Lubripedia

A 1001 Concise Illustrated Encyclopedia Reference of all Things Lubrication

RIVER PUBLISHERS SERIES IN ENERGY ENGINEERING AND SYSTEMS

Series Editor

BOBBY RAUF
Semtrain
LLC, USA

The "River Publishers Series in Energy Engineering and Systems" is a series of comprehensive academic and professional books focussing on the theory and applications behind various energy-related technologies and control systems. The series features handbooks for related technology, as well as manuals on the fundamentals and theoretical aspects of energy engineering projects.

The main aim of this series is to be a reference for academics, researchers, managers, engineers, and other professionals in related matters with energy engineering and control systems.

Topics covered in the series include, but are not limited to:

- Energy engineering;
- Machinery;
- Testing;
- HVAC;
- Electricity and electronics;
- Water systems;
- Industrial control systems;
- Web based systems;
- Automation;
- Lighting controls.

For a list of other books in this series, visit www.riverpublishers.com

Lubripedia

A 1001 Concise Illustrated Encyclopedia Reference of all Things Lubrication

Kenneth E. Bannister
Canada

River Publishers

Routledge
Taylor & Francis Group
NEW YORK AND LONDON

Published 2025 by River Publishers
River Publishers
Broagervej 10, 9260 Gistrup, Denmark
www.riverpublishers.com

Distributed exclusively by Routledge
605 Third Avenue, New York, NY 10158, USA
4 Park Square, Milton Park, Abingdon, Oxon OX14 4RN

Lubripedia: A 1001 Concise Illustrated Encyclopedia Reference of all Things Lubrication / Kenneth E. Bannister.

Routledge is an imprint of the Taylor & Francis Group, an informa business

ISBN 978-87-7004-140-9 (hardback)
ISBN 978-87-4380-802-2 (paperback)
ISBN 978-87-4380-661-5 (online)
ISBN 978-8-743-80660-8 (ebook master)

DOI 10.1201/9788743806608

Contents

Preface

Lubrication is both a science and an art, woven into the fabric of engineering, manufacturing, and daily life in ways that are often unseen but always essential. From the precision bearings of aerospace systems to the gears in industrial machinery; from the engines that power global transportation to the delicate medical instruments that safeguard health, the role of lubrication is critical to asset performance, reliability, and longevity, making the difference between seamless motion and costly breakdown. And yet, despite its importance, the language of lubrication can often feel like a maze of technical jargon, acronyms, and specialized expressions—posing challenges to students, professionals, and practitioners alike.

This concise encyclopedic dictionary of **1001 lubrication terms** has been compiled to serve as a definitive reference of all things lubrication, bridging the gap between technical precision and accessible understanding. It is designed for engineers, tribologists, maintainers, lubrication practitioners, laboratory technicians, researchers, and decision-makers who require clarity in a discipline where exactness matters. Furthermore, it is meant to serve as both a quick reference and a deeper learning tool, capturing the breadth of lubrication related vocabulary used across industries and disciplines. Each entry has been carefully selected to represent not only the fundamental vocabulary of lubrication science but also the emerging lexicon shaped by advancements in synthetic materials, nanotechnology, sustainability practices, and digital monitoring systems.

The structure of this work is deliberately encyclopedic: readers will find concise definitions enriched with illustrations, photos, context, cross-references, and—where relevant—historical notes or modern applications. Think of it as a toolbox: sometimes you'll need a simple definition, other times you'll want a fuller picture of how a term fits into the wider world of lubrication science and practice. By presenting both foundational and cutting-edge terms, ***Lubripedia*** acknowledges the evolving nature of lubrication in a world where innovation and sustainability increasingly dictate engineering solutions.

At its heart, this book reflects the simple truth that knowledge shared is knowledge strengthened. Lubrication may be a specialized subject, but it touches everyday life in countless ways. With clearer language and a common vocabulary, we can better connect ideas, improve performance, and push the boundaries of what's possible. It is my hope that this volume will become a practical companion for quick consultation, a tool for education and training, and an inspiration for deeper inquiry into the discipline of tribology. In an age when the smallest efficiencies can yield the greatest benefits, understanding the precise language of lubrication is not a luxury but a necessity.

With gratitude to the many researchers, practitioners, and educators whose contributions continue to expand the boundaries of lubrication science, I present this work to you—the reader—whether you are embarking on your first studies in the field or applying decades of expertise to the challenges of tomorrow. So, turn the page, explore, and let the words guide you through the fascinating world of lubrication…

Kenneth E. Bannister

List of Figures

List of Tables

List of Abbreviations

AAMA	American Automobile Manufacturers Association
ABMA	American Bearing Manufacturers Association
ACA	Asset criticality assessment
AES	Atomic emission spectroscopy
AFM	Atomic force microscopy
AGMA	American Gear Manufacturers Association
ANSI	American National Standards Institute
API	American Petroleum Institute
ATF	Automatic transmission fluid
AW	Antiwear
BL	Boundary layer
BSI	British Standards Institution
BTU	British thermal unit
CBL	Condition-based lubrication
CBM	Condition-based maintenance
CCS	Cold crank simulator
CDL	Commercial driver's license
CGM	Centimeter-gram-second
CLS	Certified Lubrication Specialist
CM	Condition monitoring
CMFS	Certified Metalworking Fluids Specialist
CMMS	Computerized maintenance management system
COA	Certificates of Analysis
COC	Cleveland Open Cup
COF	Coefficient of friction
CVT	Continuously Variable Transmission
CWA	Clean Water Act
DBE	Dibasic acid ester
DFMEA	Design failure modes and effects analysis
DLC	Diamond-like carbon
DLS	Dynamic light scattering
DOT	Department of Transportation

DTC	Diagnostic trouble code
EAL	Environmentally acceptable lubricant
EFL	Environmentally friendly lubricant
EG	Ethylene glycol
EHDL	Elastohydrodynamic lubrication
EHL	Elastohydrodynamic lubrication
EI	Energy Institute
EOLCS	Engine Oil Licensing and Certification System
EP	Extreme pressure
EPA	Environmental Protection Agency
EU	European Union
FIFO	First-in, first-out
FILO	First in, last out
FMEA	Failure modes and effects analysis
FRL	Filter, regulator, lubricator
FTIR	Fourier transform infrared
GC	Gas chromatography
GC-MS	Gas chromatography-mass spectrometry
GDP	Gross domestic product
GHS	Globally harmonized system
GPM	Gallons per minute
HCC	Hose connection control
HCS	Hazard communication standard
HDL	Hydrodynamic lubrication
HDPE	High-density polyethylene
HEPA	High-efficiency particulate air
HPLC	High-performance liquid chromatography
HSL	Hydrostatic lubrication
HTHS	High-temperature high-shear
HVAC	Heating, Ventilation and Air Conditioning
IBC	Intermediate bulk container
ICML	International Council for Machinery Lubrication
ICP	Inductively coupled plasma
ICP-AES	Inductively coupled plasma atomic emission spectroscopy
ICP-MS	Inductively coupled plasma mass spectrometry
IFH	Industrial fluid handling
IFT	Interfacial tension
ILF	Instantaneous lubrication failure
ILSAC	International Lubricant Standardization and Approval Committee

IR	Infrared
ISO	International Organization for Standardization
JAMA	Japan Automobile Manufacturers Association
KPI	Key performance indicator
LC	Liquid chromatography
LCA	Lifecycle cost analysis
LD	Laser diffraction
LFMEA	Lubrication failure modes and effects analysis
LLA	Laboratory Lubricant Analyst
LMMS	Lubrication maintenance management system
LMS	Lubrication management system
LOER	Lubrication operation effectiveness review
LOTO	Lockout/tagout
LWI	Load wear index
MBTLF	Mean time between lubrication failures
MIT	Massachusetts Institute of Technology
MLA	Machine Lubricant Analyst
MLT	Machine Lubrication Technician
MMR	Manufacturing maintenance readiness
MON	Motor octane number
MPC	Membrane patch colorimetry
MRV	Mini-rotary viscometer
MTBLF	Mean time between lubrication failure
NFPA	National Fire Protection Association
NLGI	National Lubricating Grease Institute
NRC	National Research Council
NTU	Nephelometric turbidity units
OLE	Overall lubrication effectiveness
OMA	Oil Monitoring Analyst
OPC	Optical particle counter
OSHA	Occupational Safety and Health Administration
PAG	Polyalkylene glycol
PAH	Polycyclic aromatic hydrocarbon
PAO	Polyalphaolefin
PAS	Publicly available specification
PBM	Perimeter-based maintenance
PD	Plastic deforming
PDI	Positive displacement injector
PEMAC	Plant Engineering Maintenance Association of Canada
PFMEA	Process failure modes and effects analysis

PFPE	Perfluorinated polyethers
PLC	Programmable logic controller
PM	Preventive maintenance
PMCC	Pensky–Martens Closed Cup
POE	Polyol ester
PPB	parts per billion
PPD	Pour point depressant
PPE	Personal protective equipment
PPE	Polyphenyl ether
ppm	Parts per million
PSI	Pounds per Square Inch
PTFE	polytetrafluoroethylene
PVC	Polyvinyl Chloride
RBOT	Rotating Bomb Oxidation Test
RCAF	Root cause analysis of failure
RCL	Reliability-centered lubrication
RCM	Reliability-centered maintenance
RCRA	Resource Conservation and Recovery Act
RMS	Root mean square
RPM	Revolutions per minute
RPN	Risk priority number
RPVOT	Rotary Pressure Vessel Oxidation Test
SAE	Society of Automotive Engineers
SDS	Safety data sheet
SEM	Scanning electron microscopy
SEM-EDX	Scanning electron microscopy with energy-dispersive X-ray spectroscopy
SI	System International
SLR	Single-line resistance
SNR	Signal-to-noise ratio
SOP	Standard operating procedures
SPL	Single-point lubricator
SSU	Saybolt second universal
STLE	Society of Tribologists and Lubrication Engineers
SUV	Saybolt universal viscosity
TAN	Total acid number
TBN	Total base number
TCP	Tricresyl phosphates
TEM	Transmission electron microscopy
TGA	Thermogravimetric analysis

TLC	Thin-layer chromatography
TOST	Turbine Oil Stability Test
ULPA	Ultra-low penetration air
VI	Viscosity index
VPI	Vapor phase inhibitor
WHMIS	Workplace Hazardous Materials Information System
WPA	Wear particle analysis
XRF	X-ray fluorescence
ZDDP	Zinc dialkyldithiophosphate

A

1. ABMA

See "American Bearing Manufacturing Association."

2. Abrasion

Abrasion is a type of mechanical wear that occurs when a hard, rough surface or particles repeatedly slide, scrape, or rub against a softer material, causing material loss or surface damage. This process can lead to erosion and pitting on the affected surfaces, often diminishing the performance and lifespan of machinery components.

Key Characteristics:

- **Surface wear**: Abrasion results in the gradual removal of material, leading to smoother or grooved surfaces.
- **Cause of friction**: Abrasive wear increases friction, potentially causing overheating and efficiency loss in machinery.
- **Affected components**: Bearings, seals, gears, and other moving parts exposed to contaminants are especially prone to abrasion.

Prevention Strategy:
Using lubricants with antiwear additives and effective filtration can help reduce abrasive wear by keeping contaminants away from critical surfaces and providing a protective film that minimizes direct contact.

3. Abrasive Wear

Abrasive wear involves the scratching, scraping, or grinding away of a surface by hard particles or rough surfaces. It can occur through two-body or three-body abrasion.

- **Two-body abrasion**: Direct contact between two surfaces where one is harder and causes wear on the other. See Figure A1.

DOI: 10.1201/9788743806608-1

- **Three-body abrasion**: Loose particles are trapped between surfaces and cause wear as they move with the surfaces. See Figure A2.

Characteristics: Surface damage appears as fine scratches or grooves, and the wear process is more gradual than cutting wear.

Common causes: Caused by the presence of hard particles such as sand, dust, or wear debris in a system. It occurs in industrial settings with exposure to abrasive materials or poor filtration systems.

Impact: Leads to a gradual loss of material and can impair the performance and efficiency of machinery if not addressed. Abrasive wear is sometimes incorrectly referred to as "cutting wear," due to the action of abrasive particles embedded in a soft surface penetrating and cutting into its mating hard surface and producing cutting wear particles.

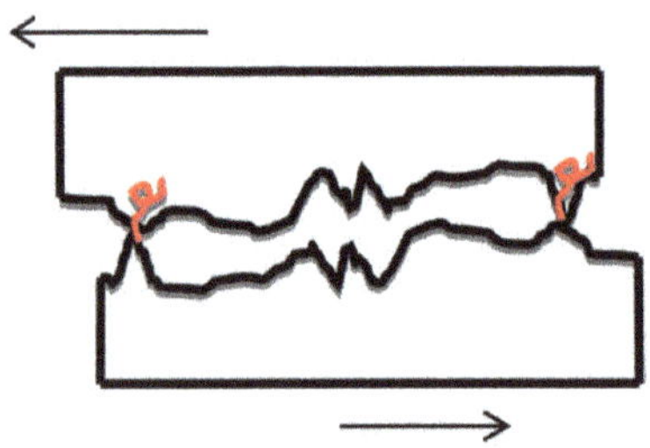

Figure A1 Two-body abrasion.

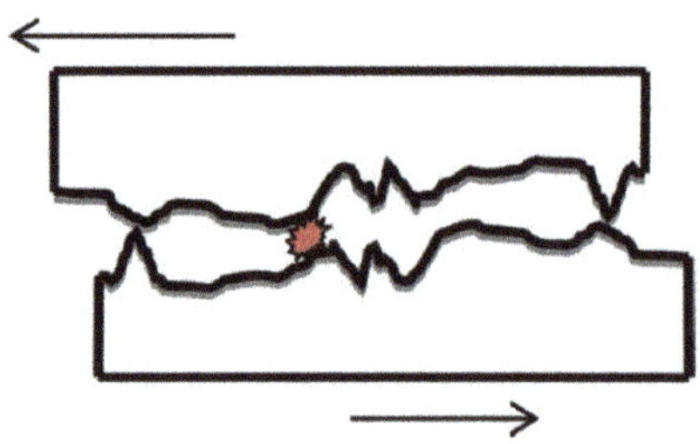

Figure A2 Three-body abrasion.

4. Absolute Rating

Absolute rating is a filtration term that defines the maximum particle size a filter can retain or block with near-complete efficiency. It indicates the filter's ability to capture and hold particles of a specific size or larger, providing a clear measure of its effectiveness in maintaining fluid cleanliness. See Table A1.

Key Characteristics:

1. **Particle size specification**: The absolute rating is expressed in microns (μm) and represents the smallest particle size that will be captured by the filter at a defined efficiency level, typically 98%–99.9%.

2. **Precise filtration**: Unlike **nominal ratings**, which give an approximate filtration efficiency, the absolute rating provides a guaranteed level of performance.

3. **Pore size uniformity**: Filters with an absolute rating are manufactured with consistent pore sizes to ensure uniform filtration throughout the medium.

Applications:

1. **Hydraulic systems**: Ensures the removal of particles that could damage sensitive components like pumps, valves, and actuators.
2. **Lubrication systems**: Maintains lubricant cleanliness to prevent wear, clogging, and damage to bearings, gears, and seals.
3. **Industrial and automotive filters**: Used in applications where precise filtration is critical to maintain performance and extend equipment life.
4. **Contamination control**: Essential for systems requiring strict cleanliness levels, such as aerospace, medical, and food-grade applications.

Importance of Absolute Rating:

1. **Equipment protection**: Prevents particles larger than the specified size from entering and damaging components.
2. **Consistent performance**: Ensures reliable filtration across the entire operating lifecycle of the filter.
3. **Compliance with standards**: Supports adherence to cleanliness and performance standards, such as ISO 4406 or SAE standards for fluid systems.

Table A1 Comparison: absolute versus nominal rating.

Feature	Absolute rating	Nominal rating
Definition	Blocks particles above a specific size with near-total efficiency	Blocks a percentage of particles of a specific size
Efficiency	Typically, 98%–99.9%	Varies, often around 50%–98%
Precision	Higher precision and reliability	Approximate and less consistent

Factors Influencing Absolute Rating:

1. **Filter media**: The type and construction of the filtration material determine the pore size and absolute rating.
2. **Fluid viscosity**: Higher viscosity fluids may require specialized filters to maintain the absolute rating under flow conditions.

3. **Operating pressure**: Consistent pressure is necessary to ensure the filter maintains its absolute rating during use.

The **absolute rating** of a filter is a critical specification for ensuring precise and reliable filtration in lubrication and hydraulic systems. By defining the maximum particle size a filter can block with high efficiency, it plays a vital role in protecting equipment, maintaining fluid cleanliness, and enhancing overall system performance.

5. Absolute Viscosity

Absolute viscosity, also known as **dynamic viscosity**, is a measure of a fluid's internal resistance to flow when subjected to an external force. It quantifies how easily a fluid flows and is a fundamental property of lubricants, oils, and other fluids. Absolute viscosity is typically expressed in units of **poise (P)** or **centipoise (cP)**, where 1 cP = 0.001 P (Water at 20 °C has an absolute viscosity of approximately 1 cP), and in the SI system as **Pascal-seconds (Pa-s)**.

Key Characteristics:

1. **Resistance to flow**: Indicates the effort required to move one layer of fluid over another. Higher absolute viscosity means the fluid resists flow more.
2. **Independent of gravity**: Unlike kinematic viscosity, absolute viscosity does not depend on the density of the fluid or gravitational force.
3. **Temperature dependency**: Absolute viscosity decreases with an increase in temperature for liquids, while it increases for gases.

Formula:

Absolute viscosity is mathematically defined as:

$$\text{Absolute Viscosity} = \frac{\text{Shear stress (force per unit area applied to the fluid)}}{\text{Shear rate (velocity gradient perpendicular to the direction of flow)}}.$$

Applications:

1. **Lubricant selection**: Absolute viscosity helps determine the suitability of a lubricant for specific applications, ensuring proper film formation and protection against wear.

2. **Industrial processes**: Used to control fluid dynamics in hydraulic systems, gearboxes, and engines to optimize performance and efficiency.
3. **Temperature performance**: Absolute viscosity guides the selection of lubricants for operations under varying temperature conditions to maintain consistent flow and protection.

Key Examples:

- **Low absolute viscosity**: Fluids like water or light oils flow easily with minimal resistance, making them suitable for high-speed, low-load applications.
- **High absolute viscosity**: Heavy oils or greases resist flow and are ideal for high-load, low-speed environments, providing strong film strength.

Importance in Lubrication:

1. **Film strength**: Ensures the lubricant forms a protective barrier between moving surfaces.
2. **Energy efficiency**: Affects the energy required to pump and move the lubricant through a system.
3. **Wear protection**: Ensures optimal lubrication under specific load and speed conditions, minimizing wear and extending component life.

Absolute viscosity is a critical property for understanding and selecting fluids in lubrication systems and industrial applications.

6. Absorbent Filter

An **absorbent filter** refers to a filtration device designed to trap and retain contaminants by **absorbing** (see Absorption) them into the filter media.

This filter type is commonly used in lubrication, hydraulic, and fuel systems to remove moisture, oils, or chemical impurities that could degrade performance. By drawing contaminants into the media rather than simply blocking them, absorbent filters promote enhanced fluid purity that can prevent damage to sensitive components.

7. Absorption

Absorption is a process where a substance is fully taken in and incorporated into the volume of another material. For example, when a sponge absorbs water, the water penetrates and spreads throughout the sponge's interior. See Figure A3.

Key Characteristics:

- **Complete integration**: The absorbed substance is fully incorporated within the absorbing material.
- **Industrial application**: Commonly refers to fluids or gases being absorbed into solids or liquids, such as moisture being absorbed by desiccants.
- **Difference from adsorption**: Absorption involves full penetration, unlike adsorption, which occurs only at the surface.

Figure A3 Absorption occurs when particles are incorporated into the filter media.

8. Acid

Acid refers to a sour, corrosive chemical substance that increases the concentration of hydrogen ions (H^+) or protons when dissolved in water, typically resulting in a pH value below 7 (see Figure A4) that will turn litmus paper red color. Acids can be classified as strong or weak, depending on the extent to which they ionize in solution.

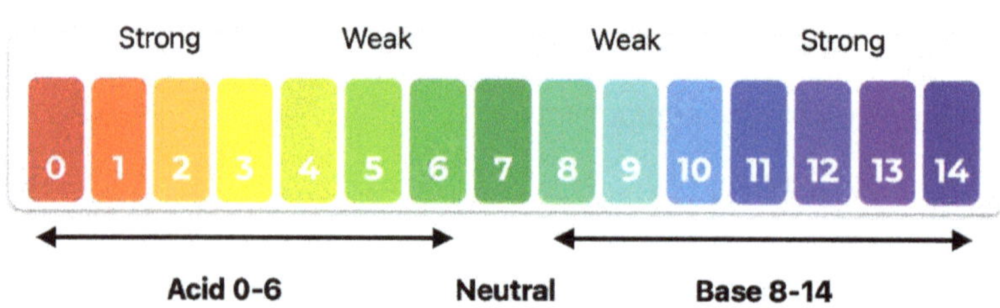

Figure A4 Typical pH scale.

In industrial and chemical systems, acids play a variety of roles, such as cleaning agents, catalysts, or reactants. In lubrication systems, the buildup of acidic compounds – often measured by the **total acid number (TAN)** – indicates fluid degradation, which can lead to corrosion of metal components and reduced system efficiency. Proper monitoring and control of acidity are crucial for maintaining performance and preventing equipment damage. See "Acidity."

9. Acid Number

See "Total Acid Number (TAN)."

10. Acidic Pitting

See "Corrosive Wear."

11. Acidity

Acidity in lubricants results from acidic compounds formed during manufacturing or as the lubricant breaks down over time. It is measured by **Total Acid Number (TAN)**, which indicates the amount of potassium hydroxide required to neutralize the acids in one gram of lubricant. High acidity can lead to metal corrosion, seal damage, and decreased lubricant performance.

Key Characteristics:

- **Total Acid Number (TAN)**: indicates the amount of potassium hydroxide required to neutralize the acids in one gram of the lubricant, which helps monitor lubricant condition.
- **Corrosion and damage**: High acidity can corrode metal components and damage seals.
- **Performance impact**: Elevated acidity reduces the lubricant's effectiveness.

Regular monitoring of **TAN** is crucial for detecting lubricant degradation and determining when corrective actions, like replacement, are needed to maintain machinery reliability and extend its service life.

12. Additive (Lubricant)

A lubricant **Additive** is a material added to a lubricant base stock to enhance its performance during use. Additives are sacrificial by nature and play a crucial role in indicating when a lubricant needs replacement. Once additives are depleted, the oil must be replaced or replenished with new additives.

Key Characteristics:

- **Additive depletion**: Monitoring and comparing the additive profile in virgin stock oil with a used oil sample (see oil analysis) helps determine the extent of depletion, assess efficacy, and identify necessary additives for restoration.
- **Functions**: Additives serve to enhance performance, introduce new properties, or suppress undesirable traits of the base oil.
- **Composition**: Additives can be organic or inorganic, dissolving in oil when submicron-sized, with larger particles remaining as suspended solids.
- **Concentration**: Additives can make up **0.1% to 30%** of the total oil volume.

Common types of additives include Antioxidant, antiwear, antifoam, corrosion inhibitors, demulsifiers, detergents, dispersants, emulsifiers, extreme pressure (EP) additives, pour point depressants, rust inhibitors, tackiness agents, and viscosity index (VI) improvers.

13. Additive Depletion

Additive depletion refers to the gradual reduction or exhaustion of chemical additives in lubricating oil due to usage, thermal stress, and contamination. Additives like **antioxidants**, **antiwear agents**, **detergents**, and **dispersants** are included in oils to enhance performance, protect equipment, and extend the lubricant's lifespan. As these additives are consumed, the oil's ability to perform critical functions – such as preventing corrosion, reducing friction, and controlling sludge buildup – diminishes.

Key Characteristics:

- **Impact on performance**: Additive depletion reduces the oil's effectiveness in protecting equipment and maintaining optimal function.
- **Monitoring importance**: Regular checks for additive depletion are essential for lubrication management to determine when oil replacement or additive replenishment is needed.

- **Consequences**: Failure to address depletion can lead to increased wear, corrosion, system inefficiencies, potential equipment failures, and higher operational costs.

Routine oil analysis helps detect additive depletion by comparing current additive levels to those in a virgin stock sample. Once additives are depleted, the oil can no longer function as intended and is considered unserviceable.

14. Additive Solvency

Additive solvency refers to a base oil or lubricant's ability to dissolve and maintain chemical additives in suspension, ensuring even distribution throughout the fluid. Proper solvency is crucial for preventing additives – such as antioxidants, detergents, dispersants, and antiwear agents – from separating from the lubricant.

Key Characteristics:

- **Even distribution**: Ensures additives remain suspended and evenly distributed to maintain lubricant effectiveness.
- **Impact of poor solvency**: Inadequate solvency can lead to the formation of deposits, sludge, or varnish, which may clog filters, impair lubrication, and reduce overall system performance.

Maintaining proper solvency is essential for optimal lubricant function and system reliability.

15. Additive Stability

Additive stability refers to the ability of chemical additives within a lubricant to maintain their effectiveness and resist degradation over time under various operational conditions. Stable additives ensure that the lubricant continues to perform essential functions, such as **reducing friction, preventing corrosion, and minimizing sludge formation**.

Key Characteristics:

- **Long-term performance**: Stable additives help maintain consistent lubricant performance throughout its service life.
- **Resistance to degradation**: Additive stability ensures that additives can withstand thermal stress, oxidation, and contamination without breaking down.

- **Impact of instability**: Unstable additives can degrade prematurely, leading to reduced lubricant performance, increased wear, and potential equipment damage.

Ensuring additive stability is crucial for maintaining the reliability and efficiency of lubricated systems.

16. Adiabatic Compression

Adiabatic compression in the context of lubrication refers to the rapid compression of air or gas within a lubricated system, such as in compressors or hydraulic systems, where no heat is exchanged with the environment. This sudden increase in pressure results in a significant temperature rise within the gas, which can impact the lubricant's performance and stability.

Key Characteristics:

- **Temperature rise**: The rapid temperature increase can lead to thermal stress on the lubricant, potentially causing oxidation, degradation, or vaporization.
- **Lubricant stability**: Lubricants must be able to withstand high temperatures without breaking down to ensure continued protection and performance.
- **Impact on equipment**: If the lubricant degrades under adiabatic compression, it can result in reduced lubrication efficiency, increased wear, and potential damage to components.

Proper lubricant selection and monitoring are essential in systems where adiabatic compression occurs to prevent issues related to high temperatures and ensure reliable operation.

17. Adhesion

Adhesion in lubrication refers to the ability of a lubricant to stick or bond to the surfaces it is applied to, forming a protective film. See also: Affinity. This property is crucial for maintaining consistent lubrication under various operating conditions, including high temperature and shock environments. This property is enhanced with the use of antiwear and extreme pressure additives and is often referred to as "metal wetting."

Key Characteristics:

- **Protective film**: Ensures continuous coverage on components, reducing friction and preventing wear.
- **Surface bonding**: Effective adhesion helps maintain lubricant position, even under challenging conditions.
- **Enhanced performance**: Essential for operations in high-stress environments to ensure reliable protection and performance.

Adhesion is balanced with other lubricant properties to optimize overall performance, especially in vertical or high-motion applications.

18. Adhesive Wear

Adhesive wear occurs during high-load sliding friction, often when the lubricant used has an incorrect viscosity or lacks the necessary extreme pressure **(EP** or antiwear **(AW)** additives. This type of wear happens when metal surfaces come into contact, leading to material transfer, tearing, and surface damage (see Figure A5).

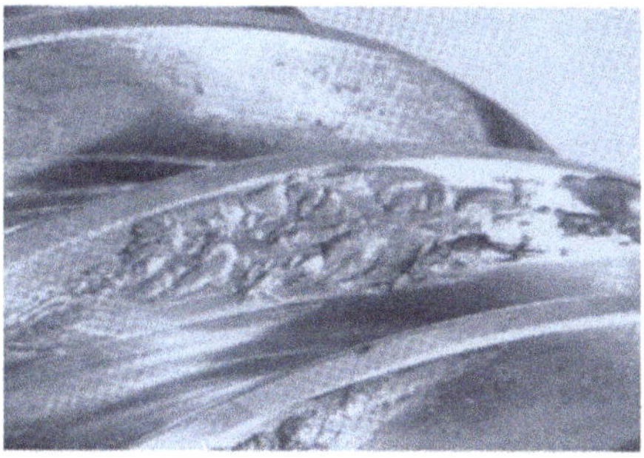

Figure A5 Metal tear due to adhesion.

Key Characteristics:

- **Causes**: Typically, due to insufficient lubricant properties, such as improper viscosity or missing additives.
- **Effects**: Results in material transfer between surfaces, causing additional wear and potential surface damage to components. This type of wear is also known as scuffing, shearing, or galling and can lead to severe mechanical failure if not addressed.

- **Impact on performance**: Can compromise equipment performance and reliability if not addressed.

To prevent adhesive wear, using the correct lubricant with appropriate viscosity and additives is essential. Regular monitoring and maintenance help ensure optimal lubrication and protect against wear in high-load applications.

19. Adsorbent Filter

Adsorbent filter refers to a filtration device designed to capture contaminants by attracting and binding them to the surface of the filter media through adsorption. Unlike conventional filters that physically block particles, adsorbent filters use materials with high surface areas – such as activated carbon, silica gel, or alumina – to trap impurities like moisture, odors, oils, or gases on their surfaces.

Key Characteristics:

- **Adsorption process**: Contaminants are attracted and held on the filter's surface rather than being blocked or absorbed.
- **High surface area media**: Utilizes materials like activated carbon and silica gel for efficient impurity capture.
- **Applications**: Effective for trapping moisture, odors, oils, and gases.

Adsorbent filters are valuable in systems where maintaining air or fluid purity is essential for preventing contamination and ensuring optimal performance.

20. Adsorption

Adsorption is a process in which molecules – such as gases, liquids, or dissolved substances – adhere to the surface of a solid or liquid without penetrating it. Unlike absorption, which involves the substance being taken into the material, adsorption occurs solely at the surface as depicted in Figure A6.

Key Characteristics:

- **Surface interaction**: Molecules bind to the surface without being absorbed internally.
- **Comparison to absorption**: Adsorption is a surface process, while absorption involves full penetration into the material.

- **Industrial use**: Commonly used to capture contaminants, moisture, or impurities from fluids or air to enhance system performance and reliability.

Adsorption plays a vital role in improving the efficiency and cleanliness of industrial systems by effectively trapping unwanted substances.

Figure A6 Adsorption where particles adhere to the surface and do not penetrate.

21. Aeration

Aeration in lubrication refers to the introduction and dispersion of air within a lubricant, which can occur during the operation of machinery or fluid systems. This process can lead to the formation of air bubbles or foam within the lubricant, potentially impacting its effectiveness.

Key Characteristics:

- **Air entrapment**: Involves the unintended mixing of air with the lubricant, leading to bubble formation.
- **Effects on performance**: Excessive aeration can reduce the lubricant's ability to form a consistent film, decrease its cooling efficiency, and lead to increased oxidation.
- **Prevention and control**: Proper system design, use of antifoam additives, and maintaining appropriate oil levels help mitigate aeration issues.

Managing aeration is crucial in lubrication systems to ensure the lubricant maintains its protective properties, supports optimal equipment function, and prevents potential damage or performance issues.

22. AES

See "Atomic Emission Spectroscopy."

23. Affinity

Affinity in lubrication refers to the tendency of a lubricant to bond or interact with surfaces and other materials it comes into contact with. This property is critical for ensuring that the lubricant adheres effectively to surfaces and maintains a consistent protective film.

Key Characteristics:

- **Surface bonding**: Indicates how well a lubricant attaches to metal surfaces or other components to prevent wear and reduce friction.
- **Compatibility**: A lubricant's affinity for certain materials impacts its effectiveness in specific applications, ensuring optimal interaction with seals, gaskets, or other system components.
- **Performance impact**: High affinity helps maintain a stable lubrication layer, enhancing equipment protection and performance, while low affinity may lead to inadequate surface coverage and increased wear.

Understanding the affinity of a lubricant is essential for selecting the right product that will provide consistent protection, reduce friction, and extend the lifespan of machinery.

24. Age Hardening

Age hardening in relation to lubrication refers to the process in which a lubricant, usually grease, or its additives undergoes chemical changes over time, resulting in a harder, more viscous, or less effective state. This typically occurs due to prolonged exposure to heat, oxidation, or contaminants, which alters the physical and chemical properties of the lubricant.

Key Points:

- **Chemical changes**: Involves the gradual hardening or thickening of the grease lubricant as its molecular structure changes over time. See Figure A7.
- **Impact on performance**: Age-hardened lubricants may lose their ability to flow properly, reducing their effectiveness in protecting and lubricating moving parts.

Figure A7 Failed needle bearing with heat-hardened grease. Courtesy: ENG**TECH** Industries Inc.

- **Causes**: Prolonged exposure to high temperatures, oxidation, contamination, or extended use can accelerate age hardening.

Managing and monitoring the condition of lubricants to prevent or mitigate age hardening is crucial for maintaining proper equipment lubrication and preventing wear or damage. Regular oil/grease analysis and timely replacement help ensure that lubricants maintain their desired properties throughout their service life.

25. Agglomeration

Agglomeration in lubrication refers to the process where small particles or droplets cluster together to form larger masses or aggregates. This can occur when contaminants such as dirt, metal particles, or degraded additives bind together, resulting in sludge, deposits, or filter blockages.

Key Characteristics:

- **Formation of aggregates**: Small particles combine to create larger clusters, impacting the lubricant's consistency.
- **Effects on performance**: Agglomeration can impair the lubricant's flow properties, increase wear, and cause system inefficiencies.
- **Prevention**: Proper filtration, the use of dispersant additives, and regular oil analysis help monitor and mitigate contamination levels.

Preventing agglomeration is essential for maintaining the effectiveness of lubricants and ensuring the smooth operation of machinery.

26. AGMA

See "American Gear Manufacturers Association."

27. Air Breather

See "Breather."

28. Air Entrapment

Air entrapment in lubrication refers to the unintended capture and suspension of air bubbles within a lubricant during operation. This can occur due to turbulence, agitation, rapid fluid movement, or improper filling of reservoirs.

Key Characteristics:

- **Causes**: Air becomes trapped in the lubricant through turbulent flow, agitation, or incorrect reservoir filling.
- **Impact on performance**: Leads to foaming, reduced film strength, inconsistent lubrication, and increased component wear.
- **Risks**: Entrapped air accelerates oxidation and raises the risk of cavitation, which can damage pumps, valves, and other equipment.
- **Prevention**: Effective system design, careful filling procedures, and the use of antifoam additives are essential for minimizing air entrapment.

Preventing air entrapment helps maintain the lubricant's effectiveness, ensuring reliable performance and protecting machinery from potential damage.

29. Air Filter

Air filter refers to a device designed to remove contaminants such as dust, pollen, dirt, and other particles from the air before it enters a system or equipment. In industrial and mechanical settings, air filters are used to protect machinery, engines, and HVAC systems by ensuring clean air intake, which helps maintain optimal performance and extends the lifespan of components. See Airborne Contaminants.

Key Points:

- **Contaminant removal**: Captures particles to prevent them from entering and damaging lubricated systems.
- **Applications**: Used in engines, compressors, HVAC systems, and industrial machinery.
- **Performance benefits**: Enhances system efficiency, reduces wear, and helps maintain consistent operation.

Air filters play a vital role in removing dust and air particulates that could end up on a lubricated open-bearing surface or lubricant reservoir that could be pulled into the lubricant and bearing creating a harmful lapping paste.

30. Airline Lubricator

Airline lubricator refers to a device used in pneumatic systems to introduce controlled amounts of lubricant into the compressed air stream. This lubrication helps reduce friction, prevent wear, and extend the lifespan of pneumatic tools and components.

Key Characteristics:

- **Function**: Adds a fine mist of lubricant to the air stream to maintain and protect pneumatic equipment.
- **Benefits**: Reduces friction, prevents corrosion, and minimizes wear on moving parts.
- **Applications**: Commonly used in air-powered tools, cylinders, and valves within pneumatic systems.

Airline lubricators ensure that pneumatic systems operate smoothly and reliably by maintaining proper lubrication and minimizing component degradation. Airline lubricators are often combined with a filter and regulator assembly – See Filter, Regulator, Lubricator.

31. Air/Oil Lubrication

Air/oil lubrication system is an automated high-pressure lubricant delivery system specifically designed to lubricate spindle-type bearings operating at speeds greater than 20,000 rpm. This system introduces lubricant into a compressed air stream at high pressure through a metering valve located prior close to the entrance of each individual bearing point as depicted in Figure A8. This lubrication system is sometimes referred to as "Spindle" lubrication due to its preferable use in lubricating high-speed spindles.

Key Characteristics:

- **Atomization**: The constant air stream atomizes the lubricant into a fine mist of microdroplets, which are then propelled through the metering valve at high velocity into the bearing cavity.
- **Efficient cooling and lubrication**: The high-speed air and micro drop lubrication effectively cool and lubricate the bearings, reducing fluid friction drag.
- **Optimized performance**: The system ensures efficient lubrication while minimizing resistance, enhancing the longevity and reliability of high-speed bearings.

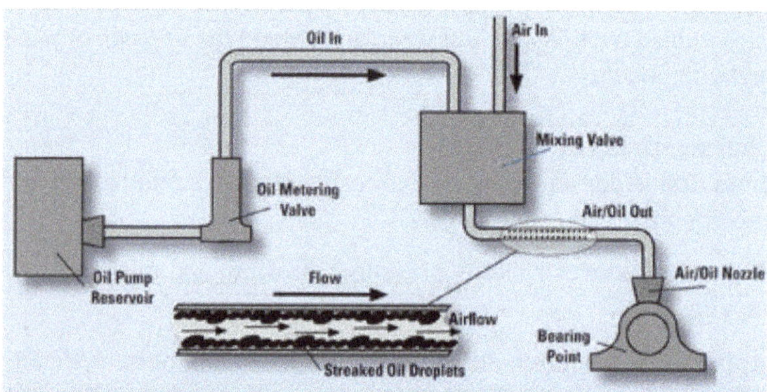

Figure A8 High-pressure air–oil system.
Lubrication Management and Technology magazine – Ken Bannister

This advanced lubrication method is ideal for applications requiring precise and efficient bearing lubrication at extremely high rotational speeds.

32. Air Release Property

Air release property refers to a lubricant's ability to **quickly separate and expel entrained air** that becomes trapped during operation. It is a critical characteristic in **hydraulic fluids, turbine oils, and circulating lubrication**

systems, where excessive air bubbles can lead to **foaming, cavitation, and reduced lubrication efficiency**.

Key Functions of Air Release Property in Lubrication

- **Prevents foam formation**: Reduces bubbles that can interfere with film formation and lubrication efficiency.
- **Enhances hydraulic and circulating system performance**: Ensures proper fluid pressure and prevents pump cavitation.
- **Minimizes oxidation and oil degradation**: Air release capability helps prevent accelerated oxidation that can lead to sludge and varnish formation.
- **Improves lubricant film integrity**: Maintains consistent film strength in high-speed and high-pressure systems, reducing wear and friction.

Factors Affecting Air Release in Lubricants

- **Viscosity and base oil type**: Higher viscosity oils trap air more easily, while synthetic base stocks generally offer better air release properties.
- **Additive formulation**: Antifoam and air-release additives help break up air bubbles and improve separation.
- **Operating temperature and pressure**: High temperatures **can increase air solubility, while rapid pressure changes can cause air entrapment and foaming**.
- **System design and flow rate**: Turbulent flow, small reservoirs, or improper venting can contribute to air entrapment issues.

33. Airborne Contaminants

Airborne contaminants refer to unwanted particles, gases, or biological agents suspended in the air that can negatively affect equipment, processes, or air quality. In industrial and mechanical systems, these contaminants may include dust, dirt, pollen, metal particles, smoke, oil mist, moisture, chemicals, and microorganisms.

Key Characteristics:

- **Sources and impact**: Airborne contaminants can enter machinery through ventilation or open spaces, causing abrasion, clogging, corrosion, and reduced efficiency.

- **Effect on lubrication**: Accelerates fluid degradation, leading to sludge or deposits that impair system performance.
- **Control measures**: Effective filtration using air filters, air/oil separators, and ventilation systems is crucial to manage and minimize contaminants.

Proper control of airborne contaminants is essential to maintain equipment reliability, ensure efficient operation, and protect system components from damage.

34. Alkali

Alkali, sometimes referred to as **Base**, refers to a chemical substance with strong basic (sometimes referred to as **nonacidic** properties that include hydroxides of ammonium, lithium, potassium, sodium, barium, and calcium. These bases are soluble in water and capable of neutralizing acids.

Key Characteristics:

- **Acid neutralization**: Alkali helps counteract acid buildup and prevent corrosive wear in machinery.
- **Lubrication use**: Commonly used in lubricating oils to protect against acid formation and oxidation byproducts.
- **Application in engines**: Essential for internal combustion engines to maintain efficiency and reduce the risk of damage from acidic conditions.

35. Alkaline

Alkaline refers to a substance with a pH value greater than 7, indicating it is the opposite of acidic (see Figure A4 – pH scale). Alkaline materials, or bases, release hydroxide ions (OH) in water and neutralize acids. Examples include sodium hydroxide (NaOH), calcium carbonate ($CaCO_3$), and ammonia (NH_3).

Key Characteristics:

- **pH greater than 7**: Alkaline substances are characterized by their ability to neutralize acids.
- **Industrial use**: Maintaining an alkaline balance in lubrication systems prevents corrosion and acid buildup.

- **Additives for protection**: Engine oils use alkaline additives, like detergents, to neutralize acidic byproducts, ensuring equipment durability and performance.

Alkalinity in lubrication helps protect machinery from wear and maintains fluid integrity for long-lasting operation.

36. Alkaline Reserve

Alkaline reserve refers to a lubricant's **ability to neutralize acidic contaminants**, primarily in engine oils and industrial lubricants. It is directly related to the **total base number (TBN)**, which measures the concentration of alkaline additives designed to counteract acid formation from fuel combustion, oxidation, and contamination.

Key Functions of Alkaline Reserve in Lubrication

- **Neutralizes acidic byproducts**: Protects metal surfaces from corrosion and prevents lubricant degradation caused by sulfuric and nitric acids in combustion engines.
- **Extends oil service life**: Helps maintain lubricant effectiveness over time, reducing the need for frequent oil changes.
- **Improves engine and component protection**: Prevents pitting, rust, and wear in engine bearings, liners, and hydraulic systems.
- **Supports long-drain intervals**: Critical for diesel engines, marine applications, and industrial systems operating under high-sulfur or extreme conditions.

Factors Affecting Alkaline Reserve

- **TBN level**: Higher TBN values indicate greater acid-neutralizing capacity, suitable for heavy-duty applications.
- **Operating conditions**: High temperatures and contamination accelerate acid buildup, requiring a strong alkaline reserve.
- **Lubricant formulation**: Additives like calcium, magnesium, and boron compounds contribute to alkaline reserve and acid neutralization.
- **Fuel quality and sulfur content**: High-sulfur fuels generate more acidic by-products, demanding higher alkaline reserves for protection.

37. Aluminum Complex Grease

Aluminum complex grease, developed in the 1950s, was designed to offer exceptional water resistance and washout protection. With a high dropping point exceeding 450 °F (230 °C), it is well-suited for automotive uses like wheel bearings and food manufacturing applications.

Key Characteristics:

- **High-temperature performance**: Effective in high-heat environments but may form a gel and lose lubricating properties at extreme temperatures.
- **Compatibility concerns**: Incompatible with most other grease types, requiring careful grease management when used alongside different greases.
- **Applications**: Ideal for demanding conditions where water resistance and heat stability are critical.

Proper handling and a disciplined lubrication program are necessary to maximize the benefits and performance of aluminum complex grease.

38. Ambient Condition Factors

See "Environmental Condition Factors."

39. Ambient Temperature Factor

Ambient temperature refers to the temperature of the **surrounding environment in which a lubricant is stored, applied, or operates**. It plays a critical role in lubricant selection, viscosity performance, and overall lubrication effectiveness. Extreme ambient temperatures, whether hot or cold, can significantly impact a lubricant's flow characteristics, oxidation stability, and protective capabilities.

Key Functions of Ambient Temperature in Lubrication

- **Affects lubricant viscosity**: High ambient temperatures thin lubricants, while low temperatures increase viscosity, potentially causing poor flow and lubrication starvation.
- **Impacts lubrication film thickness**: A lubricant must maintain sufficient film strength to prevent wear and metal-to-metal contact under varying ambient conditions.

- **Influences oxidation and degradation rates**: Higher temperatures accelerate oxidation and varnish formation, reducing lubricant lifespan.
- **Determines lubricant selection**: Lubricants must be formulated to withstand ambient temperature extremes, ensuring consistent performance in different environments.

Factors Affected by Ambient Temperature

- **Cold-temperature performance**: Low ambient temperatures can cause oil thickening, leading to startup wear, increased energy consumption, and potential lubricant solidification.
- **High-temperature performance**: Elevated ambient temperatures can lead to oil thinning, evaporation loss, increased oxidation, and film breakdown.
- **Grease behavior**: Greases can become stiff in cold environments, leading to poor flow, while high temperatures can cause oil separation and excessive softening.

Lubricant Selection Based on Ambient Temperature

- **Low-temperature lubricants**: Require low-viscosity base oils with pour point depressants, such as synthetic PAO-based oils for cold climates.
- **High-temperature lubricants**: Use high-viscosity index (VI) oils, oxidation inhibitors, and thermal stabilizers to prevent breakdown.
- **Multigrade oils**: Designed to perform in both hot and cold conditions, such as 5W-30 engine oil, which flows well in cold starts but maintains protection at high temperatures.

40. American Bearing Manufacturing Association

The American Bearing Manufacturers Association (ABMA) was established in 1917 by a small group of US bearing manufacturers to support bearing production during World War I. Today, this national organization represents 27 manufacturers of various bearing types.

Key Points:

- **Founded**: 1917, to support the bearing industry during wartime.
- **Representation**: Comprises 27 manufacturers specializing in different bearing types.

- **Contact information**: More details can be found at www.american-bearings.org.

The ABMA continues to play a crucial role in advancing the bearing industry and supporting its members through various initiatives and standards. For more information visit www.americanbearings.com

41. American Gear Manufacturers Association

The American Gear Manufacturers Association (AGMA) was established in 1916 when the RD Nuttall Company convened several gear manufacturers to develop technical standards for gear production and maintenance. AGMA remains committed to promoting and advancing standards for gears and gear lubricants.

Key Points:

- **Founded**: 1916, with a focus on standardizing gear manufacturing and maintenance practices.
- **Purpose**: Develop and maintain technical standards for gears and lubricants.
- **Contact information**: More details can be found at www.agma.org.

42. American National Standards Institute

The American National Standards Institute (ANSI) was established in 1918 as a private, not-for-profit organization to serve as the voice of the US standards and conformity assessment system. Since that time, ANSI has continued to oversee the development and use of thousands of standards and guidelines that impact a wide range of industries, from acoustical devices to energy distribution.

Key Points:

- **Founded**: 1918, to coordinate the US standards and conformity assessment system.
- **Scope**: Influences sectors such as construction, agriculture, and energy with norms and guidelines.
- **Accreditation**: Accredits conformance assessment programs, including ISO 9000 (quality) and ISO 14000 (environmental) management systems.

ANSI continues to play a pivotal role in shaping standards that promote safety, efficiency, and innovation across various sectors. For more information, visit www.ansi.org.

43. American Petroleum Institute

The **American Petroleum Institute (API)** was founded in 1919 by the US domestic oil and gas industry to aid in fuel supply for the military during World War I. Today, API represents the oil and gas industry, engaging with the public, US Congress, the executive branch, state governments, regulatory bodies, and the media.

Key Points:

- **Founded**: 1919, to support military fuel supply and now serves as an industry advocate.
- **Role**: Represents the oil and gas sector and influences public policy and industry regulations.
- **Standards**: Maintains over 500 standards and recommended practices for all segments of the oil and gas industry, including oil and grease standards.

API continues to set industry benchmarks and promote best practices to ensure safety, sustainability, and efficiency. For more information, visit www.api.org.

44. American Society for Testing Materials – ASTM International

ASTM International (formerly the American Society for Testing and Materials) was established in 1898 and is today a global organization dedicated to developing and publishing international voluntary consensus standards. These standards cover materials, products, systems, and services, including those for lubricant testing.

Key Points:

- **Founded**: 1898, with a focus on creating standards for a wide range of industries.
- **Membership**: Over 30,000 members representing more than 110 partner countries.

- **Scope**: Maintains more than 12,800 standards impacting industries worldwide.

ASTM International continues to be a leader in standardization, ensuring quality, safety, and efficiency in global markets. For more information, visit www.astm.org.

45. AN (Acid Number)

See "Total Acid Number."

46. Analytical Ferrography

Analytical ferrography is a diagnostic technique used to analyze wear particles in lubricants to evaluate a machine's condition. It involves separating and examining metal particles suspended in the lubricant, offering insights into the type, severity, and source of wear within a system.

Key Characteristics:

- **Process**: A lubricant sample is passed over a magnetic slide, attracting ferrous particles, which are then analyzed under a microscope to assess size, shape, and composition (see Ferrogram).
- **Wear mechanisms**: Identifies types of wear, such as abrasion, adhesion, fatigue, or corrosion.
- **Predictive maintenance**: Detects early signs of abnormal wear, enabling timely intervention to prevent equipment failure and extend machinery lifespan.

Analytical ferrography is a valuable tool for proactive maintenance, helping to ensure the reliability and efficiency of industrial equipment.

47. Anhydrous Lubricants

Anhydrous lubricants are lubricants that **do not contain water** in their formulation. These lubricants are specifically designed for applications where moisture can cause oxidation, corrosion, or degradation of machinery components. They are commonly used in high-temperature, high-pressure, and water-sensitive environments to ensure long-lasting protection and reliability.

Key Functions of Anhydrous Lubricants

- **Prevents corrosion and oxidation**: Since water is a major contributor to rust and oil breakdown, anhydrous lubricants help protect metal surfaces and extend component life.
- **Enhances thermal stability**: These lubricants can withstand high operating temperatures without breaking down, making them ideal for industrial machinery, bearings, and gear systems.
- **Resists moisture contamination**: In environments where humidity or water exposure is high, anhydrous lubricants ensure consistent lubrication without emulsification or dilution.
- **Improves lubricant longevity and efficiency**: Without water content, anhydrous lubricants retain their viscosity and performance properties over extended periods.

Types of Anhydrous Lubricants

- **Anhydrous grease**: Greases formulated without water, often thickened with calcium sulfonate, aluminum complex, or lithium-based thickeners for enhanced resistance to water washout.
- **Anhydrous oils**: Specialty base oils used in hydraulic systems, compressors, and high-performance machinery, ensuring moisture-free operation.
- **Dry film lubricants**: Solid lubricants like graphite, molybdenum disulfide (MoS_2), and PTFE that function without liquid carriers, ideal for extreme conditions where liquid lubricants may evaporate or degrade.

Applications of Anhydrous Lubricants

- **Food-grade and pharmaceutical equipment**: Ensures lubrication without risk of bacterial growth or contamination due to water presence.
- **Aerospace and automotive components**: Protects moving parts exposed to extreme temperatures and pressure variations.
- **Industrial bearings and gears**: Prevents moisture-related wear in steel mills, marine environments, and heavy machinery.
- **High-temperature applications**: Used in kilns, ovens, and turbines where traditional lubricants may degrade due to heat exposure.

48. Aniline Point

Aniline point is a measure of a hydrocarbon's solvent power, determined by the lowest temperature at which equal volumes of aniline (a benzene derivative) and a petroleum product become completely miscible. Lower aniline point values indicate higher solvency levels. Aromatic hydrocarbons, known for their strong solvency, have the lowest aniline points.

Key Characteristics:

- **Measurement**: Reflects the solvent power of a hydrocarbon, with lower values signifying greater solvency.
- **Aromatic hydrocarbons**: Exhibit the lowest aniline points and highest solvency.
- **Lubrication use**: Assesses oil compatibility with seals, gaskets, and other nonmetallic components, and provides insight into the lubricant's ability to dissolve impurities and additives.

The aniline point is essential in evaluating the performance and compatibility of lubricants within various systems, helping to ensure reliable and efficient operation.

49. Animal/Vegetable Base Oil

Animal and vegetable base oils played a primary role in both Industrial Revolutions which relied heavily on olive oil and rendered animal fats (see "Tallow") for lubrication until these natural lubricants proved inadequate in preventing rapid acid formation under increasing speeds and loads. The discovery of crude mineral oil addressed these limitations. An exception was sperm whale oil, (see "Sperm Whale Oil") valued for its exceptional lubricating properties and used as a base oil for automotive automatic transmission fluid (ATF) until the early 1970s.

Today, for the most part, this classification of lubricants is generally reserved for cooking purposes. However, the environmental movement has created renewed interest, research, and use of vegetable-based lubricants that are now making a comeback.

Key Points:

- **Early lubricants**: Olive oil and animal fats were primary lubricants until crude mineral oil became necessary for higher performance.

- **Sperm whale oil**: Used for its superior qualities but phased out by the 1970s due to high sperm whale mortality and the advancement of synthetic lubricants.
- **Modern trends**: There is a renewed interest in vegetable-based soluble oils for hydraulic systems. With a focus on environmental sustainability, research into vegetable oils is growing due to their natural biodegradability and potential use as alternative diesel fuel.

The shift toward eco-friendly products continues to drive innovation and research into sustainable lubrication solutions.

50. ANSI

See "American National Standards Institute."

51. Anticorrosive Additive

See "Corrosion Inhibitor Additive."

52. Antifoam Additive

Antifoam additives are chemical compounds added to lubricants, oils, and other fluids to prevent or reduce foam formation. Foam can occur when air becomes trapped in the lubricant due to agitation, turbulence, or rapid circulation, impairing the lubricant's ability to provide consistent coverage and protection.

Key Characteristics:

- **Impact of foam**: Foam accelerates oxidation, reduces lubricant effectiveness, and can lead to cavitation damage in pumps and components.
- **Function**: Antifoam additives work by destabilizing air bubbles, causing them to coalesce and rise to the surface, where they quickly collapse and dissipate.
- **Common agents**: Include silicone-based compounds and polymers, which effectively minimize foam formation.
- **Applications**: Used in industrial and automotive lubrication systems, hydraulic fluids, and engine oils to maintain optimal fluid performance and prevent inefficiencies caused by foam buildup.

Also known as defoamants, foam inhibitors, or antifoam agents, these additives are essential for ensuring the reliability and efficiency of fluid systems.

53. Antifoam Agent

See "Antifoam Additive."

54. Antifriction Bearing

See "Rolling Contact Bearings."

55. Antigalling Agent

An **Antigalling Agent** is a **lubricant additive or coating** designed to **prevent galling**, a severe form of adhesive wear that occurs when metal surfaces under high loads weld together and tear apart. Galling typically happens in fasteners, bearings, gears, and sliding metal components, especially those made from stainless steel, aluminum, and titanium due to their tendency to cold-weld.

Key Functions of an Antigalling Agent

- **Prevents metal adhesion and seizure**: Reduces friction and surface welding, preventing permanent metal deformation.
- **Minimizes wear and surface damage**: Forms a protective barrier between contact surfaces, reducing scuffing, scoring, and material transfer.
- **Enhances load-carrying capacity**: Allows high-pressure applications to operate smoothly without premature failure.
- **Improves assembly and disassembly of components**: Helps in preventing thread damage in bolts, nuts, and fasteners.

Types of Antigalling Agents

- **Solid lubricants (dry film lubricants)**: **molybdenum disulfide (MoS_2), graphite, and PTFE** provide long-lasting protection in extreme conditions.
- **Extreme pressure (EP) additives**: Additives like **sulfur, phosphorus, and zinc-based compounds (ZDDP)** reduce metal-to-metal contact under high loads.

- **Specialty antiseize compounds**: Grease-based products containing **copper, nickel, or aluminum particles** to prevent galling in high-temperature fasteners and fittings.
- **High-performance oils and coatings**: Synthetic lubricants and ceramic coatings designed for **severe-duty** applications.

56. Antioxidant Additive

Oxygen is a lubricant's primary adversary, especially at high temperatures and in the presence of contaminants like water. When oxygen reacts with oil, it causes oxidation, leading to the formation of sludge, varnish, tar, and corrosive acids (see Figure A9). These byproducts increase viscosity and cause physical damage to bearing surfaces, eventually clogging filters and impairing equipment performance, which can result in costly downtime.

Figure A9 Example of sludge buildup on a chain drive. Courtesy: ENGTECH Industries Inc.

Key Characteristics:

- **Oxidation effects**: Produces harmful byproducts that degrade lubricant quality and damage equipment.
- **Antioxidant additives**: Also known as oxidation inhibitors, these additives improve oxidation stability by neutralizing catalytic metals and breaking down hydroperoxides.

- **Common additives**: Zinc dialkyldithiophosphate (ZDDP) is a widely used antioxidant that also provides antiwear protection. Organic amines and phenols are also employed to slow oil degradation.

Regular oil analysis and timely replenishment of antioxidants are critical to maintaining lubricant integrity and extending the service life of the equipment.

57. Antirust Additive

An **Antirust additive** is a chemical component added to lubricants to **prevent rust and corrosion on metal surfaces**. These additives work by forming a **protective barrier** against moisture, oxygen, and other corrosive elements that can degrade metal components in engines, gear systems, hydraulic systems, and industrial machinery.

Key Functions of Antirust Additives in Lubrication

- **Prevents rust formation**: Creates a **thin protective film** on metal surfaces to **block moisture and oxygen exposure**.

- **Inhibits corrosion**: Neutralizes acidic contaminants and prevents **chemical reactions that cause metal deterioration**.

- **Extends equipment life**: Reduces **wear and damage caused by rust**, ensuring **longer service life for lubricated components**.

- **Enhances lubricant performance**: Helps maintain **oil stability and prevents sludge or deposit buildup**.

Types of Antirust Additives

- **Polar additives (surface-active compounds)**: These molecules **attach to metal surfaces**, forming a protective layer that repels water and corrosion. Examples: **amines, esters, and phosphates**.

- **Oil-soluble rust inhibitors**: Dissolved in oil, these additives coat metal surfaces to prevent **moisture-induced oxidation**. Common in **engine and gear oils**.

- **Emulsifiable antirust agents**: Used in **metalworking fluids and water-based lubricants**, providing **temporary corrosion protection**.

- **Vapor phase inhibitors (VPIs)**: Release a protective vapor that **forms a rust-preventive film on nearby metal surfaces**, used in **storage and shipping applications**.

58. Antiscuff Additive

Antiscuff Additives are specialized chemical compounds added to lubricants to **prevent severe surface damage under high-load, high-pressure conditions**. These additives create a protective film that reduces metal-to-metal contact, preventing scuffing, scoring, and adhesive wear in gears, bearings, camshafts, and high-pressure components.

Key Functions of Antiscuff Additives in Lubrication

- **Prevents scuffing and surface welding**: Forms a **protective film** that reduces **friction and metal transfer** during extreme operating conditions.
- **Enhances extreme pressure (EP) protection**: Helps lubricants withstand **high loads without breaking down**, ensuring **longer component life**.
- **Reduces wear in boundary lubrication conditions**: Effective when the **lubricant film thickness is reduced**, such as during **startup, shock loads, or low-speed operations**.
- **Extends lubricant and equipment life**: Protects **gears, cams, and other high-load surfaces** from damage, reducing maintenance costs.

Types of Antiscuff Additives

- **Extreme pressure (EP) additives**: Contain **sulfur, phosphorus, or chlorinated compounds** that create a **chemical reaction with metal surfaces**, forming a wear-resistant layer.
- **Solid lubricants**: **Molybdenum disulfide (MoS_2), graphite, and PTFE (Teflon)** provide a **physical barrier** to reduce friction and prevent scuffing.
- **Zinc dialkyldithiophosphate (ZDDP)**: Common in **engine oils**, it offers both **antiwear and antioxidant protection**.
- **Borate-based additives**: Used in gear and hydraulic oils, providing **thermal stability and surface protection** under extreme loads

59. Antiseize Compound

An **Antiseize compound** is a high-performance lubricant designed to **prevent metal-to-metal seizing, galling, and corrosion** in threaded fasteners, joints, and high-temperature applications. It is commonly used on bolts, studs, flanges, and press-fit assemblies to ensure easy disassembly, reduced friction, and long-term protection in extreme environments.

Key Functions of Antiseize Compounds

- **Prevents seizing and galling**: Reduces friction and metal adhesion, preventing fasteners from welding together under high loads and temperatures.
- **Enhances corrosion and oxidation resistance**: Forms a protective barrier against moisture, chemicals, and oxidation, reducing rust and corrosion in extreme environments.
- **Facilitates easy component disassembly**: Prevents threads and mating surfaces from locking up, making maintenance and repairs easier.
- **Reduces wear in extreme conditions**: Provides high-load and high-temperature protection, preventing surface damage in threaded components and metal joints.

Types of Antiseize Compounds

- **Copper-based antiseize**: Designed for high-temperature applications (up to 1800 °F/982 °C) in exhaust systems, bolts, and industrial equipment.
- **Nickel-based antiseize**: Used in extreme heat (up to 2600 °F/1427 °C) and corrosive environments, ideal for stainless steel, chemical processing, and aerospace applications.
- **Aluminum-based antiseize**: A general-purpose antiseize for moderate temperatures (up to 1600 °F/871 °C), commonly used in automotive and machinery components.
- **Graphite-based antiseize**: Provides solid lubrication for high-pressure and high-temperature environments.
- **Molybdenum disulfide (MoS_2) Antiseize**: Ideal for high-load applications, such as bearing surfaces and sliding components.
- **Food-grade antiseize**: Formulated for FDA-compliant applications in food processing, pharmaceutical, and beverage industries.

60. Antistatic Additive

Antistatic additives are chemical compounds used in fuel oils and gasoline to increase fluid conductivity and prevent hazardous static discharge. These additives are essential during transportation, pumping, and filling operations to reduce the risk of ignition caused by static electricity.

Key Characteristics:

- **Function**: Enhance fluid conductivity to safely dissipate static charges.
- **Application**: Used in fuels to prevent dangerous static discharge during handling and transfer.
- **Importance**: Critical for ensuring safety in environments where static sparks could lead to combustion.

Antistatic additives play a vital role in maintaining safe operations when handling flammable fuels.

61. Antistick-slip Additives

Antistick-slip additives are specialized chemical compounds incorporated into lubricants to minimize the stick-slip phenomenon, which occurs when there is a sudden transition between static and kinetic friction in moving parts. This effect often causes jerky or uneven motion, leading to noise, vibration, and potential damage to machinery.

Key Characteristics:

- **Function**: Reduce the difference between static and kinetic friction to ensure smooth, consistent movement of components.
- **Applications**: Commonly used in precision machinery, hydraulic systems, and applications requiring steady, controlled motion, such as in guides and slides.
- **Benefits**: Help prevent vibration, noise, and wear, enhancing the overall efficiency and lifespan of equipment.

Antistick-slip additives are crucial for applications where precise and stable motion control is required to maintain equipment performance and reliability.

62. Antiwear Additive

Antiwear Additives are chemical components added to lubricants to reduce friction and protect metal surfaces from damage under high-pressure or boundary lubrication conditions. These additives form a protective film on metal surfaces, minimizing direct metal-to-metal contact, reducing wear, and extending the life of equipment.

Key Characteristics:

- **Function**: Create a protective layer to prevent surface damage and reduce friction.
- **Applications**: Essential in engines, gearboxes, and hydraulic systems, where the lubricant film may degrade under extreme conditions.
- **Common Additives**: Zinc Dialkyldithiophosphate (ZDDP) is widely used for both antiwear and antioxidant protection. Other agents include Tricresyl Phosphates (TCP), halocarbons (chlorinated paraffins), sulfur-based compounds, and esters, which chemically bond to metal surfaces to form a protective ash-like film.

Antiwear additives are crucial for ensuring equipment reliability and performance, especially in high-stress or high-load environments.

63. API

See "American Petroleum Institute"

64. API Gravity

API Gravity is a scale developed by the **American Petroleum Institute (API)** to measure the relative density of petroleum liquids. It indicates whether a petroleum liquid is heavier or lighter than water, with values expressed in degrees.

Key Characteristics:

- **Measurement**: API gravity values range from 10 to 70 degrees. If the API gravity is greater than 10, the liquid will float on water; if less than 10, it will sink.
- **Calculation: API gravity is determined using the Formula:**

$$\text{API Gravity} = \frac{141.5}{\text{Specific Gravity at 60F}} - 131.5.$$

- **Importance**: Used to classify and evaluate petroleum liquids in terms of density and quality.

API gravity helps determine the characteristics and market value of different petroleum products, playing a crucial role in the petroleum industry.

65. API Rating

The **American Petroleum Institute (API)** establishes standards for motor oils to ensure they meet specific performance and protection criteria for various engine types. These standards are denoted by service categories, each tailored to particular engine requirements and technological advancements.

API Service Categories:
Gasoline Engines: **S Series**: Designated for spark-ignition engines, these categories include:

- **API SP**: The latest standard, introduced in 2020, offering enhanced protection against high-temperature deposits, oxidation, and wear.
- **API SN**: Introduced in 2010, providing improved high-temperature deposit protection and better wear protection.
- **API SM**: Introduced in 2004, offering enhanced oxidation stability and wear protection.
- **API SL**: Introduced in 2001, providing improved high-temperature deposit protection and wear protection. See Figure A10.
- **API SJ**: Introduced in 1996, offering improved oxidation stability and wear protection.

Diesel Engines: **C Series**: Designed for compression-ignition engines, these categories include:

- **API CK-4**: Introduced in 2016, formulated for high-speed four-stroke cycle diesel engines, providing enhanced protection against oil oxidation, viscosity loss due to shear, and oil aeration.
- **API CJ-4**: Introduced in 2006, designed to meet 2007 on-highway exhaust emission standards, offering improved protection against oil oxidation and wear.
- **API CI-4**: Introduced in 2002, formulated to sustain engine durability where exhaust gas recirculation (EGR) is used, providing improved protection against oil oxidation and wear.

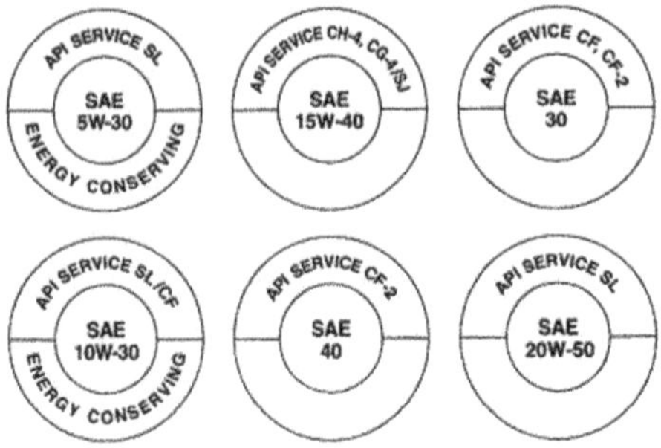

Figure A10 PI engine rating identification symbols typically found on lubricant containers.

Understanding API Ratings:
Each API service category is represented by a two-letter code:

- **First Letter**: Indicates the engine type (e.g., "S" for spark-ignition, "C" for compression-ignition).
- **Second Letter**: Denotes the performance level; higher letters signify more advanced standards.

For example, **API SP** indicates a service category for spark-ignition engines with the latest performance standards.

Selecting the Appropriate Motor Oil:
When choosing motor oil, consider the following:

- **Vehicle Manufacturer Recommendations**: Always refer to your vehicle's owner manual for the recommended API service category and viscosity grade.
- **Driving Conditions**: Severe driving conditions may necessitate oils with higher performance standards.
- **Oil Compatibility**: Ensure the oil is compatible with your engine's design and emission system.

66. Apparent Viscosity

Apparent viscosity refers to the resistance of a non-Newtonian fluid to flow under specific shear conditions, measured in poise or centipoise. Unlike Newtonian fluids, such as water or mineral oils, which have a constant viscosity, non-Newtonian fluids – like greases, polymer solutions, or slurries – experience viscosity changes based on the shear rate applied.

Key Characteristics:

- **Non-Newtonian behavior**: Viscosity varies with the speed or force of movement, causing fluids to either shear thin (decrease in viscosity) or shear thicken (increase in viscosity).
- **Industrial relevance**: Apparent viscosity is crucial for understanding fluid behavior under dynamic conditions and ensuring proper lubricant performance.
- **Lubrication management**: Helps predict a lubricant's ability to form a protective film under different operating speeds and loads, aiding in wear prevention and maintaining system efficiency.

Measuring apparent viscosity is essential in various industrial applications to optimize fluid selection and ensure reliable equipment operation.

67. Appearance – Grease

Appearance describes the visual characteristics of an undisturbed grease surface using specific terms:

Key Descriptors:

1. **Bleeding**: Presence of free oil on the surface (see Figure A11).
2. **Cracking**: Visible surface cracks.
3. **Grainy**: Small lumps or granules present.
4. **Rough**: Surface with numerous small irregularities.
5. **Smooth**: Surface without visible irregularities.

Additional Visual Characteristics:

- **Bloom**: The sheen or reflection on the surface.
- **Color**: The hue of the grease.
- **Luster**: The shine or glossiness.
- **Texture**: The feel or consistency when touched.
- **Bulk appearance**: The overall look of the grease in its container.

These descriptors and characteristics are important for assessing the quality and condition of grease in lubrication applications.

Figure A11 Old grease with an appearance of **bleeding** and a **rough surface**. Courtesy: ENGTECH Industries Inc.

68. Aqueous Lubricant

An **Aqueous lubricant** is a water-based or water-containing lubricant formulated to provide low friction, cooling, and environmental benefits while maintaining effective lubrication in industrial, biomedical, and specialty applications. Unlike traditional oil-based lubricants, aqueous lubricants are designed to reduce heat, prevent wear, and offer biodegradable or nontoxic alternatives for various industries.

Key Functions of Aqueous Lubricants

- **Reduces friction and wear**: Provides a thin lubricating film that minimizes metal-to-metal or material-to-material contact.
- **Enhances cooling and heat dissipation**: Water content absorbs and transfers heat efficiently, preventing thermal buildup in high-speed applications.
- **Environmentally friendly and biodegradable**: Many aqueous lubricants are nontoxic and eco-friendly, making them ideal for food-grade, medical, and sustainable industrial use.
- **Reduces fire hazard**: Unlike petroleum-based lubricants, aqueous lubricants have a high water content, reducing flammability risks in industrial environments.

Types of Aqueous Lubricants

- **Water-based metalworking fluids**: Used in cutting, grinding, and machining operations to provide cooling and lubrication.
- **Emulsion-based lubricants**: Oil-in-water formulations that combine oil's lubricity with water's cooling properties, often used in industrial and automotive applications.
- **Synthetic water-soluble lubricants**: Fully water-based, containing polymers or advanced additives for high-performance lubrication in aerospace and biomedical applications.
- **Biomedical and personal lubricants**: Water-based lubricants used in medical devices, surgical instruments, and pharmaceutical applications for safe, nontoxic lubrication.

69. Aromatic

Aromatic refers to a class of organic compounds characterized by one or more benzene rings or cyclic structures with alternating double and single bonds, forming a conjugated system. These compounds possess enhanced stability due to aromaticity, where electrons are shared across the ring, creating a delocalized electron cloud. Common examples include benzene, toluene, and naphthalene.

Key Characteristics:

- **Structure and stability**: Aromatics have a stable ring structure with delocalized electrons.
- **Industrial use**: Often utilized as solvents or additives in lubrication due to excellent solvency properties.
- **Reactivity**: High aromatic content can increase chemical reactivity and potentially degrade materials like seals or gaskets.

In lubrication systems, managing aromatic content is crucial to balance performance with material compatibility. Due to their high reactivity and solvency, aromatic hydrocarbons are preferred over paraffin and naphthene stocks as petrochemical building blocks for base oils. Aromatic refers to a class of organic compounds characterized by one or more benzene rings or related cyclic structures with alternating double and single bonds, known as conjugated systems. These compounds exhibit enhanced stability due to aromaticity, a property where electrons are shared across the ring structure, forming a delocalized electron cloud. Examples include benzene, toluene, and naphthalene.

70. Aromatic Base Oil Stock

Base oil stock manufactured from aromatic hydrocarbon stock, see "Aromatic."

71. Arrhenius Rule

The Arrhenius rule (or Arrhenius equation) is a mathematical formula that describes how the rate of a chemical reaction increases with temperature. Developed by Swedish scientist Svante Arrhenius, it explains that reaction rates accelerate exponentially with higher temperatures because greater thermal energy leads to more frequent and energetic molecular collisions, overcoming the activation energy barrier.

Equation:

$$k = A \cdot e^{\frac{-Ea}{RT}}.$$

where:

- k: Reaction rate constant
- A: Frequency factor, indicating the frequency of collisions
- E_a: Activation energy, the minimum energy needed for the reaction
- R: Universal gas constant
- T: Absolute temperature in Kelvin

Applications in Lubrication:

- **Oxidation prediction**: Used to estimate how temperature influences the oxidation rate of lubricants and the degradation of components.
- **Lifespan estimation**: Helps predict the lifespan of oils, greases, and materials, as a 10 °C (18 °F) increase typically doubles the oxidation rate, reducing the lubricant's lifecycle by half. Conversely, lowering the temperature can extend the lubricant's service life.

Understanding the Arrhenius Rule is crucial in lubrication management and material science to optimize maintenance intervals and prolong the lifespan of lubricants and equipment.

72. Arrhenius, Svante

Svante Arrhenius (1859–1927) was a pioneering Swedish scientist recognized as one of the founders of physical chemistry. His groundbreaking work led to significant discoveries, including the **Arrhenius equation**, (see Arrhenius rule) which describes how temperature influences chemical reaction rates. In 1903, he was awarded the **Nobel Prize in Chemistry** for his theory of electrolytic dissociation, showing that salts dissociate into ions when dissolved in water, enabling electrical conductivity.

Key Contributions:

- **Chemistry**: Developed the Arrhenius equation and advanced understanding of ionic dissociation.
- **Climate science**: In 1896, created the first quantitative model of global warming caused by rising atmospheric carbon dioxide, pioneering research on the greenhouse effect.
- **Academic leadership**: Served as the first director of the Nobel Institute for Physical Chemistry in Stockholm.

Arrhenius's work has had a lasting impact across fields such as environmental science, physical chemistry, and materials science, shaping our understanding of both chemical reactions and climate change.

73. Ash

Ash in lubricating oil refers to the noncombustible residue left after the oil is burned. This residue is primarily due to metallic-based additives in the lubricant, such as detergents, dispersants, and antiwear agents, which contain elements like calcium, magnesium, zinc, or phosphorus. These additives are crucial for enhancing oil performance by neutralizing acids, preventing sludge, and protecting against wear.

Key Points:

- **Additive contribution**: Elements in metallic additives form ash when burned.
- **Performance impact**: Excess ash content can lead to deposits on engine components, such as valves and pistons, causing issues like preignition, knocking, or fouling.
- **Guidelines**: The recommended maximum ash content for most lubricating oils is around 0.2% by weight.
- **Modern engines**: Monitoring ash levels is particularly important in low-ash oils to avoid emissions problems and maintain engine efficiency.

Proper ash management is essential to ensure optimal engine performance and longevity, especially in modern, emissions-sensitive engines.

74. Asperity

Asperity refers to the microscopic high points or roughness found on the surface of a material. Even surfaces that appear smooth to the naked eye contain asperities, which can come into contact under load, leading to friction, wear, or surface damage. Figure A12 illustrates the actual microscopic asperic high/low surface profile of a supposed smooth metal surface.

Key Points:

- **Surface roughness**: Asperities are the tiny peaks and valleys on a material's surface.
- **Friction and wear**: When two surfaces come into contact, asperities interact, affecting lubrication and potentially causing material degradation.
- **Relevance in lubrication**: Effective lubrication aims to minimize the contact between asperities, reducing friction and wear to enhance the performance and longevity of machinery.

Full-film lubrication forms a protective layer over these microscopic peaks, preventing direct surface-to-surface contact, thereby reducing friction, and extending equipment life. Understanding asperities is crucial for designing and maintaining systems that rely on smooth, efficient motion and wear resistance.

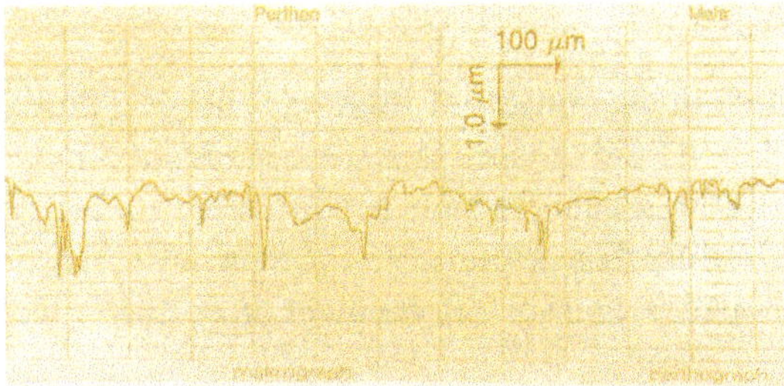

Figure A12 Ground metal surface profilogram demonstrates the surface asperity of metal that appears perfectly smooth to the naked eye.

75. ASTM

See "American Society for Testing Materials."

76. ASTM Test

ASTM lubricant test refers to a standardized testing method designated by a specific reference number established by ASTM International. These tests are used to evaluate various properties and performance characteristics of lubricants, ensuring they meet industry standards.

For example: the **ASTM test D892** describes the test procedure for the **Foaming Characteristics of Lubricating Oils**.

Reference and Resources:

- For a comprehensive list of ASTM lubricant tests, refer to **Practical Lubrication for Industrial Facilities, Fourth Edition, 2024** by Kenneth E. Bannister.
- Additional information can be found on the ASTM website at www.astm.org.

These standardized tests are critical for maintaining the quality and reliability of lubricants in industrial applications.

77. ATF

See "Automatic Transmission Fluid."

78. Atmosphere

Atmosphere (atm) in the context of fluid pressure is a unit used to describe the force exerted by a fluid on a surface. One standard atmosphere is defined as **101.325 kPa** (kilopascals), which is equivalent to **14.7 psi** (pounds per square inch) or **760 mmHg** (millimeters of mercury). This value represents the average atmospheric pressure at sea level on earth.

Key Characteristics:

- **Standard reference**: 1 atm is used as a baseline for measuring pressure.
- **Pressure measurements**: Gauge pressure is measured relative to atmospheric pressure, while absolute pressure is measured relative to a vacuum.
- **Application example**: A system with an absolute pressure of 2 atm means the fluid exerts twice the pressure of the surrounding atmosphere.

Understanding atmospheric pressure is essential for calibrating hydraulic and pneumatic systems, ensuring correct fluid flow, and maintaining efficiency under varying environmental conditions.

79. Atomic Emission Spectroscopy

Atomic emission spectroscopy (AES) is an analytical technique used to identify and quantify the elemental composition of a lubricant sample by analyzing the light emitted from excited atoms. When a sample is energized, typically using high-temperature sources such as plasma, flame, or arc, atoms absorb energy and emit light as they return to their ground state. Each element emits light at specific wavelengths, enabling identification and measurement (see Oil Analysis).

Key Characteristics:

- **Mechanism**: Atoms emit light at characteristic wavelengths, which are analyzed to determine the elements present.
- **Qualitative and quantitative analysis**: AES provides detailed information about the types and amounts of elements in a sample.
- **Applications**: Used in metallurgy, environmental monitoring, and lubrication analysis to detect trace metals, impurities, or additives.

Inductively coupled plasma atomic emission spectroscopy (ICP-AES) is a common variant that uses a plasma source for improved sensitivity and accuracy. AES is highly valued for its precision and ability to simultaneously analyze multiple elements. In lubrication analysis, it is routinely used to detect wear metals, helping predict equipment failures and optimize maintenance schedules.

80. Auto-ignition Point

Auto-ignition point is the minimum temperature at which a substance, such as a fuel or lubricant, spontaneously ignites without an external ignition source, like a flame or spark. At this temperature, the substance rapidly oxidizes, releasing heat and triggering combustion.

Key Characteristics:

- **Spontaneous ignition**: Occurs without any external spark or flame.
- **Safety importance**: Understanding the auto-ignition point is essential in industrial settings to prevent accidental combustion, especially in high-temperature environments.
- **Risk factors**: Lower auto-ignition points increase the risk of combustion, necessitating stringent safety measures. Factors like pressure, oxygen concentration, and contamination can also affect the auto-ignition temperature.

81. Automated Control

Automated control in lubrication delivery systems refers to the use of computers, electronic sensors, controllers, and pumps to precisely manage the delivery of lubricant to machine bearing points without manual intervention. These systems ensure that the correct amount of lubricant is applied at optimal intervals, minimizing the risks of overlubrication, underlubrication, or missed lubrication points (see Automatic Lubrication System).

Key Characteristics:

- **Precision delivery**: Lubricant is applied accurately to the right points, enhancing machine reliability.
- **Components**: Includes a central computer/controller/timer, pump, valves, and sensors that monitor parameters like temperature, pressure, and flow rate.
- **Real-time adjustments**: The system uses real-time data to automatically adjust lubrication, ensuring consistent and efficient performance.

Automated lubrication systems can greatly improve equipment lifespan and efficiency while reducing maintenance costs and the risk of human error.

82. Automatic Lubrication System

An **automatic lubrication system** is a self-operating, automated lubrication system designed to deliver precise amounts of lubricant to machinery components without manual intervention. These systems optimize lubrication schedules, reduce human error, and improve equipment reliability by ensuring consistent and accurate lubricant application.

Key Characteristics:

- **Delivers lubricant automatically**: Ensures **timely and precise lubricant application**, preventing **over or underlubrication**.
- **Reduces maintenance costs and downtime**: Eliminates the need for **manual lubrication**, increasing **equipment uptime and efficiency**.
- **Extends equipment life**: Maintains **consistent lubrication film**, reducing **friction, wear, and component failures**.
- **Improves workplace safety**: Reduces the need for **workers to access hazardous or hard-to-reach lubrication points**.

Types of Autonomous Lubrication Systems

- **Single-point lubricators**: **Battery- or gas-powered devices** that dispense lubricant at set intervals for bearings, chains, and small machinery.
- **Multipoint lubrication systems**: Delivers lubricant from a centralized reservoir to multiple lubrication points in industrial applications.
- **Circulating oil systems**: Continuously circulates and filters oil to ensure clean, consistent lubrication in large rotating equipment.
- **Progressive lubrication systems**: Uses metering devices to distribute precise lubricant amounts to multiple points in a sequential manner.

Benefits:

- **Improved efficiency**: Lubricates equipment while in operation, reducing maintenance interruptions.
- **Reduced maintenance costs**: Prevents overlubrication and underlubrication, reducing the risk of equipment failure.

- **Safety**: Removes the need for technicians to access hard-to-reach or hazardous areas for lubrication.

Examples of automatic lubrication systems include single-line resistance (SLR) systems, dual-line systems, and progressive lubrication systems, all tailored for various industrial and heavy machinery applications. Figure A13 depicts a dual-line parallel automated lubricant delivery system.

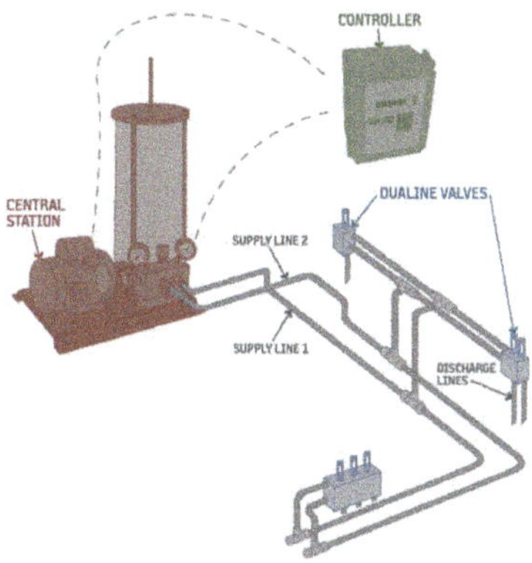

Figure A13 Dual line parallel automated lubricant delivery system. Source: Bijur–Delimon International.

83. Automatic Transmission Fluid

Automatic transmission fluid (ATF) is a specialized lubricant formulated for use in vehicle automatic transmissions. It plays several essential roles, including lubricating moving parts, cooling the transmission, reducing friction, and transmitting hydraulic pressure. ATF is also designed to prevent sludge, varnish, and other contaminant buildup, ensuring smooth gear shifts and prolonging the transmission's lifespan.

Key Characteristics:

- **Composition**: A blend of base oils and additives, such as detergents, antiwear agents, friction modifiers, and antioxidants, to meet the demands of automatic transmissions.
- **Types**: Various formulations are designed to match specific transmission requirements, such as DEXRON, MERCON, and CVT fluids.
- **Identification**: ATF is usually dyed red or orange to distinguish it from other automotive fluids.

Automatic Transmission Fluid is crucial for maintaining efficient and reliable transmission performance in a wide range of vehicles.

84. Automotive Bearing

Automotive bearing refers to a mechanical component used in vehicles to reduce friction between moving parts and support rotational or linear motion. Bearings are crucial for the smooth operation of various automotive systems, such as wheels, transmissions, engines, and drivetrains.

Key Characteristics:

- **Function**: Minimize friction and wear, facilitate smooth motion, and support loads in automotive applications.
- **Types**: Common types include ball bearings, roller bearings, and needle bearings, each designed for specific functions and load conditions.
- **Applications**: Used in wheel hubs, gearboxes, engines, gearboxes, steering columns, and other critical vehicle components.
- **Commonality**: Because of their commonality and manufactured volume, these bearings are inexpensive to purchase.

Automotive bearings comply with a specified automotive industry standard for size, fit, and quality. In doing so, they are interchangeable between manufacturers and volume manufacturing makes them inexpensive to purchase.

85. Autonomous Lubrication System

See "Automatic Lubrication System."

86. Axial Bearing

Axial bearing refers to a type of bearing designed to accommodate forces that are applied parallel to the axis of rotation, known as axial or thrust loads. These bearings are specifically engineered to handle loads that push or pull along the shaft rather than perpendicular to it.

Key Characteristics:

- **Function**: Supports and manages axial forces in rotating or moving components.

- **Types**: Includes thrust ball bearings, cylindrical thrust roller bearings, and tapered thrust roller bearings, each suited for different load capacities and applications.
- **Applications**: Commonly used in automotive transmissions, rotary tables, propeller shafts, and other machinery where axial load support is critical.

Axial bearings are essential for ensuring smooth operation and stability in systems where axial loads are present, protecting components from excessive wear and damage.

B

87. Babbitt

Babbitt is a soft, white, nonferrous alloy used as an inner bearing surface material for lining large journal bearings. Invented by Isaac Babbitt in 1839, it provides a soft, malleable surface that allows particulates to embed within the material, reducing abrasive damage to the shaft and extending the bearing's lifespan.

Key Characteristics:

- **Original composition**: Initially lead-based but now primarily made from tin-based alloys.
- **Modern composition**: Typically consists of tin, copper, and antimony to improve strength and durability.
- **Function**: Protects the shaft from wear and damage by embedding debris and maintaining a smooth operation.

Babbitt materials are crucial in heavy machinery and industrial applications, providing reliable and durable bearing surfaces. Babbitt-lined bearings, also referred to as shell bearings, are still used today in engines, turbines, compressors, and electric motors. They provide smooth operation under high loads by forming a conformable surface that compensates for slight misalignments. Figure B1 shows a failed babbitt-lined automotive shell bearing with embedded debris.

DOI: 10.1201/9788743806608-88

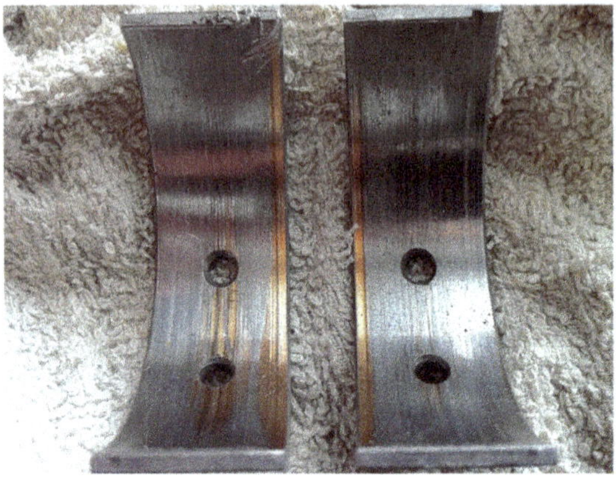

Figure B1 Worn babbitt-lined shell bearing with embedded debris. Courtesy: ENG**TECH** Industries Inc.

88. Backflow Contamination

Backflow contamination occurs when used, dirty, or contaminated lubricant flows backward into a clean system, introducing particles, moisture, or degraded oil that can cause increased wear, reduced efficiency, and potential equipment failure. This issue is common in hydraulic systems, lubrication circuits, and fluid handling applications where improper sealing, pressure imbalances, or system design flaws allow contaminants to enter unintended areas.

Causes of Backflow Contamination

- **Pressure fluctuations**: Sudden pressure drops or surges can cause the lubricant to reverse flow, carrying contaminants with it.
- **Faulty or missing check valves**: Check valves prevent backflow, and if they fail or are absent, contaminated lubricant can re-enter clean areas.
- **Improperly designed drain or return lines**: Poor system design can lead to reverse flow, especially in gravity-fed or pump-driven lubrication systems.
- **Filter saturation or clogging**: If a lubrication system filter becomes clogged, the contaminated lubricant may bypass filtration and re-enter the system.

- **External contaminant intrusion**: In open systems, water, dust, or foreign particles can enter and mix with clean lubricant due to backflow.

Effects of Backflow Contamination

- **Accelerated wear and component damage**: Contaminants such as metal particles, dirt, and water can cause abrasive and corrosive wear.
- **Reduced lubrication efficiency**: Dirty oil or grease can lose viscosity, degrade film strength, and reduce friction protection.
- **Increased downtime and maintenance costs**: Contaminated systems require frequent oil changes, filter replacements, and component repairs.
- **Potential system failures**: Hydraulic and lubrication systems can lose pressure or lubrication film integrity, leading to overheating, galling, or seizure.

89. Bactericide Additive

Bacterial additives are chemicals added to lubricants, fuels, or fluids to control and prevent bacterial growth. In systems such as hydraulic systems, cutting fluids, or storage tanks, bacterial contamination can cause problems like slime formation, corrosion, and foul odors. Also known as biocides, these additives are formulated to either inhibit bacterial growth or eliminate existing bacterial colonies, keeping fluids effective and free from contamination.

Key Characteristics:

- **Function**: Prevent or eliminate bacterial growth to maintain fluid integrity.
- **Applications**: Used in environments prone to moisture exposure, such as cutting fluids or fuel storage, where microbial growth is a risk.
- **Benefits**: Prevent microbial-induced corrosion, reduce sludge formation, extend fluid service life, and maintain system hygiene.

Bacterial additives are crucial for protecting systems from microbial damage.

90. Bag Filter

Bag filter refers to a filtration device that uses fabric or felt bags to remove solid particles from liquids or gases. In industrial and automotive applications,

bag filters are commonly used to filter lubricants, hydraulic fluids, or air, to ensure contaminants like dirt, metal particles, and sludge are removed to assure system performance.
Bag filters work by forcing the fluid or gas through the fabric of the bag, which traps contaminants while allowing the clean medium to pass through. Bag filters are able to capture contaminants down to a 1-micron rating and can be configured in single or multiple-bag housings for higher filtration efficiency. See Figure B2.

Key Points:

- **Ease of maintenance**: Bags can be replaced quickly when clogged.
- **Scalability**: Suitable for high-flow applications with multiple-bag setups.
- **Cost-effectiveness**: They provide an economical filtration solution for large systems.

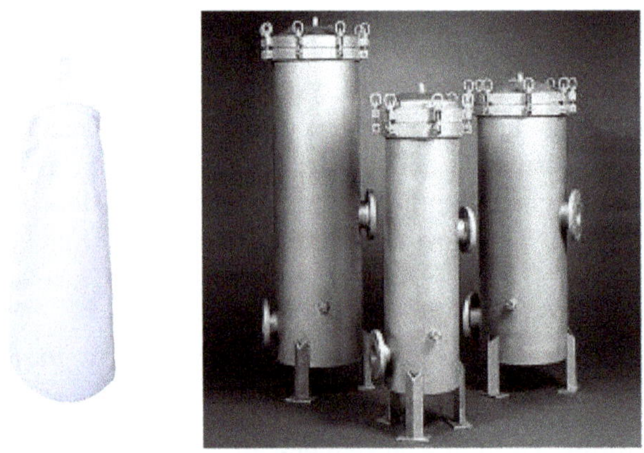

Figure B2 Fabric bag filter (left) that fits into the cylindrical bag filter housings (right).

Bag filters are widely used in industries such as oil refining, chemical processing, wastewater treatment, and food and beverage production, as a primary filter on the pressure side of large-flow circulative oil and hydraulic systems.

91. Ball Bearing

Ball bearing is an inexpensive rolling-element bearing that uses balls to maintain separation between bearing races, reducing friction and enabling smooth rotation (see Figure B3). Designed to handle both radial loads (forces perpendicular to the axis of rotation) and axial loads (forces parallel to the axis), ball bearings are known for their versatility, high speed capability, and relatively low friction compared to other types of bearings.

Key Characteristics:

- **Inner and outer races**: The balls roll between these rings, distributing the load.
- **Cage**: Holds the balls evenly spaced and prevents them from contacting each other.
- **Balls**: Provide minimal contact area, reducing friction.

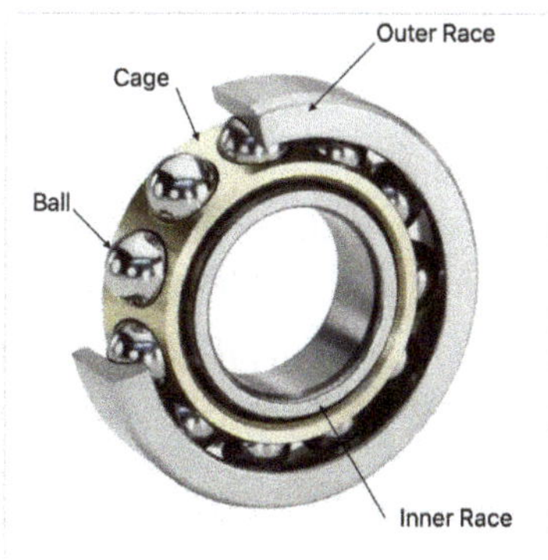

Figure B3 Ball bearing cutaway.

Ball bearings are inexpensive and widely used in applications such as motors, wheels, fans, gearboxes, and pumps, where both smooth motion and low friction are required (see Automotive Bearings). However, compared to roller bearings, they have a limited capacity for handling heavy loads.

92. Bar

Bar is a unit of pressure used to quantify the force exerted by a fluid or gas. One bar is equal to **100 kilopascals** (kPa) or approximately **14.5 pounds per square inch** (psi), making it slightly less than the average atmospheric pressure at sea level (1.013 bar).

Key Characteristics:

- **Equivalences**: 1 bar ≈ 1 atmosphere ≈ 100,000 Pascals (Pa) ≈ 14.5 psi.

- **Applications**: Commonly used in measuring tire pressures, fuel injection systems, lubrication and hydraulic pump pressure.

The bar is a convenient metric unit of pressure often referenced alongside psi for various industrial and automotive applications. See "PSI" for more information on pressure units.

93. Barium Grease

Barium grease is a lubricant formulated with barium-based thickeners, such as barium soaps or barium complex compounds. Renowned for its excellent water resistance, oxidation stability, and high-temperature performance, barium grease is ideal for challenging industrial and automotive applications.

Key Characteristics:

- **Performance**: Offers superior protection in environments exposed to moisture, saltwater, or harsh chemicals, making it valuable for marine operations, mining, and heavy machinery.
- **High-load capability**: Adheres well to metal surfaces, providing durable lubrication under high loads and pressures, and can operate at temperatures exceeding 275 °F (135 °C).
- **Water resistance**: Maintains consistency and resists emulsification in wet conditions, enhancing equipment protection and reducing maintenance.

Environmental Considerations:

Due to concerns over the environmental impact and toxicity of barium compounds, the use of barium-based greases is often restricted, and safer alternative formulations are being developed. Despite these limitations, barium grease remains a dependable choice for specialized applications requiring its unique properties.

94. Barrel

Barrel is a standard unit of liquid volume used in the US petroleum industry. One barrel of crude oil or petroleum product is defined as **42 US gallons, 35 imperial gallons, or 159 liters**.

Key Points:

- **Volume**: 1 barrel = 42 US gallons = 35 imperial gallons = 159 liters.
- **Global use**: Outside the United States, petroleum products are often measured in cubic meters or tonnes, but the barrel remains essential for pricing and trading in the global oil market.

The barrel is a crucial unit in the oil industry, providing a consistent standard for comparing and trading oil volumes worldwide.

95. Barrel Pump

Barrel pump is a mechanical/pneumatic pumping device designed to efficiently and safely transfer fluids from barrels, drums, or other containers. They are sometimes referred to as a "drum" pump. Commonly used for handling oils, chemicals, fuels, lubricants, and various other liquids, barrel pumps help prevent spillage, reduce waste, and ensure a clean transfer process without manual pouring. As seen in Figure B4, a barrel pump can also be used with a control system to feed a centralized lubrication delivery system.

Key Characteristics:

- **Design**: Consists of a cylindrical tube with a piston or plunger mechanism, operated manually, electrically, or pneumatically. The pump is inserted into the container through a bung opening, and the liquid is drawn up and dispensed through a nozzle or spout.
- **Types**: Available in hand-operated, electric, and pneumatic versions to suit different fluid types and volumes.

Applications:

- **Industrial lubrication**: Transferring oils and greases in lubrication systems.
- **Fuel handling**: Used in automotive and marine industries for safe fuel transfer.
- **Chemical processing**: Safely moving hazardous or reactive substances in industrial settings.

Barrel pumps are essential in industrial environments for the safe and efficient transfer of various liquids, reducing the risk of spills and ensuring smooth operations.

Figure B4 Barrel pump and controller hooked up to a multipoint central lubrication system. Courtesy: ENG**TECH** Industries Inc.

96. Base (Alkaline)

Base refers to a chemical substance that can accept protons (H^+ ions) or release hydroxide ions (OH^-) in an aqueous solution. Bases are the opposite of acids on the pH scale, having a pH greater than 7. They are essential in various chemical and industrial processes, including neutralizing acids and maintaining pH balance. See Figure A4 – pH scale.

Key Characteristics:

- **Proton acceptor**: Bases accept H^+ ions or release OH^- ions in water.
- **pH Scale**: Bases have a pH value greater than 7.
- **Applications**: Used in processes such as acid neutralization and pH control in industrial settings.

Bases are fundamental in chemistry and industry, playing a vital role in maintaining chemical stability and reactivity.

97. Base Oil

Base oil is the primary fluid component used in the formulation of lubricants, greases, and industrial oils. It forms the foundation of a lubricant's properties, with additives blended in to enhance specific performance characteristics. Base oils can be derived from crude oil (mineral oils) or synthetically produced through chemical processes and are classified into five groups according to the American Petroleum Institute (API) standards.

Key Characteristics:

- **Composition**: Base oils make up at least 70% of the final lubricant product and can be mineral, semisynthetic, synthetic, or derived from animal/vegetable sources.
- **API classifications**: There are five API groups that categorize base oils based on their composition and refining methods (see "Base Oil Group Classification").
- **Performance factors**: Base oils influence key properties like viscosity, volatility, and thermal stability. The choice of base oil affects performance under temperature extremes, oxidation resistance, and compatibility with seals and additives.

Synthetic base oils generally provide superior performance and a longer service life compared to mineral oils, making them suitable for high-demand applications like automotive engines and industrial machinery.

98. Base Oil Group Classification

Most modern industrial oils rely on either mineral or synthetic base oils as their primary components. These base oils are classified into five distinct groups based on the refining or manufacturing methods used.

Group Classifications:

- **Group I**: Conventional base oils produced through solvent refining of crude oil. These oils have a viscosity index (VI) between 80 and 120, a sulfur content above 0.03%, and less than 90% saturated hydrocarbons. They offer moderate performance and are cost-effective.
- **Group II**: Hydrotreated mineral oils with improved oxidation stability. The hydrotreating process involves infusing hydrogen at temperatures over 600 °F and pressures above 500 psi, resulting in oils with sulfur content below 0.03% and hydrocarbon saturation of 90% or higher.

- **Group III**: Highly refined mineral oils processed through hydrocracking, which removes impurities like sulfur and nitrogen and saturates the oil with hydrogen. This intensive process occurs at temperatures above 650 °F and pressures over 1000 psi, producing oils with a high VI above 120 and a low pour point. Group III oils are often marketed as synthetic due to their advanced performance characteristics.
- **Group IV**: Fully synthetic oils composed of polyalphaolefin (PAO) base oils. These are synthesized from small hydrocarbon molecules, offering exceptional stability and performance in extreme temperatures and conditions.
- **Group V**: A diverse category that includes all other non-PAO synthetic base oils, such as esters, silicone oils, and polyalkylene glycols (PAGs). These oils provide specialized properties for unique or demanding applications.

Each group classification plays a vital role in lubricant formulation, balancing performance, cost, and environmental compliance to meet the needs of various industrial applications.

99. Base Oil Property

Base oil property refers to the essential characteristics of base oil that influence its performance and suitability for various lubrication applications. These properties are critical for predicting how the lubricant will behave under different operating conditions, such as temperature, load, and exposure to contaminants. Key properties include:

1. **Viscosity**: Measures the oil's resistance to flow, determining its ability to form a lubricating film between surfaces, which is crucial for minimizing wear and friction.
2. **Viscosity index (VI)**: Indicates how well the oil maintains its viscosity across a temperature range. A higher VI means the oil remains stable under fluctuating temperatures, preserving its lubricating properties.
3. **Oxidation stability**: The oil's resistance to degradation when exposed to oxygen and high temperatures. High oxidation stability reduces sludge and varnish formation, extending the oil's lifespan.
4. **Pour point**: The lowest temperature at which the oil remains fluid. Oils with a lower pour point are suitable for cold environments, ensuring smooth startup and operation.

5. **Volatility**: The oil's tendency to evaporate at high temperatures. Low volatility is important for minimizing oil loss and ensuring consistent lubrication.
6. **Flash point**: The temperature at which the oil emits a vapor that can ignite. A higher flash point indicates greater thermal stability and a reduced risk of fire hazards.
7. **Compatibility**: The oil's ability to work with system components like seals, gaskets, and additives without causing degradation or leakage over time.

These properties, along with the type of base oil – whether mineral, synthetic, or bio-based – determine the oil's overall performance. Selecting the right lubricant for an application depends on understanding and prioritizing these critical properties.

100. Bath Oiler

See "Oil Bath Lubricator."

101. Battle, John R.

John R. Battle (1889–1975) was an American pioneer in lubrication engineering from St. Louis, Missouri, who made significant and lasting contributions to the field. In 1916, he authored the **Lubricating Engineer's Handbook**, a foundational text that has guided professionals in understanding and enhancing lubrication practices. Battle also invented the "Gun-Fil" lubricating device, known for its effectiveness and reliability in machinery lubrication and still used today.

Legacy:

- The **International Council for Machinery Lubrication (ICML)** honors his contributions through the annual John R. Battle Award, recognizing organizations that excel in machinery lubrication practices.
- His work laid the foundation for modern lubrication standards, and his influence continues to be celebrated and respected within the industry.

John R. Battle's innovations and writings continue to remain integral to the evolution of lubrication engineering.

102. Bearing

A **bearing** is a sacrificial mechanical component designed to support and reduce friction between moving parts in machinery. It facilitates smooth motion – whether rotational, as in wheels and motors, or linear, as in sliding elements – by maintaining alignment and evenly distributing loads. Bearings reduce wear and resistance, enhancing the efficiency and lifespan of machinery.

Common Bearing Types:

1. **Ball bearings**: Handle both radial and axial loads, commonly used in motors and wheels.
2. **Roller bearings**: Provide higher load capacity than ball bearings, making them ideal for heavy machinery.
3. **Needle bearings**: Compact roller bearings designed for applications with limited space.
4. **Tapered roller bearings**: Handle combined radial and thrust loads, often used in automotive hubs.
5. **Plain bearings**: The simplest type, featuring a sliding surface without rolling elements, suitable for low-speed operations. These can be rotating or linear sliding surfaces.

Importance of bearing lubrication: Proper lubrication is crucial for efficient bearing performance. It prevents metal-to-metal contact, reduces friction, and shields bearings from contaminants, ensuring smooth operation and extending the service life of machinery.

A bearing is a sacrificial mechanical component that supports and reduces friction between moving parts in machinery. It enables smooth motion – whether rotational (as in wheels and motors) or linear (as in sliding elements) – by maintaining alignment and distributing loads evenly. Bearings minimize wear and tear by facilitating motion with minimal resistance, improving both the efficiency and lifespan of machinery.

103. Bearing Isolator

Bearing isolator is a two-part dynamic seal consisting of a stator and a rotor attached to the shaft. According to Inpro/Seal Corporation, the rotor and stator form a noncontacting labyrinth seal with no wearing parts. The isolator is specifically designed to extend bearing life by preventing the escape of lubricant and the entry of contaminants. See Figure B5.

Key Features:

- **Labyrinth design**: The inner portion of the labyrinth captures and redirects bearing lubricant back into the bearing, preventing lubricant loss.
- **Contaminant expulsion**: The outer labyrinth traps external contaminants, which are then expelled through a port in the rotor using centrifugal force and gravity.

By effectively managing both lubricant retention and contaminant exclusion, bearing isolators play a crucial role in maintaining the performance and longevity of bearings.

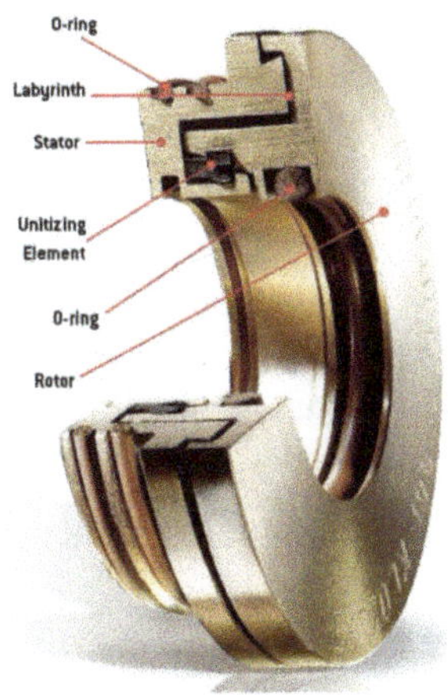

Figure B5 Typical bearing isolator. Courtesy: Impro Seal.

104. Bellows

Bellows are flexible, accordion-like components used in mechanical and industrial systems to accommodate expansion, contraction, or movement while protecting sensitive parts from external contaminants. They are made from various materials, including rubber, metal, leather, or fabric, and serve as seals or protective covers for moving parts, allowing axial, lateral, or angular flexibility.

Key Applications:

- **Sealing and protection**: Shield lubricated machinery components like actuators, shafts, and bearings from dirt, dust, and moisture.
- **Expansion joints**: Absorb thermal expansion in pipelines, reducing stress and enabling smooth operation.
- **Pneumatics and hydraulics**: Serve as flexible chambers, compressing air or fluids in valves and pumps.

Bellows are essential for ensuring the reliability and longevity of equipment by accommodating movement and preventing contamination.

105. Bentonite Grease

Bentonite grease is a nonsoap-based lubricant thickened with bentonite clay, valued for its outstanding high-temperature stability. Unlike soap-thickened greases, bentonite grease does not melt under extreme heat, making it ideal for applications where traditional greases might degrade. It is effective at temperatures above 200 °C (392 °F) and is commonly used in industrial equipment, ovens, kilns, bearings, and automotive components exposed to high heat.

Key Features:

- **Thermal stability**: Resistant to melting, making it well-suited for high-temperature environments.
- **Water resistance**: Maintains performance in moist or humid conditions, as it does not absorb water.
- **Oxidation resistance**: Minimizes the formation of varnish or sludge, enhancing reliability.
- **Long service life**: Retains consistency under high loads and heat, reducing the need for frequent relubrication.

Bentonite grease is widely used in automotive, industrial, and marine settings, particularly for bearings and equipment exposed to high heat or water contamination. However, caution is needed when using it in centralized lubrication systems, as it can harden in extreme conditions and potentially cause system blockages.

106. Beta Ratio

Beta ratio is a performance metric used to measure the efficiency of a filter by comparing the concentration of particles upstream and downstream of the filter. The Beta Ratio is determined by the filter manufacturer following the **ISO 16889:1999 standard** "multipass method for evaluating filtration performance of a fine filter element." This method quantifies how effectively a filter captures particles of a specified size or larger, making it an essential parameter in evaluating the filter's ability to remove contaminants from fluids like oil, fuel, or hydraulic fluid. Table B1 demonstrates filter efficiency as a beta ratio.

Formula:

$$\beta_x = \frac{C_u}{C_d}.$$

where:
- β_x = Beta ratio for particles of size "x" or larger.
- C_u = Particle concentration upstream.
- C_d = Particle concentration downstream.

Interpretation:
- β = 2: Indicates that 50% of particles of size "x" are removed.
- β = 200: Means that 99.5% of the particles are captured.

Table B1 Filter efficiency measured as a beta ratio. Courtesy: ENG**TECH** Industries Inc.

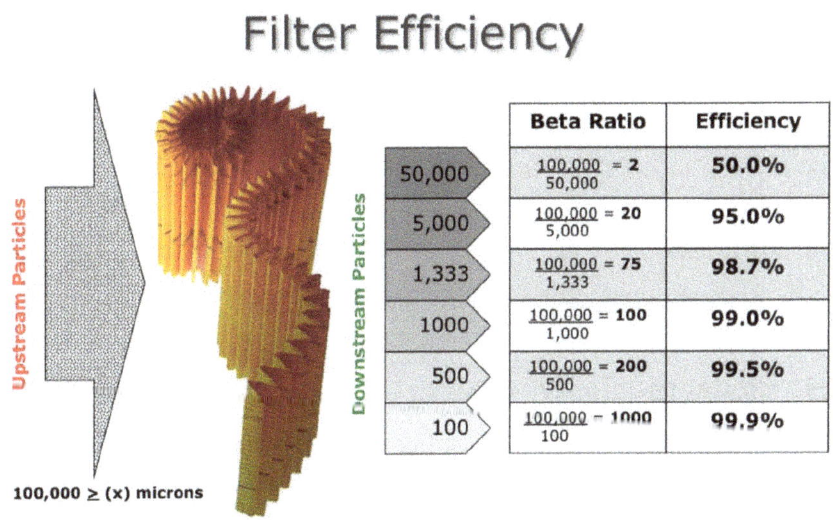

Beta Ratio	Efficiency
100,000 / 50,000 = 2	50.0%
100,000 / 5,000 = 20	95.0%
100,000 / 1,333 = 75	98.7%
100,000 / 1,000 = 100	99.0%
100,000 / 500 = 200	99.5%
100,000 / 100 = 1000	99.9%

Filters with higher beta ratios are more effective at removing contaminants, making this measurement critical for applications in hydraulic systems, automotive engines, and industrial equipment to ensure system cleanliness and performance.

107. Bijur, Joseph

Joseph Bijur was an American inventor and a key figure in the evolution of automotive lubrication in the early 20th century. In 1923, he introduced the world's first centralized lubrication system, revolutionizing vehicle maintenance. Before this innovation, car owners, or their chauffeurs, had to manually lubricate up to 50 chassis points daily, a labor-intensive and time-consuming task. Bijur's system, known as the **single-line resistance (SLR) lubrication system**, automated the effort by delivering metered oil to multiple points with a simple pull of a handle under the dashboard, significantly improving operator and maintainer efficiency while extending the lifespan of automotive components.

Key Contributions:

- **SLR system**: Simplified vehicle lubrication, reducing manual effort and enhancing reliability.
- **Impact**: Made cars more reliable and accessible for the general public by lowering maintenance demands.

Bijur's invention laid the groundwork for modern automated lubrication systems used in industries beyond automotive, including industrial machinery and transportation equipment. His legacy continues through **Bijur–Delimon International**, a leader in lubrication technology, known for their continued innovation in the field.

The simplicity and effectiveness of Bijur's system made it one of the most widely adopted lubrication solutions in history, influencing lubrication practices across multiple sectors and shaping the future of automated maintenance.

108. Biocide

See "Bactericide Additive."

109. Biodegradability

Biodegradability refers to the ability of a substance, often organic or petroleum-based, to naturally decompose through the action of microorganisms like bacteria, fungi, or algae. This process breaks down the material into simpler compounds such as water, carbon dioxide, and organic matter that can be safely reintegrated into the environment.

Key Characteristics:

- **Natural breakdown**: Microorganisms convert the substance into harmless, simple compounds.
- **Environmental impact**: Essential for reducing pollution, especially in environmentally sensitive areas like marine environments or forests.

In the context of lubricants, biodegradability is a critical factor for ensuring eco-friendly performance and compliance with environmental regulations. Biodegradable lubricants are crucial in applications where minimizing environmental impact is essential, helping to promote sustainability and align with stewardship goals.

110. Biodegradable Lubricants

Biodegradable lubricants are specially formulated lubricants designed to naturally break down through the action of microorganisms such as bacteria, fungi, or algae. These lubricants decompose into nontoxic byproducts – mainly water, carbon dioxide, and organic matter – minimizing their environmental impact. They are often made from vegetable oils, synthetic esters, or other eco-friendly base stocks.

Key Benefits:

- **Eco-friendly**: Reduce pollution risks in sensitive environments, such as marine ecosystems or forests.
- **Regulatory compliance**: Adhere to strict environmental standards and certifications.
- **Reduced toxicity**: Safer for use in areas where accidental spills could harm wildlife or water systems.

Biodegradable lubricants are commonly used in applications like offshore drilling, forestry equipment, agriculture, and hydroelectric plants, where traditional petroleum-based products could pose a contamination risk.

111. Bitumen

Bitumen in lubrication refers to a viscous, semisolid hydrocarbon material used in specialized industrial applications as a lubricant or protective coating.

See Figure B6. Although not a conventional lubricant, bitumen is valued for its strong adhesive properties, water resistance, and high-temperature stability, making it suitable for demanding conditions where traditional oils or greases may fail.

Figure B6 Bitumen is a viscous, semisolid hydrocarbon material.

Common Applications:

- **Open-gear lubrication**: Used in heavy equipment and industrial machinery to provide long-lasting protection.
- **Wire rope coatings**: Protects against corrosion and environmental exposure, extending the life of wire ropes.
- **Protective coatings**: Applied to pipelines or metal structures, serving as a lubricant and moisture barrier.

Bitumen-based lubricants are ideal for applications requiring durable, heavy-duty lubrication under harsh conditions. However, due to its high viscosity, bitumen is not suitable for high-speed machinery and is best used where durability and environmental protection are more critical than low-friction performance.

112. BL

See "Boundary Layer Lubrication."

113. Black Oil

Black oil refers to heavy lubricants or asphalt-based products formulated with tackifier agents to enhance adhesion and stickiness. Tackifiers are chemical additives that improve the bonding ability of asphaltic oils, ensuring the oil adheres effectively to surfaces even under demanding conditions. These oils provide both lubrication and protection, making them suitable for industrial gears, chains, wire ropes, and open gears in harsh environments.

Key Characteristics:

- **High viscosity**: The presence of asphaltic components gives the oil thick, long-lasting film strength, enabling it to withstand heavy loads and extreme temperatures.
- **Durable adhesion**: Tackifiers ensure the oil stays in place, resisting washout from water and reducing the frequency of reapplication.
- **Harsh environment performance**: Effective in wet, high-pressure, or extreme temperature conditions, ideal for applications like construction equipment, mining machinery, and marine operations.

Black oils are designed to form a resilient protective coating, providing reliable lubrication and reducing wear, even in the most challenging industrial settings.

114. Bleeding

Bleeding refers to the involuntary separation of oil from a grease's thickener matrix, which can occur over time or under certain operating conditions. Grease is a blend of base oil, thickener, and additives, and bleeding happens when the oil seeps out of the thickener (see Figure A11 for a bleeding example). However, a controlled amount of bleeding is necessary for proper lubrication, as the released oil forms a protective film on components. However, excessive bleeding can lead to performance issues, such as:

Potential Problems:

- **Loss of lubrication**: Reduces the grease's effectiveness, leading to increased wear or friction.
- **Contamination risks**: Separated oil can attract dirt and debris, especially in open or unsealed systems.
- **Component damage**: Excessive oil loss can leave only the thickener behind, compromising lubrication and causing damage.

Influencing factors: Temperature, load pressure, storage conditions, and improper grease formulation can all affect bleeding. Managing this involves choosing greases with suitable properties for the application and monitoring for signs of overseparation to maintain consistent equipment performance.

115. Blending

Blending in the context of lubricants refers to the process of mixing base oils and additives to create a lubricant with specific performance characteristics. This process ensures the lubricant meets the required viscosity, thermal stability, oxidation resistance, antiwear properties, and chemical balance for a particular application.

Main Methods of Blending:

1. **Batch blending**: Involves mixing a set quantity of base oil and additives in controlled batches, typically used for small or specialized production runs.
2. **Continuous in-line blending**: Components are continuously mixed during production, ideal for large-scale manufacturing and ensuring consistent product quality.
3. **Hybrid blending**: Combines batch and continuous methods, where base oils and additives are first mixed in a batch blender and then passed through a continuous mixer to ensure thorough and uniform blending.

Proper blending is crucial for achieving optimal lubrication performance across automotive, industrial, and marine applications. Formulations are tailored based on operating conditions, environmental requirements, and equipment specifications, with additives like detergents, dispersants, antioxidants, and antiwear agents carefully chosen to meet performance needs.

116. Block Grease

Block grease is a type of lubricant with an exceptionally hard consistency, classified as **NLGI Grade 6** on the National Lubricating Grease Institute (NLGI) scale. See Figure B7. This grade signifies the highest level of stiffness in grease, giving it a solid form comparable to hard wax or cheddar cheese.

Key Characteristics:

- **High consistency**: Retains a solid structure, making it ideal for applications where the lubricant must remain in place under high loads or extreme conditions.

- **Durability**: Offers long-lasting lubrication by resisting softening or displacement, even under high pressures or elevated temperatures.

Figure B7 NLGI 6 – 4 lb (approx. 2 kg) grease block.

Applications:
Block grease is used in specialized settings, such as slow-moving or heavily loaded machinery, industrial crushers, mining equipment, and exposed gears and bearings operating in severe conditions. Its ability to withstand extreme stress makes it essential for reliable performance in harsh environments.

Handling and Application:
Due to its stiffness, block grease requires manual or spot lubrication methods, as conventional grease dispensing systems may not be suitable for applying this type of grease effectively.

117. Bloom

Bloom refers to a visual phenomenon observed on the surface of lubricants or greases, characterized by a hazy or iridescent appearance. It often occurs due to the separation or migration of certain additives or oil components to the surface, typically influenced by temperature fluctuations or aging.

Key Characteristics:

- **Appearance**: Presents as a hazy film or iridescent sheen on the lubricant or grease surface.
- **Causes**: Can result from additive separation, temperature changes, or prolonged storage.
- **Impact**: Generally, does not indicate a significant degradation of the lubricant's performance but may warrant further inspection to ensure the product's integrity.

Bloom is primarily a cosmetic issue, though understanding its causes can be important for maintaining optimal lubrication conditions. In coatings, adhesives, and greases, the bloom is often monitored to maintain product consistency and performance.

Controlling bloom involves selecting compatible additives and stabilizers that stay evenly dispersed throughout the product under the intended operating conditions.

118. Blotter Spot Test

Blotter spot test is a simple diagnostic method used to assess possible machine problems through a quick visual check of the oil by placing a drop of the oil sample on absorbent paper (such as filter paper) and observing how it spreads and separates over a number of hours and days. This simple test provides a dispersion pattern of discoloration that can be compared to known photographic examples of lubricant/machine issues and alerts the user to the presence of contaminants, oxidation, fuel dilution, and water content in the oil that may require further laboratory oil analysis. See the typical blotter example shown in Figure B8.

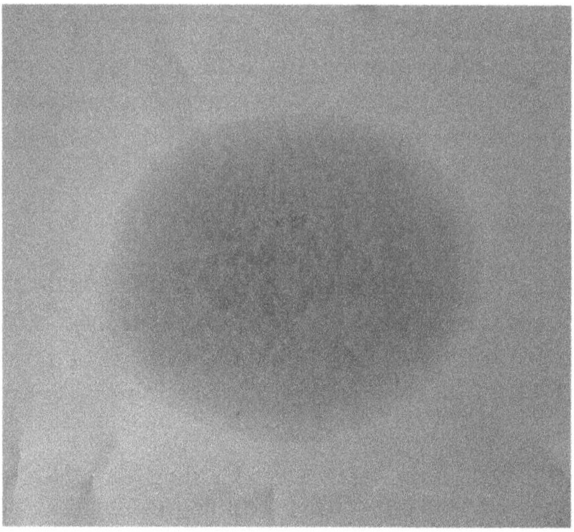

Figure B8 Typical dispersion pattern found on an oil blotter test.

Procedure and Interpretation:

1. Place a small drop of oil on a piece of **blotter paper**.

2. Observe how the oil spreads over time, noting any distinct patterns and discoloration that appear. For example:

- **Clear center with dark ring**: Indicates possible contamination or soot in the oil.
- **Irregular or discolored patterns**: May suggest oxidation, fuel dilution, or water contamination.
- **Uniform spread**: Suggests the oil is in good condition with no significant contamination.

The blotter spot test serves as a quick, qualitative field assessment tool to determine if further, more detailed analysis of the oil is necessary.

119. Boiling Point

Boiling point is the temperature at which a liquid (such as a lubricant) transitions into a gas (vaporizes) under standard atmospheric pressure (1 atmosphere or 101.3 kPa). At this temperature, the vapor pressure of the liquid equals the external pressure, causing the liquid to form vapor bubbles and convert into gas.

Importance in Lubrication and Fluid Applications:

- **High boiling point**: Indicates good thermal stability, allowing the fluid to remain in a liquid state under high temperatures. This is essential for lubricants and hydraulic fluids used in high-heat environments.
- **Low boiling point**: Fluids with lower boiling points may vaporize when exposed to heat, leading to fluid loss, increased wear, and potential component failure.

The boiling point is a crucial property for ensuring the reliable performance of lubricants and fluids under varying temperature conditions.

120. Borate Ester

Borate ester is a lubricant additive created by chemically reacting boric acid with alcohols to form boron-containing esters. These additives are prized for their antiwear, friction-reducing, and oxidation stability properties, making them particularly effective in engine oils, gear oils, and industrial lubricants.

Key Benefits:

- **Antiwear protection**: Reduces surface wear in high-stress conditions, prolonging equipment life.
- **Friction reduction**: Improves efficiency by minimizing friction between moving parts.
- **Oxidation stability**: Enhances the lubricant's resistance to oxidation, helping to maintain performance over time.

Borate esters are often used in lubricant formulations to boost performance, especially in demanding applications, ensuring equipment longevity and operational efficiency.

121. Boundary Layer Lubrication

Boundary layer lubrication (BL) is a lubrication regime where the lubricant film between two sliding surfaces is extremely thin, often just a few molecules thick. In this regime, the surfaces may come into partial contact, and the lubricant depends on specific additives to reduce friction and prevent wear, as there is not enough film thickness to fully separate the surfaces. See Figure B9.

Key Characteristics:

- **Thin lubricant film**: The film is insufficient to completely separate the surfaces, leading to partial metal-to-metal contact.
- **Additive reliance**: Boundary layer lubrication relies heavily on additives, such as antiwear agents and friction modifiers, which form a protective layer on the metal surfaces. This layer helps soften the contact points, allowing surfaces to slide without causing significant damage to the parent metal.
- **Higher friction and wear risk**: This regime typically has higher friction and wear compared to other lubrication types (like hydrodynamic or elastohydrodynamic lubrication), particularly under high load or low-speed conditions.

Boundary layer lubrication is common in applications with high loads, low speeds, or frequent stop-start cycles, where the lubricant film cannot maintain complete separation of moving parts. Examples include gears, camshafts, bearings during startup, and slow-moving or indexing machinery.

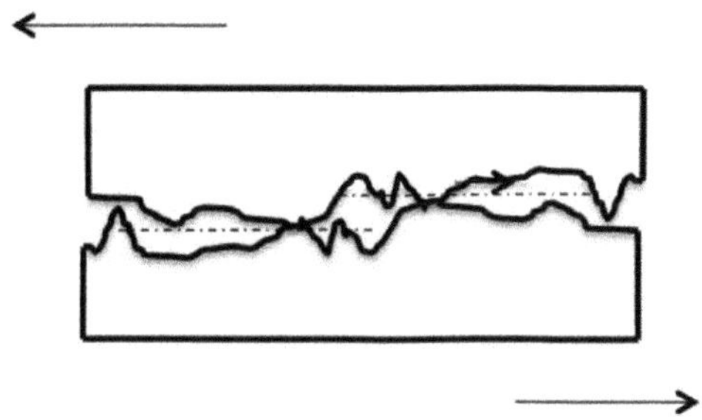

Figure B9 Boundary layer lubrication regime.

122. Box Cam Lubricator

Box cam lubricator is a mechanical device designed to deliver precise amounts of lubricant to specific points in machinery, ensuring optimal performance and reducing wear. It features a self-contained reservoir of lubricant and a cam-driven mechanism that operates a series of small pistons. As the camshaft rotates, its multiple lobes engage the pistons, drawing lubricant from the reservoir and pumping a measured amount through output lines to the lubrication points. The stroke of each piston can be adjusted to control the volume of lubricant delivered.

Key Components:

- **Reservoir**: Stores the lubricant, which can be either oil or grease, depending on the application.
- **Camshaft**: Powered by the machinery via gears, chains, or direct coupling. As it rotates, the cams lift and actuate the pumps to dispense lubricant.
- **Pumps**: Metered pumps that deliver a controlled amount of lubricant with each rotation of the camshaft.
- **Distribution lines**: Carry the lubricant from the pumps to the designated lubrication points.

Box cam lubricators were commonly used during the Industrial Revolution, such as on steam engines in cotton mills, to automate lubrication and improve the efficiency and reliability of heavy machinery. Figure B10 illustrates a typical box cam lubricator fitted to a 19th-century steam engine powering a cotton carding mill in Lancashire, England, during the Industrial Revolution – still operational today.

Figure B10 Victorian steam engine box cam lubricator. Courtesy: ENG**TECH** Industries Inc.

123. Brake Fluid

Brake fluid is a specialized hydraulic fluid used in vehicle brake systems to transmit force from the brake pedal to the brake calipers or wheel cylinders. This hydraulic action enables the brake pads or shoes to press against the rotors or drums, slowing or stopping the vehicle effectively.

Key Properties:

- **High boiling point**: Essential for operation in high-temperature environments created by braking friction. A high boiling point prevents vaporization, which can cause air bubbles in the brake lines and lead to brake fade or reduced braking performance.
- **Corrosion resistance**: Formulated to protect metal components, such as brake lines, the master cylinder, and calipers, from corrosion and degradation over time.
- **Material compatibility**: Designed to work with brake system components, such as rubber seals and hoses, to prevent swelling, deterioration, and leaks.

Brake Fluid Types:

1. **DOT 3 and DOT 4**: Glycol-based fluids suitable for most standard braking systems.
2. **DOT 5.1**: A glycol-based fluid (not to be confused with silicone-based DOT 5) that provides higher performance for extreme temperature conditions.

It is crucial to use the brake fluid type specified by the vehicle manufacturer to ensure compatibility with the system materials and to maintain reliable braking performance.

124. Breather

Breather is a component used in lubrication and hydraulic systems to allow air to flow in and out of a reservoir or enclosure, maintaining pressure equilibrium and preventing the ingress of contaminants like dust, moisture, or dirt. Breathers are vital in systems where internal pressure changes occur due to temperature fluctuations, fluid movement, or the filling and draining of fluids. See Figure B11.

Key Functions:

- **Pressure regulation**: Maintains balance between internal and external pressure, ensuring stable system operation.
- **Contamination prevention**: Equipped with filters or desiccants to trap airborne contaminants and moisture, protecting the fluid and components from degradation.
- **Extended equipment life**: By keeping the system clean and preventing pressure build-up, breathers help extend the longevity and reliability of lubricated components.

Figure B11 Dessicant style breather. Courtesy: Fluid Defense Systems.

Applications:
Breathers are commonly used in gearboxes, hydraulic systems, engines, and storage tanks. They are especially critical in heavy machinery, industrial equipment, and vehicles to maintain fluid integrity and minimize maintenance needs in environments where contamination or moisture could affect system performance.

125. Bridging

Bridging in the context of lubrication and filtration refers to the accumulation of particles or contaminants that form a bridge across filter media, pores, or openings. This build-up can obstruct the flow of lubricant, oil, or air, reducing the filter's effectiveness and potentially causing pressure drops, restricted flow, or even system failures.

Key Implications:

- **Filter clogging**: Bridging accelerates filter clogging, necessitating more frequent maintenance or replacement.
- **Flow restriction**: The particle bridge obstructs flow, reducing efficiency and potentially leading to component wear or overheating.
- **Inconsistent lubrication**: In lubrication systems, bridging can cause insufficient lubrication at critical points, increasing friction and wear.

Bridging is a concern not only in filtration systems for engines, hydraulics, and industrial machinery but also in storage tanks and piping systems where contaminants can accumulate and disrupt system performance.

126. Bright Stock

Bright stock is a heavy, high-viscosity base oil produced through solvent refining and dewaxing processes. It is derived from residual oil, which is the thick, high-molecular-weight fraction left after the lighter oils are distilled from crude oil. Bright stock is valued for its excellent thermal stability and high film strength, making it a crucial component in the formulation of heavy-duty lubricants and greases.

Common Uses:

- **Marine oils**: Provides robust protection for engines and components in marine environments.

- **Industrial gear oils**: Used in formulations to handle high-load conditions and prevent wear.
- **Engine oils and industrial lubricants**: Blended with lighter base oils to meet specific viscosity and performance requirements for engines, compressors, and heavy machinery.

Bright stock is essential for applications where strong film formation and resistance to thermal breakdown are needed to ensure equipment reliability and longevity.

127. Brinelling

Brinelling refers to a specific type of damage or deformation occurring on the surfaces of rolling-element bearings, such as ball bearings and roller bearings. This damage happens when a bearing is subjected to high static or impact loads beyond the elastic limit of the bearing race material, causing the rolling elements (balls or rollers) to create indentations or depressions on the bearing race surfaces (inner and outer rings) where contact occurs. These indentations, known as **Brinell marks** or **Brinell impressions**, are shallow, circular depressions (see Figure B12) that differentiate brinelling from other types of bearing damage, such as spalling.

Figure B12 Brinelled ball indentations. Courtesy: Barden Precision Bearings.

Key Impacts:

- **Increased friction**: Brinell marks disrupt smooth rolling, increasing friction and reducing efficiency.

- **Vibration and noise**: The indentations cause irregular movement, leading to increased vibration and noise during operation.
- **Premature failure**: If not addressed, brinelling accelerates wear and can cause early bearing failure.

Prevention: To prevent brinelling, it is crucial to ensure proper bearing installation and maintenance, avoid excessive static or impact loads, and maintain correct lubrication practices. These measures help distribute loads evenly and protect the bearing surfaces, preserving overall bearing integrity.

128. British Thermal Unit

British thermal unit (BTU) is a unit of measurement for heat energy, specifically relevant in industrial lubrication for understanding heat transfer and energy dissipation. One BTU is defined as the amount of heat needed to raise the temperature of one pound of water by one degree Fahrenheit at sea level. In lubrication contexts, BTUs are crucial for evaluating the heat generated and managed by lubricants in machinery and industrial systems.

Key Characteristics:

- **Energy conversion**: 1 BTU is approximately equal to 1055 joules, a useful measure for calculating energy dissipation in lubricated equipment.
- **Applications**: BTUs are used to quantify the heat management capacity of lubricants in industrial machinery, HVAC systems, and large-scale equipment where heat control is critical.
- **Heat management**: Understanding the BTU ratings of lubricants and cooling systems helps engineers design efficient lubrication solutions to prevent overheating, ensure optimal viscosity, and maintain equipment performance.

In industrial lubrication, BTUs provide a standardized way to assess and manage heat generated in equipment, ensuring reliable operation and effective temperature control in high-demand environments.

129. Bromine Number

Bromine number is a measure used to assess the degree of unsaturation in the base oil used to formulate lubricants. It quantifies the amount of bromine that reacts with the unsaturated molecules in the base oil, similar to how unsaturation is measured in oils and fats. The bromine number is important in lubricant formulation because it influences key properties such as oxidative stability, thermal stability, and resistance to degradation.

Key Characteristics:

- **Degree of unsaturation**: A higher bromine number indicates more unsaturated molecules, which are generally more prone to oxidation and thermal breakdown. This can reduce lubricant performance and shorten service life.
- **Formulation impact**: Lubricant manufacturers use the bromine number to select suitable base oils and additives, optimizing formulations to meet specific performance requirements.
- **Customization**: By controlling unsaturation levels, manufacturers can create lubricants that withstand various conditions, such as extreme temperatures, heavy loads, or extended service intervals, ensuring optimal performance and durability.

The bromine number is a critical tool in the development of high-performance lubricants, helping to balance stability and efficiency in demanding industrial and automotive applications.

130. Brookfield Viscometer

Brookfield viscometer is a type of rotational viscometer used to measure the viscosity of fluids by determining their resistance to flow under applied shear stress. Developed by Brookfield Engineering Laboratories, Inc (see Figure B13), this instrument operates by immersing a spindle or rotor into the fluid and rotating it at a set speed. The resistance encountered by the rotating spindle is measured and converted into viscosity units, typically reported in centipoise (cP) or millipascal-seconds (mPa·s).

Key Features:

- **Rotational measurement**: Delivers accurate viscosity readings by assessing the torque needed to rotate the spindle.
- **Versatile application**: Suitable for measuring various fluids, including oils, greases, and lubricants, especially at low temperatures.
- **Standardized testing**: Used in **ASTM method D2983** for evaluating low-temperature viscosity, crucial for assessing flow characteristics in industrial applications.

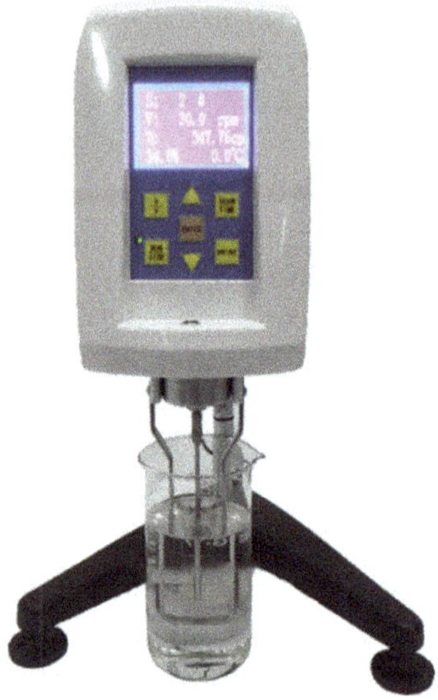

Figure B13 Brookfield viscometer.

The Brookfield viscometer is extensively used in the lubricant, food, pharmaceutical, and chemical industries to ensure consistent fluid performance under specific conditions. It aids in optimizing formulations for their intended operating environments, ensuring product reliability and efficiency.

131. Brookfield Viscosity

Brookfield viscosity refers to the measurement of a fluid's resistance to flow under applied shear stress, especially at low temperatures. It forms the basis of **ASTM Method D2983** and is determined using a **Brookfield viscometer**, (see entry 119 – an instrument developed by Brookfield Engineering Laboratories, Inc. This rotational viscometer works by immersing a spindle or rotor into the fluid and rotating it at a constant speed. The resistance encountered by the spindle during rotation is measured and converted into viscosity units, typically expressed in centipoise (cP) or millipascal-seconds (mPa·s).

Importance:

- **Low-temperature evaluation**: Critical for assessing the flow properties of lubricants and other fluids in cold environments.

- **Performance testing**: Ensures that fluids maintain suitable viscosity levels under specific operating conditions, which is essential for reliable and efficient performance.

Brookfield viscosity measurements are widely used in the lubricant industry to verify that products will function effectively in applications where temperature fluctuations could impact fluid behavior.

132. BTU

See "British Thermal Unit."

133. Bulk Appearance (Grease)

Bulk appearance in lubricating grease refers to the visual characteristics observed when the grease is in bulk form, such as in a container or storage vessel. This term describes the overall look, texture, and consistency of the grease, which can vary based on its composition, formulation, manufacturing process, and intended application.

Key Characteristics:

- **Consistency**: The firmness or softness of the grease, which can range from fluid-like to stiff or solid at room temperature.
- **Texture**: The feel of the grease surface, which may be smooth and homogeneous or coarse and lumpy, depending on the distribution of additives and thickeners.
- **Color**: Can vary from white, translucent, amber, brown, and black to specific identification colors (e.g., red, green, or blue) that may signify particular additives or base oils.
- **Clarity**: Some greases are clear or transparent, while others are opaque or cloudy, often due to the presence of solid additives or impurities.
- **Homogeneity**: High-quality grease should appear uniform, with no visible separation of components or large aggregates of particles.
- **Identification**: The bulk appearance can serve as a visual check to ensure the correct grease is being used for a particular application.

The bulk appearance of grease provides critical insights into its formulation and suitability for specific uses, helping operators verify that they are applying the appropriate product for optimal lubrication performance.

134. Bulk Modulus of Elasticity

Bulk modulus of elasticity, often referred to as the **bulk modulus**, measures a lubricant's resistance to uniform compression. It quantifies how much a material deforms when subjected to hydrostatic pressure (pressure applied equally in all directions).

Mathematically, the bulk modulus **(K)** is defined as the ratio of the applied pressure change **(P)** to the resulting relative volume change **(ΔV/V)**, given by

$$K = -\frac{P}{\Delta V / V}.$$

where:

- K = Bulk modulus of elasticity
- P = Applied pressure change
- $\Delta V/V$ = Relative volume change (change in volume divided by the original volume)

The **bulk modulus** is typically expressed in units of pressure, such as **pascals (Pa)** or **gigapascals (GPa)**, because it represents the pressure required for a unit change in volume. Materials with a high bulk modulus are highly resistant to compression and exhibit minimal volume changes under pressure, while materials with a low bulk modulus deform more easily and show significant volume changes when subjected to pressure.

In lubrication, understanding the bulk modulus is crucial for applications where fluids experience high pressures, as it influences the lubricant's performance and ability to maintain protective film strength under varying load conditions.

135. Bulk Oil

Bulk oil refers to oil that is supplied, stored, and transported in large quantities, typically without individual packaging, to meet the needs of high-volume users. It is widely utilized in manufacturing plants, automotive service centers, agricultural operations, and transportation fleets where significant amounts of lubricating or hydraulic oil are regularly required. Bulk oil is dispensed directly from large storage tanks into equipment or smaller containers as needed, ensuring a continuous supply and enhancing operational efficiency.

Key Characteristics:

- **Storage**: Kept in tanks, drums, or large containers (see Figure B14 Totes) to allow easy access and efficient distribution.
- **Cost-effective**: Purchasing oil in bulk reduces packaging costs, providing a more economical solution for operations with high oil consumption.
- **Environmental benefit**: Minimizes packaging waste, making bulk distribution a more sustainable option.

Bulk oil is commonly used across industrial, automotive, and commercial applications, contributing to streamlined operations and reduced downtime in high-demand environments.

Figure B14 On-site bulk oil tote containers. Courtesy: ENG**TECH** Industries Inc.

136. Bulls Eye Sight Gauge

A **bull's eye** in lubrication and fluid management refers to a circular sight glass that allows operators to visually check the oil level and condition in equipment such as gearboxes, reservoirs, and engines. Positioned on the exterior of the machinery (see Figure B15), the bull's eye provides a clear view of the lubricant without needing to open the system, facilitating quick and convenient monitoring.

Figure B15 Modern 3D "Bull's-eye" sight gauge. Copyright: Des-Case Corporation, 2022.

Key Characteristics:

- **Oil level monitoring**: Enables operators to easily confirm that the oil is at the correct level, ensuring proper lubrication and preventing equipment damage.
- **Condition check**: Offers a visual assessment of the oil's color and clarity, helping identify issues like contamination, oxidation, or the need for an oil change.
- **Simple maintenance**: Minimizes the need to disassemble equipment for routine checks, improving maintenance efficiency and saving time.

The term "Bull's Eye" was originally used as a brand name for a Detroit hydrostatic side-feed steam locomotive lubricator, and it remains an essential tool for modern fluid management and equipment maintenance.

137. Burst Pressure Rating

Burst pressure rating is the maximum pressure that a fluid-carrying or containment component, such as a hose, pipe, or tank, can endure before it ruptures or fails. This rating defines the upper-pressure limit for the material and design of the component, determined under controlled testing conditions.

Key Characteristics:

- **Safety margin**: The burst pressure is usually several times greater than the component's normal working pressure, providing a safety buffer to handle unexpected pressure spikes.
- **Application-specific**: Crucial for selecting appropriate materials and components in high-pressure systems like hydraulics, pneumatics, and fuel lines, where pressure fluctuations are common.
- **Failure prevention**: Exceeding the burst pressure can result in dangerous ruptures, leaks, or catastrophic failures, making this rating essential for maintaining safe operation.

In high-pressure systems such as hydraulic equipment, industrial machinery, and aerospace applications, the burst pressure rating ensures the durability and safety of components under extreme conditions, helping to prevent accidents and system failures.

138. Bushing

See "Plain Bearing."

139. Button Head Grease Fitting

A **Button head grease fitting** is a **large-diameter grease fitting designed for high-volume grease applications in heavy-duty industrial and off-road equipment**. Unlike standard grease nipples (Zerk fittings), button head fittings have a flat, wide surface that allows for better grease retention and high-pressure grease delivery, making them ideal for slow-moving, high-load bearings and joints in construction, mining, and heavy machinery applications. See Figure B16.

Key Functions of a Button Head Grease Fitting

- **Delivers large volumes of grease efficiently**: Supports high grease flow rates required for large bearings and heavy-duty components.
- **Ensures secure coupling with grease guns**: Designed to work with a button head coupler, preventing leakage or grease wastage.

- **Prevents contamination and protects lubrication points**: Features a secure sealing mechanism that blocks dirt and moisture from entering.
- **Supports high-load, low-speed applications**: Used in slow-moving bearings, pivot joints, and bushings that require heavy lubrication.

Types of Button Head Grease Fittings

- **Straight button head fitting**: Standard type for **easily accessible lubrication points**.
- **Angled button head fitting (45° or 90°)**: Designed for **hard-to-reach lubrication points in confined spaces**.
- **High-pressure button head fitting**: Used in **applications requiring extreme-pressure lubrication**.
- **Flush-mount button head fitting**: Installed flush with the surface for **protection in environments prone to impact or debris accumulation**.

Figure B16 Button head grease fitting and positive lock grease gun connector. Courtesy: ENGTECH Industries Inc.

Applications of Button Head Grease Fittings

- **Construction and mining equipment**: Used in **excavators, bulldozers, and loaders** for **high-load pivot points and joints**.

- **Heavy-duty industrial machinery**: Applied in **presses, rollers, and large rotating assemblies**.
- **Off-road and agricultural equipment**: Lubricates **steering joints, suspension systems, and large bearings**.
- **Railroad and marine applications**: Ensures **reliable lubrication in high-pressure, high-contamination environments**.
- **Military equipment**: Applied to military vehicles, large field guns and Ship guns

Factors Affecting Button Head Grease Fitting Performance

- **Grease gun coupler compatibility**: Requires a **button head coupler to ensure proper grease delivery**.
- **Operating load and lubrication frequency**: High-load components require **consistent greasing intervals** to prevent wear.
- **Environmental contamination risks**: Properly sealed fittings **prevent dirt, water, and debris infiltration**.
- **Grease viscosity and flow rate**: Heavy greases with **high viscosity** work best for **button head applications**.

140. Bypass Filtration

Bypass filtration is a method of filtering a portion of the lubricant flow through a separate, finer bypass filter while the rest of the fluid continues to circulate normally without passing through this filter. Unlike full-flow filtration, which processes the entire fluid volume, bypass filtration allows a small amount of the total lubricant flow to pass through a high-efficiency bypass filter media. These filters are capable of capturing extremely fine contaminants, often down to submicron sizes that full-flow filters may not remove.

Key Benefits:

- **Enhanced filtration**: Bypass filters capture smaller contaminants, gradually ensuring cleaner oil and better fluid quality.
- **Extended equipment life**: Cleaner oil reduces wear on components, helping to extend the lifespan of machinery and equipment.
- **Reduced maintenance costs**: Maintaining oil cleanliness minimizes the frequency of oil changes and lowers overall maintenance expenses.

Applications:

Bypass filtration systems are widely used in heavy machinery, engines, hydraulic systems, and industrial equipment. They are especially valuable in applications where maintaining high fluid cleanliness can enhance performance and reliability over long operating periods.

141. Bypass Indicator

Bypass indicator is a device used to monitor the condition of a bypass filter and notify operators or maintenance personnel when the filter requires servicing or replacement. This ensures the filtration system operates efficiently, preventing contaminants from bypassing the filter and causing equipment damage or performance degradation. See Figure B17.

Key Features:

- **Clog detection**: Identifies when the filter is nearing clogging, alerting that maintenance is needed to maintain filtration effectiveness.
- **Visual and electronic alerts**: Depending on the setup, bypass indicators may feature gauges or sight glasses to display differential pressure across the filter element, or electronic sensors and alarms for immediate warnings.

Figure B17 Filter bypass indicator. Courtesy: ENGTECH Industries Inc.

- **Improved maintenance**: Facilitates timely servicing, ensuring continuous protection and reducing the risk of contaminants reaching sensitive components.

Applications:
Bypass indicators are crucial in industrial, automotive, and hydraulic systems where efficient filtration is vital. They support proactive maintenance, enhance system reliability, extend equipment life, and ensure consistent performance.

142. Bypass Valve

Bypass valve is a mechanical component used in filtration systems to allow fluid to bypass the filter element when it becomes clogged or if the system pressure exceeds a predetermined threshold. This valve ensures continuous fluid flow, preventing pressure build-up that could damage equipment or interrupt operation. See Figure B18.

Key Features:

- **Pressure relief**: Automatically opens when the filter is blocked, allowing unfiltered fluid to bypass and maintain system pressure.
- **Flow continuity**: Ensures that critical components, such as hydraulic systems or engine parts, continue to receive lubrication or hydraulic fluid even if the filter is obstructed.
- **Protection against damage**: Prevents overpressurization, protecting both the filter element and the overall system from potential failure or damage.

How a Bypass or "Unloading" Filter Works

1. **Filtration process**:
   - Oil or lubricant flows through the **filter element**, capturing **particles, dirt, and contaminants**.
2. **Pressure regulation**:
   - If the pressure rises due to flow surges or clogging, the **bypass/unloading valve activates**, redirecting excess flow.
3. **Automatic reset**:
   - Once pressure returns to normal, the **filter resumes standard operation**, preventing unnecessary bypassing.

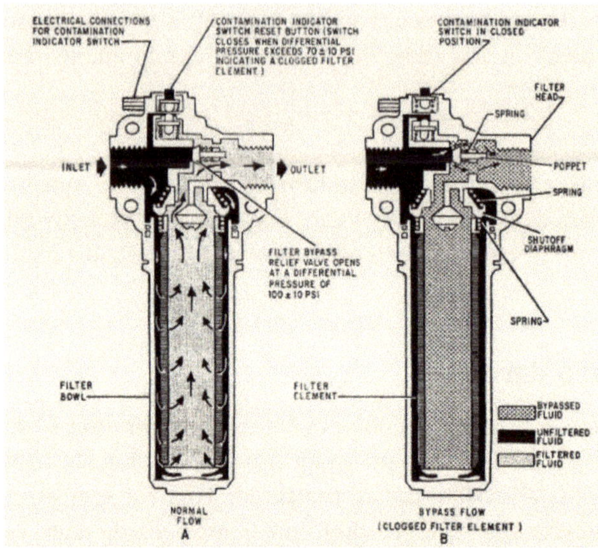

Figure B18 Hydraulic bypass filter. Source: UFI Hydraulic Filters.

Applications:
Bypass valves are widely used in hydraulic systems, engines, fuel systems, and lubrication systems where maintaining continuous fluid flow is essential. They play a crucial role in protecting sensitive equipment by enabling unfiltered flow, when necessary, thus supporting uninterrupted operation and safeguarding the system from damage.

C

143. Calcium Grease

Calcium grease is a lubricating grease thickened with calcium soaps or calcium compounds, known for its excellent water resistance and strong adhesion properties. It is particularly effective in applications exposed to moisture and wet conditions, making it a popular choice for marine, agricultural, and outdoor equipment.

Key Characteristics:

- **Water resistance**: Highly effective at resisting washout and maintaining performance in wet or humid environments.
- **Good adhesion**: Adheres well to metal surfaces, providing reliable and continuous lubrication.
- **Moderate temperature stability**: Performs effectively in low to moderate temperatures but is less suitable for high-temperature applications compared to lithium or complex greases.

Applications:
Calcium grease is ideal for chassis lubrication, wheel bearings, marine equipment, and agricultural machinery where exposure to water or humidity is prevalent. However, it has limitations, including a tendency to stiffen in very cold temperatures, which can affect its pumpability and flow characteristics. As a result, it is often not recommended for use in extremely low-temperature environments.

144. Capillarity

Capillarity, also known as **capillary action** or **capillary rise**, is the phenomenon where liquids move within narrow spaces or tubes (capillaries) against the force of gravity. This movement is driven by the interplay of cohesive and adhesive forces and is influenced by the properties of the liquid and the surrounding surfaces.

DOI: 10.1201/9788743806608-145

Key Characteristics:

- **Cohesion**: The attraction between molecules of the same substance. In capillarity, cohesive forces pull liquid molecules together, aiding upward movement within the capillary.
- **Adhesion**: The attraction between molecules of different substances. Adhesive forces between the liquid and the capillary surface cause the liquid to "wet" and adhere to the walls, encouraging upward motion.
- **Surface tension**: The property of a liquid surface to minimize its area. Surface tension helps draw the liquid upward, working alongside cohesion and adhesion to counteract gravity.

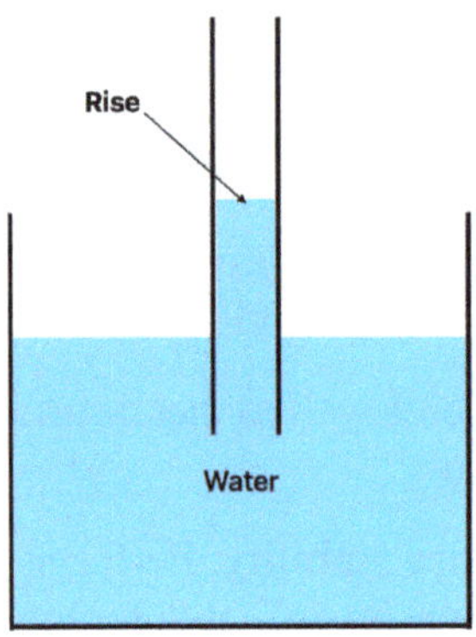

Figure C1 Capillary rise.

Capillary rise: The height to which a liquid rises in a capillary tube depends on factors like surface tension, density, viscosity of the liquid, and the tube's diameter. Narrower capillaries generally produce a greater capillary rise because of the larger surface area available for adhesion and the stronger resistance to gravitational pull. See Figure C1.

Applications: Capillarity is crucial in fluid transport systems, such as lubrication systems and porous materials, where efficient movement of fluids in narrow spaces is essential. It enables effective distribution and penetration of lubricants in small gaps and interfaces.

145. Carbon

Carbon is a chemical element with the symbol **C** and atomic number **6**, known as a fundamental building block of life and a key element in organic chemistry. Carbon atoms can form stable bonds with other carbon atoms and with elements like hydrogen, oxygen, and nitrogen, enabling the creation of a vast array of complex compounds, including hydrocarbons and essential biological molecules.

Key Forms:

- **Graphite**: A soft, slippery form of carbon used as a lubricant and in pencils. It is valued for its low friction and high thermal resistance.
- **Diamond**: A hard, highly organized form of carbon used in cutting tools and jewelry, known for its exceptional hardness.
- **Amorphous carbon**: Found in materials like coal, soot, and charcoal, used in various industrial processes and applications.
- **Carbon nanostructures**: Forms such as graphene and carbon nanotubes, utilized in advanced technologies for their remarkable strength, electrical conductivity, and versatility.

Importance in Lubrication:
Carbon-based materials like graphite and certain synthetic carbon compounds are highly valued in lubrication applications. They provide stability, thermal resistance, and low friction properties, making them ideal for high-temperature and high-load environments. Carbon's versatility extends to numerous fields, including materials science, electronics, and energy storage, where its unique properties enable groundbreaking advancements.

146. Carbonization

Carbonization is the process in which organic materials decompose under high temperatures in the absence of oxygen, resulting in carbon-rich residues. During carbonization, volatile compounds are released, and the remaining material transforms into solid carbon forms such as charcoal, coke, or carbon deposits. This process is widely used in coal processing, biomass conversion, and petroleum refining.

Key Characteristics:

- **High temperature decomposition**: Occurs at elevated temperatures without oxygen to avoid combustion.
- **Formation of carbon residue**: Produces stable, solid carbon with increased purity and density.

- **Byproducts**: Releases volatile compounds like gases and tars, which can be collected and utilized separately.

In lubrication systems, carbonization can result in carbon deposits or sludge formation, particularly in engines and high-temperature equipment. These deposits can reduce efficiency, increase wear, and cause overheating. To mitigate these issues, anticarbonization additives are included in lubricants to minimize deposit formation and improve performance, extending the life of equipment.

Factors Contributing to Carbonization:

- **Thermal stress**: High operating temperatures cause the breakdown of hydrocarbon molecules, leading to carbon-rich residues.
- **Oxidation**: The reaction of lubricants with oxygen at elevated temperatures forms carbonaceous byproducts.
- **Contaminants and impurities**: Foreign particles in the lubricant can accelerate the carbonization process.

Hardened carbon deposits can score mating parts, resulting in surface friction that causes excessive wear, reduced machinery efficiency, and preventable damage. Proper lubricant selection and maintenance practices are crucial to managing carbonization in high-temperature applications.

147. Carcinogen

In the context of lubricants, a **carcinogen** refers to any ingredient or byproduct in the lubricant formulation that has the potential to cause cancer through prolonged or repeated exposure. Carcinogenic substances in lubricants may include certain aromatic hydrocarbons, polycyclic aromatic compounds, or heavy metals that can be present in base oils or additives, especially in older or improperly refined products.

Key Concerns:

- **Exposure risks**: Workers can be exposed to carcinogenic components through skin contact, inhalation of fumes, or accidental ingestion, posing significant health risks.
- **Regulation**: Agencies like **OSHA** and the EU's **REACH** enforce strict guidelines to limit or ban carcinogenic substances in lubricants, requiring safety measures, proper labeling, and disclosure of hazardous ingredients.

- **Modern formulations**: Advances in lubricant technology have led to the development of low-toxicity and eco-friendly alternatives, reducing or eliminating known carcinogens and making lubricants safer for applications in automotive, manufacturing, and metalworking industries.

Minimizing exposure to carcinogens in lubricants involves implementing appropriate protective equipment, ventilation, and safe handling procedures. These practices help ensure compliance with safety regulations and promote healthier work environments.

148. Cartridge Filter

A lubricant **cartridge filter** is a filtration device used to remove contaminants, particles, and impurities from lubricants. It consists of a cylindrical cartridge filled with a filtration medium, such as paper, cellulose, synthetic fibers, or metal mesh. The cartridge is housed within a protective casing or canister, designed to allow lubricant to flow through the medium, trapping contaminants and enabling clean lubricant to pass through. See Figure C2.

Figure C2 Typical automotive-style cartridge filters in a bolt-on and spin-on housing. Source: ENG**TECH** Industries Inc.

Key Features:

- **Filtration medium**: The type, porosity, and design of the medium determine the filter's efficiency in capturing particles and impurities.
- **Replaceable cartridge**: The cartridge is easily replaceable, simplifying maintenance. When clogged or at the end of its service life, the cartridge can be swapped out to restore filtration performance.
- **Durable housing**: Provides structural support and protects the filtration medium from external damage.

Application:
Cartridge filters are widely used in industrial machinery, automotive engines, and hydraulic systems to ensure lubricant cleanliness, minimize wear, and extend equipment life through effective and reliable filtration.

149. Castor Oil

Castor oil is a natural lubricant derived from the seeds of the castor plant (**Ricinus communis**), known for its unique properties and versatility. Rich in ricinoleic acid, a distinctive fatty acid, castor oil is characterized by a specific odor and exceptional thermal stability.

Key Properties:

- **High-temperature operability**: Withstands high temperatures, making it suitable for demanding lubrication applications.
- **Thermal stability and adhesion**: Adheres well to surfaces, ensuring long-lasting lubrication even under intense heat.
- **Biodegradability**: Environmentally friendly, as it is a naturally derived and readily biodegradable product.

Applications:
Before synthetic lubricants became widespread, castor oil was a popular choice in competitive motorsport engines due to its reliable performance at high temperatures. At the time it was affectionately called "Bean Oil" and was immediately recognizable by its smell as the race vehicle passed by. However, a significant drawback is its tendency to oxidize and form gummy residues over time, especially in high-heat environments. This oxidation can require periodic cleaning and maintenance to prevent buildup and ensure optimal engine performance.

150. Catalyst

Catalyst in lubrication refers to a substance that initiates, accelerates, or facilitates a chemical reaction within the lubricant without being consumed in the process. While catalysts are commonly associated with industrial chemical reactions, they also play a crucial role in lubricant formulations by promoting reactions that enhance or preserve the lubricant's performance.

Key Applications:

- **Antiwear additives**: Catalysts in antiwear additives help form protective films on metal surfaces, reducing friction and wear, thereby extending the lifespan of machinery.
- **Antioxidants**: Catalysts aid in stabilizing the lubricant by slowing down oxidation, which prolongs the lubricant's life and maintains its effectiveness.
- **Detergents and dispersants**: Catalytic action enhances the ability of detergents and dispersants to manage and neutralize contaminants, keeping the lubricant clean and efficient.

151. Catastrophic Failure

Catastrophic failure in lubrication terms refers to a sudden and severe breakdown of equipment caused by the complete loss or extreme degradation of lubricant/lubrication. This failure occurs when the lubricant can no longer protect, cool, or reduce friction between moving parts, leading to rapid wear, overheating, or the seizure of components.

Key Causes:

- **Lubricant depletion**: Insufficient lubricant levels due to leaks, evaporation, or improper maintenance, leaving surfaces unprotected.
- **Contamination**: The presence of dirt, moisture, or metal particles that compromise the lubricant's effectiveness, increasing the risk of damage.
- **Thermal breakdown**: Excessive temperatures that degrade the lubricant's properties, causing it to lose viscosity or form harmful deposits.
- **Leakage**: Rapid oil loss due to a broken line, or slow lubricant loss over time where fluid levels go unnoticed by maintenance.

Consequences:

Catastrophic failure can cause irreversible damage to critical components such as bearings, gears, or seals, often necessitating extensive repairs or even complete equipment replacement. To prevent such failures, it is crucial to implement regular lubrication checks, effective contamination control, and timely maintenance. These measures help ensure the reliability and longevity of equipment.

152. Cavitation

Cavitation is a phenomenon in which rapid pressure changes within a fluid cause the formation of vapor-filled cavities or bubbles. These bubbles collapse with significant force, generating shock waves that can damage components in fluid-handling systems, such as pumps, hydraulic systems, propellers, and turbines. Cavitation reduces system efficiency, accelerates wear, and can even lead to catastrophic failure if not managed properly.

Types of Cavitation in Pressurized Oil Systems:

1. **Hydraulic cavitation**: Occurs in pumps, valves, and hydraulic systems due to pressure variations caused by flow rate changes or obstructions. These fluctuations create bubbles that collapse, causing damage.
2. **Venturi cavitation**: Happens in areas where fluid is constricted, such as nozzles, venturi tubes, or damaged lines. Increased fluid velocity lowers pressure, leading to bubble formation and subsequent collapse.

Key Impacts:

- **Surface erosion**: The collapse of bubbles can erode metal surfaces, causing pitting and significant surface damage.
- **Efficiency reduction**: Pressure fluctuations and the presence of bubbles reduce overall system performance and energy efficiency.
- **Noise pollution**: The collapse of cavitation bubbles produces noise, adding to operational disturbances.

Preventing cavitation involves proper system design, efficient flow management, and regular maintenance.

153. Cavitation Wear

Cavitation wear, also known as **cavitation erosion**, refers to the material damage or loss that occurs when vapor-filled bubbles in a fluid collapse, producing intense localized forces. This violent collapse generates shock waves and high temperatures, leading to erosion and mechanical damage to the surfaces exposed to the fluid.

Key Characteristics:

- **Localized damage**: The impact of collapsing bubbles causes pitting and surface degradation in specific areas, often appearing as small, crater-like indentations.
- **High temperatures and shock waves**: The collapse produces both extreme heat and forceful shock waves, accelerating the wear process and weakening material surfaces.
- **Accelerated erosion**: Prolonged exposure to cavitation wear can lead to significant material loss, jeopardizing the structural integrity of affected components.

Applications Affected:

Cavitation wear is a major concern in pumps, hydraulic systems, impellers, propellers, turbines, and other equipment where fluid pressure variations can induce cavitation. See Figure C3. Effective design, proper flow management, and regular maintenance are essential strategies to mitigate cavitation wear, extend component life, and maintain optimal system performance.

Figure C3 Typical cavitation wear found on the axial piston hydraulic pump.Source: http://hydro.tg.free.fr/reparation_pompe/cavitation/cavitation.jpg

154. CBL

See "Condition Based Lubrication."

155. CBM

See "Condition Based Maintenance."

156. CCS

See "Cold Crank Simulator."

157. Cellulose

Cellulose is a natural polysaccharide that serves as the primary structural component of plant cell walls, providing strength and rigidity to plants. Made up of long chains of glucose molecules, cellulose is the most abundant organic polymer on Earth, prized for its biodegradability, renewability, and structural strength.

Key Properties:

- **High tensile strength**: Essential for applications like paper production, textiles, and bio-based materials.
- **Biodegradable and renewable**: Being plant-derived, cellulose is environmentally friendly and naturally decomposes, making it a sustainable resource.
- **Insoluble in water**: Offers structural stability, which is beneficial in applications requiring durability and resistance to water.

Industrial Uses:
Cellulose is commonly used as a filtration medium in lubricants to remove impurities efficiently. It is also a base material for producing biofuels and biodegradable plastics, playing a significant role in supporting sustainability efforts across various industries.

158. CentiPoise

Centipoise (cP) is a unit of **dynamic viscosity** used to measure a fluid's **resistance to flow**. One centipoise is equal to one-hundredth of a poise and is equivalent to one millipascal-second (mPa·s) in the **SI system**.

Key Characteristics:

- **Viscosity measurement**: Centipoise is commonly used to express fluid viscosity, indicating how easily or slowly a fluid flows when subjected to an applied force.
- **Reference point**: Water at 20 °C has a viscosity of approximately 1 cP, serving as a standard benchmark for comparing the viscosity of other fluids.
- **Applications**: Used to evaluate the flow properties of lubricants, oils, paints, syrups, and other industrial and consumer products where understanding fluid behavior is essential.

In lubrication, knowing the centipoise value is critical for selecting the appropriate lubricant for a given application, ensuring effective flow and the formation of a protective film to reduce friction and wear.

159. Centistoke

Centistoke (cSt) is a unit of **kinematic viscosity**, measuring a fluid's resistance to flow under the influence of gravity. One centistoke equals one-hundredth of a **stoke** and is equivalent to one square millimeter per second (mm^2/s) in the **SI system**.

Key Characteristics:

- **Kinematic viscosity**: Unlike dynamic viscosity (measured in centipoise), centistoke takes into account the fluid's density, offering a measure of how easily the fluid flows compared to water.
- **Reference point**: Water at 20 °C has a kinematic viscosity of approximately 1 cSt, providing a useful standard for comparing the flow characteristics of other fluids.

Widely used for classifying lubricants, oils, and fuels, especially in scenarios in which understanding the flow behavior of fluids under gravity and effective protective film formation at operating temperatures is crucial. This helps to minimize wear and enhance the reliability and longevity of equipment.

160. Centralized Lubrication

Centralized lubrication is a system designed to deliver metered amounts of lubricant to multiple points within machinery or equipment from a single, centralized source. This ensures that each lubrication point receives the correct amount of lubricant, minimizing manual labor and enhancing equipment performance. Centralized lubrication systems can be fed by an automated pumping system, a manual pump, or a grease gun (see Figure C4). Centralized systems can operate while the machinery is running, thereby reducing downtime and increasing productivity. They also improve safety by eliminating the need for maintenance personnel to access hazardous areas for manual lubrication.

Key Characteristics:

- **Metered delivery**: Lubricant is automatically supplied to designated points, reducing the risk of under or overlubrication.

- **Efficiency**: Saves time and resources by eliminating the need for manual lubrication, especially beneficial for large or complex machinery.
- **Enhanced equipment lifespan**: Provides consistent lubrication, which helps prevent wear, reduce friction, and extend the life of components.

Applications:
Centralized lubrication systems are commonly used in manufacturing, automotive, agricultural, and heavy-duty industrial equipment where reliable, continuous lubrication is crucial to minimize downtime and maintain optimal performance.

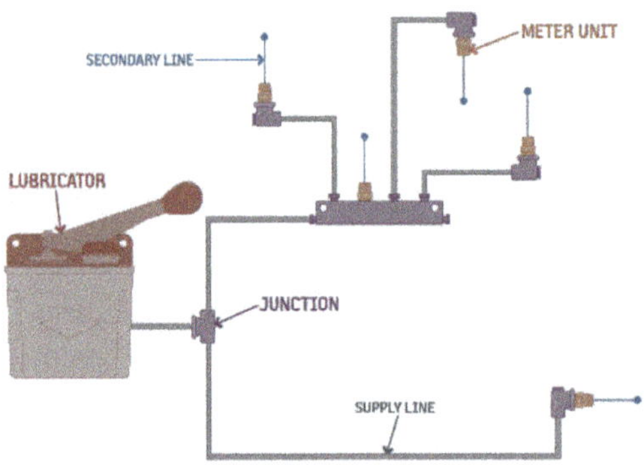

Figure C4 Centralized lubrication delivery system with a manual oil pump.Courtesy: Bijur–Delimon International.

161. Centrifuge

A **centrifuge** is a device that uses centrifugal force to separate substances of different densities or particle sizes within a mixture. By spinning the mixture at high speeds, the centrifuge forces denser components to move outward toward the perimeter, while lighter components remain closer to the center. See Figure C5.

Key Features:

- **Separation of substances**: Commonly separates solids from liquids or two immiscible liquids, such as oil and water.
- **High-speed rotation**: Generates powerful centrifugal force to achieve rapid separation.

- **Versatile applications**: Centrifuges play a crucial role in maintaining the cleanliness and performance of lubricants in industrial and automotive applications, helping to extend the life of both the lubricant and the equipment it lubricates.

Applications in Lubrication:
In lubrication and fluid processing, centrifuges are often used to remove contaminants like metal particles, dirt, and water from oils and other lubricants, maintaining fluid purity and enhancing equipment performance.

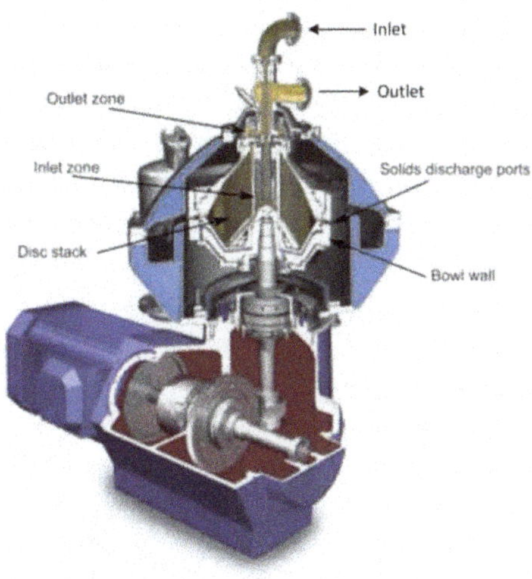

Figure C5 Oil centrifuge. Source: Dolphin Centrifuge, 2012.

162. Chain Conveyor Lubricant

Chain Conveyor Lubricant is a specialized lubricant formulated to reduce friction, wear, and corrosion on chains, sprockets, and other linked mechanical components. It ensures smooth movement and prolongs the life of the chain by creating a protective layer that withstands environmental conditions.

Key Characteristics:

- **Adhesion**: Designed to stick to the chain and resist fling-off, especially in high-speed applications.
- **Penetration**: Able to penetrate into chain links and pins, ensuring complete lubrication of all moving parts.

- **Resistance to contaminants**: Often includes properties that repel water, dust, and dirt to prevent corrosion and buildup.

Applications:
Chain lubricants are widely used for bicycles, motorcycles, industrial machinery, agricultural equipment, and conveyor system chains. Selecting the appropriate chain lubricant for the specific environment and load conditions is essential for optimal performance and minimal maintenance.

163. Chain Conveyor Lubricator

A Chain Conveyor Lubricator is a system designed to automatically deliver oil or grease to chain link bearings, conveyor truck bearings, slide rails, and other moving components. Commonly used in industrial and automotive manufacturing, mining, warehousing, and food processing environments, these lubricators ensure consistent and precise lubrication to maintain smooth equipment operation.

Key Features:
- **Automated lubrication**: Uses pneumatic pumps to apply the correct amount of lubricant, either by injecting oil into bearing spaces or dispensing grease into moving bearing nipples. See Figure C6.

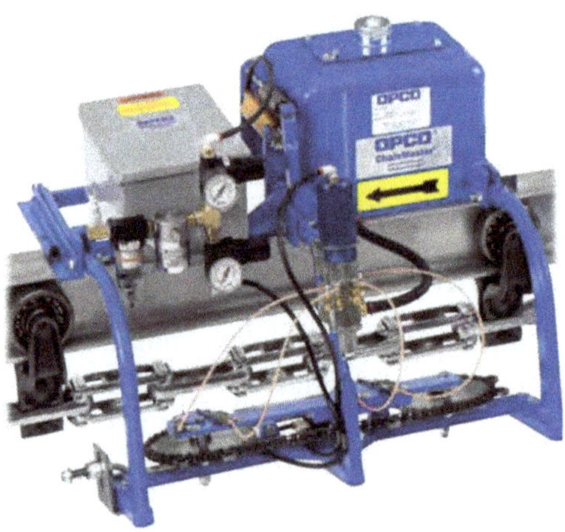

Figure C6 Opco OP-4a automatic conveyor chain pin lubricator. Source: Opco Lubrication Systems Product Catalog, 2020.

- **Efficiency**: Reduces manual lubrication efforts, ensuring all critical components are properly lubricated and minimizing the risk of over or underlubrication.

- **Enhanced equipment lifespan**: By reducing friction and wear, the system extends the life of machinery, lowers energy consumption, and prevents unexpected breakdowns.

Chain/conveyor lubricators are essential in environments where reliable and continuous lubrication is critical. They support higher productivity, maintain efficient operation, and improve safety standards, even in demanding or high-load conditions.

164. Channeling (Grease)

Channeling in grease refers to a phenomenon where a low-pressure path or groove forms within the grease as mechanical components, such as bearings, move through it. This path allows the moving parts to rotate without pushing new grease into the contact zone, potentially reducing lubrication effectiveness over time. See Figure C7.

Key Characteristics:

- **Formation of grooves**: Repeated motion creates channels, causing the grease to stay in place rather than replenishing the lubricated area.
- **Reduced lubrication**: Channeling can prevent fresh grease from moving into the contact area, leading to inadequate lubrication and increased wear.
- **Temperature influence**: Greases with higher stiffness or those used in low temperatures are more prone to channeling.

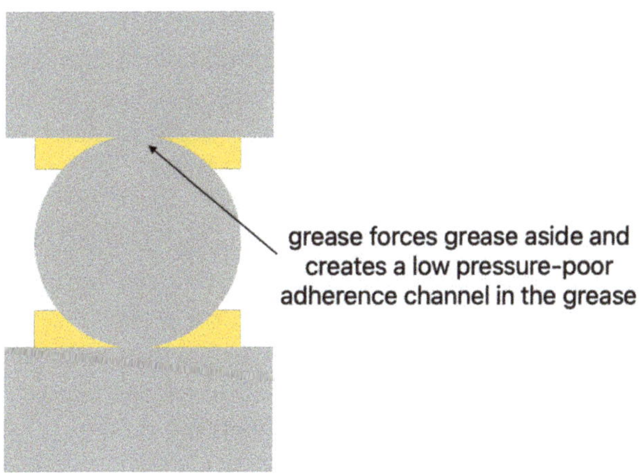

Figure C7 Grease channeling.

Implications:
Channeling may reduce the effectiveness of lubrication, increase friction, and cause premature wear of components. Selecting the appropriate grease consistency and ensuring proper application can help mitigate channeling, maintaining consistent lubrication and optimal equipment performance.

165. Chemical Wear

Chemical wear in the context of lubrication refers to the deterioration or removal of material from a surface due to chemical reactions between the lubricant, additives, and the materials being lubricated. This type of wear occurs when the chemical components of the lubricant react with the metal surfaces or contaminants in the system, leading to surface degradation.

Key Characteristics:

- **Oxidation and corrosion**: Chemical wear often involves reactions such as oxidation or corrosion, where acids or reactive compounds in the lubricant attack the surface.
- **Additive interactions**: Certain additives in lubricants, such as extreme pressure (EP) or antiwear additives, can cause controlled chemical reactions to form protective layers. However, improper use or depletion of these additives can lead to detrimental chemical wear.
- **Temperature influence**: Elevated temperatures can accelerate chemical reactions, increasing the risk of chemical wear.

Implications:
Chemical wear can lead to pitting, surface roughening, and a reduction in component lifespan. Ensuring the right formulation of lubricant with stable chemical properties and effective additive packages is essential for preventing chemical wear and maintaining machinery performance.

166. Chlorinated Wax

Chlorinated wax is a chemical compound produced by chlorinating paraffin wax, resulting in a substance with varying levels of chlorine content, typically between 30% and 70%. Chlorinated waxes are valued for their flame-retardant properties, plasticizing capabilities, and use as extreme pressure **(EP)** additives in lubricants and metalworking fluids.

Key Characteristics:

- **High chlorine content**: Provides enhanced thermal stability and flame retardancy.
- **Lubricant additive**: Used as an EP additive to improve lubrication performance by forming a protective film on metal surfaces, reducing friction and wear.
- **Plasticizer**: Serves as a plasticizer in PVC and other polymers to enhance flexibility and durability.

Chlorinated wax is commonly used in metalworking fluids, cutting oils, and industrial lubricants to improve performance under high pressure and temperature conditions. It is also used in flame-retardant coatings and plastics manufacturing. However, due to environmental and health concerns, the use of chlorinated waxes is regulated in some regions, prompting the development of safer alternatives.

167. Chromatography

Chromatography, pioneered by Russian–Italian botanist **Mikhail Tsvet** in 1908, is an analytical technique used to separate, identify, and quantify components in a mixture. It works by passing the mixture through a medium (**stationary phase**) while being carried by a solvent or gas (**mobile phase**).

Different components of the mixture move at varying speeds, allowing them to be separated based on their chemical properties such as polarity, size, or affinity to the stationary phase.

Key Types:

- **Gas chromatography (GC)**: Used for volatile compounds, where the mobile phase is a gas.
- **Liquid chromatography (LC)**: Uses a liquid mobile phase and is suitable for nonvolatile substances.
- **Thin-layer chromatography (TLC)**: Utilizes a thin stationary phase on a flat substrate, often used for quick qualitative analysis.
- **High-performance liquid chromatography (HPLC)**: A highly precise form of liquid chromatography that operates under high pressure for better resolution.

Chromatography is widely used in pharmaceuticals, environmental testing, food safety, lubrication, and petrochemical analysis to separate complex mixtures and identify individual components. In lubrication, chromatography can help analyze the composition of oils and detect the presence of contaminants or additives, ensuring the quality and stability of the lubricant.

168. Churning (Grease)

Churning in grease refers to the **mechanical working** of grease within a bearing or machinery as it is subjected to **repetitive motion**. This action can create a high resistance and drag that can affect the distribution and consistency of the grease, influencing its ability to provide effective lubrication. See Figure C8.

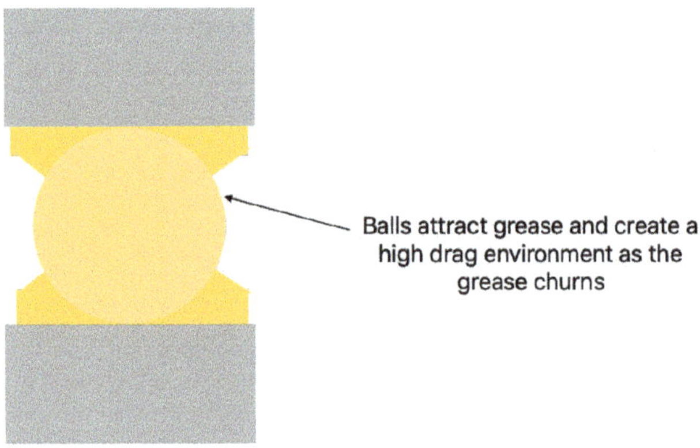

Figure C8 Grease churning.

Key Characteristics:

- **Mechanical mixing**: Continuous movement of the grease as components move, often causing changes in texture and performance.
- **Temperature rise**: Overfilling the bearing cavity or use of an incorrect higher viscosity grease can lead to increased friction, resulting in higher operating temperatures and potential grease degradation.
- **Air entrapment**: Churning may introduce air into the grease, potentially reducing its effectiveness and leading to oxidation or foaming.

Implications:
Churning can lead to grease breakdown, causing a loss of lubrication properties and increased wear on moving parts. To minimize churning, it is essential to use grease with the **appropriate consistency** (**NLGI grade**) for the

application and ensure the correct amount of grease is applied to avoid excessive working. Proper grease selection and maintenance help maintain effective lubrication and extend the life of machinery.

169. Circulating Oil System

A **circulating oil system** is a type of lubrication system in which oil is **continuously circulated** from a central reservoir to machinery components and back. This process ensures that critical parts are consistently lubricated, cooled, and cleaned, maintaining optimal performance and preventing wear or damage.

Key Characteristics:

- **Continuous lubrication**: Oil flows through the system, providing a constant supply of lubricant to moving parts.
- **Cooling effect**: The circulation helps dissipate heat generated by machinery, preventing overheating and maintaining temperature control.
- **Contaminant removal**: As the oil returns to the reservoir, it can pass through filters to remove contaminants, ensuring clean oil is recirculated.

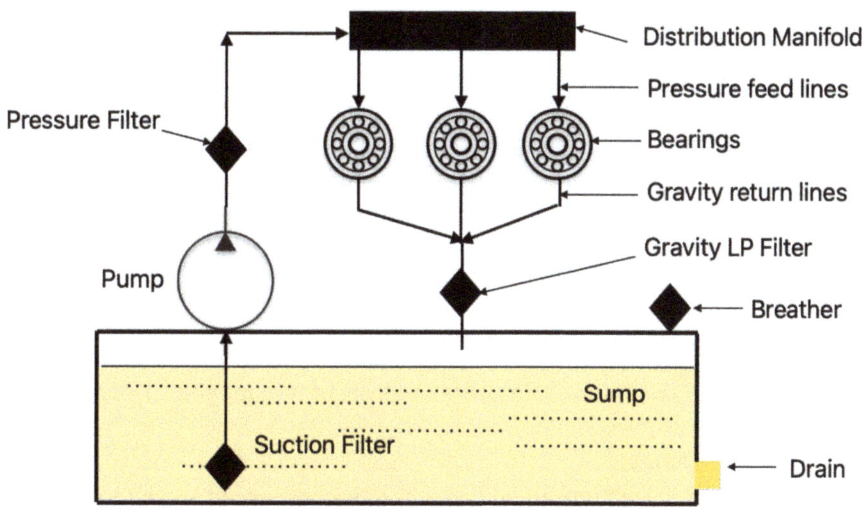

Figure C9 Circulating oil system – pressurized.

Circulating oil systems, similar to the one shown in Figure C9 are commonly used in industrial machinery, large gearboxes, turbines, pumps, and compressors, where consistent lubrication and temperature management are crucial.

This system helps extend the life of components by reducing friction, wear, and operating temperatures, promoting reliable and efficient machinery operation.

170. Class A Lubricant

Class A lubricant refers to a **category of lubricant** that meets high standards of purity, performance, and stability, suitable for use in critical machinery and industrial applications where reliability is paramount. These lubricants are formulated to provide superior protection against wear, oxidation, and corrosion while maintaining their properties under a wide range of operating conditions.

Key Characteristics:

- **High-quality base oils**: Typically made from premium base oils, ensuring low impurities and high thermal stability.
- **Advanced additives**: Contains additives that enhance properties like antiwear, antioxidation, and corrosion protection.
- **Performance standards**: Meets or exceeds industry standards for viscosity, film strength, and temperature resistance.

Class A lubricants are used in high-performance engines, turbines, compressors, and industrial machinery where consistent lubrication is essential for maintaining operational efficiency and preventing equipment failure. These lubricants are chosen for their ability to operate effectively under demanding conditions and for extended periods, reducing downtime and maintenance costs.

171. Class B Lubricant

Class B lubricant is a **category of lubricant** that denotes a performance level ranked below Class A lubricants in terms of quality and capability.

Key Characteristics:

- **Quality and performance**: Class B lubricants are of moderate quality, offering reliable lubrication with fewer advanced properties compared to Class A. They provide basic antiwear, oxidation resistance, and corrosion protection but may not be suitable for extreme or high-stress environments.

- **Applications**: Used in less demanding industrial applications and machinery that do not operate under severe conditions. Common in general machinery, light-duty motors, and noncritical equipment where high performance is not essential.
- **Cost**: More cost-effective than Class A lubricants, making them a practical choice for applications where premium lubrication is not required.

Class B lubricants are chosen based on the operational demands, offering a balance between performance and cost.

172. Class C Lubricant

Class C lubricant is a lubricant category that denotes lubricants with different performance levels and applications, typically ranked below Class A and B lubricants in terms of quality and capability.

Key Characteristics:

- **Quality and performance**: Class C lubricants are considered lower-grade and provide basic lubrication with minimal additives. They may lack enhanced properties such as high thermal stability or superior antiwear performance.
- **Applications**: Suitable for low-demand equipment, intermittent use machinery, or applications where lubrication requirements are minimal and exposure to extreme conditions is unlikely.
- **Cost**: Typically, the most economical option, used where the primary goal is simply to reduce friction and provide basic protection.

Class C lubricants are best suited for simpler applications where high performance is not necessary.

173. Clean Room

A **Clean room**, sometimes referred to as a **white room**, is a specially designed, HEPA-filtered/controlled environment where airborne particles, contaminants, and pollutants are minimized to ensure a high level of cleanliness. Clean rooms are classified according to the number and size of particles permitted per volume of air, as defined by standards such as **ISO 14644-1**.

Key Characteristics:

- **Controlled environment**: Clean rooms are equipped with HEPA (high-efficiency particulate air) or ULPA (ultra-low penetration air) filters to regulate air quality and control particulate contamination.
- **Temperature and humidity control**: Maintained within strict parameters to meet specific operational requirements.
- **Positive pressure**: Keeps contaminants out by ensuring that air flows out of the room when doors or access points are opened.
- **Clean protocols**: Includes practices such as wearing specialized clothing (e.g., cleanroom suits) to prevent contamination from people.

Clean rooms are essential in industries such as bearing manufacturing, pharmaceuticals, semiconductor manufacturing, biotechnology, aerospace, and medical device production, where even the smallest contaminants can compromise product integrity or safety. The classification levels (e.g., ISO Class 1 to Class 9) indicate the degree of cleanliness, with Class 1 being the most stringent, permitting the fewest particles.

174. Cloud Point

Cloud point is the **temperature** at which a liquid, such as oil or fuel, begins to form solid wax crystals and becomes **cloudy or hazy**. See Figure C10. This is an important characteristic, especially for fuels and lubricants used in cold environments, as it indicates the temperature at which the fluid starts to lose its fluidity and may begin to clog filters or impede flow.

Key Characteristics:

- **Indicates wax formation**: The cloud point signals the temperature at which wax components in the fluid begin to crystallize and separate from the liquid.
- **Affects flow properties**: At or below the cloud point, the fluid's viscosity increases, potentially impacting its ability to flow and be pumped efficiently.
- **Critical for cold-weather performance**: Essential for applications involving diesel fuel, biodiesel, and certain lubricants, where maintaining fluidity at lower temperatures is necessary for proper equipment operation.

Understanding the cloud point is vital for industries using engines, hydraulic systems, or machinery in colder climates, as it helps in selecting suitable fluids that remain effective at low temperatures to ensure reliable operation and prevent system blockages.

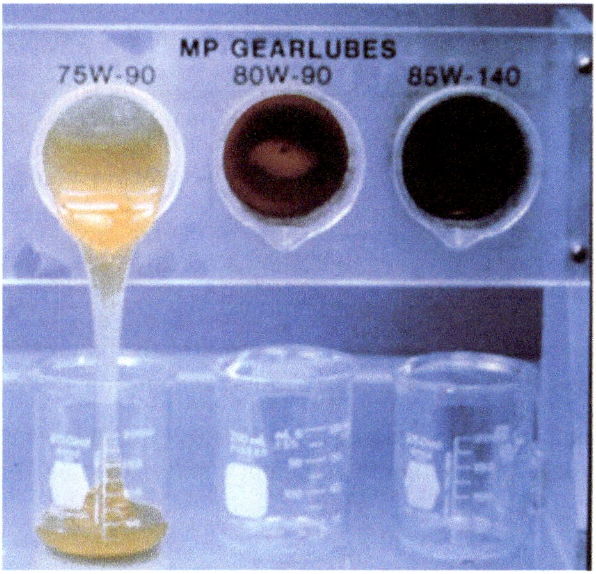

Figure C10 Clouded 75W90 gear oil at –30 °C (–22 °F). Courtesy: Petro Canada.

175. CLS Designation (STLE)

The **Certified Lubrication Specialist (CLS)** designation is awarded by the **Society of Tribologists and Lubrication Engineers (STLE)**. A CLS may be recognized as a "Lubrication Engineer" by their employer and is responsible for various tasks related to lubrication management and maintenance.

Key Responsibilities:

- **Lubricant evaluation and selection**: Chooses the appropriate lubricants for specific applications and recommends changes when needed.
- **Cost-effective purchasing**: Identifies interchangeable products and suppliers to assist in cost-effective lubricant procurement.
- **Lube surveys and specifications**: Conducts lubrication surveys and specifies suitable lubricants and procedures for new equipment purchases or installations.
- **Training and work management**: Trains lubricators and develops work lists, establishing lubrication frequencies for all equipment.

- **Dispensing system design**: Designs or modifies lubricant dispensing systems to improve efficiency and reliability.
- **Quality assurance programs**: Develops and oversees quality assurance programs and used lubricant analysis to maintain equipment health.
- **Waste management**: Manages the collection, reclamation, or disposal of oily waste and implements bulk usage where economical.
- **Inventory management**: Consolidates lubricant inventory to minimize the number of products used.
- **Record keeping**: Maintains detailed records of lubricant applications throughout the facility.
- **Troubleshooting**: Diagnoses lubrication-related failures, identifies root causes, recommends corrective actions, and ensures follow-up.

This comprehensive role ensures the effective use and management of lubricants, promoting operational efficiency and extending the lifespan of equipment.

176. CM

See "Condition Monitoring."

177. CMFS Designation (STLE)

The **Certified Metalworking Fluids Specialist (CMFS)** designation is awarded by the **Society of Tribologists and Lubrication Engineers (STLE)**. A CMFS is an individual who has met the minimum standards of experience, knowledge, and written examination requirements set by the **STLE Metalworking Fluids Certification Committee** to provide technical consultation in metalworking fluids management.

Key Responsibilities:

- **Evaluation and selection**: Assesses and selects the appropriate metalworking fluids and fluid control measures for various applications.
- **Inventory management**: Manages and maintains fluid inventories to ensure consistent supply and usage.
- **Surveys and training**: Conducts surveys and trains operating personnel on proper fluid use and maintenance procedures.

- **Operational recommendations**: Recommends optimal operating parameters and fluid handling equipment to enhance performance and efficiency.
- **Troubleshooting**: Diagnoses and resolves issues related to metalworking fluid applications on the shop floor.
- **Communication facilitation**: Acts as a liaison between fluid suppliers, end users, peers, and other stakeholders to ensure effective communication and collaboration.

Knowledge Base:
A **CMFS** possesses a **foundational understanding** of fluid chemistry, machining processes, tooling, filtration, metallurgy, and waste treatment, enabling them to provide comprehensive support and guidance in metalworking fluid management.

178. CMMS

See "Computerized Maintenance Management System."

179. Coalescor

A **coalescer** is a device used to separate and **remove liquid contaminants** from another liquid or gas stream by coalescing smaller dispersed droplets into larger ones, which can then be more easily separated. Coalescers are commonly used in **filtration systems** to enhance the purity and quality of fluids/lubricants by removing water, oil, or other immiscible liquid contaminants.

Key Characteristics:

- **Coalescing medium**: The device uses a specialized medium or filter material that encourages smaller droplets to merge into larger ones.
- **Gravity separation**: Once the droplets grow larger, they can be separated from the main fluid stream by gravity or mechanical means.
- **Application flexibility**: Can be used for both liquid-liquid and liquid-gas separation processes.

Coalescers are widely employed in oil and gas processing, hydraulic systems, compressed air systems, and fuel filtration to remove water or other contaminants. This ensures better performance and reliability of equipment by maintaining clean, contaminant-free fluids.

180. Coefficient of Friction (COF)

Coefficient of friction (**COF**) is a measure that **quantifies the resistance to sliding between two contacting surfaces** and is represented by the Greek letter **mu (μ)**. It is expressed as a dimensionless value that indicates the ratio between the force required to move one surface over another and the normal force pressing them together. COF values can range from **nearly 0 to well over 1**, with lower values indicating less resistance and easier movement. See Table C1 for both dry and lubricated values for popular bearing materials.

Table C1 Typical dry and wet coefficient of friction values.

Materials		Dry μ	Lubricated μ
Brass	Steel	0.35	0.19
Sintered Bronze	Steel	0.16	Self lubricated
Cast Iron	Oak	0.49	0.075
Graphite	Steel	0.1	Self lubricated
Iron	Iron	1.0	0.2
Teflon	Steel	0.05-0.2	No lube required
Teflon	Teflon	0.04	No lube required
Wood	Concrete	0.62	
Copper Lead	Steel	0.22	
Tire	Dry Road	1.0	0.6 (wet)
Tire	Snow	0.3	

Source: Plant Lubrication for Industrial Facilities – Fourth Edition, Kenneth E. Bannister, 2024.

Key Points:

- **Static COF**: Refers to the measure of friction when an object is stationary, representing the force needed to initiate movement.
- **Kinetic (dynamic) COF**: Refers to the measure of friction when an object is already in motion, usually lower than static COF.
- **Material and fluid variability**: The COF differs based on the materials and fluids involved, affecting how easily one surface moves over another.

Significance:
Lower COF values are preferred in applications where movement and work are required, as they reduce the energy expenditure necessary to achieve motion. For example, a motor will require less amperage, or an engine will consume less fuel, to produce the desired performance with lower COF. Without proper lubrication, surfaces in contact would require significant energy to overcome large static and kinetic forces. Introducing a fluid film between moving parts reduces these forces and facilitates smoother, more efficient movement.

COF is crucial in industries such as automotive, manufacturing, and material handling, impacting the efficiency, performance, and lifespan of machinery. Effective lubrication plays a key role in lowering COF, leading to reduced energy consumption, less heat generation, and extended equipment life.

181. Cold Flow Property

Cold flow property refers to a lubricant's ability to **remain fluid and pumpable at low temperatures** without thickening, gelling, or solidifying. It is a critical characteristic for engine oils, hydraulic fluids, gear oils, and greases operating in cold climates or low-temperature environments. A lubricant with good cold flow properties ensures consistent lubrication, easy startup, and reduced wear in cold conditions.

Key Functions of Cold Flow Property in Lubrication

- **Ensures proper lubrication in cold temperatures**: Prevents oil thickening, ensuring smooth startup and lubrication film formation.
- **Maintains pumpability and flowability**: Allows the lubricant to circulate efficiently in hydraulic systems, fuel injectors, and engines even in subzero conditions.
- **Prevents component wear and startup damage**: Reduces friction and metal-to-metal contact by ensuring lubricant reaches critical parts quickly.
- **Improves energy efficiency**: Reduces drag and resistance caused by high-viscosity oil in cold environments.

Factors Affecting Cold Flow Property

- **Viscosity index (VI)**: Lubricants with a higher VI exhibit better viscosity stability across temperature changes.

- **Base oil type**: Synthetic oils have superior cold flow properties compared to conventional mineral oils, which tend to thicken at low temperatures.
- **Pour point and cloud point**: A lower pour point (the temperature at which oil stops flowing) and cloud point (temperature where wax crystals form) indicate better cold flow performance.
- **Additive formulation**: Pour point depressants (PPDs) and low-temperature modifiers help maintain fluidity at extremely cold temperatures.

182. Cohesion

Cohesion in lubrication terms refers to the **force of attraction** between molecules within the same lubricant. This property determines the **internal strength** and the ability of the lubricant to stay intact under stress, contributing to its overall performance in creating a continuous lubricating film.

Key Characteristics:

- **Film strength**: High cohesion helps maintain a stable, continuous lubricant film between surfaces, preventing metal-to-metal contact.
- **Viscosity contribution**: Cohesion is a factor influencing the viscosity of a lubricant, as it affects how the molecules interact and flow.
- **Resistance to separation**: Strong cohesive forces mean the lubricant can withstand mechanical stresses without breaking apart or thinning excessively.

Cohesion is crucial in maintaining the integrity of the lubricant film under load and during high-speed operation. Proper cohesive properties ensure that the lubricant provides consistent protection, reduces friction, and prevents wear in machinery and moving components.

183. Cohesive Lubricant

A **Cohesive lubricant** is a type of lubricant characterized by **strong intermolecular forces** that help maintain its integrity and consistency under mechanical stress. These cohesive forces ensure that the lubricant can form a continuous, stable film that adheres to surfaces and resists separation, which is crucial for effective lubrication.

Key Characteristics:

- **Film stability**: A cohesive lubricant maintains a consistent film thickness, preventing metal-to-metal contact and reducing wear.
- **Resistance to shear**: Strong cohesion helps the lubricant withstand shear forces without breaking down, ensuring reliable performance under load.
- **Effective protection**: Provides long-lasting protection by preventing lubricant thinning or displacement during operation.

Cohesive lubricants are ideal for use in high-load machinery, gears, bearings, and other moving parts where maintaining a continuous film is critical for reducing friction and wear. This type of lubricant is essential for equipment that operates under **severe conditions**, such as high speeds or heavy loads, where maintaining a stable lubrication layer is necessary to protect components.

184. Cold Cranking Simulator (CCS)

A **cold cranking simulator (CCS)** is a device used to measure the low-temperature viscosity of engine oils under conditions that simulate a cold engine start. The CCS is designed to replicate the stress that engine oil experiences when an engine is cranked at low temperatures, typically during winter or in cold climates. This test provides insight into how well an oil flows and lubricates during initial startup when the oil is thick and the engine components are most vulnerable to wear.

Key Characteristics:

- **Simulates cold start conditions**: Tests the oil's performance at low temperatures to ensure it can effectively flow and protect engine parts during startup.
- **Viscosity measurement**: Results are expressed in units of centipoise (cP), indicating the oil's resistance to flow.
- **Standardized testing**: Commonly used in industry-standard tests (e.g., ASTM D5293) to classify oils according to their SAE viscosity grades for cold performance.

Application:

Using a cold cranking simulator helps manufacturers and engineers ensure that engine oils meet the necessary viscosity specifications for cold weather performance. This is also a useful test for gear oils used in outdoor applications, critical for maintaining engine and gearbox reliability and reducing wear during the most demanding part of any mechanical device/ engine operation – start-up in cold temperatures.

185. Colloidal

A **colloidal** is a type of **fluid mixture** in which tiny particles of one substance are evenly dispersed throughout another substance. These particles are typically between 1 and 1000 nanometers in size and do not settle out or separate under normal conditions, distinguishing colloids from solutions and suspensions.

Key Characteristics:

- **Heterogeneous mixture**: Colloids consist of two phases – a dispersed phase (the particles) and a continuous phase (the medium in which particles are distributed).
- **Stable dispersion**: The particles in a colloid are small enough to remain suspended indefinitely without settling.
- **Tyndall effect**: Colloids can scatter light, making a light beam visible when passed through them (e.g., fog or milk).

In lubrication, **colloidal suspensions** might include greases where thickeners like soaps are dispersed in a continuous oil phase, or additives, for example, molybdenum disulphide suspended in oil to enhance performance. The particles in a colloidal suspension are larger than those in a true solution but small enough that they do not settle out or separate under normal conditions. Colloids ensure that lubricants maintain their properties and function effectively over time, providing consistent protection and performance for machinery and equipment.

186. Color (Oil/Grease)

Color in the context of oil and grease refers to the **visual appearance or hue** of the lubricant, which can indicate various characteristics, such as its composition, condition, and contamination level. The color of oil and grease can range from clear or light amber to dark brown or even black, and specialized lubricants may be dyed in specific colors for identification purposes.

Key Points:

- **Indicator of quality and additives**: The color can signify the type of base oil used and the presence of certain additives. For example, some high-performance or synthetic oils may have a unique tint to distinguish them.

- **Contamination and degradation**: Changes in color over time can indicate contamination (e.g., darkening due to soot or particulate matter) or oxidation, which suggests the oil or grease is breaking down and may need to be replaced.
- **Identification**: In industrial and automotive applications, color-coding lubricants help prevent cross-contamination and ensure the correct product is used.

Applications:
The color of lubricants is assessed visually or with instruments and can be part of routine maintenance checks. Dark or discolored oil and grease may indicate the presence of impurities, water, or metal particles, signaling potential issues that require further inspection or action.

187. Color Antirust

Color antirust refers to a type of protective coating or lubricant that is specifically formulated to prevent corrosion while also incorporating a colored dye or pigment. This dual-purpose product provides rust protection for metal surfaces and allows for easy visual identification due to its color.

Key Characteristics:

- **Corrosion prevention**: Contains additives that form a protective barrier on metal surfaces, inhibiting rust formation.
- **Colored appearance**: The inclusion of color helps users visually confirm application and coverage, making it easier to ensure that all necessary areas are protected.
- **Lubrication properties**: Some color antirust products also serve as lubricants, providing both rust prevention and friction reduction.

Color antirust coatings are used in industrial equipment, automotive parts, machinery, and metal structures exposed to moisture or corrosive environments.

188. Color Coding (Lubricant)

Color coding of lubricants is a systematic method used to **identify** and **differentiate** lubricants based on their type, application, or specific properties.

Colors are often combined with shapes for identification clarity. See Table C2. This practice helps prevent cross-contamination, reduces misapplication errors, and streamlines maintenance processes by allowing for easy visual identification of lubricants.

Table C2 Lubricant ID control system chart.

Lubricant ID Control System Chart		
Symbol	**Color**	**Lubricant Description**
	Red	Oil, Gear, Mineral, ISO 220, XYZ Company, Gearlube-2000
	Blue	Oil, Hydraulic, ISO 68, XYZ Company, Hydroblue
	Green	Grease, Lithium, EP2, XYZ Company, Crasso 2
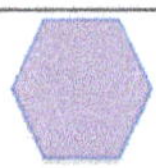	Purple	Grease, Lithium, EP1, XYZ, Crasso 1

Source: Plant Lubrication for Industrial Facilities – Fourth Edition, Kenneth E. Bannister, 2024.

Key Points:

- **Visual identification**: By assigning specific colors to various lubricants, maintenance staff can quickly and easily recognize the correct product for a particular application.
- **Safety and efficiency**: Color coding enhances safety by ensuring that the proper lubricant is used, reducing the risk of equipment damage due to incorrect lubricant application.
- **Standardized system**: Many organizations develop standardized color-coding schemes tailored to their equipment and lubricant types, using color-coded tags, labels, or containers.

Common Applications:

- **Industrial facilities**: Use color coding to distinguish between different types of lubricants such as gear oils, hydraulic fluids, greases, and coolants.

- **Automotive maintenance**: Garages and service centers may use color-coded systems to differentiate between various engine oils, transmission fluids, and other automotive lubricants.
- **Food-grade operations**: Color coding helps identify food-safe lubricants that comply with strict industry regulations, preventing contamination with nonfood-grade products.

Implementing a color-coding system for lubricants improves maintenance efficiency, reduces errors, and helps ensure that each piece of equipment receives the correct lubricant for optimal performance and longevity. Note: Colors depicted in Table C2 are example only and do not infer a color standard is in use.

189. Color Coding (Lubrication Delivery point ID)

Color coding of lubrication points is a maintenance strategy used to **visually identify** and differentiate the various lubrication points on equipment or machinery. This practice helps ensure that the correct type of lubricant is applied to each specific point, preventing cross-contamination and misapplication, and supporting efficient maintenance operations.

Key Points:

- **Visual cues**: Each lubrication point is marked with a distinct color that corresponds to a specific type of lubricant, making it easy for maintenance personnel to quickly identify which lubricant to use.
- **Organized maintenance**: Reduces the chance of human error during lubrication routines, ensuring that all points receive the proper attention.
- **Increased equipment longevity**: Helps maintain consistent and appropriate lubrication, which can extend the life of machinery and optimize performance.

Common Practices:

- **Color-coded tags or labels**: Placed near lubrication points to indicate the required lubricant type. For example, red for grease, blue for gear oil, and yellow for hydraulic fluid. See Figure C11.
- **Standardized schemes**: Companies may create specific color-coding systems based on their unique lubricant inventory and application needs.

- **Training and reference**: Maintenance staff are trained to recognize and adhere to the color-coding system, sometimes supported by reference charts or guides.

Figure C11 Color coded lube label, grease fitting collars, tape for lube point identification Courtesy: ENG**TECH** Industries Inc.

190. Combination Friction

Combination friction refers to the simultaneous presence of sliding and rolling contact friction on a bearing surface, a phenomenon that is especially characteristic of meshing gears. This occurs due to the dynamic nature of the gear tooth profile, where gear teeth slide against each other until their opposing pitch surfaces align (see Figure C12). At this point, rolling friction predominates as the teeth disengage and transfer the motion smoothly.

Key Features:

- **Dynamic friction transition**: In gear systems, friction shifts from sliding to rolling as the teeth come into contact and then separate. This transition requires the lubricant to handle both high-friction sliding and lower-friction rolling effectively.
- **Increased sliding friction in certain gears**: Specific types of gears, such as hypoid gears and worm gears, experience a much higher degree

of sliding friction due to their design. The angled contact and continuous sliding in these gears create significant frictional forces, necessitating specialized lubricants that can withstand intense heat and wear.

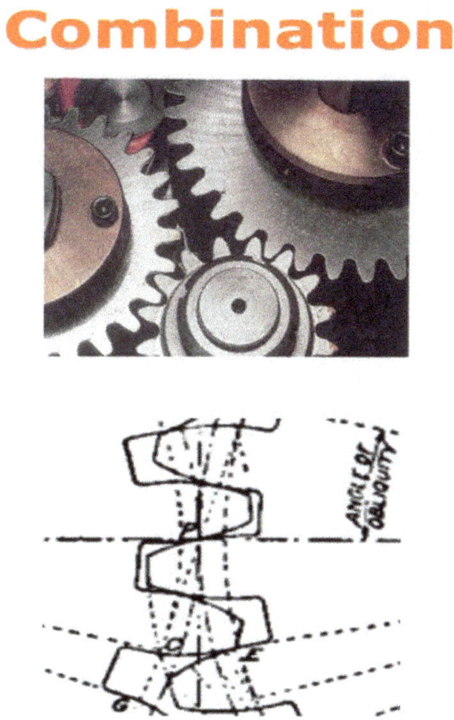

Figure C12 Typical gear application depicting rolling and sliding friction. Courtesy: ENGTECH Industries Inc.

Applications and Considerations:

- **Lubrication requirements**: Gears operating under combination friction need lubricants with high film strength, effective antiwear additives, and the ability to withstand varying frictional conditions.
- **Gear Types**:
  - **Hypoid gears**: Found in automotive differentials, they have an offset design that produces a greater sliding motion, leading to increased friction and heat.
  - **Worm gears**: Common in machinery for heavy torque applications, these gears rely heavily on sliding contact, making lubrication crucial for reducing wear and maintaining efficiency.

Understanding and managing combination friction is essential for optimizing gear performance, reducing energy losses, and extending the lifespan of gear systems. The right lubricant formulation ensures that both sliding and rolling contacts are adequately protected, enabling smooth operation under varying load and speed conditions.

191. Complex Soap

Complex soap refers to a type of **thickener u**sed in the formulation of greases, providing enhanced performance characteristics compared to conventional soap-based greases. Complex soap thickeners are created by combining a metallic soap (such as lithium, calcium, or aluminum) with a complexing agent, such as an organic acid or a short-chain dicarboxylic acid, to create a more intricate structure.

Key Characteristics:

- **Higher drop point**: Greases made with complex soaps have a higher drop point (the temperature at which the grease starts to liquefy), making them suitable for high-temperature applications.
- **Improved performance**: Offers better thermal stability, water resistance, and mechanical stability than simple soap-based greases.
- **Enhanced load-carrying capacity**: Provides strong protection under high load conditions, reducing wear and extending equipment life.

Common Types:

- **Lithium complex grease**: Known for excellent high-temperature performance and versatility, used in automotive, industrial, and heavy machinery applications.
- **Calcium complex grease**: Offers good water resistance and is often used in marine and outdoor environments.
- **Aluminum complex grease**: Provides excellent water resistance and adhesion, ideal for applications exposed to moisture.

Complex soap greases are widely used in automotive wheel bearings, industrial machinery, and high-load equipment where performance under high temperatures and challenging conditions is required.

192. Compatibility (of Lubricants)

Compatibility in lubrication refers to the ability of different lubricants, additives, and materials to function together without causing negative chemical or physical interactions. It is essential when mixing oils or greases, selecting lubricants for seals and gaskets, or transitioning between different lubrication products. Incompatibility can lead to lubrication failure, component damage, and increased maintenance costs.

Key Aspects of Compatibility in Lubrication

- **Lubricant-to-lubricant compatibility**: Ensures that different base oils, additives, and viscosity grades can mix without separation or degradation.
- **Grease compatibility**: Determines whether different thickener types, base oils, and additives can work together without causing hardening, softening, or oil separation.
- **Seal and elastomer compatibility**: Ensures that the lubricant does not cause swelling, shrinking, or degradation of seals, O-rings, and gaskets.
- **Material compatibility**: Prevents lubricants from reacting with metals, plastics, or coatings, which could lead to corrosion, wear, or loss of lubrication film.

Factors Affecting Compatibility

- **Base oil type**: Mineral, synthetic, and bio-based oils may have different solvency properties that affect mixing behavior.
- **Additive package**: Some antiwear, antioxidation, and detergent additives may be chemically incompatible with others, leading to deposit formation or performance loss.
- **Grease thickener type**: Lithium, calcium, polyurea, and other thickener types may not blend well, resulting in softening, separation, or hardened deposits.
- **Operating conditions**: Temperature, pressure, and speed variations can impact lubricant stability and compatibility.

Testing Compatibility

- **ASTM D6185 (grease compatibility test)**: Evaluates whether different grease formulations can mix without causing performance issues.
- **ASTM D741 Seal swell test**: Measures how lubricants affect elastomer expansion and durability.
- **Oil compatibility testing**: Analyze how two different oils mix, separate, or degrade over time.

193. Compounded Oil

Compounded oil is a type of lubricating oil that consists of mineral or synthetic base oil blended with specific amounts of **fatty oil** (such as animal or vegetable oils) additives to enhance its lubricating properties. This combination improves the oil's ability to adhere to surfaces, providing better lubrication under certain conditions.

Key Characteristics:

- **Enhanced lubricity**: The addition of fatty oils improves the oil's lubricating film strength, making it suitable for applications where superior boundary lubrication is needed.
- **Adhesion**: Compounded oils adhere better to metal surfaces, reducing friction and wear in environments where consistent lubrication is crucial.
- **Improved wetting**: These oils have better wetting properties, allowing them to spread evenly and provide thorough coverage on surfaces.

Compounded oils are commonly used in steam cylinder lubrication, worm gears, and applications involving high humidity or water exposure, such as in textile machinery and industrial steam engines. Their ability to provide strong, reliable lubrication in wet or high-pressure conditions makes them essential for equipment where standard mineral oils might not perform adequately.

194. Compounding

Compounding refers to the process of **blending** base oils with various additives or enhancing agents to create a lubricant with specific properties tailored for particular applications. This process modifies the performance characteristics of the oil or grease to meet specific operational requirements, such as increased lubricity, improved temperature stability, or enhanced protection against wear and corrosion.

Key Characteristics:

- **Customization**: Compounding allows for the formulation of lubricants that cater to unique environmental or mechanical conditions.
- **Enhanced properties**: The addition of fatty oils, antiwear additives, detergents, or oxidation inhibitors improves the overall performance of the base oil.

- **Application-specific**: Different types of compounding produce lubricants suitable for varied conditions, such as high-load environments, extreme temperatures, or exposure to water and steam.

Compounding is widely used in the production of industrial lubricants, automotive oils, gear oils, and hydraulic fluids. For example, compounded oils with added fatty oils are used in steam cylinder lubrication to improve adhesion and reduce friction in wet or high-temperature conditions.

195. Compressed Air

Compressed air is air that has been **pressurized** to a level higher than atmospheric pressure, stored or conveyed for use as a **power source** or in various industrial applications. The process of compressing air concentrates its energy, allowing it to perform mechanical work when released.

Key Characteristics:

- **Stored energy**: Compressed air acts as a reservoir of potential energy that can be used to power tools, machinery, and processes.
- **Pressure levels**: Measured in pounds per square inch (psi) or bars, with varying levels depending on the application.
- **Moisture content**: Air compression often results in moisture condensation, necessitating the use of dryers or moisture separators to prevent corrosion and maintain equipment efficiency.

Common Application:

- **Lubrication**: Many grease and oil lubricant pumps use compressed air as their power source.
- **Industrial**: Powering pneumatic tools, conveyor systems, paint sprayers, and automated lubrication pumps.
- **Transportation**: Used in braking systems and chassis lubrication pumps for trucks and trains.
- **Maintenance and cleaning**: Employed to blow debris off machinery and parts or clear dust from surfaces.
- **Manufacturing processes**: Provides energy for functions like assembly line automation, packaging, and controlling robotic arms.

Benefits:

- **Versatile**: Compressed air can be used across various industries, including construction, manufacturing, automotive, and food processing.
- **Safe power source**: Generally safer than other energy sources like electricity, as there is no risk of sparks in flammable environments.

196. Compressibility (of Lubricants)

Compressibility of a lubricant refers to its ability to decrease in volume when subjected to pressure. While lubricants are generally considered incompressible, they do exhibit slight compressibility under extreme pressure conditions, which can influence film thickness, load-carrying capacity, and overall lubrication performance.

Key Functions of Lubricant Compressibility:

- **Maintains film thickness under pressure**: Compressibility affects hydrodynamic and elastohydrodynamic lubrication (EHL) by influencing the formation and stability of the lubricating film.
- **Impacts high-pressure system performance**: In hydraulics and high-load applications, excessive compressibility can lead to delayed pressure response and energy loss.
- **Influences load-carrying capacity**: More compressible lubricants may reduce efficiency in bearings and gears, leading to higher friction and potential wear.
- **Affects cavitation resistance**: Lubricants with low compressibility help reduce the formation of vapor bubbles (cavitation) in pumps and hydraulic systems.

Factors Affecting Lubricant Compressibility:

- **Base oil type**: Synthetic lubricants, such as polyalphaolefins (PAOs) and esters, generally have lower compressibility than mineral oils, improving performance in high-pressure applications.
- **Operating pressure**: Higher pressures cause greater lubricant volume reduction, affecting film formation in hydraulic, aerospace, and high-load systems.
- **Temperature**: Increased temperatures typically reduce compressibility, improving lubricant stability under load.

- **Viscosity and additives**: Higher-viscosity lubricants and extreme pressure (EP) additives enhance film formation and load capacity, counteracting compressibility effects.

Measurement of Lubricant Compressibility:

- **Bulk modulus** (β): A measure of a fluid's resistance to compression, expressed in Pascals (Pa) or psi. Higher bulk modulus values indicate lower compressibility.
- **High-pressure tribology tests**: Used to assess lubricant behavior under extreme pressure conditions, ensuring adequate film thickness and wear protection.

Applications Where Lubricant Compressibility is Critical:

- **Hydraulic systems**: Compressibility affects fluid efficiency, pressure stability, and actuator responsiveness.
- **High-speed bearings and gears**: Impacts EHL film thickness and rolling element lubrication.
- **Aerospace and automotive engineering**: Determines lubricant performance in extreme pressure and temperature conditions.
- **Deep-sea and high-pressure equipment**: Lubricants must remain stable under high hydrostatic pressure conditions.

197. Compressor Lubricants

Compressor lubricants are specially formulated oils designed to lubricate, cool, and protect the internal components of air, gas, and refrigeration compressors. They help reduce wear, dissipate heat, prevent deposit formation, and enhance energy efficiency in high-pressure environments.

Key Functions of Compressor Lubricants

- **Reduces friction and wear**: Forms a protective film on moving parts, preventing metal-to-metal contact and extending component life.
- **Dissipates heat**: Helps remove excess heat generated by compression, preventing overheating and oil degradation.
- **Prevents deposit and varnish formation**: Minimizes carbon buildup, oxidation, and sludge, which can reduce compressor efficiency.
- **Seals and protects components**: Ensures tight sealing in piston rings and rotary mechanisms, improving compression efficiency.

- **Resists oxidation and contamination**: Designed to withstand high temperatures, moisture, and chemical exposure without breaking down.

Types of Compressor Lubricants

1. **Based on Base Oil Type**
   - **Mineral oil-based compressor lubricants**: Traditional and cost-effective, suitable for general-purpose applications with moderate operating conditions.
   - **Synthetic compressor lubricants**: Offer better oxidation resistance, thermal stability, and extended service life, commonly used in high-performance and extreme temperature applications.

2. **Based on Compressor Type**
   - **Reciprocating compressor lubricants**: Must handle high pressures, piston friction, and heat generation, often formulated with antiwear and deposit control additives.
   - **Rotary screw compressor lubricants**: Require low-foaming, oxidation-resistant oils to maintain proper lubrication in continuous-duty cycles.
   - **Rotary vane compressor lubricants**: Designed for consistent viscosity and wear protection in high-speed vane compressors.
   - **Centrifugal compressor lubricants**: Typically use low-viscosity synthetic oils with excellent oxidation resistance to reduce friction in high-speed impellers.
   - **Refrigeration compressor lubricants**: Specially designed for compatibility with refrigerants (R-134a, ammonia, CO_2, etc.), low-temperature stability, and moisture resistance.

198. Computerized Maintenance Management System

A **computerized maintenance management system (CMMS)** is a **software solution** designed to help organizations plan, track, manage, and optimize maintenance and lubrication activities and assets. CMMS platforms centralize maintenance and lubrication management information and streamline processes to improve the efficiency of maintenance and lubrication operations.

Key Features:

- **Work order management**: Allows for the creation, assignment, and tracking of maintenance work orders.
- **Asset management**: Keeps detailed records of equipment and assets, including maintenance history, specifications, and operational status.
- **Preventive maintenance and lubrication scheduling**: Automates the scheduling of routine maintenance and lubrication tasks to prevent equipment failures and extend asset life.
- **Inventory control**: Manages spare parts and lubrication supplies to ensure that necessary items are available when needed.
- **Reporting and analysis**: Generates reports and analytics on maintenance and lubrication activities, costs, and equipment performance for better decision-making.

Key Benefits:

- **Increased efficiency**: Streamlines maintenance workflows, reduces downtime, and improves response times.
- **Enhanced equipment life**: Helps prevent unplanned breakdowns and prolongs the life of machinery through timely maintenance.
- **Cost savings**: Reduces labor costs and material waste by optimizing maintenance schedules and resource allocation.
- **Compliance**: Assists in meeting regulatory requirements by maintaining accurate records and maintenance logs.

CMMSs are widely used in industries such as manufacturing, facilities management, healthcare, transportation, and utilities, where maintaining equipment reliability and efficiency is critical. The system enables maintenance teams to better organize their work, manage resources, and maintain optimal operational performance.

199. Condition Base Lubrication (CBL)

Condition-based lubrication (CBL) is a proactive **lubrication work strategy** that involves monitoring the condition of machinery and applying lubrication only when specific indicators show that it is needed. Unlike routine or time-based lubrication schedules, CBL aims to **optimize lubrication intervals** by assessing the **real-time state** of the equipment to prevent overlubrication or underlubrication.

Key Characteristics:

- **Data-driven**: Relies on data collected from sensors, oil analysis, vibration analysis, and other monitoring tools to determine the need for lubrication.
- **Optimized maintenance**: Reduces unnecessary lubricant application, extending the life of both the lubricant and the machinery.
- **Prevents wear and failure**: Helps identify the precise moment when lubrication is necessary, reducing the risk of component wear and unexpected breakdowns.

Key Benefits:

- **Improved equipment performance**: Ensures machinery operates efficiently by maintaining optimal lubrication levels.
- **Cost-effective**: Minimizes lubricant waste and reduces maintenance costs by only servicing equipment when needed.
- **Extended component life**: Reduces the risk of overlubrication, which can attract contaminants, and underlubrication, which can increase friction and wear.

Condition-based lubrication is used in industries with critical machinery, such as manufacturing, power generation, oil and gas, and transportation, where maintaining optimal performance and minimizing downtime are essential. It leverages predictive maintenance technologies to ensure that lubrication efforts are targeted and effective, enhancing overall reliability and operational efficiency.

200. Condition-based Maintenance (CBM)

Condition-based maintenance (CBM) is a **proactive maintenance strategy** that involves monitoring **the real-time condition** of equipment and **performing maintenance** only when indicators suggest that a failure or performance decline is imminent. This approach aims to maintain equipment efficiency and reliability while minimizing maintenance costs and preventing unexpected downtime.

Key Characteristics:

- **Real-time monitoring**: Utilizes data from sensors, inspections, and diagnostic tools to assess equipment condition. Parameters such as vibration, temperature, pressure, and lubricant analysis are commonly tracked.
- **Data-driven decisions**: Maintenance activities are based on actual equipment conditions rather than on predefined schedules, optimizing the timing of repairs and interventions.
- **Targeted interventions**: Maintenance is performed only when necessary, reducing the risk of overmaintenance and minimizing unplanned failures.

Key Benefits:

- **Increased equipment life**: Ensures that maintenance is performed at the right time, extending the lifespan of equipment and components.
- **Reduced downtime**: Helps prevent unexpected equipment failures, enhancing overall operational uptime.
- **Cost savings**: Decreases unnecessary maintenance activities, reducing labor costs and extending the use of resources.
- **Improved safety**: Prevents accidents and equipment malfunctions by addressing issues before they escalate.

Condition-based maintenance is widely used in industries such as manufacturing, transportation, energy, and aerospace, where critical machinery requires reliable performance and minimized downtime. CBM is supported by predictive technologies and advanced analytics, making it an integral part of predictive maintenance programs to optimize maintenance practices and operational efficiency.

201. Condition Monitoring (CM)

Condition monitoring is the practice of continuously or periodically assessing the current condition of equipment or machinery to detect potential issues early and prevent failures. By measuring and analyzing various parameters, condition monitoring helps identify signs of wear, deterioration, or abnormal operation, enabling timely maintenance actions to be taken.

Key Characteristics:

- **Real-time and periodic data collection**: Uses sensors and diagnostic tools to gather data on factors like vibration, temperature, pressure, sound, lubrication quality, and electrical performance.

- **Noninvasive**: Monitors equipment without interfering with its regular operation, ensuring minimal disruption to production.
- **Predictive insights**: Helps forecast potential failures or performance declines, supporting predictive maintenance strategies.

Key Benefits:

- **Increased reliability**: Early detection of issues prevents unexpected equipment failures and reduces downtime.
- **Cost savings**: Avoids unnecessary maintenance by targeting specific components that need attention, and optimizing repair schedules.
- **Enhanced safety**: Identifies potential hazards before they cause accidents or damage.
- **Extended equipment life**: Maintains optimal operation and minimizes the wear and tear of components by addressing issues proactively.

202. Cone Penetration Test

A **cone penetration test**, also known as a **cone test**, is a method used to **measure the consistency or stiffness** of a material, such as grease or soil, by determining how far a standardized cone penetrates into the sample under specific conditions. See Figure C13.

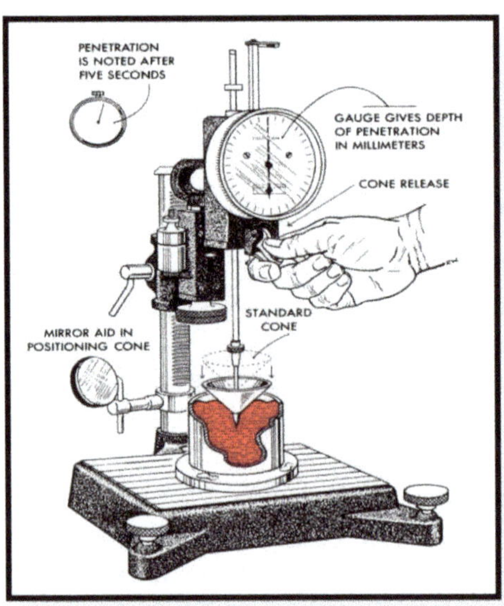

Figure C13 ASTM D-217 cone penetration test apparatus. Source: Gulf Oil.

In lubrication, the cone test is essential for evaluating the workability and grade of grease.

Key Characteristics:

- **Standardized procedure**: A cone of known weight and dimensions is allowed to sink into the grease for a set time (usually 5 seconds) at a specific temperature.
- **Penetration measurement**: The depth of penetration is measured in tenths of a millimeter. The result classifies the grease according to the **NLGI (National Lubricating Grease Institute) scale**, which ranges from **000 (very soft) to 6 (very hard). See "NLGI Grease Scale.**"
- **Consistency indicator**: The test helps determine the grease's consistency, which affects its application and performance under different operational conditions.

203. Consistency (Grease)

Consistency, when referring to **grease**, specifies the measure of a **grease's firmness or stiffness**, indicating its ability to maintain its structure and stay in place under varying conditions. It is an essential property that determines how the grease behaves during application and use, impacting its performance in lubricating and protecting machinery.

Key Characteristics:

- **Measured by cone penetration test**: The consistency is determined by the depth a standardized cone penetrates the grease under specific conditions, with results expressed in tenths of a millimeter.
- **NLGI classification**: Grease consistency is classified by the NLGI scale, which ranges from 000 (very soft) to 6 (very hard). Most general-purpose greases fall between NLGI 1 and NLGI 3, with NLGI 2 being the most common.
- **Impact on application**: The consistency affects the grease's pumpability, load-carrying capacity, and ability to stay in place without dripping or leaking.

Importance: The consistency of grease is a critical factor in ensuring effective lubrication. It must be matched to the specific application to provide optimal protection, stability, and performance under varying loads, temperatures, and operational conditions. For example, softer greases (lower NLGI grades) are more suited for low-temperature or high-speed applications, while harder greases (higher NLGI grades) are better for high-load, low-speed applications.

204. Consolidation of Lubricants

See "Lubricant Consolidation Program."

205. Constant Level Lubricator

Constant level lubricator is an automatic device designed to maintain a consistent lubricant level in machinery reservoirs, such as gearboxes, bearings, or pump housings. It ensures that the appropriate amount of lubricant is supplied as needed, without the need for manual intervention. This type of lubricator helps prevent overfilling or underfilling, both of which can lead to equipment inefficiency or damage. See Figure C14.

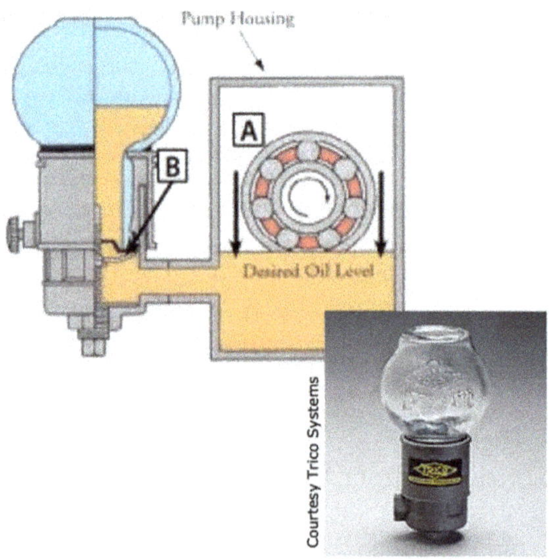

Figure C14 Constant level lubricator. Courtesy: Trico Systems.

How It Works:

- A reservoir within the constant-level lubricator holds a supply of lubricant.
- When the lubricant level in the machinery's housing drops below the set level, air enters the lubricator's reservoir, allowing lubricant to flow into the housing.

- The flow stops automatically when the lubricant reaches the predetermined level, as the air entry is sealed off, creating a vacuum.

Key Features:

1. **Automatic level control**: Ensures a steady supply of lubricant without the need for frequent manual checks.
2. **Transparent reservoir**: Many constant-level lubricators have a clear reservoir, allowing operators to visually monitor lubricant levels.
3. **Adjustable level**: Some models allow the level to be easily adjusted to meet the specific needs of the equipment.

Applications:

- **Bearings**: To maintain proper lubrication in rotating equipment bearings.
- **Gearboxes**: Ensures gears are continuously lubricated to reduce wear and improve efficiency.
- **Pumps and compressors**: Keeps critical moving parts adequately lubricated, even during extended operation.

Key Benefits:

1. **Reduced maintenance**: Minimizes the need for manual lubrication checks and top-offs.
2. **Consistent lubrication**: Prevents over or underlubrication, reducing wear and extending equipment life.
3. **Improved reliability**: Helps maintain optimal operating conditions, reducing the risk of downtime or failure.

206. Containment Berm

A **Containment berm** is a **protective barrie**r designed to contain spills, leaks, or accidental releases of lubricants and other fluids found in lubricant storage areas and rooms These berms are typically constructed from durable, chemical-resistant materials and are placed around storage areas, dispensing equipment, or containers to prevent fluid from spreading and causing contamination or environmental damage.

Key Characteristics:

- **Spill containment**: Acts as a first line of defense to capture and contain fluids, preventing them from spreading to surrounding areas.

- **Material composition**: Made from rubber, plastic, metal, or other durable, nonporous materials that resist chemicals and oils.
- **Modular and portable options**: Can be fixed structures or portable for easy repositioning and flexibility in the lube room layout.

Importance in a Lube Room:

- **Safety**: Protects workers by reducing slip hazards associated with spilled lubricants.
- **Environmental protection**: Prevents the contamination of soil and water sources by containing potential leaks or spills.
- **Regulatory compliance**: Helps facilities meet local, state, or federal regulations regarding spill containment and environmental safety.

Containment berms are used in industrial facilities, workshops, automotive service centers, and any location where bulk lubricant storage or fluid transfer takes place. They are a critical component of spill prevention plans, ensuring that any accidental release of fluids is contained, making clean-up easier, and mitigating potential harm to the environment and equipment.

207. Containment System

A **containment system**, as it pertains to lubricant loss and the environment, refers to a series of protective measures and structures designed to capture and prevent the unintended release of lubricants and other fluids into the surrounding environment. These systems play a crucial role in environmental protection and spill prevention, ensuring that potential leaks or spills are controlled and do not cause contamination or harm to ecosystems.

Key Characteristics:

- **Primary containment**: The main vessel or container (e.g., tanks, drums, or equipment) that holds the lubricant.
- **Secondary containment**: An additional layer of protection, such as berms, drip trays, or dikes, that captures and contains any lubricant that may escape from the primary container. See Figure C15.
- **Tertiary measures**: Backup systems, such as spill response equipment and absorbent materials, to address containment failures.

Functions:

- **Leak and spill prevention**: Containment systems are designed to hold fluids in the event of a leak or spill, preventing them from reaching soil, water bodies, or other sensitive areas.
- **Environmental compliance**: Helps facilities meet regulatory requirements related to environmental safety and spill containment.
- **Workplace safety**: Reduces slip hazards and exposure risks to workers by controlling lubricant flow and preventing uncontrolled spreading.

Figure C15 Typical lubricant spill containment pallet. Courtesy: ENG**TECH** Industries Inc.

Types of Containment Systems:

- **Berms and barriers**: Structures that surround storage and handling areas to contain spills.
- **Drip pans and trays**: Used under machinery or containers to catch any leaks or drips.
- **Spill containment pallets**: Platforms with built-in containment areas to capture spills from storage drums or tanks.

Containment systems are essential in industrial facilities, workshops, and storage areas where lubricants are used or stored in bulk. They protect the environment, promote sustainable practices, and ensure that facilities comply with environmental safety regulations by preventing the release of hazardous substances.

208. Contamination

Contamination refers to the presence of **unwanted** or **foreign substances** within a system, material, or environment that can compromise its quality, performance, or safety. In the context of lubricants and lubrication systems, contamination involves particles, moisture, chemicals, or other impurities that infiltrate the lubricant, potentially leading to degraded performance and increased wear of machinery.

Types of Contaminants:

- **Solid particles**: Dirt, metal shavings, or other debris that can cause abrasion and accelerate wear.
- **Moisture**: Water can lead to rust, corrosion, and the breakdown of lubricant additives.
- **Chemical contaminants**: Reactive substances, such as acids or fuel, that can alter the lubricant's chemical composition.
- **Sources**: Contamination can enter through leaks, poor sealing, improper storage, cross-contamination, or as a byproduct of equipment operation (e.g., metal particles from wear).
- **Impact on machinery**: Increases friction, wear, and potential equipment failure, leading to reduced efficiency and higher maintenance costs.

Prevention and Control:

- **Filtration systems**: Using high-quality filters to remove particles and impurities.
- **Proper storage**: Ensuring lubricants are stored in sealed, clean containers to prevent external contaminants from entering.
- **Regular monitoring**: Conducting oil analysis and inspections to detect contamination early and take corrective action.

Contamination control is essential for maintaining the reliability and longevity of machinery, as clean lubricants help ensure optimal performance and minimize maintenance issues.

209. Contamination Control

Contamination control refers to the set of **practices, strategies, and systems** designed to prevent, detect, and manage contamination within machinery and lubricant systems. Effective contamination control is essential for maintaining the quality and performance of lubricants, minimizing wear on equipment, and extending machinery life.

Key Measures Required:

- **Filtration systems**: Installing high-quality filters in lubricant and hydraulic systems to trap and remove solid particles and moisture.
- **Sealing and enclosures**: Ensuring that machinery is well-sealed to prevent the ingress of dust, water, or other contaminants. This includes using appropriate gaskets, seals, and protective covers.
- **Proper storage and handling**: Storing lubricants in clean, sealed containers and keeping them in designated areas to avoid exposure to contaminants.
- **Cleanliness**: Using clean transfer equipment (e.g., pumps and funnels) to handle lubricants and prevent cross-contamination.
- **Breathers**: Using desiccant breathers on storage containers and machinery to prevent moisture and airborne particles from entering while allowing airflow.
- **Regular monitoring and testing**:
    - **Conducting** oil analysis to check for the presence of contaminants like metal particles, dirt, or water.
    - Using **contamination detection tools** such as particle counters and moisture sensors.
- **Routine maintenance**: Implementing regular inspection schedules to identify and address contamination sources in machinery and lubricant storage areas.

- **Training and education**: Educating personnel on proper lubricant handling, storage procedures, and contamination prevention methods to reinforce best practices.

Importance:

- **Improved equipment reliability**: Reduces the risk of unexpected breakdowns by maintaining clean lubricants that perform efficiently.
- **Extended lubricant and equipment life**: Prevents wear and damage caused by contaminants, leading to longer intervals between maintenance and reduced operational costs.
- **Compliance and safety**: Helps facilities meet industry standards and regulations related to contamination control, promoting a safer working environment.

Contamination control is a critical part of lubrication management and predictive maintenance strategies, ensuring that machinery runs smoothly and efficiently while minimizing the risk of costly repairs or downtime.

210. Continuous Delivery Lubrication System

See "Circulating Oil System."

211. Control Device

A **lubrication control device** is an apparatus or system used to regulate, monitor, and manage the application and distribution of lubricants to machinery and equipment. These devices help ensure that the correct type and amount of lubricant is delivered to critical points, maintaining optimal machinery performance, preventing overlubrication or underlubrication, and reducing wear and tear.

Key Characteristics:

- **Precision delivery**: Provides accurate control over the amount of lubricant dispensed, ensuring consistent application and preventing waste.
- **Monitoring capability**: Many devices come equipped with sensors or indicators that monitor lubrication levels, pressure, or flow rates.
- **Automation**: Advanced lubrication control devices are part of automated systems that deliver lubricant at scheduled intervals without manual intervention.

- **Integration**: Can be integrated with centralized lubrication systems to manage the distribution of lubricants across multiple points in a complex machine or facility.

Types of Lubrication Control Devices:

- **Metering valves**: Manage the flow rate of lubricant to ensure it is distributed accurately and evenly.
- **Pressure valves**: Control the pressure of the lubricant being delivered to maintain optimal flow.
- **Timers and controllers**: Automate lubrication intervals and schedules based on machinery needs.
- **Monitoring sensors**: Detect and alert users to changes in lubricant levels or flow disruptions.

Lubrication control devices are widely used in industrial machinery, automotive systems, manufacturing plants, and heavy equipment where precise lubrication is critical to prevent breakdowns, minimize friction, and extend the life of components.

Benefits:

- **Enhanced equipment performance**: Maintains the proper lubrication needed for smooth operation.
- **Reduced maintenance costs**: Prevents overlubrication and lubricant waste, leading to cost savings.
- **Improved reliability**: Helps prevent equipment failures due to inadequate lubrication.
- **Safety and compliance**: Assists in meeting industry standards for safe and effective lubrication practices.

212. Control Unit

A **control unit** is a continuous **metering valve** utilized in a single-line resistance (SLR) centralized lubrication system. Designed by Joseph Bijur and pioneered by the original Bijur Lubricating Corporation in 1924, this control unit is responsible for **restricting and metering** the delivery of **lubricating oil** in a continuous delivery lubrication system.

How an SLR Meter Unit Works

Control units can be positioned at each individual bearing point or in an upstream manifold. The continuous lubricant flow is controlled at the pump and the individual continuous flow to each bearing is apportioned by an engineered screw device inside of the control unit.

Each control unit has a number representing its apportioning capability. The higher the number, the more the flow, for example, a CSA-2 control unit will deliver twice the flow of the number CSA-1 sized control unit. The next size-up CSA-3 control unit is designed to deliver twice that of a CSA-2 and four times the flow of a CSA-1, and visa-versa as you go down in number and size. The stamped arrow on the unit designates the flow direction as the valves are unidirectional.

Although similar in look and size to the Bijur meter unit (see Meter Unit), the control unit differs in that it has an internal screw that controls the fluid flow by changing the screw pitch allowing the lubricant to continuously free flow at differing rates, unlike the meter unit that is designed for only cyclic lubrication operations. See Figure C16.

Key Characteristics:

- **Automated lubrication management**: Controls the volume of lubricant flowing to each point, reducing the need for manual intervention.
- **Integration with centralized systems**: Works seamlessly with other components of centralized lubrication systems, including pumps, reservoirs, and filters

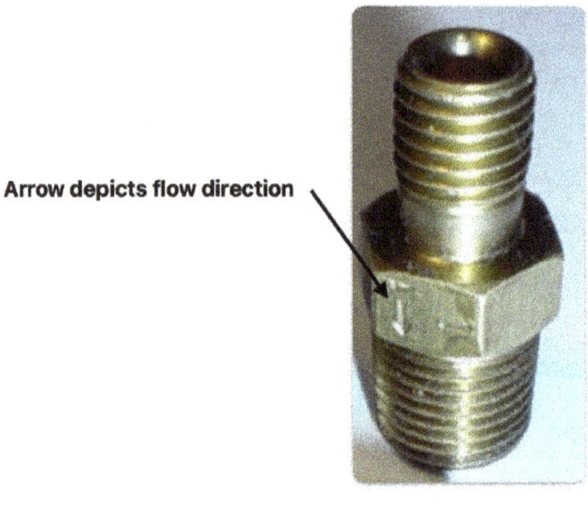

Figure C16 Typical CSA control unit. Courtesy **ENGTECH** Industries Inc.

Applications:

- **Industrial equipment**: Used in manufacturing plants for lubricating metalworking machines, gearboxes, etc.
- **Automotive and heavy machinery**: Provides consistent lubrication to reduce wear and tear on critical components such as gears, bearings, and chains.
- **Textile and food processing**: Ensures smooth operation of machinery that requires precise lubrication to prevent contamination or downtime.

213. Conventional Base Oil

See "Base Oil Group" (Group I).

214. Conversion Factor

A **conversion factor** in lubrication refers to a **numerical value** used to translate measurements from one unit to another to ensure consistency and accuracy in calculations related to lubricants and lubrication management. These factors assist in converting data such as viscosity, volume, pressure, and flow rates between different unit systems (e.g., metric to imperial) or measurement standards.

Key Characteristics:

- **Standardization**: Allows for the consistent comparison and analysis of lubricant properties, ensuring that measurements align with industry or regional standards.
- **Unit conversion**: Commonly used to convert measurements such as gallons to liters, centistokes to centipoise, or pounds per square inch (psi) to bar.
- **Application in formulas**: Essential in calculations involving lubricant flow, film thickness, and application rates, where accurate unit matching is crucial.

Importance:

- **Global communication**: Facilitates clear communication between manufacturers, suppliers, and users across different regions with varying unit systems.

- **Precision in maintenance**: Ensures maintenance personnel apply the correct amount and type of lubricant, which is vital for machinery performance and longevity.
- **Accurate data interpretation**: Helps engineers and operators understand and compare specifications, such as viscosity grades and flow rates, without ambiguity.

Example:
If a lubrication system specification is given in liters per minute (**L/min**), but the operational standard requires gallons per minute (**GPM**), a conversion factor of **1 L/min ≈ 0.264 GPM** would be applied to translate the measurement accurately.

Conversion factors are essential tools in lubrication management for maintaining consistency, ensuring compatibility with equipment specifications, and supporting effective communication and maintenance practices across global operations.

215. Conveyor Lubrication

See Chain/Conveyor.

216. Copper Corrosion Test

The **copper corrosion test** is a method used to assess the **corrosive properties** of a lubricant, fuel, or other liquid on copper surfaces. This test helps determine whether a substance has a tendency to cause corrosion that could affect metal components in engines, machinery, or industrial systems. The test is conducted according to standardized procedures, such as **ASTM D130** for petroleum products.

Key Characteristics:

- **Procedure**: A polished copper strip is immersed in the test liquid at a specified temperature for a defined period. After the test duration, the copper strip is removed, cleaned, and visually compared to a standard rating chart (see Figure C17) to assess the level of corrosion.
- **Rating scale**: The results are rated on a scale from **1 to 4**, where:

  1 (**slight tarnish**)**:** Indicates minimal or no corrosion.

2 (**moderate tarnish**): Shows slight but noticeable tarnishing.

3 (**dark tarnish**): Indicates significant tarnish.

4 (**corrosion**): Indicates severe corrosion or pitting.

This test is commonly used to evaluate the potential corrosiveness of engine oils, hydraulic fluids, fuels, and other lubricants.

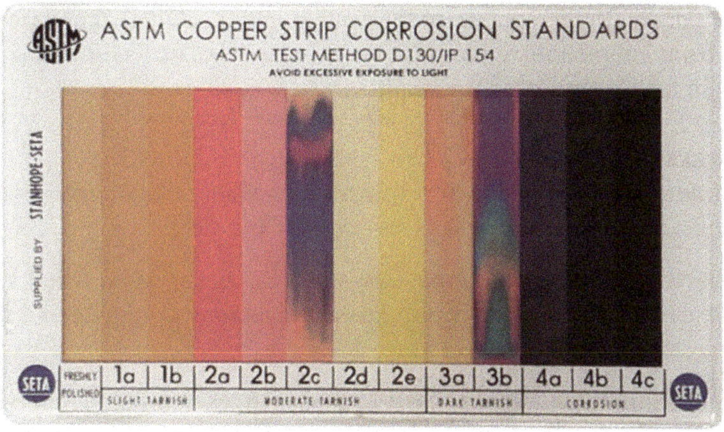

Figure C17 ASTM D130 standard color rating chart. Source: Stanhope-Seta

Importance:

- **Equipment protection**: Ensures that lubricants or fluids used in machinery do not cause detrimental corrosion to copper or copper-alloy components, which could lead to equipment damage and failure.
- **Quality control**: Helps manufacturers formulate products that meet industry standards and specifications for corrosion resistance.
- **Maintenance planning**: Provides valuable information for selecting the right lubricants and fluids to extend the life of components and reduce maintenance costs.

Benefits:

- **Enhanced reliability**: Identifying and using noncorrosive lubricants prevents premature failure of metal parts and increases the longevity of machinery.

- **Regulatory compliance**: Assists in meeting industry regulations and standards for safe and effective fluid use.

217. Corrosion

Corrosion is the gradual **deterioration** or **decomposition** of a material, typically metal, due to **chemical** or **electrochemical** reactions with its environment. This process leads to the weakening and eventual failure of structures or components. In the context of lubrication and machinery, corrosion often results from interactions with moisture, oxygen, chemicals, or contaminants present in the operating environment.

Key Characteristics:

- **Oxidation**: A common type of corrosion where metal reacts with oxygen, forming oxides such as rust on iron and steel.
- **Electrochemical reactions**: Corrosion can occur when metal surfaces are exposed to an electrolyte, causing metal ions to migrate and degrade the material.
- **Environmental influences**: Factors such as humidity, temperature, salinity, and pollutants can accelerate corrosion processes.

Types of Corrosion:

- **Uniform corrosion**: Even deterioration across the surface of a metal.
- **Pitting corrosion**: Localized damage that results in small, deep pits, often difficult to detect.
- **Galvanic corrosion**: Occurs when two dissimilar metals are in contact in the presence of an electrolyte, leading to accelerated corrosion of the more reactive metal.
- **Crevice corrosion**: Develops in confined spaces where the access of a fluid is limited, such as joints or under gaskets.

218. Corrosion Inhibitor Additive

Corrosion inhibitor additives are chemical compounds formulated as base oil additives, also known as **"rust inhibitors"** that create a protective shield on metal surfaces to prevent and mitigate the formation of corrosion and rust. These additives are crucial in protecting ferrous metals, as well as copper, tin, and lead-based metals, by safeguarding them from the damaging effects of water and corrosive acids.

Key Characteristics:

- **Protective barrier formation**: Corrosion inhibitors work by forming a film on metal surfaces that blocks the interaction with moisture, oxygen, and corrosive substances.
- **Versatile application**: Suitable for use in a range of lubricants, including engine oils, hydraulic fluids, and industrial lubricants, to maintain the integrity of metal components.
- **Compatibility**: Designed to work effectively with various base oils without compromising other lubricant properties.
- **Dual role with detergent additives**: Detergent additives, developed to neutralize corrosive acids in lubrication systems, also serve as effective corrosion inhibitors due to their acid-neutralizing properties. This dual function enhances the overall protection provided by the lubricant, ensuring cleaner and more reliable operation.

Corrosion inhibitor additives are widely used in automotive, industrial, marine, and aerospace applications where exposure to moisture, chemical contaminants, and acidic byproducts is common.

219. Corrosive Acid

Corrosive acid refers to an acidic substance that can cause significant damage to metals, organic tissues, or other materials through a chemical reaction. These acids have a **low pH value**, typically below pH 4, and their strong reactive nature enables them to break down substances and lead to corrosion or material degradation.

Key Characteristics:

- **Reactivity**: Corrosive acids react with metals, resulting in the formation of metal salts and hydrogen gas. This reaction can weaken the structure of the metal, leading to rust, pitting, and general material loss.
- **Examples**: Common corrosive acids include sulfuric acid (H_2SO_4), hydrochloric acid (HCl), nitric acid (HNO_3), and phosphoric acid (H_3PO_4).
- **Impact on machinery**: In lubrication systems, corrosive acids can form as a byproduct of oxidation, contamination, or degradation of

lubricants. They can attack metal components, leading to corrosion, reduced performance, and increased wear.

- **Prevention and control**: Use of additives: Lubricants often contain detergent additives and corrosion inhibitors to neutralize acids and protect metal surfaces.
- **Regular monitoring**: Conducting oil analysis and monitoring pH levels can help detect and manage the presence of corrosive acids.
- **Proper filtration and maintenance**: Ensuring that contaminants do not enter the system and that lubricants are maintained can reduce the formation of corrosive acids.

Corrosive acids are particularly problematic in industrial environments, automotive engines, and marine applications, where exposure to moisture and contaminants can accelerate acid formation.

220. Corrosive Wear

Corrosive wear refers to the damage that occurs when a bearing contact surface becomes **acid-etched** due to an **oxidative chemical reaction**. This reaction is accelerated in the presence of moisture contamination and heat, leading to the gradual deterioration of the surface. Corrosive wear is commonly known as **acidic pitting** and manifests as small pits or etched marks on metal surfaces. See Figure C18.

Figure C18 Close-up image of a corrosive worn surface. Courtesy ENGTECH Industries Inc.

Key Causes:

- **Moisture-contaminated lubricant**: The presence of water in the lubricant promotes oxidation and the formation of corrosive acids.

- **Additive depletion**: Occurs when the protective additives in the lubricant, such as corrosion inhibitors, are consumed or deteriorate over time.
- **Lack of corrosion inhibitor**: Using a lubricant that does not contain a corrosion inhibitor additive increases the risk of surface damage due to unchecked acid formation.

Characteristics:
- **Surface damage**: The affected area typically shows pitting or etched patterns that weaken the structural integrity of the bearing.
- **Performance impact**: Corrosive wear leads to increased friction, reduced efficiency, and potential failure of the bearing or machinery.

Corrosive wear can be managed and prevented by using high-quality lubricants with corrosion inhibitor additives, maintaining proper moisture control, and regularly monitoring lubricant conditions to ensure additives remain effective.

221. Coulomb Friction

See "Dry Friction."

222. Coupler Agent

A **coupler agent**, also known as a **coupling agent**, is a **chemical additive** used to enhance the compatibility between two or more substances that would not typically mix or bond well, such as oil and water or dissimilar polymers. In the context of lubricants and industrial applications, coupling agents help create a stable mixture or improve the adhesion between materials.

Key Characteristics:
- **Compatibility enhancement**: Facilitates the uniform blending of substances with different polarities, ensuring a consistent product.
- **Stability improvement**: Helps maintain the stability of emulsions or solutions by preventing phase separation.
- **Adhesion promotion**: Improves the bonding between surfaces or components, often used in formulations where adhesion is critical for performance.

Applications:

- **Lubricants**: Coupler agents are used to stabilize emulsified lubricants, ensuring that oil-based and water-based components remain mixed for effective application in cutting fluids and coolants.
- **Coatings and adhesives**: Enhances the adhesion properties of coatings to metal or plastic surfaces.
- **Polymer blends**: Helps mix polymers with varying chemical properties to create materials with combined beneficial traits.

The use of a coupler agent ensures that the final product maintains consistent performance, reduces separation, and provides effective lubrication or bonding properties. An agent or additive added to primarily water-based lubricants to achieve emulsion stability by allowing the solubilization of polar additives within nonpolar carriers such as those found in food-grade lubricants.

223. Crackle Test

The **crackle test** is a simple, qualitative method used to detect the presence of **water contamination** in lubricating oils or fuels. This test is often performed in the field as a quick screening tool to assess whether further, more detailed laboratory analysis is needed.

Key Characteristics:

- **Procedure**: A small sample of oil is placed on a hot surface (such as a hot plate) heated to a temperature above the boiling point of water but below the oil's flash point. If water is present, it will vaporize and create a "crackling" sound as steam bubbles form and burst.
- **Visual and auditory cues**: The presence of water is indicated by the crackling noise and visible bubbling or frothing in the oil sample.
- **Sensitivity**: The test can detect free or emulsified water content but is not quantitative and cannot measure the exact amount of water present.

Applications:

- **Field testing**: Commonly used by maintenance personnel to check oil quality on-site in industrial machinery, engines, and hydraulic systems.
- **Preliminary screening**: Helps determine if more precise, laboratory-based tests (e.g., Karl Fischer titration) are necessary for accurate water content measurement.

Importance:

- **Prevention of damage**: Identifying water contamination early helps prevent equipment corrosion, wear, and reduced lubrication performance.
- **Maintenance planning**: Supports proactive maintenance practices by indicating when the oil needs to be treated or replaced.

The crackle test is a quick and effective way to identify water contamination,

224. Creep Resistance Property

Creep resistance refers to a lubricant's ability to **stay in place and resist migration** or displacement under mechanical stress, temperature changes, or prolonged load conditions. It is a critical property in greases, oils, and solid lubricants, ensuring consistent lubrication, reduced leakage, and improved wear protection in bearings, gears, chains, and high-speed applications.

Key Functions of Creep Resistance in Lubrication

- **Prevents lubricant migration**: Ensures that oil or grease remains at the lubrication point rather than being forced out due to movement, vibration, or pressure.
- **Maintains film integrity**: Helps in forming a consistent lubricating layer, reducing friction and wear over time.
- **Reduces leakage and contamination risks**: Prevents oil or grease from escaping and contaminating adjacent components.
- **Enhances lubrication efficiency**: Ensures longer retention of lubrication film, reducing relubrication frequency and maintenance costs.

Factors Affecting Creep Resistance

- **Base oil viscosity**: Higher-viscosity oils tend to have better creep resistance than low-viscosity oils, which may migrate more easily.
- **Thickener type (for greases)**: Some thickeners, like polyurea and lithium complex, provide superior adhesion and resistance to creep.
- **Additives**: Tackifiers and adhesion promoters help lubricants cling to surfaces and resist creep.
- **Temperature and pressure**: High temperatures and mechanical loads can cause lubricants to thin out or be displaced, affecting their ability to stay in place.

- **Surface roughness and material composition**: Lubricants behave differently depending on the surface texture and material properties of the components they protect.

Testing Creep Resistance

- **ASTM D1831 (grease roll stability test)**: Evaluates grease consistency changes under mechanical stress.
- **ISO 11007 (Creep Test)**: Measures the tendency of lubricants to migrate under continuous load conditions.

225. Critical Speed Lubrication

Critical speed lubrication refers to the lubrication strategies used to ensure proper lubricant film formation and stability when rotating components approach or exceed their **critical speed** – the speed at which natural resonance causes excessive vibration and potential system instability. This is particularly important in bearings, shafts, turbines, and high-speed machinery, where improper lubrication can lead to excessive wear, overheating, and mechanical failure.

Key Functions of Critical Speed Lubrication

- **Prevents excessive heat generation**: Proper lubrication reduces **frictional heating** and ensures **effective cooling** in high-speed rotating elements.
- **Ensures lubricant film integrity**: Maintains **a stable lubricating film** to avoid **metal-to-metal contact and premature wear**.
- **Minimizes vibration-induced lubrication breakdown**: Prevents **lubricant film collapse due to oscillations and excessive loads at critical speeds**.
- **Reduces risk of bearing and shaft failure**: Helps bearings and rotating components function **smoothly at high rotational speeds** without seizing or excessive wear.

Factors Affecting Critical Speed Lubrication

- **Viscosity selection**: Lubricants must have the **right viscosity to form a stable film** without creating excessive drag or friction.

- **Lubrication method**: High-speed applications require **mist, circulating oil, or air–oil lubrication** to ensure proper lubricant distribution.
- **Temperature and heat dissipation**: High speeds generate heat, requiring **thermally stable lubricants with oxidation resistance**.
- **Bearing type and load**: Lubrication needs to vary based on **bearing design (rolling element vs. fluid film bearings)** and the **applied load**.
- **Centrifugal forces**: At high speeds, **grease may be forced away from the contact area**, necessitating specialized lubricants with **tackifiers or high-speed grease formulations**.

226. Crude Oil

Crude oil is a naturally occurring, unrefined petroleum product composed of hydrocarbon deposits and other organic materials. It is extracted from underground reservoirs and is a primary source of fuel and raw material for the petrochemical industry. Crude oil is processed in refineries to produce a variety of petroleum products, including gasoline, diesel, kerosene, lubricants, and asphalt.

Key Characteristics:

- **Complex mixture**: Composed mainly of hydrocarbons, which include alkanes, cycloalkanes, aromatic hydrocarbons, and trace amounts of sulfur, nitrogen, and oxygen compounds.
- **Varied composition**: The specific makeup of crude oil varies significantly depending on its geographical source, which affects its physical properties, such as viscosity, color, density, and sulfur content.
- **Grades**: Classified based on its density (light, medium, heavy) and sulfur content (sweet or sour). Light, sweet crude is more desirable as it yields more high-value products and requires less refining. See also Napthenic Crude Oil.

Extraction and Processing

- **Extraction**: Obtained through drilling wells into underground reservoirs. Advanced techniques, such as offshore drilling and fracking, are used to access deeper or unconventional reserves.

- **Refining**: Crude oil is refined through processes like distillation, cracking, and reforming to separate and convert it into usable products. This process helps isolate components based on their boiling points and rearranges chemical structures to create more efficient fuels and lubricants.

Uses:

- **Energy production**: The primary source for producing fuel (e.g., gasoline, diesel, and jet fuel) that powers vehicles, aircraft, and other machinery.
- **Petrochemical feedstock**: Provides the raw materials for the production of plastics, synthetic rubber, and chemical products.
- **Lubricants**: Refined into base oils that are further processed into various lubricants used in machinery, vehicles, and industrial applications.
- **Other products**: Includes asphalt for road construction and bitumen for waterproofing.

Crude oil remains a **cornerstone of the global economy** due to its diverse applications and role in energy production, despite its significant environmental challenges and the push for more sustainable alternatives. Crude oil is a **naturally occurring fluid** composed of hydrogen and carbon formed over millions of years deep below the earth's surface through the application of heat and pressure to organic-rich mud. Crude oil is sometimes found above ground in oil seeps and tar pits. Once fractioned or refined to remove impurities and separate the oil molecules by size and type, crude oil provides a variety of petroleum-based products that include lubricating oil base stocks. **Well over 90% of lubricating fluids are crude oil-based**.

227. cSt

See "Centistoke."

228. Cutting Wear

Cutting wear occurs when hard particles or asperities remove material from a surface by a cutting action. This type of wear resembles the action of a cutting tool where the material is sliced or chipped away, leading to a loss of the original surface. Figure C19 shows a **ferrographic** analysis of used oil depicting evidence that cutting wear has taken place.

Key Characteristics:

- **Cut or gouged surface**: Surface shows clear evidence of material removal with defined cuts or gouges. This wear is typically associated with sharp-edged particles that can plow through the surface.
- **Severity and mode**: Cutting wear results in more severe, noticeable material removal compared to the more gradual surface degradation seen with abrasive wear.

Common Causes:

- **Cutting particles**: Cutting wear is often due to sharp, cutting particles, while abrasive wear results from scratching by hard particles, either embedded in a surface or loose between surfaces. This often arises in situations where hard particles (e.g., metal shavings or debris) are trapped between moving parts or are introduced in environments where machining, drilling, or metal-on-metal contact occurs.
- **Impact**: Can lead to significant material loss and damage, requiring maintenance or replacement of affected parts.

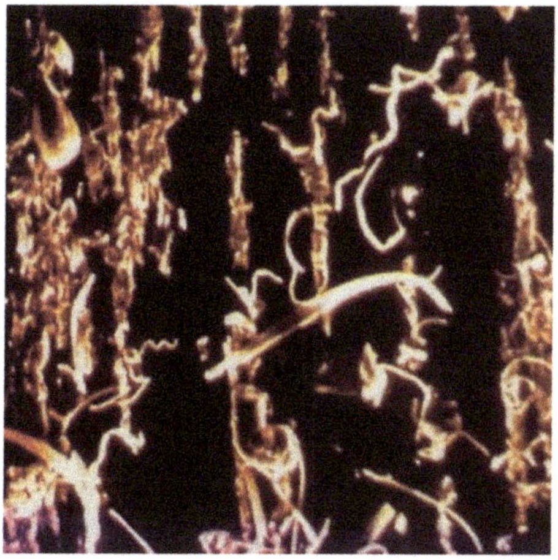

Figure C19 Example of cutting wear particles at 500× magnification using analytical ferrography. Courtesy: ENG**TECH** Industries Inc.

Appearance:

Cutting wear produces clean cuts or chips, while abrasive wear results in fine scratches or worn surfaces.

229. Cycle Count Indicator

A **cyclic count indicator** is a monitoring device used in lubrication delivery systems to track and display the number of lubrication cycles completed. It ensures that the system delivers the appropriate amount of lubricant at specified intervals, providing a way to verify that the lubrication system is functioning correctly. Indicators can be manually activated by pump or meter valve action or can be electronically triggered by the controller.

Key Functions:

- **Cycle monitoring**: Records and displays the total number of lubrication cycles performed, helping operators maintain accurate lubrication schedules.
- **Performance verification**: Ensures that the system operates as programmed, which is crucial for maintaining consistent lubrication and protecting machinery from wear and tear.
- **Maintenance planning**: Aids in planning preventive maintenance by indicating when service or inspection of the lubrication system is needed.

Cyclic count indicators are essential in automated lubrication systems, especially in industrial or heavy-duty applications, where precise and reliable lubrication delivery is critical for equipment longevity and performance.

230. Cyclic Lubrication Delivery System

A **cyclic lubrication delivery system** is an automated system designed to dispense lubricant at regular, programmed intervals, ensuring consistent lubrication of machinery components. Instead of providing a continuous flow of lubricant, the system delivers precise amounts in cycles, based on time settings or the operational needs of the equipment. Figure C20 shows a typical dual-line system that can only run in cyclic mode.

Key Features:

- **Timed or programmed cycles**: The system is configured to release lubricant at specific intervals, which can be adjusted based on machinery requirements.
- **Precision**: Dispenses exact amounts of oil or grease to reduce waste and ensure optimal lubrication without over or underlubricating components.

- **Automated operation**: Minimizes manual intervention, improving efficiency and ensuring machinery is lubricated even during continuous or high-demand operations.

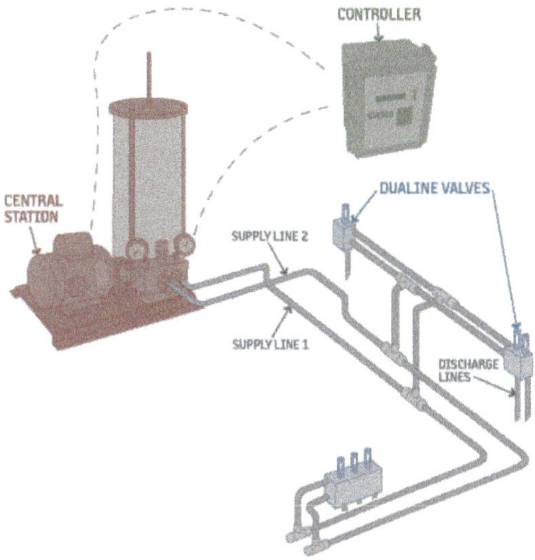

Figure C20 Dual-line lubrication cyclic delivery system. Courtesy: Bijur–Delimon International.

Cyclic lubrication delivery systems are widely used in industrial machinery, conveyors, and heavy-duty equipment where maintaining consistent lubrication is essential to reduce wear, extend equipment life, and optimize performance.

231. Cylinder Oil

Cylinder oil is a high-viscosity lubricant specifically formulated for lubricating the cylinders and pistons of large, slow-speed engines, such as those found in marine diesel engines or industrial steam engines. It is designed to withstand high temperatures and pressures, providing a protective film that reduces friction and wear on engine components.

Key Characteristics:

- **High viscosity**: Ensures effective lubrication under extreme conditions within engine cylinders.

- **Thermal stability**: Maintains performance and resists breakdown at high operating temperatures.
- **Corrosion protection**: Contains additives to neutralize acidic byproducts from combustion, protecting cylinder liners and pistons from corrosion.

Cylinder oil is primarily used in marine and industrial engines where heavy-duty lubrication is required to handle high loads in extreme operating environments, ensuring engine efficiency and longevity.

D

232. DBE

See "Dibasic Acid Ester."

233. da Vinci, Leonardo

Leonardo da Vinci (1452–1519), the renowned Italian polymath, is considered one of the early pioneers in the field of **tribology**, which is the study of friction, lubrication, and wear between interacting surfaces in motion. Though tribology as a formal discipline was established much later, da Vinci's observations and research laid the groundwork for understanding these concepts.

Key Contributions:

- **Study of friction**: Leonardo conducted detailed experiments on friction and was the first to formally describe the fundamental laws of friction. He observed that the force of friction is proportional to the load pressing two surfaces together and independent of the apparent contact area, insights that are still foundational in modern tribology (see Figure D1 and "**Friction**" entry).

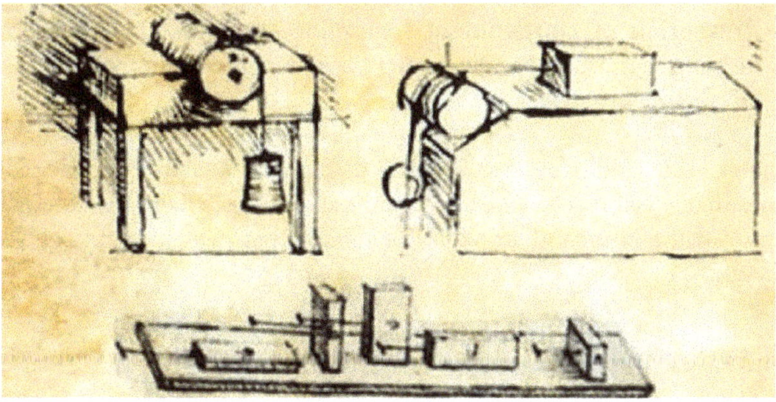

Figure D1 De Vinci's friction experiment sketches.

DOI: 10.1201/9788743806608-235

- **Lubrication concepts**: He explored the role of lubricants in reducing friction between moving parts. His sketches and notes included mechanisms where lubrication could be used to improve the efficiency of machines, such as gears and bearings.
- **Engineering designs**: Many of Leonardo's mechanical designs, such as ball bearings, gear systems, and rotating shafts, incorporated principles of reducing friction to enhance performance. His ideas foreshadowed modern lubrication practices.

Leonardo da Vinci's meticulous studies and sketches have had a lasting impact on the understanding of mechanical efficiency and the importance of lubrication, influencing both historical and contemporary engineering practices.

234. Decomposition Additive

Decomposition additive refers to a chemical compound added to lubricants or fluids to prevent or slow down the breakdown (decomposition) of the base oil or other components under high temperature, pressure, or oxidative conditions. These additives enhance the stability and longevity of lubricants by inhibiting chemical reactions that could lead to the formation of harmful byproducts, such as acids, sludge, or varnish.

Key Characteristics:

- **Oxidation inhibition**: Helps prevent the oil from reacting with oxygen, which can lead to the formation of acidic and corrosive substances.
- **Thermal stability**: Improves the lubricant's ability to withstand high temperatures without degrading or forming deposits.
- **Prevention of contaminant formation**: Reduces the likelihood of forming harmful byproducts that can damage equipment or reduce lubrication efficiency.

Decomposition additives are essential in high-stress applications, such as engines, industrial machinery, and hydraulic systems, where maintaining the integrity of the lubricant is crucial for reliable performance and equipment protection.

235. Defoamant

See "Antifoam Additive."

236. Degree Engler Viscosity

Degree Engler viscosity is a unit of measurement used to express the viscosity of a fluid, based on the time it takes for a specified volume of the fluid to flow through a standardized **Engler viscometer** at a controlled temperature. The result is compared to the flow time of the same volume of water at the same temperature, typically **20 °C (68 °F)**.

Formula and Concept:

- **Engler degree** = (Time taken by the fluid to flow/Time taken by water to flow).
- The Engler viscosity is reported as the ratio of these two times, giving an indication of how much more viscous the tested fluid is compared to water.

Applications:
Degree Engler is an older viscosity measurement system, primarily used in the petroleum and lubrication industries, particularly in Europe. Although it has been largely replaced by more standardized units like centistokes (cSt) and centipoise (cP), it is still referenced in certain legacy specifications and industrial practices.

Understanding Engler viscosity helps in comparing and selecting appropriate fluids for lubrication, ensuring that they meet the required flow characteristics for specific applications.

237. Demulsability Test

Demulsibility test is a procedure used to measure a lubricant's ability to separate from water. This test evaluates how effectively a lubricant, such as oil, can release water that has been mixed or emulsified with it. The demulsibility property is crucial for lubricants used in systems where water contamination is a risk, such as in hydraulic, turbine, and gear systems.

Key Aspects:

- **Purpose**: To ensure that the lubricant can maintain performance and protect machinery by separating from water, preventing emulsions that could lead to corrosion, reduced lubrication efficiency, or system damage.
- **Procedure**: The lubricant and water are mixed under controlled conditions, and the time it takes for a clear separation layer to form between

the oil and water is measured. The results are usually reported in terms of the time required and the clarity of the separated phases.

- **Performance indicator**: A lubricant with good demulsibility will quickly and cleanly separate from water, whereas poor demulsibility can lead to persistent emulsions.

The **demulsibility test** is critical for selecting lubricants for equipment that may be exposed to water, ensuring the lubricant's effectiveness and the reliability of the machinery.

238. Demulsification

Demulsification is the process by which a lubricant, such as oil, separates from water that has become mixed with it. In lubrication and fluid management, demulsification is a crucial property, especially for oils used in machinery that may be exposed to water, such as turbines, hydraulic systems, and marine equipment.

Key Points:

- **Purpose**: Effective demulsification ensures that water can be easily separated and removed from the oil, preventing the formation of emulsions that can impair the lubricant's performance and lead to corrosion, wear, or system failure.
- **Mechanism**: Demulsification occurs when water droplets in the oil coalesce and separate from the oil layer, often aided by the lubricant's formulation or additives designed to enhance this property.

239. Demulsifier

A **demulsifier** is a chemical base oil additive used to break emulsions, specifically to separate water from oil in lubricants, fuels, or other hydrocarbon mixtures. These additives work by altering the surface tension between the oil and water, causing water droplets to coalesce and separate more easily from the oil phase.

Key Functions:

- **Water separation**: Facilitates the quick and efficient removal of water from oil, preventing emulsion formation that can harm machinery and reduce lubricant effectiveness.

- **Contamination control**: Helps maintain the purity of lubricants and fuels by removing water, which can lead to corrosion, oxidation, and reduced lubrication performance.
- **Application**: Commonly used in the petroleum industry, power plants, and any application where water contamination in oil-based systems is a concern.

Demulsifiers are essential for ensuring the efficiency and longevity of lubricated systems.

240. Density

Density in lubrication terms refers to the mass of a lubricant per unit volume, typically expressed in units like kilograms per cubic meter (kg/m^3) or grams per cubic centimeter (g/cm^3). It indicates how compact or heavy the lubricant is in relation to the space it occupies.

Key Aspects:

- **Importance**: Density affects how a lubricant behaves under various pressure and temperature conditions. It can influence the film strength, load-carrying capacity, and efficiency of the lubricant in different applications.
- **Temperature dependence**: The density of lubricants generally decreases as temperature increases, which can impact the performance and flow characteristics of the lubricant.
- **Application**: Understanding the density is important for applications where precise fluid measurement and compatibility with different system components are critical, such as in hydraulic systems or engines.

241. Department of Transport (DOT)

The **Department of Transportation**, also known as **DOT**, is a US government agency responsible for regulating and overseeing transportation safety and standards across various sectors, including automotive, aviation, maritime, and railroads.

In the context of lubrication and fluid standards, "DOT" commonly refers to specifications set by the Department of Transportation for brake fluids, such as DOT 3, DOT 4, and DOT 5. These classifications indicate the performance and chemical composition of brake fluids used in vehicles. See "Brake Fluid".

242. Depth Filter

A **depth filter** is a type of filtration device designed to capture contaminants throughout the entire thickness of the filter medium, rather than just on the surface. These filters consist of multiple layers of porous materials, such as fibrous or granular substances, which trap particles as the fluid passes through. See Figure D2.

Key Features:

- **Multilayer filtration**: Particles of various sizes are captured at different depths within the filter material, providing a high contaminant-holding capacity.
- **Effective particle removal**: Depth filters are efficient at removing both large and fine particles from fluids, making them ideal for applications with high levels of contamination.
- **Long service life**: Due to the ability to trap contaminants throughout the filter medium, depth filters generally have a longer lifespan compared to surface filters.

Depth filters are commonly used in lubrication systems, hydraulic systems, and industrial processes to ensure fluid cleanliness, protect equipment, and maintain efficient operation. They are especially useful in environments where the fluid contains a wide range of particle sizes or high contamination loads.

Figure D2 Typical depth filter element. Source: ENG**TECH** Industries Inc.

243. Dessicant

Desiccant is a **hygroscopic** substance used to absorb and remove moisture from the surrounding environment, helping to keep products and systems dry. In lubrication and industrial applications, desiccants are often used in air breathers, filters, or storage containers to prevent moisture contamination, which can lead to rust, corrosion, or degradation of lubricants and machinery.

Common Types:

- **Silica gel**: A widely used desiccant known for its high moisture-absorbing capacity and nonreactive properties.
- **Activated alumina**: Effective at absorbing moisture and used in industrial applications requiring high moisture control.
- **Molecular sieves**: Provide precise moisture absorption and are suitable for applications requiring very low humidity levels.

Applications:
Desiccants are used in various settings, including industrial lubrication systems, packaging, air compressors, and electrical enclosures, to protect against moisture damage and ensure reliable performance of equipment and materials.

244. Dessicant Breather

A **desiccant breather** is a device used to protect equipment, such as hydraulic systems, gearboxes, storage tanks, or lubrication reservoirs, from moisture and particulate contamination. It contains a desiccant material, typically silica gel, that absorbs moisture from the air entering the system, ensuring that only dry, clean air passes through.

Key Functions:

- **Moisture control**: The desiccant absorbs humidity from incoming air, preventing moisture from contaminating the lubricant or hydraulic fluid, which can cause corrosion, oxidation, or fluid degradation.
- **Particle filtration**: In addition to moisture absorption, desiccant breathers also have built-in filters to capture airborne contaminants like dust and dirt.
- **Indicator**: Many desiccant breathers change color when the desiccant material is saturated, signaling the need for replacement.

Applications:
Desiccant breathers are commonly used in industrial machinery, storage tanks, lubricant reservoirs, and hydraulic systems where maintaining fluid purity and preventing water contamination are critical for efficient operation and equipment longevity. See Figure D3.

Figure D3 Cross section of a modern desiccant breather in which media changes color when saturated. Courtesy: Des-Case Corporation.

245. Detergent Additive

Detergent additives are a type of base oil additive primarily used in applications where combustion occurs, such as in engine and compressor oils. These additives function as **chemical cleaners**, helping to keep combustion surfaces free from harmful deposits and neutralizing combustion acids that can form during operation.

Key Characteristics:

- **Deposit control**: Detergents help prevent the buildup of sludge, varnish, and other deposits on engine and machinery surfaces, ensuring cleaner operation and preventing performance issues.
- **Acid neutralization**: They possess alkaline properties, which allow them to neutralize acidic byproducts of combustion, protecting internal metal surfaces from corrosive damage.

- **Composition**: Detergent additives are typically formulated from overbased (alkaline) organic metallic soaps, such as those containing barium, calcium, and magnesium.

Dual Role:
In addition to their primary function as cleaning agents, detergents also serve as **corrosion inhibitors** due to their ability to neutralize corrosive acids. This helps to protect metal components from chemical degradation and enhances the overall lifespan and performance of the lubricant.

Detergent additives are specifically developed for use in crankcase oils and compressor oils, where maintaining clean, deposit-free surfaces and preventing acid-related corrosion is essential for reliable operation and efficiency. They are a vital component in engine oils used in automotive, industrial, and marine applications, ensuring optimal performance under varying combustion conditions.

246. Dielectric Grease

Dielectric grease is a type of nonconductive grease used to protect electrical connections from moisture, corrosion, and contaminants. It is formulated from silicone or silicone-based compounds and is known for its excellent insulating properties, which prevent electrical arcing and enhance the durability of electrical components.

Key Characteristics:

- **Nonconductive**: Does not conduct electricity, making it ideal for protecting electrical connections without causing short circuits.
- **Waterproof**: Provides a moisture-resistant barrier that helps keep water and contaminants out, preventing corrosion and degradation of connections.
- **Temperature resistant**: Remains stable over a wide temperature range, making it suitable for use in various environmental conditions.
- **Lubricative properties**: Eases the insertion and removal of connectors by reducing friction.

Applications:

- **Automotive**: Used in spark plug boots, battery terminals, and electrical connectors to prevent corrosion and moisture intrusion.
- **Electrical components**: Applied to protect connectors, terminals, and circuit boards in household and industrial equipment.

- **Marine and outdoor equipment**: Helps shield electrical connections from saltwater and other harsh environmental conditions.

Benefits:

- **Prevents electrical failures**: Protects against moisture and oxidation, reducing the risk of electrical malfunctions and arcing.
- **Increases component longevity**: Helps maintain the integrity of electrical connections over time, reducing wear and the need for replacements.
- **Easy application**: Typically applied to connectors and terminals before assembly to ensure a protective coating.

Dielectric grease is essential for maintaining reliable electrical connections, particularly in environments exposed to moisture, dirt, and temperature fluctuations, by providing long-lasting protection and insulation.

247. Dielectric Property Test

Dielectric property test measures a lubricant's (oil or grease) ability to act as an electrical insulator by assessing its dielectric properties, such as dielectric strength, dielectric constant, and dielectric loss. These tests are crucial for determining the suitability of materials for applications where electrical insulation is required, such as in lubricants, greases, and insulating fluids.

Key Aspects of Dielectric Property Tests:

- **Dielectric strength**: The maximum electric field a material can withstand without breaking down. This test measures the material's ability to resist electrical flow and is expressed in volts per unit thickness (e.g., kV/mm).
- **Dielectric constant (relative permittivity)**: Indicates how well a material can store electrical energy in an electric field. A higher dielectric constant means the material can store more energy.
- **Dielectric loss (dissipation factor)**: Represents the energy lost as heat in an insulating material when it is subjected to an electric field. This property is important for understanding the efficiency of insulation.

Testing Procedure:

1. **Sample preparation**: The material, such as a lubricant or insulating fluid, is prepared according to standardized specifications for testing.

2. **Voltage application**: A controlled electric field is applied across the sample until the dielectric breakdown point is reached (for dielectric strength tests).

3. **Measurement equipment**: Instruments like dielectric analyzers or breakdown voltage testers are used to apply the voltage and record the results.

Applications:

- **Lubricants and greases**: Dielectric property tests ensure that dielectric greases and other lubricants used in electrical and electronic applications provide effective insulation.
- **Insulating fluids**: Tests on transformer oils and other dielectric fluids help maintain the reliability of power transmission and electrical equipment.
- **Material quality control**: The test results help manufacturers assess the quality and suitability of materials for insulation purposes, ensuring they meet industry standards and safety requirements.

Dielectric property tests are fundamental for applications where electrical insulation is critical, ensuring that materials can withstand and operate efficiently in electrical and electronic environments.

248. Dibasic Acid Ester (DBE)

Dibasic acid ester (DBE) is a type of synthetic ester formed by the reaction between a dibasic acid (an acid containing two carboxylic acid groups) and alcohol. These esters are valued in the lubrication industry for their unique properties, which make them suitable for a wide range of applications, especially in high-performance lubricants and synthetic oils.

Key Characteristics:

- **Thermal stability**: DBEs exhibit excellent stability at high temperatures, making them ideal for applications where high thermal resistance is needed.
- **Low volatility**: They have a lower tendency to evaporate under high temperatures, contributing to reduced lubricant consumption.
- **Lubricity**: Offers enhanced lubricating properties, providing a protective film that reduces friction and wear on moving parts.
- **Good solubility**: Can dissolve and disperse additives efficiently, enhancing the performance of formulated lubricants.

- **Biodegradability**: Many dibasic acid esters are environmentally friendly and biodegradable, making them suitable for use in eco-friendly lubricant formulations.

Applications:
- **High-performance engine oils**: Used as a base fluid or co-base in synthetic oils for automotive and industrial engines.
- **Compressor oils**: Ideal for applications requiring low volatility and excellent thermal stability.
- **Hydraulic fluids**: Provide reliable performance in systems that operate under high pressure and temperature conditions.
- **Aviation lubricants**: Suitable for jet engines and other aircraft components where extreme temperatures and reliability are crucial.

Common dibasic acids used to create DBEs include adipic acid, sebacic acid, and azelaic acid. When reacted with specific alcohols, these acids form esters with properties tailored for specific lubrication needs.

249. Diluent Solvent

A **diluent solvent** is a fluid used to reduce the viscosity or concentration of a substance, making it easier to handle, apply, or mix. In lubrication and industrial applications, diluent solvents are added to oils, greases, or other formulations to adjust their flow properties or to aid in the uniform dispersion of additives.

Key Characteristics:
- **Viscosity reduction**: Helps lower the viscosity of thick fluids, improving their ability to flow or spread evenly.
- **Ease of application**: Makes it simpler to apply coatings, lubricants, or other substances, especially in processes that require spraying or pumping.
- **Evaporation**: Many diluent solvents are designed to evaporate after application, leaving the primary substance to perform its intended function.

Diluent solvents are used in formulating lubricants, paints, coatings, and cleaning solutions, providing the necessary fluidity for efficient use and application. In lubrication systems, they ensure that additives are properly mixed and distributed throughout the base oil.

250. Dilution Ratio

A **dilution ratio** is a proportion used to mix a concentrated substance with a diluent, such as a solvent or water, to achieve a desired concentration or consistency. It is typically expressed as a ratio of the amount of concentrate to the amount of diluent.

Common Representation:

- A dilution ratio of **1**:**10** means one part of the concentrate is mixed with ten parts of the diluent.
- A ratio of **1**:**5** indicates one part concentrate to five parts diluent.

Applications:
In lubrication and industrial processes, dilution ratios are important for preparing cleaning solutions, adjusting the viscosity of oils or greases, and ensuring the correct concentration of additives. Using the proper dilution ratio ensures the desired performance and effectiveness of the product while preventing issues such as overconcentration or insufficient lubrication.

251. Dipstick

A **dipstick** is a measuring tool used to check the level and condition of fluid, such as engine oil, transmission fluid, or hydraulic oil, within a reservoir or system. It is typically a long, narrow metal or plastic rod marked with minimum and maximum indicators to show the acceptable fluid level range. See Figure D4.

Key Functions:
• **Fluid level check**: Allows operators to quickly and easily check if the fluid is within the appropriate range for proper operation.

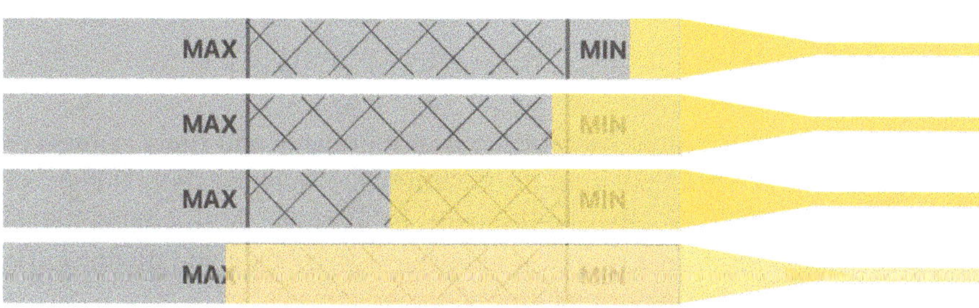

Figure D4 Dipstick level markers showing underfilled to overfilled.

- **Fluid condition assessment**: By observing the color and consistency of the fluid on the dipstick, users can get an indication of the fluid's condition and whether it may need to be changed.

Dipsticks are commonly found in automotive engines, transmissions, and industrial machinery. Regular use of a dipstick helps ensure that equipment has sufficient lubrication, preventing damage or inefficiencies caused by low or degraded fluid levels.

252. Dispersibility

Dispersibility refers to the ability of a substance, such as an additive or lubricant, to uniformly distribute and remain evenly suspended within a fluid or mixture. In lubrication terms, dispersibility is crucial for ensuring that contaminants, soot, or other particles are effectively dispersed throughout the lubricant, preventing them from settling or agglomerating, which can lead to deposits, sludge, or component wear.

Key Aspects:

- **Additive performance**: Dispersant additives are used in lubricants to maintain dispersibility, helping to keep contaminants suspended and preventing buildup on engine or machinery surfaces.
- **Fluid stability**: Good dispersibility contributes to the overall stability and effectiveness of the lubricant, especially in engines or systems exposed to combustion byproducts or contamination.

Dispersibility is important in automotive and industrial lubricants, where maintaining clean and efficient operation is essential. It helps prevent the formation of harmful deposits, extending the life of both the lubricant and the machinery.

253. Dispersant

A **dispersant** is a chemical additive used in lubricants and fuels to prevent the agglomeration and deposition of contaminants, such as soot, sludge, or other insoluble particles. Dispersants work by keeping these contaminants suspended in the fluid, ensuring they are evenly distributed and do not settle or form deposits on engine or machinery components.

Key Characteristics:

- **Contaminant suspension**: Dispersants help keep particles like soot or dirt finely dispersed within the oil, preventing them from clumping together or settling out.
- **Engine cleanliness**: By preventing the buildup of sludge and deposits, dispersants help maintain engine cleanliness, optimize performance, and extend the life of the lubricant.
- **Compatibility**: Dispersants are designed to work effectively with other additives in the lubricant, providing balanced protection for the equipment.

Dispersants are commonly used in engine oils, hydraulic fluids, and industrial lubricants where contaminant control is critical. They are especially important in automotive and diesel engines, which generate soot and combustion byproducts that need to be managed to maintain efficient and reliable operation.

254. Disposable

Disposable refers to items or components designed for single-use or limited use, intended to be discarded after serving their purpose. In the context of lubrication and industrial applications, disposable products include items like oil filters, grease cartridges, or single-use wipes and gloves used for maintenance tasks.

Key Characteristics:

- **Single-use design**: Made for convenience and efficiency, eliminating the need for cleaning or maintenance of the item itself.
- **Hygiene and safety**: In maintenance and lubrication, disposable items help prevent contamination and ensure the safe handling of chemicals or lubricants.
- **Cost-effective**: While disposable items may be more affordable up front, they are meant for quick, easy use and are then discarded to maintain cleanliness and efficiency.

Disposable products are commonly used in industrial maintenance/lubrication shops, and automotive service centers where cleanliness and contamination control are crucial.

255. Dissolved Air

Dissolved air refers to air that is mixed into and held within a liquid at a molecular level, typically under normal atmospheric pressure or through mechanical agitation. In lubrication systems, dissolved air can be a concern because it can come out of solution under certain conditions, such as changes in pressure or temperature, leading to the formation of air bubbles or foam that can quickly exhaust antifoam additives within the lubricating oil.

Key Concerns:

- **Foam formation**: When dissolved air comes out of the lubricant, it can cause foaming, which reduces the lubricant's effectiveness and may lead to inadequate lubrication and increased wear on components.
- **Oxidation risk**: Dissolved air contains oxygen, which can accelerate the oxidation of the lubricant, degrading its properties and shortening its service life.
- **Performance impact**: The presence of dissolved air can affect the viscosity and load-carrying capacity of the lubricant, impacting the performance of hydraulic systems or machinery.

Proper system design, deaeration techniques, and the use of antifoaming additives can help minimize the effects of dissolved air in lubrication systems, ensuring optimal performance and longevity of the equipment.

256. Distribution Block

A **distribution block** is a component used in lubrication systems to evenly distribute lubricants, such as oil or grease, from a central source to multiple lubrication points. It ensures that each point receives the correct amount of lubricant, which is critical for maintaining proper operation and reducing wear on machinery. See Figure D5.

Key Features:

- **Multiple outlets**: Equipped with multiple outlets to supply lubricant to different areas or components of a machine.
- **Controlled flow**: Designed to deliver a precise amount of lubricant to each point, ensuring consistent and efficient lubrication across the system.

- **Versatility**: Used in various manual and automated lubrication systems, including single-line, dual-line, and progressive lubrication setups.

Figure D5 Typical 6-point progressive divider distribution block. Courtesy: ENG**TECH** Industries Inc.

Distribution blocks are commonly found in industrial machinery, heavy equipment, and vehicles where multiple bearings require regular metered lubrication. They are essential for improving maintenance efficiency, extending equipment lifespan, and ensuring reliable and uniform lubrication throughout complex systems.

257. Dn factor

See "Speed Factor."

258. DOT

See "Department of Transport."

259. Downtime

Downtime, based on **ineffective lubrication** refers to periods when machinery or equipment is nonoperational due to failures or malfunctions caused by inadequate or improper lubrication. This type of downtime can have significant impacts on productivity, operational efficiency, and maintenance costs.

Common Causes:

- **Insufficient lubrication**: When machinery does not receive the correct amount of lubricant, leading to increased friction, overheating, and accelerated wear.
- **Contaminated lubricants**: The presence of dirt, water, or other contaminants can degrade the lubricant's effectiveness, causing component damage.
- **Incorrect lubricant selection**: Using an inappropriate type or grade of lubricant for a specific application can lead to equipment failure. Also, mixing of two different lubricant types can cause premature lubrication and bearing failure.
- **Degraded lubricant**: Over time, lubricants can break down or oxidize, losing their protective properties and leading to component damage.
- **Over/under lubrication**: More prevalent in manual grease lubrication where standardization and training are lacking.

Impacts:

- **Increased maintenance costs**: Frequent repairs or replacements of parts that have been damaged due to poor lubrication.
- **Reduced equipment lifespan**: Consistent exposure to improper lubrication can lead to premature failure of machinery components.
- **Loss of productivity**: Downtime caused by equipment failures disrupts production schedules and reduces overall efficiency.
- **Safety risks**: Malfunctioning machinery can pose hazards to workers, increasing the likelihood of workplace accidents.

To minimize downtime from ineffective lubrication, regular maintenance practices should include proper lubricant selection, routine lubrication checks, contamination control, and monitoring of lubricant condition through analysis. Automated lubrication systems can also help ensure consistent and effective lubrication across all critical components.

260. Drain Port Sampling

Drain port sampling refers to the process of collecting a fluid sample, such as oil or coolant, from the drain port of machinery or equipment for analysis. This method is commonly used in lubrication and maintenance programs to monitor the condition of the fluid and assess the health of the equipment.

Key Aspects:

- **Sampling location**: The sample is taken from the drain port, usually located close to the bottom of the oil reservoir or sump. This can help capture sediment or contaminants that may have settled, providing insights into potential issues. Note in Figure D6 that the drain sampling port is shown ideally located at least two inches off the sump bottom to avoid any sludge that may have built up in the reservoir.

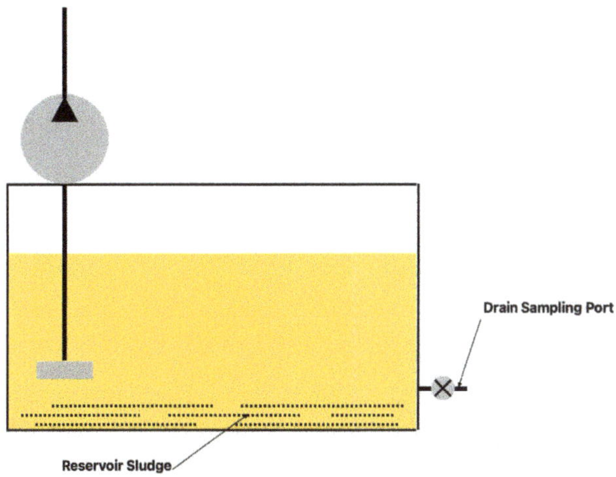

Figure D6 Typical location for a reservoir drain sampling port.

- **Fluid analysis**: The collected sample is analyzed for properties such as viscosity, contamination levels, wear metals, water content, and chemical degradation. This information helps determine if the fluid needs to be replaced or if maintenance is required.
- **Routine monitoring**: Drain port sampling is often part of a regular maintenance schedule to ensure that equipment is running efficiently and to detect problems early.

Sampling Best Practices:

- **Clean sampling tools**: Ensure that the sampling container and tools are clean to avoid contamination.
- **Proper sampling procedure**: Take the sample after the equipment has been running to ensure the fluid is well-mixed, but wait for a short period to avoid taking hot or aerated samples.

- **Consistent sampling**: Collect samples from the same location and under similar conditions each time to ensure accurate trend analysis.

Drain port sampling is an essential practice for proactive equipment maintenance, helping to extend the life of machinery, prevent unexpected failures, and optimize lubricant performance.

261. Drip Lubricator

A **drip lubricator** is a vintage lubrication device still employed today to supply a controlled, continuous flow of lubricant, such as oil, to machinery components, at minimal cost. It operates by allowing lubricant to drip at a regulated rate onto specific parts, such as bearings, chains, or gears, to reduce friction and wear.

Key Characteristics:

- **Adjustable flow rate**: The drip rate can be manually adjusted to ensure the appropriate amount of lubricant is delivered based on the equipment's needs.
- **Simple design**: Consists of a reservoir that holds the lubricant and a drip mechanism that controls the flow, making it a straightforward and reliable lubrication solution.
- **Gravity-fed system**: Utilizes gravity to deliver the lubricant, often used in low-pressure systems where precise lubrication is required.

Figure D7 Typical drip lubricator. Courtesy: **ENGTECH** Industries Inc.

Drip lubricators are commonly used in industrial machinery, engines, and other equipment where components require a steady and controlled supply of lubricant. They are particularly effective in applications where automatic lubrication is needed but a simple, low-cost solution is preferred.

262. Drip Tray

A **drip tray** is a shallow, flat container or tray used to catch and collect any leaking or spilled fluids, such as oil, grease, or coolant, from machinery or equipment. It serves as a temporary containment measure to prevent spills from spreading and to keep the work area clean and safe.

Key Features:

- **Leak containment**: Collects drips and spills, preventing fluids from contaminating the environment or posing safety hazards. **Warning!** Once a leak has been recognized it should be investigated and repaired as soon as possible. A drip tray should **not** become a "stop gap" leak repair method.
- **Material**: Often made of durable materials like plastic, stainless steel, or coated metal to withstand exposure to various chemicals and oils.
- **Portability**: Many drip trays are lightweight and easy to move, making them convenient for placement under different equipment or for temporary use.

Drip trays are widely used in industrial settings, automotive workshops, laboratories, and maintenance areas. They are placed under machines, engines, or storage containers to capture leaking fluids and help with spill management. By using drip trays, facilities can minimize environmental impact, maintain cleanliness, and ensure compliance with safety and environmental regulations.

263. Drop Tube Sampling

Drop tube sampling is a method used to collect fluid samples, such as oil or fuel, from a specific depth within a reservoir or tank. This technique ensures that the sample represents the fluid condition in the middle or lower sections of the tank, rather than just at the surface or near the bottom where contaminants may accumulate.

Key Features:

- **Sampling precision**: The drop tube, a narrow, rigid tube, is lowered into the tank through the fill port to the desired depth, allowing a sample to be drawn from a specific layer of fluid (see Figure D8).
- **Contamination avoidance**: By sampling from a controlled depth, this method reduces the risk of picking up contaminants, sediment, or water that may have settled at the bottom, providing a more accurate representation of the fluid's condition.
- **Equipment**: The drop tube is typically used with a vacuum pump or sampling device to draw the fluid into a container for analysis.

Figure D8 Typical drop tube sampling unit. Source: Trico Corporation.

Drop tube sampling is commonly used in oil analysis, fuel quality testing, and hydraulic system monitoring. It is an essential practice for proactive maintenance,

264. Dropping Point (Grease)

Dropping point is a term used to describe the temperature at which a lubricating grease transitions from a semisolid state to a liquid under specific test conditions. It is a critical indicator of the grease's thermal stability and defines the upper-temperature limit at which the grease maintains its structure before starting to flow.

Key Characteristics:

- **Thermal stability**: A higher dropping point indicates that the grease can withstand higher temperatures without melting or losing its consistency, making it suitable for high-temperature applications.
- **Performance indicator**: The dropping point helps in selecting the appropriate grease for machinery operating under different temperature conditions, ensuring reliable lubrication and protection.

It is important to note that the dropping point is not the maximum operating temperature of the grease but rather a measure of when it begins to lose its solid form.

265. Drum

See "Barrel."

266. Drum Lid

A **drum lid** is a circular hard cover (usually molded plastic or metal) designed to seal and protect the contents of a storage drum, such as those used for holding oil, lubricants, chemicals, or other industrial fluids. See Figure D9. The lid helps prevent contamination, spillage, and evaporation while ensuring the drum's contents remain secure.

Figure D9 Typical plastic drum or barrel lid. Source: Enpac.

Key Features:

- **Material**: Drum lids are commonly made from durable materials like metal (steel) or high-density plastic to withstand harsh environments and protect against corrosion or chemical exposure.
- **Sealing mechanism**: Many drum lids come equipped with gaskets or locking rings to create a tight seal, preventing leaks and keeping contaminants out.
- **Accessibility**: Some drum lids are designed with openings or ports for easy access, allowing for the attachment of pumps, faucets, or ventilation systems without removing the entire lid.

Drum lids are used in industrial settings, warehouses, and transport operations where safe and secure storage of bulk fluids or chemicals is necessary. Proper use of drum lids helps maintain fluid quality, reduce environmental spill risks, and ensure safe handling during storage and transport.

267. Drum Pump

See "Barrel Pump."

268. Drum Storage

Drum storage refers to the practice of safely storing and organizing drums that contain liquids such as lubricants, chemicals, oils, or other industrial fluids. Proper drum storage is essential to maintain the quality of the contents, prevent leaks or spills, and ensure a safe working environment.

Key Considerations:

- **Storage conditions**: Drums should be stored in a cool, dry, and well-ventilated area to prevent the contents from degrading and to minimize the risk of corrosion on the drum surfaces.
- **Orientation**: Drums can be stored upright (with bungs on top) to reduce the risk of leakage, or horizontally on racks designed to keep the bungs accessible and above the liquid level.
- **Secondary containment**: Spill containment pallets or bunds are often used to catch any leaks or spills, preventing environmental contamination and ensuring compliance with safety regulations.

- **Labeling and organization**: Drums should be clearly labeled with their contents and organized for easy access, with hazardous materials properly marked and stored according to safety guidelines.

Drum storage is widely used in industrial facilities, manufacturing plants, warehouses, and transportation hubs. See Figure D10. Proper drum storage practices help to protect the contents from contamination, ensure workplace safety, and comply with environmental and health regulations.

Figure D10 Typical outdoor drum storage area. Courtesy: ENG**TECH** Industries Inc.

269. Dry Film Lubrication

Dry film lubrication refers to a method of reducing friction and wear between surfaces by applying a thin, solid lubricant film that does not require the presence of liquid oil or grease. These dry lubricants are typically made from materials like graphite, molybdenum disulfide (MoS_2), polytetrafluoroethylene (PTFE), or other solid compounds that provide a low-friction barrier between contact surfaces.

Key Characteristics:

- **Low friction**: Dry film lubricants create a smooth, low-friction surface, reducing wear and extending the life of components even under extreme pressure or high-temperature conditions.
- **Temperature stability**: These lubricants can operate effectively at temperatures where traditional liquid lubricants may evaporate, degrade, or lose their effectiveness.

- **Clean application**: Because they are dry, these lubricants do not attract dust, dirt, or debris, making them suitable for applications where cleanliness is crucial.

Dry film lubrication is used in aerospace, automotive, industrial machinery, and military applications where traditional lubricants may not perform well or could pose contamination risks. Examples include coating moving parts in engines, gears, bearings, and components in extreme temperature or vacuum environments, such as in space applications.

270. Dry Friction

Dry friction, also known as **Coulomb friction**, refers to the resistance to motion that occurs when two solid surfaces come into direct contact without any lubrication between them. It is the type of friction that opposes the relative lateral movement of the surfaces and is a significant factor in mechanical wear and energy loss.

Key Aspects:

- **Static friction**: The force that must be overcome to initiate the movement of one surface over another. It is usually higher than kinetic friction.
- **Kinetic friction**: The resistance encountered when one surface is already sliding over another. It is generally lower than static friction.
- **Heat generation**: Dry friction generates heat due to the energy lost as the surfaces slide against each other, which can lead to material deformation or damage over time.

Figure D11 depicts the two frictional aspects experienced. Once static friction is overcome, the two surfaces are in motion and are said to be moving in the kinetic friction region where the applied force lessens considerably as the surfaces continue in motion.

Implications:

Dry friction can cause wear, surface damage, and inefficiencies in mechanical systems. In many applications, lubricants are used to reduce dry friction, minimize wear, and improve the longevity and performance of machinery components. However, in certain applications, such as braking systems or clutches, dry friction is essential for proper operation.

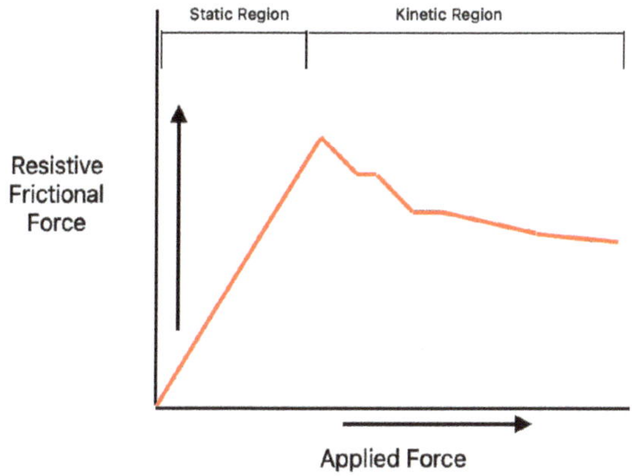

Figure D11 Dry friction has two distinct frictional regions. Source: Plant Lubrication for Industrial Facilities – Fourth Edition, 2024 – Kenneth E. Bannister

271. Dual-line Injection Lubrication System

A **dual-line injector system** is a type of automated lubrication system designed to deliver precise amounts of lubricant, such as oil or grease, to multiple lubrication points in large or complex machinery. The system operates using two main supply lines that alternately pressurize to deliver lubricant through injectors to designated points, ensuring even and consistent lubrication.

Key Features:

- **Two supply lines**: The system uses two main lines that are pressurized alternately. When one line is pressurized, the other is used to relieve pressure, allowing the injectors to reset for the next cycle.
- **Injectors**: Specialized components that dispense a metered amount of lubricant to each lubrication point when the supply line is pressurized. The amount can be adjusted to match the needs of different components.
- **Reliable for large systems**: Ideal for heavy industrial equipment, such as in mining, steel production, and large manufacturing plants, where a high number of lubrication points need to be serviced over long distances (see Figure D12).

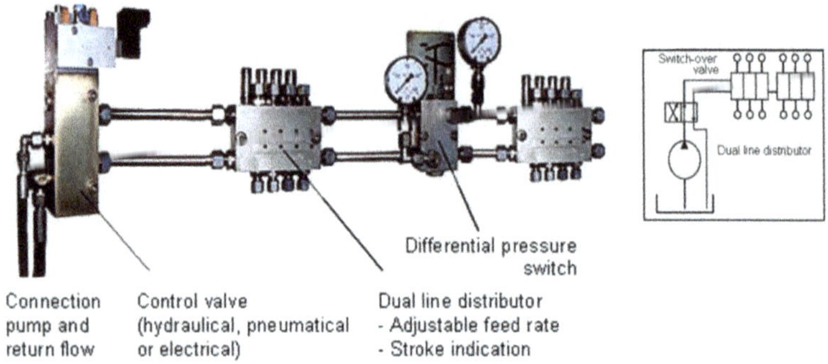

Figure D12 Dual-line injector lubrication system. Source: Plant Lubrication for Industrial Facilities – Fourth Edition, 2024 – Kenneth E. Bannister

Advantages:

- **Precise lubrication**: Delivers accurate amounts of lubricant to each point, minimizing waste and ensuring optimal equipment performance.
- **System monitoring**: Dual-line systems often include monitoring features to detect blockages or system malfunctions, helping to prevent equipment damage.
- **Versatility**: Suitable for harsh environments and heavy-duty applications, as the system can handle high-pressure and long-distance lubrication requirements.

Dual-line injector systems are commonly used in heavy machinery, industrial presses, steel mills, and other applications where reliability and consistent lubrication are critical to minimizing wear and maximizing equipment uptime.

272. Dye

Dye in the context of lubrication and industrial fluids refers to a colored substance added to oils, fuels, or other fluids for identification, leak detection, or quality control purposes. Dyes do not alter the chemical properties or performance of the fluid but serve as a visual aid.

Key Uses:

- **Identification**: Different colored dyes help differentiate between various types of fluids, such as engine oils, hydraulic fluids, or transmission fluids, to prevent cross-contamination and ensure proper use.

- **Leak detection**: Bright, fluorescent dyes are often added to fluids to help identify leaks in engines, hydraulic systems, or pipelines when illuminated with UV light or under normal lighting.
- **Quality control**: Dyes can indicate fluid degradation or contamination; for instance, a color change may signify oxidation or the presence of contaminants.

Applications:
Dyes are commonly used in automotive, industrial, and manufacturing settings. For example, antifreeze is often dyed bright colors like green or orange for easy identification, and hydraulic systems may use dyed fluids for monitoring purposes.

273. Dynamic Friction

See "Kinetic Friction."

274. Dynamic Viscosity

See "Absolute Viscosity."

E

275. EAL/EFL Lubricants

An **environmentally acceptable lubricants (EAL)**, often used interchangeably with **EFL – environmentally friendly lubricants** are lubricants formulated to minimize environmental impact in the event of a spill or leakage. EAL/EFLs are characterized by three primary properties:

1. **Biodegradability**: Lubricants designed to break down naturally over time, reducing their persistence in the environment. This property ensures that, in the event of a spill, the lubricant decomposes without causing long-term environmental harm.
2. **Low aquatic toxicity**: Lubricants are formulated to be minimally toxic to aquatic organisms, such as fish and algae. This characteristic is crucial for protecting marine life in case of accidental discharge into water bodies.
3. **Low bioaccumulation potential**: Lubricants designed to have a low potential for bioaccumulation, meaning they do not concentrate in the tissues of organisms over time. This property prevents harmful substances from entering the food chain.

The US Environmental Protection Agency (EPA) defines EAL/EFLs as lubricants that are biodegradable, exhibit low toxicity to aquatic organisms, and have a low potential for bioaccumulation. These lubricants are particularly important in industries such as marine transportation, forestry, and mining, where equipment operates in or near water bodies.

It is important to note that while EAL/EFL offer environmental benefits, they must also meet the performance standards required for their specific applications. Therefore, selecting an EAL involves balancing environmental considerations with the technical requirements of the machinery.

DOI: 10.1201/9788743806608-279

276. Effective Lubrication

Effective lubrication refers to the optimal application and performance of a lubricant, ensuring that machinery or equipment operates smoothly with minimal friction, wear, and heat generation. Proper lubrication protects moving parts, extends equipment life, and enhances efficiency and reliability.

Key Elements of Effective Lubrication:

- **Correct lubricant selection**: Using the right type and grade of lubricant suited to the specific operating conditions, such as temperature, load, and speed.
- **Adequate lubricant quantity**: Applying the appropriate amount of lubricant to avoid issues like overlubrication (which can lead to excess heat and energy loss) or underlubrication (which can cause increased friction and wear).
- **Consistent lubrication**: Ensuring the lubricant is regularly reapplied or monitored to maintain its protective properties over time.
- **Contamination control**: Keeping the lubricant-free from contaminants like dirt, water, or metal particles that can degrade its performance and harm machinery components.

Benefits:

- **Reduced friction and wear**: Minimizes the direct contact between moving surfaces, lowering the risk of damage and extending the life of components.
- **Heat dissipation**: Helps remove excess heat generated by friction, preventing overheating and maintaining optimal operating temperatures.
- **Corrosion protection**: Forms a protective barrier on metal surfaces, reducing the risk of corrosion and rust.
- **Improved efficiency**: Enhances the overall performance of machinery, reducing energy consumption and operational costs.

Effective lubrication is crucial in a wide range of industries, including automotive, manufacturing, aerospace, and heavy machinery, where the reliable performance of equipment is vital for productivity and safety. Proper lubrication practices, including regular monitoring and maintenance, are essential for achieving long-term equipment reliability.

277. Elastohydrodynamic Lubrication (EHDL)

Elastohydrodynamic lubrication (EHDL) is a type of lubrication regime that occurs when a **thin film** of lubricant separates two surfaces under **high pressure and load**, while the surfaces undergo elastic deformation. This regime is common in high-contact stress areas, such as rolling-element bearings, gears, and cam-follower systems. An elastohydrodynamic lubrication regime or state is unique to rolling contact surfaces and the friction experienced in ball and roller-style bearings and meshing gears. When a ball is rolling in a race and comes under full load, the mating surfaces will momentarily deform trapping the lubricant in the deformed area known as the **Hertzian contact area**. See Figure E1.

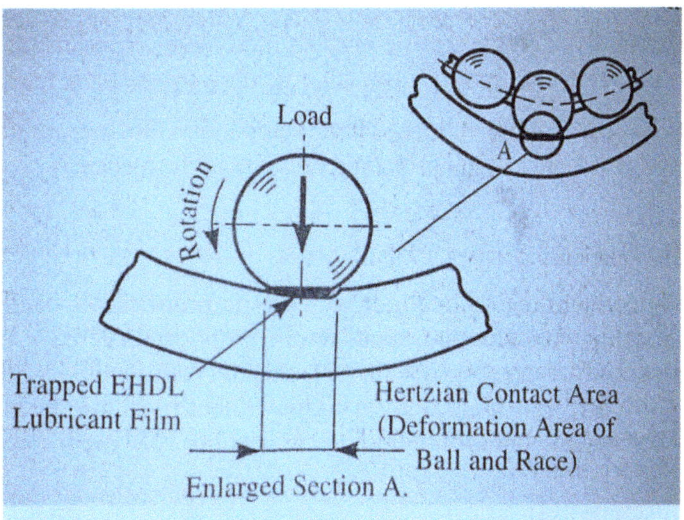

Figure E1 EHDL deformation. Source: Machinery Handbook 27th Edition, Kenneth E Bannister.

Under deformation pressure, the lubricant viscosity rapidly rises and the lubricant changes state from a liquid to a solid providing full protection to the rolling surfaces. As the ball moves out of the Hertzian load zone, the lubricant returns back to its original viscosity and fluid state.

Key Characteristics:

- **High pressure**: Under the high-pressure conditions of EHDL, the lubricant's viscosity increases significantly, forming a robust film that prevents metal-to-metal contact.

- **Elastic deformation**: The surfaces in contact deform slightly in an elastic manner, which helps distribute the load over a larger area and reduces the risk of surface damage.
- **Thin lubricant film**: The lubricant film is extremely thin yet strong enough to maintain separation between surfaces, minimizing friction and wear.

Importance:

- **Protects components**: EHDL ensures that even under extreme loads, components like gears and bearings are protected from wear and surface fatigue.
- **Enhances durability**: By preventing direct contact between moving surfaces, EHDL increases the lifespan of mechanical components.
- **Reduces friction**: Effective elastohydrodynamic lubrication reduces friction, contributing to the overall efficiency of machinery.

EHDL is critical in applications where components are subjected to high speeds and loads, ensuring reliable and efficient performance.

278. Elastomer

An **elastomer** is a type of polymer with elastic properties, meaning it can stretch significantly under stress and return to its original shape when the stress is removed. Elastomers are characterized by their flexibility, resilience, and ability to undergo large deformations without breaking, making them ideal for sealing and vibration-damping applications.

Key Characteristics:

- **High elasticity**: Can stretch and recover repeatedly without permanent deformation.
- **Durability**: Resistant to wear, abrasion, and environmental factors like temperature and humidity, depending on the formulation.
- **Versatile applications**: Elastomers can be engineered to have specific properties, such as oil resistance, chemical resistance, or high-temperature stability.

Applications in Lubrication and Industry:

- **Seals and gaskets**: Elastomers are commonly used in seals, O-rings, and gaskets to prevent leaks and keep lubricants contained within systems like engines, pumps, and hydraulic machinery.

- **Vibration dampening**: Used to absorb shock and reduce vibration in machinery, protecting components from damage.
- **Flexible couplings**: Employed in mechanical systems where flexibility is needed to accommodate movement or misalignment between parts.

Common elastomers include natural rubber, silicone rubber, nitrile rubber (NBR), and fluorocarbon rubber (FKM), each suited to different environments and operating conditions.

279. Electrostatic Lubrication

Electrostatic lubrication is an advanced lubrication technique that **uses electrically charged lubricant particles** to enhance lubrication efficiency, adhesion, and distribution. By applying an electrostatic charge to the lubricant, it can be attracted to metal surfaces, ensuring uniform coverage and reducing lubricant waste. This method is particularly useful in high-speed, high-precision, and contamination-sensitive applications, such as automotive assembly lines, metalworking, and food processing machinery.

Key Functions of Electrostatic Lubrication

- **Enhances lubricant adhesion**: Charged lubricant droplets are attracted to surfaces, ensuring uniform and controlled film formation. Lubricant is less likely to drip and contaminate the product as in automotive assembly and paint, and food processing lines.
- **Reduces lubricant consumption**: By precisely directing lubricant to where it is needed, waste is minimized, leading to cost savings and reduced contamination.
- **Improves lubrication distribution**: Ensures even coverage on moving parts, preventing dry spots that lead to wear and failure.
- **Reduces contamination and overspray**: Unlike traditional lubrication methods, electrostatic lubrication reduces misting and excess lubricant application, keeping surrounding areas cleaner.

How Electrostatic Lubrication Works

1. Charging the lubricant: Lubricant particles are electrically charged using an electrostatic generator.
2. Attraction to surfaces: The charged lubricant is drawn toward grounded metal surfaces, ensuring precise application and reduced waste.

3. Lubrication film formation: A thin, controlled lubricant layer forms on the surface, providing friction reduction and wear protection.
4. Continuous replenishment: As components move, new lubricant is attracted to exposed areas, maintaining consistent lubrication.

Applications of Electrostatic Lubrication

- **Metal forming and stamping**: Ensures even lubricant distribution on metal sheets and tools, preventing die wear and scoring.
- **Automotive assembly lines**: Used in bearing, chain, and conveyor lubrication for precise oil application.
- **Textile and paper industries**: Reduces fiber contamination and improves roller lubrication.
- **Food-grade lubrication**: Provides precise, contamination-free lubrication in food processing machinery.
- **Industrial chains and conveyors**: Ensures lubricant reaches internal links for reduced wear and extended service life.

280. Electrostatic Separation

Electrostatic separation is a process used to separate particles or droplets from a fluid using electrostatic forces. This technique is commonly applied in oil purification, recycling, and the separation of solids from liquids. The principle behind electrostatic separation is based on the difference in electrical charge or conductivity between the different components present in a mixture. In the first stage particles or water droplets in the fluid are induced with an electric charge as they pass through an electrostatic field. Due to the induced charge, particles with different electrical properties are attracted or repelled by electrodes, causing them to migrate to separate areas. The separated contaminants, such as water droplets or solid particles, are then collected and removed from the fluid.

By separating water and particles from oil, the purity and quality of the oil are maintained, enhancing equipment performance and longevity.

281. Element

An **element** in scientific terms refers to a pure substance consisting of only one type of atom, characterized by a specific number of protons in its nucleus, known as the atomic number. Elements cannot be broken down into simpler substances by chemical means and serve as the basic building blocks of all matter.

Key Characteristics:

- **Unique atomic structure**: Each element has a distinct number of protons, electrons, and, often, a specific arrangement of neutrons.
- **Periodic table**: Elements are organized in the periodic table based on their atomic number and properties, with well-known examples including hydrogen (H), oxygen (O), iron (Fe), and carbon (C).
- **States of matter**: Elements can exist in various states (solid, liquid, or gas) depending on environmental conditions such as temperature and pressure.

In the lubrication field:

- Elements like **carbon** and **sulfur** are crucial in formulating lubricants and additives.
- **Metallic elements** such as zinc, calcium, and molybdenum are used in antiwear and anticorrosion additives.
- **Nonmetal elements**, such as fluorine, are used in the production of specialty lubricants like PTFE (Teflon).

Understanding the properties of different elements is fundamental in designing and selecting materials for various industrial and lubrication applications, ensuring optimal performance and durability.

282. Elemental Analysis

Elemental analysis refers to a technique used to determine the elemental composition of a substance by identifying and quantifying the individual elements present. In the context of lubrication and oil analysis, elemental analysis is commonly used to detect wear metals, contaminants, and additive elements in lubricants, (see Table E1) providing valuable insights into the health and performance of machinery.

Key Applications:

- **Wear metal detection**: Identifies the presence of metals like iron, copper, or aluminum, which may indicate component wear or impending failure.

- **Contamination monitoring**: Detects elements such as silicon (dust/dirt), sodium, or potassium, which could signify the presence of contaminants like coolant or external debris.
- **Additive monitoring**: Measures elements like calcium, phosphorus, or zinc to ensure that the lubricant's additive package is intact and functioning correctly.

Table E1 Typical detectable lubricant elements.

Contaminant or additive	Contaminant or wear metal	Coolant or additive	Oil additive	Wear metal
Silicon	Vanadium	Boron	Barium	Aluminum
		Sodium	Calcium	Chromium
			Magnesium	Copper
			Molybdenum	Iron
			Phosphorous	Lead
			Zinc	Nickel
				Silver
				Tin

Elemental analysis can be performed using **emission spectroscopy**, a common method that uses plasma to excite atoms and measure the emitted light spectrum to identify and quantify elements, or through the use of **X-ray fluorescence (XRF)**. XRF is a nondestructive technique that uses X-rays to analyze the elemental composition of materials.

Elemental analysis is an essential tool in condition monitoring and reliability programs for industries that rely on heavy machinery, such as manufacturing, transportation, and power generation.

283. Emission Spectroscopy

Emission spectrometry is an analytical technique used to identify and quantify the elemental composition of a sample by analyzing the light emitted from excited atoms or ions. When a sample is subjected to high energy, such as heat or an electric discharge, the atoms within the sample become excited and emit light at specific wavelengths as they return to their ground state. Each element emits light at characteristic wavelengths. A polychromatic sensor then converts the light energy into a digital signal that can be interpreted by the spectrometer's software and used to print out a sample report in which element levels are compared to a known virgin sample and the equipment being monitored to determine if and what failure is occurring.

- **Energy source**: Commonly used sources include **inductively coupled plasma (ICP)** where energy is applied to the sample using an argon flame, or via an electric arc as in the **Arc Emission** method.
- **Spectrometer**: An instrument that measures the intensity of the emitted light at different wavelengths, providing a spectrum that can be analyzed to determine the elements present and their concentrations.

Emission spectrometry is capable of detecting trace levels of elements, making it highly effective for monitoring wear and contamination in lubricants. It provides rapid results, allowing for quick decision-making in maintenance and reliability programs.

Making it a vital tool in predictive maintenance and condition monitoring, helping to extend the life of machinery, optimize performance, and prevent costly failures in industrial and automotive applications.

284. Emulsification (Grease)

Emulsification in industrial grease refers to the process by which water becomes uniformly dispersed throughout the grease matrix, forming a stable or semistable mixture. This can occur due to mechanical agitation, exposure to high humidity, or direct contact with water in industrial environments.

Key Characteristics:

- **Formation of emulsions**: Water interacts with the grease's thickener and base oil, creating a mixture that may appear milky or cloudy.
- **Performance impact**: While some greases are designed to tolerate water or controlled emulsification, excessive emulsification can compromise the grease's ability to provide effective lubrication.
- **Stabilization by thickeners**: Certain thickener types (e.g., lithium or calcium-based) may stabilize water within the grease, while others may allow easier water separation.

In industrial applications, managing emulsification in grease is critical to ensuring equipment reliability and performance. Be sure to always select greases formulated for water resistance or those with good demulsibility properties to minimize emulsification risks. Ensure machinery is equipped with effective seals to limit water ingress, and regularly inspect grease conditions ensuring any contaminated grease is replaced to maintain optimal performance.

285. Emulsion

An **emulsion** is a mixture of two immiscible liquids where one liquid (such as oil) is dispersed in the form of tiny droplets within another liquid (such as water). See Figure E2.

Emulsions are stabilized by the introduction of emulsifying agents, designed to help keep the droplets from coalescing and separating over time.

Types of Emulsions:

1. **Oil-in-water (O/W) emulsion**: Oil droplets are dispersed in water. Common in food products (like milk) and certain industrial applications.
2. **Water-in-oil (W/O) emulsion**: Water droplets are dispersed in oil. Typical in lubricants and greases that need to manage moisture.

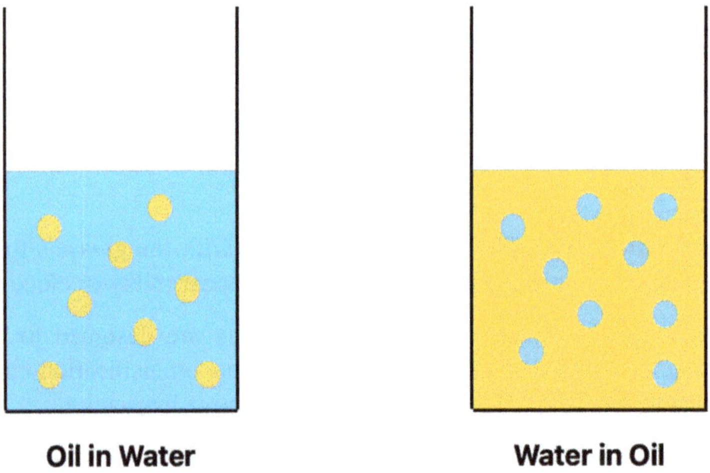

Figure E2 Emulsions.

In lubrication systems, the formation of emulsions between oil and water is usually undesirable because it can impair the lubricant's ability to reduce friction, cause corrosion, and promote the formation of sludge or deposits. Some lubricants and greases are sod already formulated with additives to resist emulsification, ensuring that water is effectively separated and does not compromise the performance of the lubricant.

Applications and Management:

- **Emulsifying agents**: Used in some industrial fluids and cutting oils to intentionally create stable emulsions for applications that require cooling and lubrication.

- **Separation**: In systems where emulsions are problematic, separators and filtration methods are used to remove water from the oil and restore lubricant performance.

Understanding how emulsions form and their effects on lubrication is crucial for maintaining the reliability and efficiency of mechanical systems, especially in environments where exposure to water is a concern.

286. Energy Institute

The **Energy Institute (EI)** is a global professional organization dedicated to the energy sector, offering resources, guidance, and training to professionals worldwide. Established in 2003 through the merger of the Institute of Petroleum (founded in 1913) and the Institute of Energy (founded in 1925), the EI serves as a central hub for energy professionals, providing access to industry knowledge, best practices, and professional development opportunities.

Key Functions and Offerings:

1. Professional development: The EI offers a range of training courses and qualifications designed to equip individuals with the knowledge and skills necessary for a just global energy transition. These programs cater to various experience levels and are accessible through the EI Academy.

2. Knowledge resources: Through its Knowledge Service, the EI provides accurate and balanced energy information to support work or studies.

3. Membership and professional registration: The EI offers membership options that provide access to exclusive resources, networking opportunities, and professional recognition. Members can apply for professional status as Member, Technician Member, or Fellow of the EI.

4. Global network: With an international membership of about 20,000 individuals and 200 companies, the EI fosters a global community of energy and lubrication professionals, facilitating collaboration and knowledge exchange across 120 countries.

By joining the Energy Institute, professionals can enhance their expertise, stay informed about industry developments, and contribute to the advancement of the energy sector.

287. Entrained Air

Entrained air refers to the presence of small air bubbles dispersed within a fluid, such as oil or lubricant. These bubbles are suspended throughout the fluid and can reduce the load-carrying capacity of the lubricant, leading to metal-to-metal contact and increased wear systems if not properly managed. In hydraulic systems, entrained air can cause cavitation, where the collapse of air bubbles generates shock waves that damage components or can result in erratic or sluggish (sponginess) movement of actuators. Air exposure can accelerate the oxidation of the lubricant, degrading its quality and shortening its service life.

Typical Entrained Air Causes:

- **Agitation**: Excessive turbulence, such as in pumps or during rapid fluid flow, can mix air into the lubricant.
- **Improper filling**: Introducing air into the system during the filling process or from leaks in seals and connections.
- **Low fluid levels**: When the fluid level in a reservoir is too low, air may be drawn into the system.

Prevention and Management:

- **Proper system design**: Minimize turbulence and use air-release devices to reduce the risk of entraining air.
- **Deaeration**: Allow time for air bubbles to rise and escape from the fluid before use, or use special equipment designed to remove entrained air.
- **Regular maintenance**: Check for and repair leaks, and ensure that fluid levels are adequate and that the system is free from contamination.

288. Entrainment

Entrainment in lubrication refers to the process **by which a moving surface pulls a lubricant into a contact zone**, such as between rolling or sliding components. This phenomenon is essential for maintaining a stable lubricant film in bearings, gears, and other machine elements operating under hydrodynamic or elastohydrodynamic (EHL) lubrication conditions.

Key Functions of Entrainment in Lubrication

- **Forms a lubricant film between surfaces**: Ensures continuous lubrication, reducing friction and wear.
- **Enhances load-carrying capacity**: In elastohydrodynamic lubrication (EHL), entrainment helps create a pressurized oil film, supporting high loads in rolling contacts.
- **Improves heat dissipation**: The lubricant carries away excess heat generated by friction, preventing thermal degradation and component damage.
- **Reduces metal-to-metal contact**: Prevents adhesive wear and scuffing by maintaining a consistent oil film thickness.

Types of Entrainments in Lubrication

1. **Rolling entrainment**: Occurs when lubricant is drawn into a rolling interface, such as in bearings and gear teeth.
2. **Sliding entrainment**: Happens when lubricant is pulled into a contact zone due to sliding motion, as in pistons and cams.
3. **Mixed entrainment**: A combination of rolling and sliding, seen in helical gears, ball screws, and journal bearings.

289. Environmental Condition Factors

Environmental condition factors refer to external elements or conditions that can influence the performance, effectiveness, and longevity of lubricants, as well as the overall operation of machinery and equipment. These factors must be carefully considered when selecting or managing lubrication systems to ensure optimal protection and performance.

Key Environmental Condition Factors:

1. **Temperature extremes**:
    - **High temperatures**: Can cause lubricants to oxidize, thin out, or lose viscosity, leading to reduced film strength and accelerated wear.

- ○ **Low temperatures**: Can increase lubricant viscosity, making it thicker and more difficult to flow, potentially leading to inadequate lubrication at startup.

2. **Humidity and moisture**:
   - ○ High humidity or direct exposure to water can lead to emulsification of lubricants, corrosion of metal parts, and microbial growth in biodegradable oils.
3. **Contaminants**:
   - ○ **Dust and dirt**: Particulate contamination can lead to abrasive wear and degradation of lubricant performance.
   - ○ **Chemicals**: Exposure to harsh chemicals or corrosive substances can degrade the lubricant and damage machinery surfaces.
4. **Pressure and load**:
   - ○ Extreme pressures or heavy loads can require specialized lubricants with high film strength and load-carrying capabilities to prevent metal-to-metal contact and surface fatigue.
5. **Vibration and shock**:
   - ○ Environments with significant vibration or shock loading may require lubricants with enhanced adhesion properties to stay in place and provide consistent protection.
6. **Exposure to environmental elements**:
   - ○ **UV radiation**: Prolonged exposure to sunlight can degrade some lubricants, especially certain synthetic and biodegradable oils.
   - ○ **Salty or marine environments**: Saltwater exposure accelerates corrosion and may require specialized marine-grade lubricants for protection.

Understanding these factors is critical when designing and maintaining lubrication systems. Proper lubricant selection, protective measures (like seals and enclosures), and regular monitoring help ensure equipment remains reliable and efficient despite environmental challenges.

290. EP Additive

See "Extreme Pressure Additive."

291. EPA

The **EPA**, or **Environmental Protection Agency**, is a US federal agency responsible for protecting human health and the environment. Established in 1970, the EPA creates and enforces regulations to manage and reduce pollution, ensure clean air and water, and oversee the safe handling of chemicals and hazardous materials.

Key EPA Responsibilities:

- **Regulation and compliance**: Establishes standards for environmental protection, such as emissions limits for pollutants and regulations for waste disposal.
- **Environmental programs**: Develops and promotes initiatives to clean up contaminated sites, protect natural habitats, and improve energy efficiency.
- **Research and monitoring**: Conducts research and monitors environmental conditions to inform policy decisions and public awareness.

Relevance to Lubrication and Industry:

- **Regulations on lubricants**: The EPA sets guidelines for the formulation, use, and disposal of lubricants to minimize environmental impact, such as restrictions on hazardous chemicals and the promotion of biodegradable lubricants in sensitive areas.
- **Emissions standards**: Implements regulations to reduce emissions from industrial machinery and vehicles, influencing the development of cleaner, more efficient lubricants.
- **Spill prevention and control**: Enforces measures to prevent and manage oil spills, including requirements for proper storage and handling of lubricants and other petroleum products.

The **EPA** plays a crucial role in ensuring that industrial and manufacturing processes, including lubrication practices, align with environmental protection goals and sustainability efforts.

292. Ergonomics

Ergonomics is the scientific study of designing and arranging workspaces, tools, and equipment to fit the needs and capabilities of the human body. The goal of ergonomics is to optimize comfort, safety, and efficiency, minimizing the risk of injury and fatigue while enhancing productivity. Many machines that have no centralized lubrication systems in place often have many bearings that are difficult or dangerous to access for manual greasing and as a consequence are often ignored until they fail.

Key Principles of Lubrication Ergonomics:

1. **Workplace design**: Arranging easy access to machine lubrication reservoirs and manual lubrication points to reduce unnecessary motion and make lubrication tasks easier to perform. This can often be achieved by piping all remote lubrication points into a safe central manifold or area that can be accessed even when the machine is running. See **PBM** – Perimeter-based Maintenance.

2. **Tool and equipment design**: Lubrication tools and machinery should be designed to minimize the strain on workers, such as using long-handled grease guns to reduce bending or ergonomically shaped containers for easy handling.

3. **Workstation layout**: Arranging lubrication stations so that all necessary tools and supplies are within easy reach, reducing the need for awkward positions or excessive stretching. See Figure E3. Ensuring that heavy lubrication drums or containers are stored and handled in a way that minimizes the risk of back injuries, such as using lifts or hoists.

4. **Posture support**: Encouraging natural, comfortable body postures to prevent musculoskeletal disorders, such as back pain or repetitive strain injuries.

5. **Adjustability**: Providing adjustable chairs, desks, and machinery to accommodate different body sizes and preferences.

6. **Minimizing repetitive motion**: Designing tasks to reduce repetitive actions that can lead to strain or injury over time.

Benefits of Ergonomics:

- **Increased safety**: Reduces the risk of workplace injuries and accidents.
- **Improved productivity**: Workers are more efficient when they are comfortable and have well-designed tools and workspaces.
- **Enhanced job satisfaction**: Comfortable and user-friendly environments can lead to higher morale and reduced fatigue.

In industrial and maintenance environments, incorporating ergonomic principles can lead to a healthier workforce, fewer work-related injuries, and more efficient lubrication and maintenance practices.

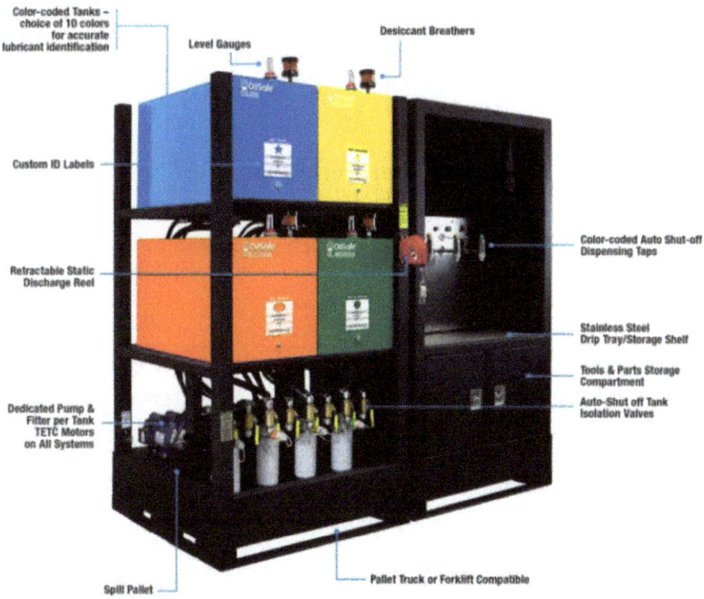

Figure E3 Ergonomically designed lubricant dispensing station. Courtesy: Fluid Defense Systems.

Preventive measures, such as using wear-resistant coatings, optimizing the flow direction, or employing erosion shields, can be employed to mitigate erosive wear and extend the lifespan of machinery and components.

293. Ester-based Lubricants

Ester-based lubricants are synthetic lubricants formulated from **ester molecules**, which are chemical compounds derived from the reaction of acids and alcohols. These lubricants offer high thermal stability, excellent oxidation resistance, superior lubricity, and biodegradability, making them ideal for high-performance, environmentally friendly, and extreme-condition applications.

Key Functions of Ester-based Lubricants

- **Provides superior high-temperature stability**: Ester-based lubricants resist oxidation and thermal breakdown, making them ideal for high-speed and high-heat applications.

- **Enhances lubricity and wear protection**: Offers lower friction and better load-carrying capacity compared to mineral oils.
- **Resists oxidation and sludge formation**: High oxidation resistance extends oil life and prevents varnish buildup in equipment.
- **Maintains performance across temperature ranges**: High viscosity index (VI) ensures stable performance in both high and low temperatures.
- **Environmentally friendly and biodegradable**: Some ester-based lubricants are readily biodegradable, making them suitable for eco-sensitive applications.

Types of Ester-based Lubricants

1. **Diester-based lubricants:** Used in compressors, bearings, and high-speed applications, known for excellent detergency and low deposit formation.
2. **Polyol ester (POE) lubricants:** Common in aviation and high-performance synthetic oils, offering superior thermal and oxidation resistance.
3. **Complex ester lubricants:** Combines properties of diesters and polyol esters for enhanced lubrication in extreme environments.
4. **Natural esters (vegetable-based oils):** Used in biodegradable lubricants for environmentally friendly applications.

Applications of Ester-based Lubricants

- **Aviation and aerospace oils**: Used in **jet engines**, **turbines**, **and hydraulic systems** due to high thermal stability.
- **Automotive engine and transmission oils**: Found in **synthetic motor oils** for better wear protection and fuel efficiency.
- **Industrial gear and compressor oils**: Provides **long-lasting lubrication** in **high-speed and high-load conditions**.
- **Biodegradable hydraulic and marine lubricants**: Used where **environmental impact is a concern**, such as **forestry and marine applications**.
- **Metalworking fluids and greases**: Reduces friction in **precision machining and metal cutting operations**.

294. Ethylene Glycol Lubricants

Ethylene glycol lubricants are specialized water-based or glycol-based lubricants formulated with ethylene glycol (EG) as the primary base fluid. These lubricants offer excellent thermal stability, low-temperature performance, fire resistance, and anticorrosion properties, making them ideal for high-performance, high-temperature, and environmentally sensitive applications.

Key Functions of Ethylene Glycol Lubricants

- **Provides superior heat transfer and cooling**: Ethylene glycol has high thermal conductivity, making it effective in cooling and heat dissipation applications.
- **Enhances fire resistance and safety**: Nonflammable under normal operating conditions, making it ideal for fire-resistant hydraulic and industrial systems.
- **Prevents corrosion and scale formation**: Formulated with corrosion inhibitors to protect metal surfaces in cooling and lubrication systems.
- **Maintains low-temperature fluidity**: Offers excellent cold flow properties, making it suitable for subzero operating conditions.
- **Reduces friction and wear**: Provides boundary and hydrodynamic lubrication, preventing metal-to-metal contact in high-speed applications.

Types of Ethylene Glycol Lubricants

1. **Water-glycol hydraulic fluids:** Used in fire-resistant hydraulic systems, commonly formulated with ethylene glycol, water, and antiwear additives.
2. **Ethylene glycol-based coolants and antifreeze:** Used in automotive and industrial cooling systems, providing thermal regulation and corrosion protection.
3. **Glycol-based compressor and gear oils:** Formulated for high-temperature, high-speed compressors and gears where oxidation resistance is critical.
4. **Metalworking fluids and cutting lubricants:** Ethylene glycol-based lubricants reduce heat and friction in machining, grinding, and cutting operations.

5. **Biodegradable and food-grade glycol lubricants:** Used in food processing and environmentally sensitive applications, ensuring nontoxicity and regulatory compliance.

295. Evaporative Loss

Evaporative loss refers to the amount of a lubricant or fluid that evaporates into the air over time, especially when exposed to elevated temperatures. This phenomenon can lead to a reduction in the lubricant's volume and affect its overall performance, potentially causing increased wear, reduced film strength, or compromised protection of machinery components.

Key Factors Influencing Evaporative Loss:

1. **Temperature**: Higher operating temperatures accelerate evaporation, particularly in environments where equipment generates a lot of heat.
2. **Volatility**: Fluids with higher volatility are more prone to evaporative loss. The base oil type and the presence of light fractions in the lubricant formulation contribute to volatility.
3. **Exposure surface area**: The greater the surface area exposed to air, the higher the potential for evaporation.

As a lubricant evaporates, its viscosity and ability to provide an effective lubricating film may diminish and can lead to an increased concentration of oxidation-prone substances, accelerating lubricant degradation. The need to replenish lubricant more frequently can result in higher operational costs and downtime.

Applications

- **High-temperature systems**: Applications such as engines, turbines, and industrial machinery are particularly susceptible to evaporative loss. Using high-quality, thermally stable lubricants can help reduce the issue.
- **Sealed systems**: Employing sealed lubrication systems can minimize the exposure of lubricants to air, thereby reducing evaporative loss.

296. Extreme Pressure Additive

EP (extreme pressure) additive is a chemical compound used in lubricants to enhance their ability to withstand high pressure and prevent wear or damage in heavily loaded mechanical systems. Under high-pressure/

high-temperature conditions, EP additives react with the metal surfaces to form a protective film or layer that can handle harsh operating conditions without breaking down.

EP additives are critical in applications where metal surfaces come into close contact under significant pressure, such as in gears, bearings, and hydraulic systems.

Common Types of EP Additives:

1. **Sulfur–phosphorus compounds**: These additives are widely used in gear oils and provide excellent wear protection by forming a durable film on metal surfaces.
2. **Chlorinated compounds**: Less common due to environmental concerns but effective in preventing scuffing and seizing in high-pressure environments.
3. **Zinc dialkyldithiophosphate (ZDDP)**: An additive that provides both antiwear and EP properties, often used in engine oils.

EP additives are a crucial component in lubrication technology, enhancing the durability and efficiency of mechanical systems that operate under severe conditions. EP additives must be compatible with the base oil and other additives in the formulation to ensure optimal performance. Be aware that some EP additives, especially those containing sulfur, can be corrosive to yellow metals (like copper or brass) if not properly formulated.

297. Extreme Pressure Property

Extreme pressure (EP) property refers to the ability of a lubricant to prevent wear and surface damage when metal components are subjected to high loads or extreme pressure conditions. This property is essential for lubricants used in applications such as heavy-duty gears, bearings, and machinery, where components are at risk of coming into direct contact under significant force.

Key Characteristics of Extreme Pressure Property:

- **Protection under load**: The EP property allows the lubricant to form a protective layer or film that prevents metal surfaces from welding or seizing together under extreme pressure.
- **Additives role**: EP properties are often imparted by EP additives, which chemically react with metal surfaces to create a sacrificial, protective film. This film minimizes wear, reduces friction, and prevents surface damage.

- **High-temperature resistance**: As extreme pressures generate significant heat, the protective film must withstand high temperatures without breaking down.

Common Applications:

- **Gearboxes and transmissions**: In automotive and industrial applications where gears experience high contact pressure.
- **Heavy machinery**: Equipment used in mining, construction, and agriculture, where loads are substantial, and protection against wear is critical.
- **Bearings**: Specifically in cases where bearings are under high loads and require enhanced protection against metal-to-metal contact.

F

298. Failure Mode Effect Analysis (FMEA)

Failure modes and effects analysis (FMEA) is a systematic method used to identify and evaluate potential failures in a product, process, or system and analyze the impact of these failures to prioritize risk-reduction actions. FMEA helps organizations proactively address potential issues and improve the reliability and safety of their operations.

Key Components of FMEA:

- **Failure mode**: Describes how a process, product, or system component might fail. Examples include mechanical failure, material degradation, or human error.
- **Effects analysis**: Evaluates the consequences of each failure mode on the overall system, product performance, or user safety.
- **Severity**: Assesses the seriousness of the failure's impact.
- **Occurrence**: Estimates the likelihood of the failure occurring.
- **Detection**: Evaluates the ability to detect the failure before it affects the end user.
- **Risk priority number (RPN)**: The RPN is calculated by multiplying the severity, occurrence, and detection ratings. It helps prioritize which failure modes require immediate action:

$$\text{RPN} = \text{Severity} \times \text{Occurrence} \times \text{Detection}$$

Lower RPN values indicate **lower risk**, while **higher values** signal the **need for corrective measures**.

Types of FMEA:

1. **Design FMEA (DFMEA)**: Focuses on potential failures in the design of a product or system.

DOI: 10.1201/9788743806608-303

2. **Process FMEA (PFMEA)**: Analyzes failures related to manufacturing or assembly processes.
3. **System FMEA**: Considers failures within complex systems, including interactions among components.
4. **Lubrication FMEA (LFMEA)**: Focuses lubrication-related failures and potential failures.

FMEA is widely used across industries such as automotive, aerospace, manufacturing, and healthcare to ensure high-quality and dependable products or processes. Benefits include risk mitigation, cost reduction, and improved reliability and safety.

299. Failure Code

Failure code refers to a unique identifier or alphanumeric sequence assigned to specific types of equipment failures or malfunctions in systems or machinery. These codes are used in lubrication management, maintenance management, and diagnostic systems to quickly communicate the nature of a fault and aid in the troubleshooting, repair, and documentation process.

Key Aspects of Failure Codes:

- **Standardization**: Failure codes are often standardized within an organization or industry to ensure consistency in reporting and understanding issues. For example, automotive diagnostic systems use standardized OBD-II (on-board diagnostics) codes.
- **Efficiency**: By using a concise code, maintenance personnel can quickly identify issues without lengthy descriptions, streamlining repair processes.
- **Integration with CMMS**: Many computerized lubrication/maintenance management systems (LMMS/CMMS) use failure codes to track and analyze maintenance data, helping organizations understand common failure modes and improve reliability.

Examples of Use:

- **Industrial equipment**: Failure codes are used in manufacturing plants to indicate problems with machinery, such as bearing failure, lubricant loss, hydraulic pressure loss, or motor overheating.

Automotive industry: Diagnostic trouble codes (DTCs) used in vehicles to pinpoint engine or system malfunctions.

Benefits:

- **Improved troubleshooting**: Quick identification of issues leads to faster repairs and less downtime.
- **Data analysis**: Tracking failure codes over time can reveal patterns and help in preventive maintenance planning.
- **Documentation and reporting**: Clear documentation of failures aids in compliance and performance tracking.

Failure codes are essential tools in maintenance and diagnostics, helping teams work more efficiently and maintain the reliability of equipment and systems.

300. Falex Test

The **Falex test** is a standardized method used to evaluate the wear, friction, and extreme pressure properties of lubricants, oils, and greases. This test is performed using a **Falex Pin and Vee Block Tester**, a device that applies a controlled load to a rotating pin placed between two stationary Vee blocks. The goal is to assess the lubricant's ability to protect surfaces under varying loads and conditions.

How the Test Works – Figure F1:

1. **Setup**: A pin and two Vee blocks are immersed in the test lubricant housed in the oil cup.
2. **Load application**: The Vee blocks are pressed against the rotating journal pin with gradually increasing force.
3. **Measurement**: The amount of load the lubricant can handle before the surfaces seize or wear excessively is recorded. Additionally, wear scars on the pin and blocks are analyzed to determine the lubricant's effectiveness.

Key Metrics:

- **Friction coefficient**: Measures the lubricant's ability to reduce friction between metal surfaces.
- **Wear scar analysis**: Examines the extent of wear on the pin and Vee blocks.

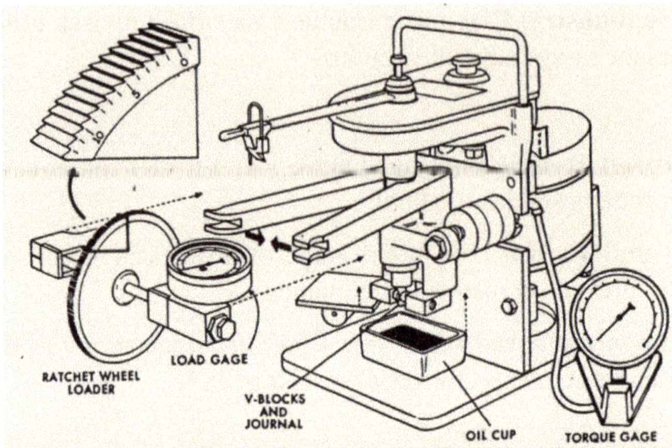

Figure F1 Falex Vee-Block and rotating pin diagram. Source: Falex Pin and Vee Block Sales Catalog.

- **Load-carrying capacity**: Indicates the maximum load the lubricant can sustain without causing damage.

The **Falex test**, originally developed by Shell Oil Company in the 1960s, is widely used in the development and quality control of lubricants, particularly for evaluating the performance of lubricants used in high-pressure or high-friction environments, such as gears, bearings, and engines. It helps manufacturers and engineers select suitable lubricants for specific applications and ensures that products meet performance and durability requirements. This test method provides valuable data for comparing different lubricants and understanding their protective capabilities under mechanical stress.

301. False Brinelling

False brinelling refers to damage or wear that occurs on the surfaces of rolling-element bearings, such as ball or roller bearings, due to vibration or oscillation while the bearing is stationary. This phenomenon results in dents or grooves on the bearing raceways that resemble those caused by brinelling (permanent deformation from static overload), but the mechanism behind false brinelling is different.

Causes of False Brinelling:

1. **Vibrations**: Repeated vibrations or small oscillations cause the rolling elements to move slightly back and forth, leading to wear patterns on the raceways.

2. **Oscillatory motion**: Bearings exposed to limited movement without completing a full rotation are prone to this type of wear.
3. **Lack of lubricant flow**: When bearings are stationary, the lubricant does not flow to replenish the protective film, exacerbating the wear.

Effects:

- **Wear marks**: Characteristic depressions or grooves form on the raceways where the rolling elements make contact. See Figure F2.
- **Reduced performance**: The wear can lead to increased friction, noise, and reduced service life of the bearing.
- **Potential failure**: Prolonged false brinelling can result in complete bearing failure, requiring replacement and causing potential damage to the surrounding equipment.

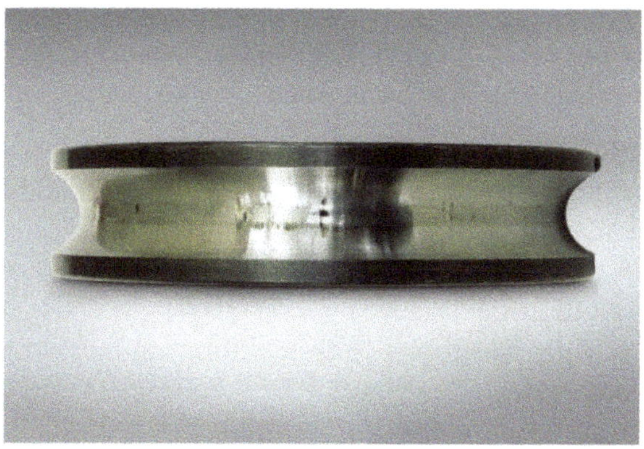

Figure F2 False brinelling of an inner race. Source: NSK Europe.

Prevention:

- **Vibration isolation**: Use dampeners or isolation pads to minimize vibrations when equipment is not in use.
- **Bearing preloading**: Applying a small, constant load to bearings can help keep the rolling elements from moving excessively under vibration.
- **Lubrication management**: Ensuring that bearings have adequate and appropriate lubrication can help mitigate the effects of oscillatory wear.

False brinelling is commonly observed in equipment that is frequently transported, such as wind turbine components or vehicles, or in machinery that remains idle but is exposed to ambient vibrations. Understanding and addressing the root causes of false brinelling are essential to maintaining bearing health and equipment reliability.

302. Fatigue Wear

Fatigue wear refers to the progressive damage that occurs on material surfaces as a result of repeated or cyclic loading. Over time, even if the stresses are below the material's ultimate strength, the repeated application of loads can lead to the formation of micro-cracks and the eventual removal of material fragments. This type of wear is common in components subjected to fluctuating mechanical forces, such as bearings, gears, and shafts.

Mechanism of Fatigue Wear:

1. **Cyclic stress**: Repeated loading and unloading cause microscopic cracks to initiate at surface imperfections or stress concentration points.
2. **Crack propagation**: As the cyclic loading continues, these cracks grow and eventually lead to the detachment of material particles from the surface.
3. **Material loss**: The accumulated effect of this crack growth results in surface pitting, spalling, or flaking, leading to wear and degradation of the component.

Characteristics of Fatigue Wear:

- **Surface pitting**: Small pits or craters form on the surface as material particles break away. See Figure F3.

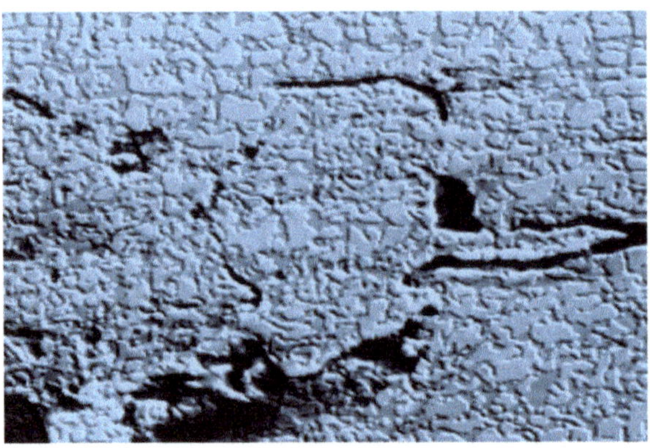

Figure F3 Fatigue worn surface. Source: Plant Lubrication for Industrial Facilities – Fourth Revision, 2024 Kenneth E. Bannister.

- **Spalling**: In severe cases, larger pieces of material may flake off the surface, causing a rough and uneven texture.
- **Progressive nature**: Fatigue wear tends to worsen over time, especially if the stress cycles are frequent and intense.

Common Causes:
Fatigue commonly occurs in rolling friction surfaces that experience repeated long-term load cycles and stress that can result in elastic deformation to the surfaces as in EHD lubrication. This long-term action eventually results in small surface and subsurface cracking, which eventually travel up to and across the bearing surface resulting in surface delamination and surface pitting. Insufficient or degraded lubrication can exacerbate fatigue wear by increasing friction and surface stress.

Preventive Measures:
- **Improved material selection**: Using materials with higher fatigue strength or surface hardness can reduce the likelihood of fatigue wear.
- **Proper lubrication**: Maintaining adequate lubrication reduces surface stress and minimizes frictional forces.
- **Surface treatments**: Techniques like case hardening or shot peening can improve the surface's resistance to crack initiation.

Fatigue wear is a critical consideration in the design and maintenance of mechanical systems. Regular monitoring and preventive measures are essential to ensure the longevity and reliability of components exposed to cyclic loading.

303. Fatty Acids

Fatty acids are organic compounds made up of a hydrocarbon chain and a carboxylic acid group (-COOH). They are fundamental components of lipids and are commonly found in natural oils and fats. Fatty acids can vary in length and degree of saturation, which influence their properties and uses in different applications, including lubrication and chemical manufacturing. There are two basic types of fatty acids:

1. **Saturated fatty acids**: These have no double bonds between carbon atoms, making them more stable and solid at room temperature (e.g., stearic acid and palmitic acid).

2. **Unsaturated fatty acids**: These contain one or more double bonds, which make them more reactive and typically liquid at room temperature (e.g., oleic acid and linoleic acid).

Role in Lubrication:
Fatty acids are used as key components in **lubricant formulations** because of their polar nature, which gives them excellent lubricity and surface-active properties. They are often used in the production of **metalworking fluids**, **biodegradable lubricants**, and **grease** formulations. Their role in lubricants is three-fold:

- **Antiwear properties**: Fatty acids can form a protective film on metal surfaces, reducing wear and friction.
- **Thickening agents**: In greases, fatty acids react with metallic bases (like lithium or calcium) to form soap-based thickeners.
- **Corrosion inhibitors**: The polar head of the fatty acid molecule can bond to metal surfaces, providing a protective barrier against moisture and corrosive elements.

Fatty acids are essential in various industrial applications due to their versatility, biodegradability, and effectiveness in enhancing lubrication performance.

304. Feedability

Feedability refers to the ease with which a material, fluid (lubricant), substance, or component can be fed or processed through a system, such as a dispensing mechanism, manufacturing equipment, or lubrication system. The concept is commonly used in industrial and manufacturing contexts where consistent and smooth feeding of materials is critical for efficiency and precision.

Key Factors Affecting Feedability:

1. **Material properties**: Characteristics such as viscosity, texture, particle size, and density impact how easily a substance can be dispensed or moved through a system.
2. **Equipment design**: The configuration of machinery or dispensing equipment can influence feedability. Lubrication systems must be tailored to accommodate the properties of the material being processed.

3. **Environmental conditions**: Temperature, humidity, and other environmental factors can alter the behavior of materials, affecting how well they flow or are fed through a lubrication delivery system.

In lubrication management, feedability is crucial for ensuring that lubricants are delivered consistently to the intended points without clogging or interruptions. Improving feedability often involves careful lubricant choice based on operational and environmental factors.

305. Ferrography

See "Analytical Ferrography."

306. Ferrous

Ferrous refers to any material or component, for example, bearings or lubricated components, that contain iron as a primary component. The term is often used to describe metals or alloys that have significant iron content and exhibit magnetic properties, as well as a tendency to oxidize and corrode in the presence of moisture.

Key Characteristics:

- **Magnetism**: Ferrous materials are usually magnetic, which is an important property for applications like transformers, electric motors, and magnetic storage media.
- **Corrosion**: They are prone to rusting and corrosion when exposed to moisture and air unless they are alloyed with elements that improve corrosion resistance, such as in stainless steel.
- **Strength and durability**: Ferrous metals are known for their high tensile strength and durability, making them essential in construction, automotive, and heavy machinery industries.

Common Ferrous Materials:

1. **Carbon steel**: A versatile and widely used material in lubricated machinery components and bearing material.
2. **Stainless steel**: An iron-based alloy with added chromium to resist corrosion.
3. **Cast iron**: Known for its excellent machinability and wear resistance, commonly used in engine blocks and piping.

Figure F4 Ferrogram image exposing a large piece of cast iron in the lubricant sample. Source: Polaris Laboratories.

In the field of lubrication and wear analysis, ferrous particles detected in oil samples using ferrography (see Figure F4) testing that will indicate wear of iron-based components, prompting maintenance or further investigation.

307. Fiber

Fiber in the context of lubrication generally refers to fibrous materials that are either added to lubricants or used in components associated with lubrication systems. These fibers can serve multiple purposes, such as enhancing the structural properties of greases or improving the filtration of lubricants.

Applications of Fiber in Lubrication:

1. **Thickening agents**: Fibers are sometimes used as a thickening agent in greases to improve consistency and performance under heavy loads. These fibers can be synthetic or organic and contribute to the grease's ability to stay in place and provide lasting lubrication.

2. **Reinforcement**: Fibers can reinforce the lubricating material, giving it added strength and stability, especially in extreme operating conditions like high temperature or pressure.

3. **Filtration**: Fiber-based materials are also used in filtration systems to remove contaminants from lubricants. These filters ensure that lubricants maintain their cleanliness and effectiveness over time, protecting

equipment from wear and tear. Figure F5 shows how fibers can present an obstacle to any contaminants trying to pass through the filter.

4. **Sealing and packing**: Fibers can be used in seals and packing materials to enhance their durability and resistance to fluid degradation, often in combination with lubricants to ensure minimal leakage and friction.

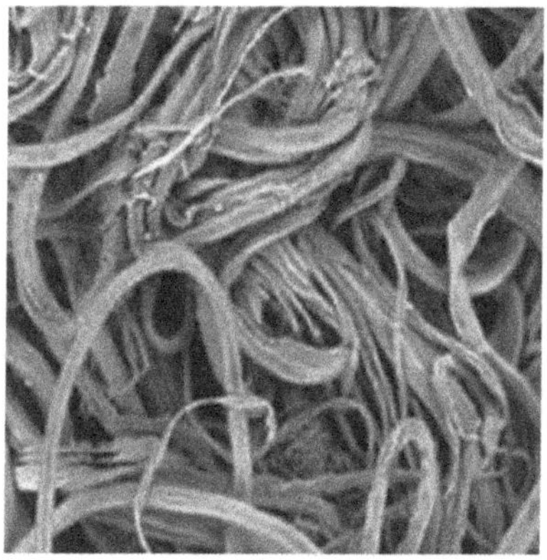

Figure F5 Magnified image of fibers in an industrial filter. Source: Unknown.

Types of Fibers Used:

- **Organic fibers**: Such as cellulose, used in biodegradable or environmentally friendly greases and filtration systems.
- **Synthetic fibers**: Like aramid or PTFE (Teflon), providing enhanced temperature and chemical resistance in demanding applications.

Fibers play a crucial role in lubrication technology, especially in applications requiring enhanced mechanical stability, controlled flow properties, or effective contaminant removal.

308. Fibril

Fibril refers to a fine, thread-like structure or fiber, which can be present in various materials or substances. In the context of lubrication, **fibrils** are often microscopic fibers that can be used to enhance the physical properties of lubricants or grease.

Role of Fibrils in Lubrication:

1. **Reinforcement**: Fibrils may be added to greases or other lubricating substances to improve their structural integrity and performance,

especially under extreme pressure or temperature conditions. They provide a network-like structure that increases the lubricant's load-bearing capacity and stability.

2. **Thickening agents**: In certain lubricating greases, fibrils act as thickening agents, helping the lubricant maintain its consistency and adhere better to surfaces.

3. **Wear resistance**: The addition of fibrils can also enhance wear resistance by providing a durable, fibrous layer that protects metal surfaces from direct contact and reduces friction.

Common Applications:

- **High-performance greases**: Used in industries such as automotive, aerospace, and heavy machinery, where lubricants are required to perform under harsh conditions.
- **Seals and gaskets**: Fibril-reinforced materials are used in sealing applications to ensure minimal fluid leakage and optimal durability.

Fibrils are crucial in advanced lubrication formulations, as they enhance mechanical properties and ensure reliable performance in demanding environments.

309. Fischer, Karl

Karl Fischer was a German chemist best known for developing the **Karl Fischer titration test**, a method for accurately measuring the water content in various substances. His groundbreaking work in analytical chemistry revolutionized the way water content is detected and quantified in liquids, solids, and gases.

Karl Fischer worked in the early 20th century, a time when accurate moisture analysis was a critical need in many scientific and industrial applications. His contribution provided a solution that was far more precise than existing methods. In 1935, Fischer introduced his **titration method**, which involved the chemical reaction of iodine, sulfur dioxide, and water in an alcohol medium with a base. This reaction forms the basis of what is now called Karl Fischer titration and is used globally for accurate water content analysis. See Karl Fischer Titration Method.

Since that time his method is still the standard for moisture determination due to its accuracy, reliability, and broad application range. It is extensively used in industries such as pharmaceuticals, lubricants, petrochemicals, food, and chemicals.

310. FIFO

See "First In, First Out."

311. Filler

Filler in lubrication terms refers to a substance added to grease or lubricants to modify or enhance their physical properties. Fillers are typically nonlubricating materials that help adjust the consistency, texture, or other characteristics of the lubricant without significantly altering its primary function.

Types and Functions of Fillers:

1. **Thickening agents**: In grease formulations, fillers like clays, silica, or organic thickeners are used to achieve the desired consistency and ensure the grease stays in place under varying conditions.
2. **Performance enhancement**: Fillers can improve properties such as load-carrying capacity, water resistance, or thermal stability. For example, solid fillers like graphite or molybdenum disulfide may be used to provide additional protection in high-load applications.
3. **Cost reduction**: Some fillers are added to reduce the overall cost of the lubricant by extending the volume, though this must be carefully balanced to avoid compromising performance.

Common Fillers Used:

- **Clays**: Used in nonmelt greases to provide stability at high temperatures.
- **Silica**: Acts as a thickener and stabilizer in certain lubricant formulations.
- **Molybdenum disulfide (MoS_2)**: A solid lubricant filler that enhances load-carrying capacity and reduces friction.

Fillers play an essential role in tailoring lubricants to meet the specific needs of various industrial and mechanical applications. The choice and amount of filler must be carefully controlled to maintain the lubricant's effectiveness and avoid unwanted side effects, such as excessive thickening, reduced lubricity, or separation.

312. Film Strength

Film strength refers to the ability of a lubricant to maintain an unbroken film between surfaces under high pressure or heavy load conditions. It is a critical property in lubrication that determines how well the lubricant can prevent metal-to-metal contact, thereby reducing friction and minimizing wear. A lubricant with high film strength can withstand extreme conditions without the film rupturing or thinning out.

Key Points about Film Strength:

1. **Protective barrier**: The lubricant film acts as a barrier that prevents direct contact between moving surfaces, reducing wear and heat generation.
2. **Load carrying ability**: High film strength ensures that the lubricant can handle heavy loads and high-pressure environments, such as in gears, bearings, and engines.
3. **Extreme pressure additives**: EP additives are often used to improve film strength, especially in situations where the lubricant might otherwise be unable to handle the load. These additives chemically react with metal surfaces to form a protective layer.

Factors Affecting Film Strength:

- **Viscosity**: Higher viscosity lubricants generally have greater film strength, but they may also increase energy consumption due to higher resistance to flow.
- **Additives**: Antiwear and extreme pressure additives enhance film strength, especially under boundary lubrication conditions.
- **Temperature**: Film strength can decrease at high temperatures, as the lubricant becomes thinner and less effective at maintaining separation between surfaces.

Applications:

- **Automotive engines**: To protect engine parts that experience high loads and temperatures.
- **Industrial machinery**: For heavy-duty equipment where metal surfaces are under constant stress and require reliable lubrication.

Film strength is essential for maintaining the longevity and efficiency of mechanical components in various applications. Selecting a lubricant with appropriate film strength is crucial for protecting equipment and ensuring smooth operation.

313. Film Thickness Ratio (Lambda)

Film thickness ratio (Lambda, λ**)** is a dimensionless parameter used to evaluate the adequacy of the lubricating film between two surfaces in relative motion. It represents the ratio of the lubricant film thickness to the combined surface roughness of the contacting surfaces. Essentially, the Lambda value helps determine whether the lubrication regime is in the boundary, mixed, or full-film (hydrodynamic or elasto-hydrodynamic) lubrication range.

Formula:

$$\lambda = \frac{h}{\sigma}.$$

where:

- h = Lubricant film thickness
- σ = Combined surface roughness of the interacting surfaces, calculated as the root sum square of the surface roughness values

Interpretation of lambda values (see Figure F6):

1. $\lambda < 1$: **Boundary lubrication** – The lubricant film is insufficient to fully separate the surfaces, leading to frequent metal-to-metal contact and increased wear.

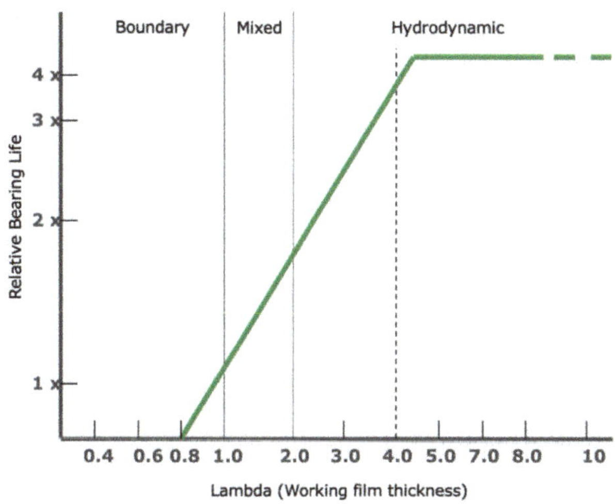

Figure F6 Lambda relationship diagram. Courtesy: ENGTECH Industries Inc.

2. **1 < λ < 3: Mixed lubrication** – Partial film separation occurs, with some surface asperities in contact. There is a combination of fluid film lubrication and boundary interactions.

3. **λ > 3: Full-film lubrication** – The surfaces are completely separated by the lubricant film, minimizing wear and reducing friction.

Importance:

- A high Lambda value indicates a robust lubricant film that can prevent surface contact, essential for minimizing wear and ensuring the longevity of components.
- Engineers and lubrication specialists use the Lambda ratio to select appropriate lubricants and optimize machine performance, especially in high-load or high-speed applications.

Understanding and calculating the film thickness ratio is crucial in the design and maintenance of machinery to ensure adequate lubrication and prevent premature failure of mechanical components.

314. FILO

See "First In, Last Out."

315. Filter

Filter in lubrication refers to a device or component designed to remove contaminants, such as dirt, metal particles, and other impurities, from lubricating oil or grease. Filtration is a crucial aspect of lubrication systems because contaminants can cause wear, corrosion, and damage to machinery components, significantly reducing equipment life and performance.

Types of Lubrication Filters:

1. **Full-flow filters**: These filters process all the lubricant passing through the system, ensuring continuous protection of components. They are commonly used in engines and hydraulic systems.

2. **Bypass filters**: A portion of the lubricant flow is diverted through a finer filter, which can remove smaller particles than full-flow filters. Bypass filters are used to provide higher filtration efficiency over time.

3. **Magnetic filters**: These filters use magnetic fields to capture ferrous metal particles suspended in the lubricant, providing additional protection.

4. **Wedge wire filters**: These filters are wedge-shaped wire coils used for trapping contaminants and particulates suspended in grease. See Figure F7.

Key Functions:

- **Contaminant removal**: Filters prevent particles from circulating through the lubrication system, reducing wear on components.
- **Maintaining lubricant quality**: By keeping the lubricant clean, filters help maintain its viscosity and protective properties, extending the service life of both the lubricant and the equipment.
- **Protecting equipment**: Effective filtration minimizes the risk of machinery failure and reduces maintenance costs.

Applications:

Filters are used in various lubrication systems, including:

- **Engines**: To keep engine oil free from debris and contaminants.
- **Hydraulic systems**: To ensure hydraulic oil remains clean and maintains system efficiency.
- **Industrial machinery**: To protect critical components in high-stress environments.

Figure F7 Wedge wire grease filter. Courtesy: ENG**TECH** Industries Inc.

316. FIlter Cart

A **filter cart** is a portable filtration system designed for removing contaminants from lubricants, hydraulic oils, or other fluids in machinery. It serves as a supplemental filtration unit, providing a convenient and efficient way to clean and maintain fluid cleanliness in industrial equipment, mobile machinery, or stationary hydraulic systems.

Key Features:

- **Portability**: Equipped with wheels and a handle, filter carts can be easily transported around a facility to service different equipment.
- **Dual filtration**: Many filter carts are equipped with two filters in series – a coarse filter to remove large particles and a finer filter for smaller contaminants, ensuring thorough filtration. Figure F8 demonstrates a filter cart that employs bag filters (see Bag Filter)
- **Versatile use**: Suitable for a variety of fluids, including hydraulic oils, gear oils, and lubricants, making them versatile tools for maintenance.

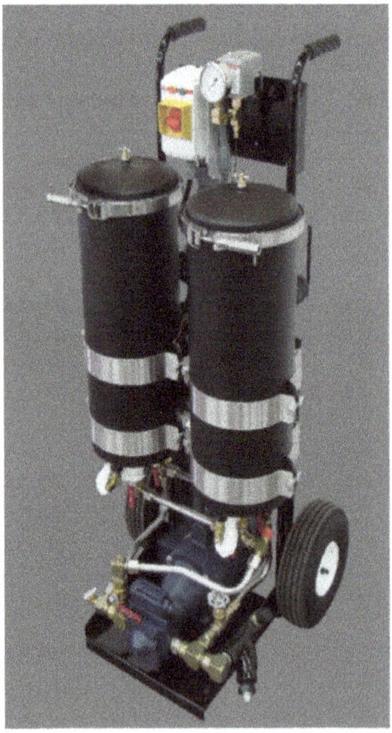

Figure F8 Typical portable filter cart. Courtesy: Fluid Defense Systems.

Filter carts can be used for many lubrication tasks that can include:

1. **Fluid transfer**: Used when transferring fluids from storage containers to machinery, ensuring that clean oil is introduced into the system.

2. **Oil maintenance**: Employed for periodic oil cleaning and conditioning, helping to extend the life of both the oil and the equipment.
3. **System flushing**: Effective for flushing and cleaning lubrication and hydraulic systems, especially for initial fill/flushing of a new or repaired machine after a component failure, or as a routine maintenance task.

By removing contaminants, filter carts help prevent wear and damage, leading to reduced maintenance costs and longer equipment life. Clean oil ensures optimal performance of machinery, minimizing the risk of breakdowns and enhancing operational efficiency.

Regular use of a filter cart reduces the frequency of oil changes and lowers the overall cost of lubricant consumption.

317. Filter Efficiency

Filter efficiency refers to the effectiveness of a filter in removing contaminants from a fluid, such as oil or air. It is a critical parameter in lubrication and hydraulic systems, where maintaining fluid cleanliness is essential for preventing wear, minimizing equipment failure, and extending machinery life.

Measuring Filter Efficiency:

- **Beta ratio**: A common metric used to express filter efficiency. It is calculated by comparing the number of particles of a certain size upstream of the filter to the number downstream. A higher beta ratio indicates better filtration efficiency. For example, a beta ratio of 200 means that 99.5% of particles of a given size are removed. See **Beta Ratio** entry.
- **Percentage efficiency: This is calculated using the Formula:**

$$\text{Efficiency }(\%) = \left(1 - \frac{\text{Number of particles downstream}}{\text{Number of particles upstream}}\right) \times 100$$

For example, if a filter has an efficiency of 98%, it means that it successfully removes 98% of the specified contaminant particles (see Table B1: Filter efficiency).

Factors Affecting Filter Efficiency:

- **Filter media**: The type and quality of the material used in the filter determine how well it captures contaminants. Fine, high-density media provide better filtration but may restrict flow more than coarse media.

- **Particle size**: Filters are rated based on the size of particles they can capture, typically measured in microns. Efficiency may vary depending on the size of the contaminants.
- **Flow rate**: The speed at which fluid flows through the filter affects how well contaminants are trapped. Higher flow rates can reduce filtration efficiency.

Importance in Lubrication Systems:

- **Protecting equipment**: Efficient filters prevent abrasive particles from damaging components such as bearings, gears, and pumps.
- **Extending oil life**: By removing contaminants, filters help maintain the quality of lubricating oils, reducing the need for frequent oil changes and lowering operational costs.
- **System reliability**: Effective filtration ensures that hydraulic and lubrication systems operate smoothly, minimizing the risk of breakdowns and increasing productivity.

318. Filter Housing

Filter housing refers to the casing or enclosure that contains and supports the filter element in a lubrication, hydraulic, or fluid filtration system. The housing is designed to securely hold the filter element while allowing fluid to flow through it, ensuring contaminants are effectively removed before the fluid continues through the system.

Key Components and Features:

1. **Inlet and outlet ports**: Openings that allow fluid to enter and exit the filter housing. The design and size of these ports are critical to ensure proper flow and minimize pressure drops.
2. **Sealing mechanism**: To prevent leaks and ensure that all fluid passes through the filter element, housings are equipped with seals or gaskets.
3. **Pressure rating**: The housing must be designed to withstand the operating pressure of the system. In high-pressure applications, robust materials such as stainless steel or reinforced polymers are used.
4. **Access for maintenance**: Filter housings often feature removable lids, clamps, or bolts to make it easy to replace the filter element. Some designs include a drain port for easy removal of trapped contaminants. Note: “Spin-on” single cartridge housings, popular in automotive

applications, are one-time use throw away filter housings as the filter cartridge cannot be replaced without destroying the housing.

Types of Filter Housings:

- **Single cartridge housing**: Holds a single filter element, commonly used in smaller systems or applications with lower flow rates.
- **Multicartridge housing**: Contains multiple filter elements to handle larger volumes of fluid, typically used in industrial applications.
- **Bag filter housing**: Designed to hold filter bags rather than cartridges, used for filtering larger particles in high-flow applications.

Filter housings are critical for ensuring that the filtration process is effective, leak-free, and able to withstand the environmental and mechanical stresses of the application. They also play a role in maintenance efficiency by allowing easy access to the filter element for cleaning or replacement.

319. Filter Media

Filter media refers to the material within a filter element that captures and removes contaminants from fluids, such as lubricants, hydraulic oil, fuel, or air. The effectiveness of a filter depends heavily on the properties and design of the filter media, which can vary widely depending on the specific application and the type of contaminants to be removed (see: **Fiber**).

Types of Filter Media:

1. **Cellulose (paper)**: Made from plant fibers, cellulose media are cost-effective and commonly used in lower-pressure systems. They are efficient for capturing larger particles but have limited durability and chemical resistance.
2. **Synthetic fibers**: These include materials like polyester or fiberglass. Synthetic media are more durable than cellulose, offer higher filtration efficiency, and can withstand higher pressures and temperatures.
3. **Metal mesh**: A durable, reusable option, metal mesh media are often used in applications that require the filtration of large particles and in systems that are regularly cleaned and reused.
4. **Microglass fibers**: Ultra-fine fibers that provide high filtration efficiency and are commonly used in hydraulic and high-performance applications. They are capable of capturing very small particles while maintaining good flow characteristics.

5. **Activated carbon**: Used for filtering out chemical impurities and odors. Activated carbon media are common in air and fuel filters, especially when odor removal is needed.

Key Considerations for Filter Media Selection:

- **Particle size**: The filter media should be capable of capturing the specific size range of particles present in the fluid. This is often expressed in microns.
- **Flow rate**: The media must allow the fluid to pass through efficiently without causing a significant pressure drop or flow restriction.
- **Durability**: In high-pressure or high-temperature environments, the media must be robust enough to maintain its integrity over time.
- **Chemical compatibility**: The filter media should be compatible with the fluid being filtered to prevent degradation or contamination.

320. Filter, Regulator, Lubricator (FRL)

Filter, **regulator**, **and lubricator** are three essential components used in pneumatic systems to ensure the proper operation of air-driven tools and equipment. Together, an FRL unit conditions the compressed air supply by removing contaminants, regulating pressure, and introducing lubrication to maintain efficiency and prolong the life of pneumatic lubrication pumps and other devices.

Components of an FRL Unit:

1. **Filter**: Removes particulates, dirt, and moisture from the compressed air to prevent contamination and damage to the equipment.
2. **Regulator**: Controls and maintains a consistent air pressure, ensuring that the pneumatic tools receive the correct amount of pressure for optimal performance.
3. **Lubricator**: Adds a fine mist of oil into the compressed air to lubricate internal moving parts of pneumatic tools and air-powered lubrication pumps, reducing wear and friction.

Functions and Benefits:

Clean, well-regulated, and lubricated air helps reduce wear and tear on pneumatic components. By regulating pressure and providing proper lubrication,

FRL units help maintain the consistent operation of air-driven equipment and prevent damage and malfunctions caused by contaminants or inadequate lubrication, minimizing asset downtime.

Applications:
FRL units are widely used in manufacturing, automotive repair, assembly lines, and any industry where pneumatic tools or machinery are essential to control and clean the air supply to any pneumatic lubrication pump. See Figure F9.

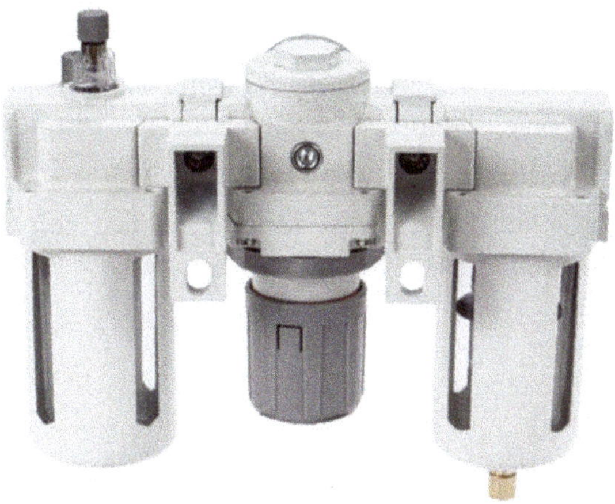

Figure F9 Typical FRL unit. Source: Unknown.

321. Filterability Test

Filterability test in the context of lubrication refers to the assessment of a lubricant's ability to pass through a filter without forming blockages or causing a reduction in flow rate. This test is essential for determining how well a lubricant can perform in systems with filtration requirements, such as hydraulic systems, where clogging could lead to decreased efficiency, increased wear, or equipment failure.

Key Aspects of the Filterability Test:

- **Purpose**: The test measures a lubricant's capacity to remain fluid and filterable over time, even in the presence of potential contaminants, oxidation byproducts, or water content.
- **Procedure**: A sample of the lubricant is passed through a standardized filter, and the test measures either the time taken for a set volume of

oil to pass through the filter or the amount of residue left on the filter. Sometimes, contaminants are intentionally added to the sample to evaluate how the lubricant handles them.

- **Parameters evaluated**: The test looks at the tendency of the lubricant to form deposits or gels that can block filters, and it assesses the potential impact of water contamination on filterability.

Importance of Filterability:
- **Lubricant reliability**: Good filterability indicates that the lubricant will not clog filters, ensuring consistent lubrication and fluid flow.
- **Prevents maintenance issues**: By choosing a lubricant with high filterability, equipment operators can reduce the risk of filter blockages and subsequent maintenance downtime.
- **Water and contaminant resistance**: The test helps identify lubricants that maintain performance even when exposed to water or other contaminants

322. Filtergram

A **filtergram** in lubrication refers to an analytical technique used to assess the presence and characteristics of particulate contaminants within a lubricant. This test involves passing a sample of the lubricant through a specialized filter membrane and then analyzing the particles trapped on the membrane. The process helps to evaluate contamination levels, larger wear debris, and the overall cleanliness of the lubricant, which are critical factors for maintaining equipment health and reliability. Figure F10 shows how filtergrams are useful in discovering larger wear particle debris and how it act as a companion analysis method for other debris detection techniques.

Key Aspects:
- **Visual and microscopic analysis**: The particles collected on the filter membrane are examined using microscopes or imaging techniques to determine their size, shape, composition, and possible source.
- **Contamination assessment**: The test identifies contaminants such as dirt, metal wear particles, or oxidation byproducts, helping to pinpoint potential issues with the lubrication system or the equipment being lubricated.

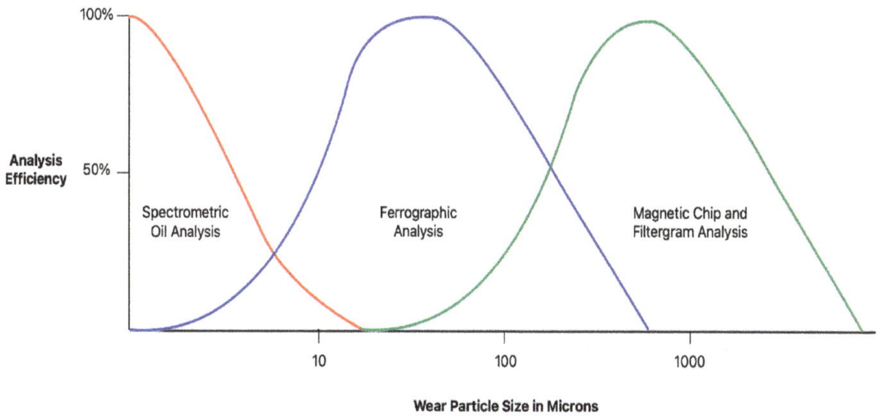

Figure F10 Filtergrams are part of a trio of wear detection methods.

Applications:

- **Condition monitoring**: A filtergram is often used as part of a comprehensive oil analysis program to monitor the condition of lubricants in machinery, hydraulic systems, or engines.
- **Preventive maintenance**: By identifying abnormal wear or contamination early, maintenance teams can take corrective action to prevent equipment failure.

The filtergram technique is a valuable diagnostic tool in predictive maintenance strategies, offering insights into lubricant and machinery health.

323. Filtration

Filtration in lubrication refers to the process of removing contaminants, such as dirt, metal particles, and debris, from a lubricant or hydraulic fluid. Effective filtration is crucial to maintaining the purity of the fluid, ensuring the efficient performance and longevity of the lubricant, machinery, and equipment. Contaminants in lubricants can cause wear, corrosion, and damage to critical components, ultimately leading to equipment failure or reduced efficiency.

Key Elements of Filtration:

1. **Filter media**: The material used in a filter to trap and remove contaminants. It can be made of cellulose, synthetic fibers, metal mesh, or microglass fibers, depending on the application's requirements.

2. **Filtration efficiency**: Measured by the filter's ability to capture and retain contaminants, typically expressed in terms of particle size (in microns) that the filter can effectively remove.

3. **Flow rate**: The rate at which the lubricant passes through the filter. The filtration system must balance high efficiency with an adequate flow rate to ensure consistent lubrication.

Types of Filtration Systems:

- **Full-flow filtration**: The entire fluid volume passes through the filter, commonly used in engines and hydraulic systems.
- **Bypass filtration**: Only a portion of the fluid is filtered at a time, using finer filters to capture smaller particles and contaminants missed by full-flow systems.
- **Magnetic filtration**: Uses magnetic fields to remove ferrous metal particles from the lubricant.

324. Fire Point

Fire point refers to the temperature at which a lubricant or other flammable liquid produces enough vapor to sustain continuous combustion when exposed to an open flame. It is a critical safety parameter that indicates the fluid's potential fire hazard, especially in high-temperature applications.

Key Points:

- **Measurement**: The fire point is determined under controlled laboratory conditions, typically following standard test methods such as ASTM D92 (Cleveland Open Cup method).
- **Higher than flash point**: The fire point is always higher than the flash point, which is the minimum temperature at which the fluid produces enough vapor to ignite briefly. While the flash point only indicates a momentary ignition, the fire point signifies the temperature at which the fluid continues to burn.
- **Safety and risk management**: Understanding the fire point of a lubricant helps in designing safer operating environments, especially in industries where heat and flammable fluids are present, such as manufacturing, automotive, and aerospace.

Applications:

- **Lubricant selection**: Choosing lubricants with appropriate fire points is essential for equipment exposed to high temperatures to minimize the risk of fire.
- **Fire prevention**: The fire point is used to assess the risk of fire in storage, handling, and operational scenarios, guiding the implementation of safety measures.

By considering the fire point in lubricant management, industries can enhance safety protocols and ensure that fluids are used within safe temperature ranges.

325. Fire Cabinet

Fire cabinet refers to a secure storage unit designed to safely house flammable and combustible liquids, including lubricants, chemicals, and fuels. These cabinets are constructed to meet strict safety standards and help prevent the ignition of flammable materials by containing potential fires and shielding the contents from external heat or flames.

Key Features:

- **Fire-resistant construction**: Typically made of heavy-duty steel with double-walled designs and air gaps to insulate against heat.
- **Self-closing doors**: Many fire cabinets are equipped with self-closing mechanisms to minimize the risk of fire spreading.
- **Ventilation**: Cabinets may include built-in vents to prevent the buildup of flammable vapors.
- **Spill containment**: Designed with leak-proof sump trays at the bottom to contain spills and prevent chemical reactions.

Safety Standards and Regulations:

- Fire cabinets must comply with standards set by regulatory agencies, such as the National Fire Protection Association (NFPA) and the Occupational Safety and Health Administration (OSHA).
- The color and labeling of the cabinet may also be regulated, with red and yellow being a common color for flammable liquid storage. See Figure F11 where the lubricant storage cabinet is used for small oil

cans, grease tubes and guns, and paint. The yellow bin on the left is for temporary storage of used oil-soaked rags.

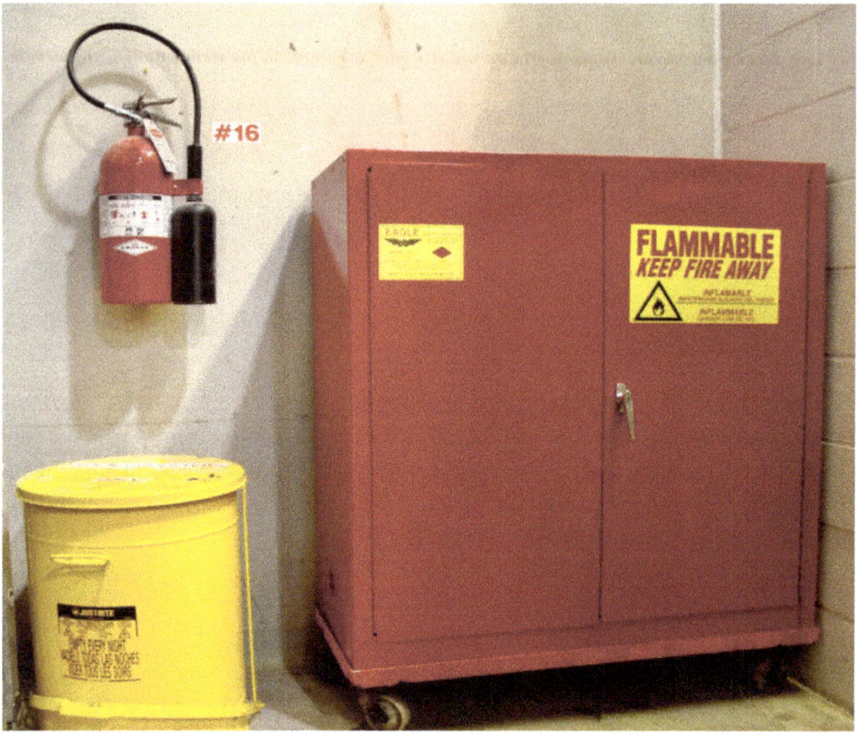

Figure F11 Lubricant and dirty oil rag storage in a food processing plant. Courtesy: ENG**TECH** Industries Inc.

Applications:
Fire cabinets are widely used in industries like manufacturing, automotive, laboratories, and any environment that handles or stores hazardous substances. Proper use of fire cabinets ensures compliance with safety regulations and minimizes the risk of fire-related incidents. By storing flammable materials in a fire cabinet, organizations can enhance workplace safety and reduce the risk of accidental fires or explosions.

326. Fire-resistant Fluid

Fire-resistant fluid refers to a class of specialized hydraulic and lubrication fluids formulated to reduce the risk of ignition and to prevent the spread of fire in high-temperature or hazardous environments. These fluids are used in applications where flammable fluids could pose a significant safety hazard, such as in steel mills, mining operations, and aviation. These fluids have a higher ignition point compared to conventional oils, reducing the risk of catching fire. They are designed to slow down or prevent the spread of fire if exposed to a flame or high heat.

Types of Fire-resistant Fluids:

1. **Water-based fluids**:
   - **HFA (high water content fluids)**: Contains over 90% water, mixed with oil or additives to improve lubrication and corrosion protection.
   - **HFB (water-in-oil emulsions)**: Emulsions with a high percentage of water dispersed in oil, providing both fire resistance and lubrication.
   - **HFC (water glycol fluids)**: Composed of water, glycol, and thickeners, offering good fire resistance and lubrication properties, commonly used in hydraulic systems.
2. **Nonwater-based fluids**:
   - **Phosphate esters**: Synthetic fluids known for their excellent fire resistance and stability at high temperatures, often used in aviation and power generation.
   - **Polyol esters**: Offer good fire resistance and are used in high-performance applications where thermal stability is required.

Applications:

- **Industrial settings**: Used in steel production, foundries, and glass manufacturing, where hot surfaces and open flames are common.
- **Mining equipment**: Essential for hydraulic systems in underground mining to minimize the risk of fire.
- **Aviation**: Utilized in aircraft hydraulic systems to ensure safety in case of leaks near engines or other heat sources.

Fire-resistant fluids are critical in industries where fire hazards are a major concern, providing an added layer of protection for both equipment and personnel. Proper selection and maintenance of these fluids are essential for optimal safety and performance.

327. First In, First Out (FIFO)

First in, first out (FIFO) for lubricants and greases refers to a rotational inventory management practice used to ensure that older stock of lubricants and greases are used **before** newer stock. This method is critical in maintaining product effectiveness and preventing issues related to the aging or degradation of lubricants.

Importance of FIFO in Lubricant and Grease Management:

1. **Shelf-life management**: Lubricants and greases have a finite shelf life, during which their chemical and physical properties remain optimal. Over time, additives can settle or degrade, and the lubricant may lose its effectiveness. Using the oldest stock first helps maintain lubrication quality.

2. **Preventing contamination**: Proper stock rotation minimizes the risk of contamination that can occur from older, poorly stored products.

3. **Consistency in performance**: Ensuring that lubricants are used within their intended lifespan helps keep machinery running efficiently, reducing the risk of equipment failure due to inadequate lubrication.

4. **Cost efficiency**: FIFO helps avoid the financial loss associated with discarding expired or degraded lubricants, making inventory management more cost-effective.

How to Implement FIFO for Lubricants and Greases:

- **Labeling**: Clearly label lubricant containers with the date they were received and the date they should be used by.

- **Organized storage**: Arrange storage areas so that older stock is the most accessible, and new stock is placed behind the existing inventory that is then rotated to the front once the older stick has been used.

- **Regular checks**: Conduct periodic inventory reviews to ensure that FIFO procedures are being followed.

Challenges and Considerations:

- **Storage conditions**: Even with FIFO, improper storage conditions (such as exposure to extreme temperatures or humidity) can affect lubricant quality. Ensure proper storage practices are in place.

- **Monitoring shelf life**: Keep track of manufacturers' recommended shelf life for each lubricant type to ensure timely usage. If the lubricant is out of date, it should be checked by the supplier for "fitness for use."

FIFO is a best practice for any organization that handles large volumes of lubricants and greases, helping maintain product quality and operational efficiency.

328. First In, Last Out (FILO)

First in, last out (FILO) for lubricant products is an inventory management method less commonly used compared to FIFO but might be relevant in specific scenarios when maintaining certain product characteristics or ensuring freshness is not the primary concern. FILO refers to a system where the newest stock is used or accessed first, while the oldest stock remains in storage until later use.

Application of FILO in Lubricants and Greases:

1. **Low-risk applications**: In some industrial applications where the lubricant is stored offsite or for lubricants that have long lead times and minimal degradation or where additives and properties remain stable over extended periods, FILO may be more acceptable.
2. In industries where the logistics of storage make it difficult to access older stock first, such as large-scale industrial settings with complex storage systems.

Potential Challenges of Using FILO:

- **Quality degradation**: Lubricants and greases can degrade over time, even under ideal storage conditions. Additives may separate, and the lubricant may become less effective, potentially leading to equipment failure or inefficiencies.
- **Inconsistent performance**: Using older stock that has aged longer than newer batches may lead to inconsistent lubrication performance, affecting machinery operation.

Mitigation Strategies:

- **Periodic testing**: Regularly test the quality of the lubricant in storage to ensure it meets performance standards.

- **Proper storage conditions**: Maintain optimal storage conditions (e.g., controlled temperature and humidity) to slow the degradation of lubricant products.

While FILO is generally not the preferred method for managing lubricant inventories, understanding its implications can help organizations implement appropriate measures to safeguard lubricant quality and equipment reliability.

329. Flash Point

Flash point is the lowest temperature at which a lubricant or other flammable liquid produces enough vapor to ignite when exposed to an ignition source, such as a flame or spark. It is a critical property for safety in handling, storing, and using lubricants, especially in environments with high heat or open flames.

Key Points:

- **Measurement methods**: Flash points are typically measured using standardized test methods such as the **Cleveland Open Cup (COC)** or **Pensky–Martens Closed Cup (PMCC)**. The open-cup method tends to give slightly higher values compared to the closed-cup method, as it allows for greater vapor dispersion.
- **Safety implications**: Knowing the flash point of a lubricant helps in assessing the fire risk associated with its use and informs appropriate storage and handling practices. A higher flash point indicates greater fire safety.
- **Lubricant performance**: Lubricants with higher flash points are preferable in high-temperature applications, as they are less likely to form flammable vapors.

Applications:

- **Engine oils and hydraulic fluids**: In engines and hydraulic systems, the flash point helps determine the operating temperature range and the risk of vapor ignition.
- **Industrial safety**: In manufacturing or industrial settings, flash points guide fire safety measures, such as ventilation and fire suppression systems.

Practical Example:

- **Engine oil**: If an engine oil has a flash point of 220 °C (428 °F), it should not be used in applications where operating temperatures could approach or exceed this level, as it would increase the risk of fire or explosion

- **Paraffinic oils** have higher flash points than **napthenic oils**.

330. Flocculate

Flocculate refers to the process by which fine particles in a fluid aggregate or clump together to form larger clusters or flocs. This occurs when particles suspended in a liquid bind together due to chemical or physical forces, making them easier to separate or filter out. Flocculation is often induced through the addition of flocculating agents, which can be chemicals that encourage particle aggregation.

Applications in Lubrication and Filtration:

- **Contaminant removal**: In the context of lubrication, flocculation can occur when impurities or contaminants aggregate in the lubricant, which can then be removed through filtration.
- **Oil–water separation**: In industrial processes, flocculation is used to separate water from oil or to clarify fluids by aggregating and removing unwanted particles.

Factors Influencing Flocculation:

- **Chemical agents**: The addition of certain chemicals, such as surfactants or electrolytes, can enhance the flocculation process.
- **Temperature and pH**: Changes in temperature or pH can affect the stability of particles in suspension, promoting or inhibiting flocculation.

331. Flow Improver

Flow improver, also known as a viscosity modifier or pour point depressant, is an additive used in lubricants and oils to enhance their flow characteristics, especially under cold temperature conditions. These additives help maintain optimal fluidity and lubrication efficiency by lowering the temperature at which the lubricant thickens or becomes too viscous to flow.

Key Functions:

- **Viscosity reduction**: Flow improvers reduce the tendency of oil molecules to cluster together in low temperatures, ensuring smoother flow and easier pumping.
- **Cold weather performance**: By improving flow, these additives are crucial for engines and machinery that need to operate efficiently in

cold environments. They prevent issues like oil starvation or poor circulation.

- **Fuel applications**: In the fuel industry, flow improvers are used in diesel fuels to prevent gelling and maintain fuel flow in cold weather.

Applications:

- **Automotive oils**: Enhances the cold-start performance of engines by keeping oil viscous enough to circulate quickly.
- **Hydraulic fluids**: Ensures hydraulic systems function smoothly in colder climates.
- **Pipeline transport**: Used to improve the flow of crude oil in pipelines, reducing pumping costs and minimizing blockages.

Flow improvers are essential for maintaining the reliability and performance of lubricants and fuels in varying temperature conditions, extending the operational range of both machinery and vehicles.

332. Flow Laminar

Laminar flow refers to a smooth, orderly fluid movement where the fluid flows in parallel layers along its shear planes as in Figure F12, with minimal mixing between the layers. In this type of flow, the fluid velocity is constant at any given point and moves in a streamlined fashion, which is typical for lower flow velocities or in situations where the fluid's viscosity dominates over inertia forces. See also Fluid Friction.

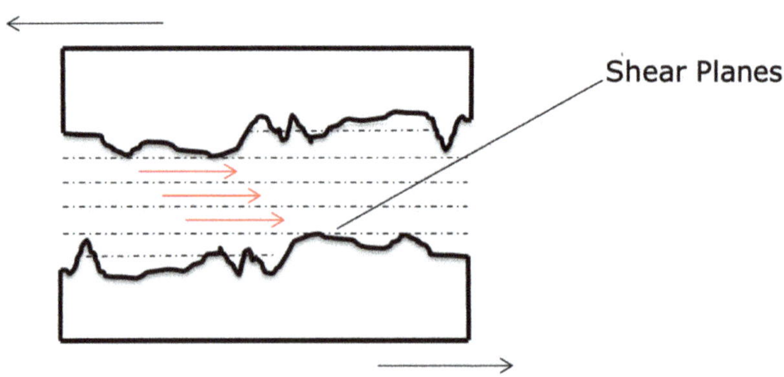

Figure F12 Lubricant laminar flow along shear planes. Courtesy: **ENGTECH** Industries Inc.

Key Characteristics of Laminar Flow:

- **Predictable flow patterns**: The flow is well-structured, and each particle of the fluid follows a specific path, creating a stable and predictable movement.
- **Low Reynolds number**: Laminar flow generally occurs when the Reynolds number (a dimensionless number used to predict flow patterns) is below 2000. The Reynolds number is calculated based on fluid velocity, characteristic length, and kinematic viscosity.
- **Minimal turbulence**: There is little to no turbulence in laminar flow, and energy loss due to internal friction is relatively low compared to turbulent flow.

Applications in Lubrication:

- **Efficient lubrication**: In lubrication systems, maintaining laminar flow ensures a consistent and even distribution of the lubricant, reducing friction and wear on machine components.
- **Hydraulic systems**: Laminar flow is desirable in hydraulic systems to maintain precise control of fluid movement and minimum pressure loss

333. Flow Meter

A **flow meter** is a device used to measure the flow rate or quantity of a fluid moving through a pipe or system. It is an essential tool in various industrial and commercial applications, including lubrication systems, hydraulic systems, and fuel monitoring, where accurate measurement of fluid flow is crucial for optimal operation and efficiency.

Types of Flow Meters:

1. **Positive displacement flow meters**: Measure the volume of fluid passing through by trapping and releasing it in measured amounts. These are ideal for high-viscosity fluids like lubricating oils.
2. **Turbine flow meters**: Use a rotor placed in the path of the fluid flow; the speed of the rotor's rotation is proportional to the flow rate. These are commonly used in water and fuel systems. See Figure F13.
3. **Ultrasonic flow meters**: Use sound waves to measure the velocity of fluid flow, suitable for both clean and dirty fluids.

4. **Magnetic flow meters**: Use a magnetic field to measure flow rate, applicable for conductive fluids like water and certain chemicals.

Applications in Lubrication:

- **Monitoring lubricant flow**: Ensures that the correct amount of lubricant is delivered to machinery, preventing overlubrication or underlubrication, which can lead to equipment damage.
- **Hydraulic systems**: Flow meters help maintain optimal fluid movement and pressure in hydraulic applications.
- **Fuel systems**: Used to monitor fuel consumption and efficiency in engines and industrial machinery.

Figure F13 Typical digital turbine flow meter. Source: Unknown.

334. Flow Rate

Flow rate refers to the volume of fluid that passes through a point in a system over a given period. It is commonly measured in units such as liters per minute (L/min), gallons per minute (GPM), or cubic meters per second (m^3/s), depending on the application.

Formula for Flow Rate:

The flow rate can be calculated using the equation:

$$Q = A \times v.$$

where:

- Q = Flow rate
- A = Cross-sectional area of the pipe or channel
- v = Velocity of the fluid

Importance of Flow Rate in Lubrication:

- **Proper lubricant distribution**: Ensures that the right amount of lubricant reaches all parts of a system, preventing wear and extending equipment life.
- **System performance**: Monitoring flow rate helps maintain optimal operation in hydraulic and lubrication systems, preventing overheating or inefficiencies.
- **Safety**: In processes involving fuel or chemical flows, accurate flow rate monitoring is crucial for avoiding hazardous conditions.

Applications:

- **Industrial machinery**: Used to maintain the correct flow of lubricants and coolants.
- **Hydraulic systems**: Essential for controlling the movement and force generated by hydraulic fluids.
- **Pumping systems**: Ensures efficient fluid transfer in pipelines and storage tanks.

335. Fluid Film

Fluid film in the context of lubrication refers to a thin layer of lubricant that separates two moving surfaces, reducing friction and preventing direct metal-to-metal contact. This film forms due to the lubricant's viscosity and the relative motion of the surfaces, creating a hydrodynamic or elastohydrodynamic effect, depending on the load and speed.

Types of Fluid Film Lubrication:

1. **Hydrodynamic lubrication**: Occurs when a thick fluid film completely separates the moving surfaces, typically under high-speed and low-load conditions. The pressure within the fluid film supports the load, preventing surface contact.

2. **Elastohydrodynamic lubrication (EHL)**: This type of lubrication occurs in conditions of high load and moderate to high speeds, where the surfaces slightly deform elastically under pressure, and a thin lubricant film still manages to separate them.

3. **Boundary lubrication**: In situations where the fluid film is very thin, boundary additives are required to prevent surface contact, as fluid film lubrication cannot be maintained.

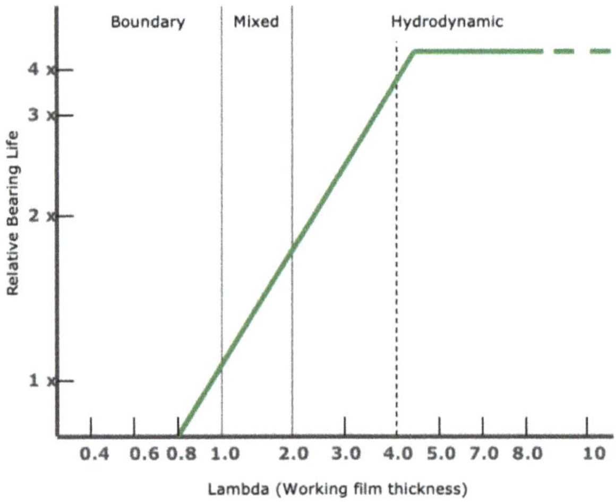

Figure F14 Working film thickness versus relative working life. Courtesy: ENGTECH Industries Inc.

Figure F14 demonstrates that only four-micron film thickness is required to achieve a full separation hydrodynamic state.

Importance of Fluid Film Lubrication:

- **Reduces friction and wear**: The fluid film minimizes contact between surfaces, reducing friction and preventing wear.

- **Heat dissipation**: The lubricant film helps in transferring and dissipating heat generated by friction.

- **Protects against corrosion**: By keeping surfaces separated and reducing exposure to contaminants, fluid film lubrication helps protect against corrosion.

336. Fluid Friction

Fluid friction refers to the resistance that occurs when layers of a fluid move relative to one another. It is a type of friction that arises within a fluid itself (internal friction) or when a fluid flows past a solid surface (external friction). Lubricants or fluid films are not themselves "frictionless" as they rely on an action known as "shearing" in which fluid friction takes place between the molecular shear planes of the lubricant as they move across one another when the load moves. See Figure F15.

Fluid, or "wet" friction increases as a lubricant's viscosity gets thicker, or as the lubricant becomes oxidized and dirty. Although a small amount of energy is required to overcome fluid friction, it is negligible in comparison to having no fluid film present. Fluid friction plays a significant role in various applications, especially in lubrication and fluid dynamics.

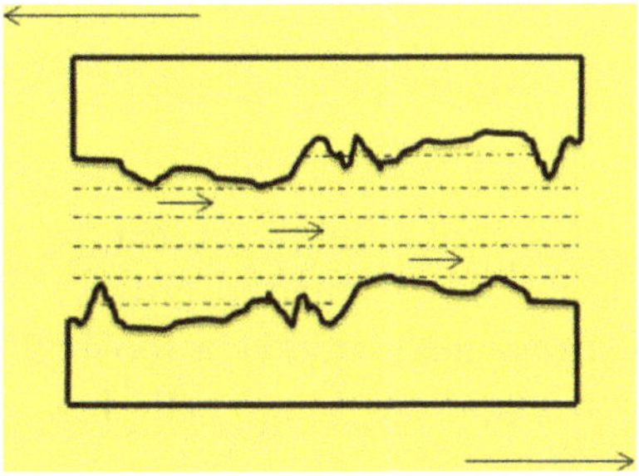

Figure F15 Fluid friction is present when the fluid shears along its shear planes. Courtesy: ENG**TECH** Industries Inc.

Key Aspects of Fluid Friction:

- **Internal friction**: Occurs between layers of the fluid as they slide past each other. The viscosity of the fluid determines the level of internal friction. For example, honey has higher internal friction compared to water because it is more viscous.
- **External friction**: Takes place when a fluid moves over a solid surface, such as oil flowing through a pipeline or water flowing over a boat's hull. The roughness of the surface and the viscosity of the fluid affect the external fluid friction.

Role in Lubrication:

- **Reducing wear**: In lubrication systems, fluid friction is managed to ensure that moving parts are adequately separated, reducing wear and energy loss.
- **Efficiency**: Fluid friction can affect the efficiency of machines and engines. Minimizing excessive friction through appropriate lubricants helps improve energy efficiency and the lifespan of components.

Factors Affecting Fluid Friction:

- **Viscosity**: Higher viscosity fluids experience greater internal friction.
- **Flow velocity**: As the velocity of the fluid increases, the frictional forces may increase, especially in turbulent flow conditions.
- **Surface roughness**: A rough surface increases the friction experienced by the fluid in contact with it.

337. Fluid Grease

Fluid grease is a type of semifluid lubricant that has a consistency between oil and traditional grease. It is designed to provide effective lubrication in applications where traditional grease might be too thick and oil might not provide adequate protection. Fluid grease typically has a National Lubricating Grease Institute (NLGI) consistency grade between 00 and 0, which allows it to flow more easily compared to standard greases.

Key Characteristics:

- **Flowability**: Fluid grease is less viscous than regular grease, allowing it to flow under moderate pressure, making it suitable for automatic lubrication systems and enclosed gearboxes.
- **Adhesion**: It provides good adhesion to metal surfaces, ensuring that the lubricant stays in place, even under moderate pressure and temperature fluctuations.
- **Protection**: Offers protection against wear, corrosion, and contamination, making it suitable for applications with high loads and harsh environments.

Applications:

- **Centralized lubrication systems**: Fluid grease is ideal for use in centralized lubrication systems that distribute grease to multiple points within a piece of machinery.

- **Enclosed gearboxes**: It is often used in porous casting gearboxes that require lubrication but where oil would leak out or not provide sufficient film strength
- **Automotive and industrial equipment**: Used in applications like heavy-duty trucks, construction equipment, and certain industrial machinery where a consistent and flowable lubricant is needed.

338. Fluid Opacity

Fluid opacity refers to the degree to which a fluid blocks light from passing through it, affecting its transparency or clarity. In lubrication and fluid management contexts, opacity can be an indicator of contamination, degradation, or the presence of suspended particles within the fluid.

Key Points About Fluid Opacity:

- **Clarity as an indicator**: In many industrial applications, the clarity or opacity of a fluid can provide visual clues about its condition. For example, fresh lubricants are generally clear, while used or contaminated oil may appear cloudy or dark.
- **Causes of increased opacity**: The opacity of a fluid may increase due to the accumulation of contaminants such as dirt, metal particles, oxidation byproducts, or water. The presence of emulsified water or microbial growth can also contribute to a milky or cloudy appearance.
- **Measuring opacity**: Opacity is often assessed visually or using optical sensors to determine whether a fluid needs replacement or further testing.

Importance in Lubrication:

- **Monitoring fluid health**: Regularly checking the opacity of lubricants can help in the early detection of issues like contamination or oil breakdown, enabling proactive maintenance and preventing equipment damage.
- **Quality control**: Ensuring that fluids meet specific opacity standards is essential for applications where cleanliness and performance are critical.

339. Fluid Target Cleanliness Level

Fluid target cleanliness levels refer to specific standards or benchmarks set for the acceptable level of particulate contamination in a lubricant or hydraulic fluid. Maintaining the appropriate cleanliness levels is crucial for the performance and longevity of machinery, as contamination can lead to increased wear, component damage, and system inefficiencies.

Key Concepts:

- **ISO Cleanliness Codes**: One of the most common ways to specify target cleanliness levels is through the ISO 4406 standard, which classifies fluid cleanliness based on the number of particles of specific sizes (usually ≥4 μm, ≥6 μm, and ≥14 μm) per milliliter of fluid. For example, as depicted in Figure F16, a cleanliness code of 18/16/13 means:
    - **18**: Number of particles ≥4 μm per milliliter.
    - **16**: Number of particles ≥6 μm per milliliter.
    - **13**: Number of particles ≥14 μm per milliliter.

Target cleanliness levels are useful for evaluating equipment based on the following:

1. **Equipment sensitivity**: The required cleanliness level depends on the sensitivity of the components. For instance, high-precision hydraulic systems or servo valves require cleaner fluid than less sensitive systems.
2. **Operating environment**: Machinery operating in dirty or harsh environments may require more stringent filtration to achieve the desired cleanliness levels.
3. **Manufacturer recommendations**: Many equipment manufacturers specify target cleanliness levels to ensure optimal operation and warranty compliance.

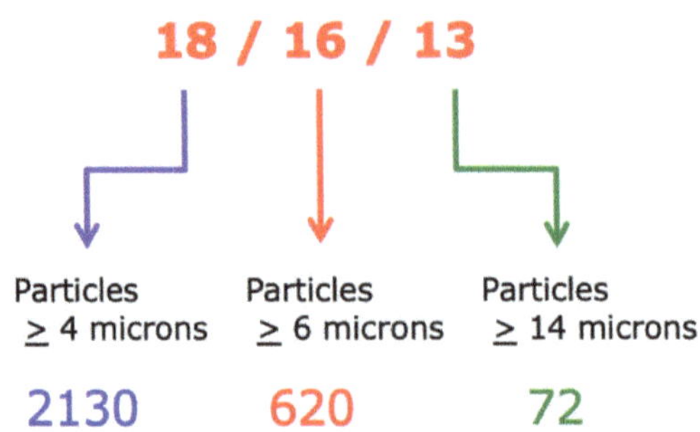

Figure F16 ISO 4406 cleanliness code interpretation.

These levels differ based on the system or equipment being analyzed, for example:

- **Hydraulic systems**: Typical target levels might range from **16/14/11 to 18/16/13**, depending on the system's pressure and precision requirements.
- **Turbine lubrication**: Often requires cleanliness levels around **17/15/12** or better to prevent bearing and gear damage.
- **Gearboxes**: May have a target cleanliness level of **19/17/14**, depending on the load and speed.

340. Fluidity

Fluidity refers to the ability of a fluid to flow easily. It is essentially the measure of how effortlessly a liquid can move and spread under the influence of external forces, such as gravity or pressure. Fluidity is inversely related to viscosity, which is the resistance of a fluid to flow; a higher fluidity means a lower viscosity and vice versa.

Key Points About Fluidity:

- **Low viscosity, high fluidity**: Fluids like water and light oils have high fluidity because they flow freely and quickly.
- **High viscosity, low fluidity**: Substances like honey or grease have low fluidity, as they resist flow and move more slowly.

Importance in Lubrication:

- **Temperature dependence**: The fluidity of lubricants often changes with temperature. As temperature increases, lubricants generally become more fluid (less viscous), which can be beneficial or detrimental depending on the application.
- **Performance**: High fluidity is essential for lubricants that need to reach tight, intricate areas quickly. Conversely, lower fluidity may be preferred for applications requiring a more stable film to reduce wear.

341. Flushing Oil

Flushing oil is a special type of oil used to clean and remove sludge, deposits, contaminants, and residual oil from oil lubrication delivery systems, oil reservoirs, gearboxes, engines, or other machinery. It is often used during maintenance procedures, particularly before refilling a system with fresh oil, to ensure the internal components are free from harmful buildup. It is also used to clean out newly installed lubrication delivery systems to flush out any internal debris prior to initial start-up.

Key Characteristics:

- **Light viscosity**: Flushing oil is usually less viscous than standard lubricating oils, allowing it to flow easily and flush out contaminants.
- **Detergent additives**: Often contains detergents and dispersants to break down and suspend sludge and deposits, facilitating effective removal.

In Industrial applications flushing oils are used in gearboxes, hydraulic systems, and other machinery where contamination or sludge buildup could impact performance. They are also used in automotive engines to cleanse internal parts before an oil change, especially in high-mileage vehicles or engines that have experienced oil degradation.

Typical Flushing Procedure:

1. **Drain the old oil**: The old or contaminated oil is completely drained from the system.
2. **Add flushing oil**: The flushing oil is introduced and the engine or machinery is run for a specified amount of time to circulate the oil and clean internal components.
3. **Drain and replace**: The flushing oil, along with suspended contaminants, is drained, and fresh lubricant is added.

Benefits:

- **Enhanced performance**: Removes sludge and contaminants that can hinder machinery efficiency.
- **Prolonged equipment life**: By keeping internal components clean, flushing oil helps prevent wear and extend the lifespan of engines and machinery.
- **Preparation for fresh oil**: Ensures that the new oil operates in a clean environment, providing optimal protection and performance.

342. FMEA

See "Failure Mode Effect Analysis."

343. Foam

Foam in the context of lubrication refers to the formation of air bubbles trapped in the oil or lubricant, creating a frothy or foamy layer on the surface. This phenomenon can lead to significant issues in lubrication systems, as it reduces the efficiency of the lubricant, impairs its ability to form a consistent protective film, and can cause cavitation or damage to equipment.

Causes of Foam Formation:

1. **Air entrapment**: Air can become entrained in the oil due to agitation, rapid fluid movement, or leaks in the system, leading to bubble formation.
2. **Contaminants**: The presence of water, dirt, or incompatible additives can exacerbate foaming.
3. **High viscosity oils**: Thicker oils are more prone to trapping air and forming foam.

Problems Caused by Foam:

- **Reduced lubrication efficiency**: Foam can prevent proper oil film formation, increasing the risk of wear and overheating in machinery.
- **Oxidation**: The presence of air in the oil accelerates oxidation, leading to sludge and varnish formation.
- **Cavitation**: In hydraulic systems, foam can cause cavitation, which may damage pumps and other components.

Solutions to Prevent or Control Foam:

- **Antifoam additives**: These additives help break down air bubbles, reducing foam formation and promoting quick release of trapped air.
- **Proper system design**: Ensuring appropriate oil levels, minimizing turbulence, and using proper reservoir venting can help reduce foam.
- **Regular maintenance**: Monitoring for leaks and contaminants, and ensuring proper oil types are used can help prevent foaming issues.

344. Foam Inhibitor

See "Antifoam Additive."

345. Foam Tendency

Foam tendency refers to the propensity of a lubricant or fluid to form foam when subjected to mechanical agitation or turbulent conditions. It is an important characteristic to evaluate, as excessive foaming in a lubricant can lead to performance issues and damage to machinery.

Factors Influencing Foam Tendency:

1. **Viscosity**: Higher viscosity fluids are more likely to trap and retain air bubbles, leading to greater foam formation.
2. **Contamination**: The presence of water, dirt, or incompatible additives can increase the likelihood of foam formation.
3. **Additive formulation**: Some additives can stabilize or destabilize the foam. Antifoam additives are often used to minimize foam tendency.
4. **Temperature**: Elevated temperatures can increase foam tendency, as the oil becomes thinner and more prone to aeration.

Measuring Foam Tendency:

Foam tendency is typically assessed using standardized tests, such as the **ASTM D892** test method. This test evaluates the foam formation characteristics of a lubricant at different temperatures and under controlled conditions.

Impact of High Foam Tendency:

- **Reduced lubrication**: Foam can prevent the formation of a continuous lubricant film, leading to increased wear and friction.
- **Oxidation acceleration**: The increased exposure of the oil to air promotes oxidation, degrading the lubricant more rapidly.
- **Pump cavitation**: In hydraulic systems, foam can cause cavitation, leading to damage and reduced efficiency.

Solutions to Minimize Foam Tendency:

- **Use antifoam additives**: These additives help to reduce foam formation and promote the quick release of entrained air.
- **Proper maintenance**: Regularly inspect and maintain systems to prevent contamination and ensure that seals and filters are functioning correctly.

- **System design**: Design the system to minimize turbulence and optimize oil reservoir venting.

346. Follower Plate (Grease)

A **follower plate** is a component used in lubricant reservoir pump systems to assist in the efficient dispensing of grease or other heavy lubricants from a container, such as a drum or pail. The follower plate sits directly on top of the grease and moves downward as the lubricant is pumped out, maintaining pressure and ensuring a continuous, air-free flow.

Key Functions of a Follower Plate:

- **Maintains pressure**: The follower plate (see lower disc plate in Figure F17) applies consistent pressure to the grease, ensuring that it is evenly distributed and preventing the formation of air pockets, which could disrupt the lubrication process.
- **Prevents contamination**: By covering the grease, the follower plate acts as a protective barrier that keeps dirt, dust, and moisture from contaminating the lubricant.
- **Ensures complete usage**: The plate helps to push all of the grease out of the container, minimizing waste and ensuring that every bit of lubricant is utilized.

Applications:

- **Industrial and heavy machinery**: Commonly used in grease pumps for large industrial equipment, construction machinery, and centralized lubrication systems that require reliable, high-pressure grease distribution.
- **Automated grease systems**: Follower plates are often used in automated or centralized lubrication systems where precise grease application is critical.

Advantages:

- **Efficient dispensing**: The even pressure provided by the follower plate ensures that the grease is dispensed smoothly and consistently.
- **Reduced air entrainment**: By eliminating air pockets, the follower plate helps maintain the integrity of the grease and ensures effective lubrication.

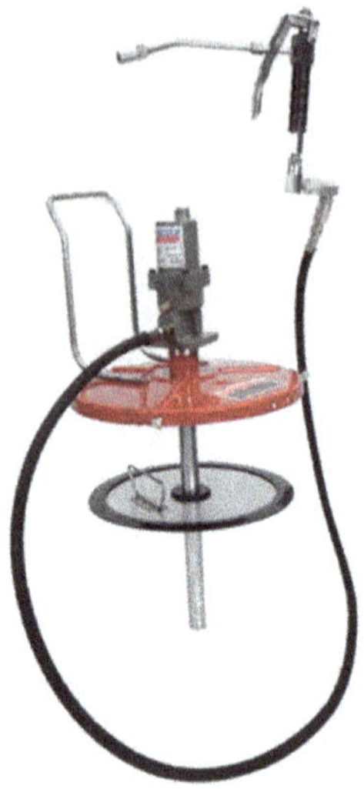

Figure F17 Lincoln lube HP grease bucket pump with follower plate. Source: LPT System Product Catalog.

347. Food Grade Lubricant

Food grade lubricant refers to a type of lubricant that is specially formulated for use in food processing and handling equipment where incidental contact with food may occur. These lubricants are designed to meet strict safety standards and regulations to ensure they are safe for use in environments where contamination could pose a risk to human health.

Key Characteristics of Food Grade Lubricants:

- **Nontoxic and safe**: Formulated to be safe if they come into incidental contact with food, food-grade lubricants are typically made from FDA-approved ingredients and meet standards set by organizations like NSF International.
- **NSF classifications**:
  - **H1**: Lubricants approved for incidental food contact.
  - **H2**: Lubricants used in areas where there is no chance of contact with food.
  - **H3**: Soluble oils used to prevent rust on hooks, trolleys, and similar equipment that may come into contact with food.

- **Odorless and tasteless**: They do not affect the taste or smell of the food if incidental contact occurs.
- **High performance**: Despite being nontoxic, these lubricants are formulated to provide excellent performance, including resistance to heat, cold, water, and other harsh conditions typically found in food processing environments.

Applications:
- **Food processing equipment**: Used in mixers, conveyors, packaging machines, and other equipment where lubricant could potentially come into contact with food.
- **Beverage manufacturing**: Applied to components in bottling and canning lines.
- **Pharmaceuticals and cosmetics**: Sometimes used in manufacturing processes to ensure product purity.

Regulations and Standards:
- **NSF International**: Food-grade lubricants are often registered under NSF H1, meaning they are safe for incidental food contact.
- **FDA compliance**: Ingredients must comply with FDA regulations outlined in 21 CFR for lubricants with incidental food contact.

348. Force-feed Lubrication

Force-feed lubrication is a system used to deliver lubricants under pressure to various parts of a machine or engine. Unlike gravity-fed or splash lubrication systems, force-feed systems ensure that oil is consistently and evenly distributed to components, even under high loads or extreme operating conditions. This is crucial in high-performance machinery where reliable lubrication is essential to prevent friction and wear. See also "Automatic Lubrication System" entry – Figure A10.

How It Works:
- **Pump mechanism**: A pump forces the lubricant through a series of channels or pipes, directing it to specific points where lubrication is needed, such as bearings, gears, or other moving parts.

- **Pressurized delivery**: The lubricant is delivered under pressure, ensuring it reaches areas that are difficult to access, even against gravity or through narrow spaces.
- **Recirculation**: In some systems, the lubricant is collected and recirculated back to the reservoir, filtered, and reused.

Advantages:

1. **Consistent lubrication**: Ensures that critical components receive a continuous supply of lubricant, reducing the risk of wear or overheating.
2. **Effective in high-speed applications**: Suitable for engines and machinery that operate at high speeds, where splash or gravity systems might be insufficient.
3. **Precise distribution**: Can be tailored to provide lubrication exactly where it is needed, improving the efficiency and longevity of the equipment.

349. Four Ball Method

Four ball method is a widely used test procedure for evaluating the antiwear and extreme pressure (EP) properties of lubricants. This method involves the use of four steel balls to simulate and measure the lubricant's ability to protect metal surfaces under high-pressure contact conditions. The test results are essential for understanding the performance of lubricants in applications that involve heavy loads or severe operating conditions.

How It Works:

1. **Setup**: Three steel balls are clamped together in a fixed position, forming a cradle, while a fourth ball is placed on top and pressed against the three stationary balls. The fourth ball rotates, creating a high-pressure contact point. See Figure F18.
2. **Antiwear test**: The lubricant is applied to the contact area, and the test apparatus rotates the top ball under a specified load, speed, and duration. After the test, the wear scar diameter on the stationary balls is measured to determine the lubricant's antiwear performance.
3. **Extreme pressure (EP) test**: In the EP test, the load on the balls is gradually increased until the lubricant film fails, and the metal surfaces weld together. The test measures the lubricant's load-carrying capacity, providing insights into its EP properties.

Key Parameters Measured:

- **Wear scar diameter**: Indicates the effectiveness of the lubricant in preventing wear. Smaller wear scars suggest better antiwear performance.
- **Weld point**: The load at which the balls weld together, indicating the maximum pressure the lubricant can withstand before failing.
- **Load wear index (LWI)**: A measure of the lubricant's ability to withstand increasing loads without failing.

Applications:

- **Engine oils and gear lubricants**: The four ball method helps determine the suitability of lubricants for protecting engine and gearbox components under extreme conditions.
- **Greases and hydraulic fluids**: Used to evaluate the performance of greases and hydraulic fluids in heavy machinery and industrial applications.

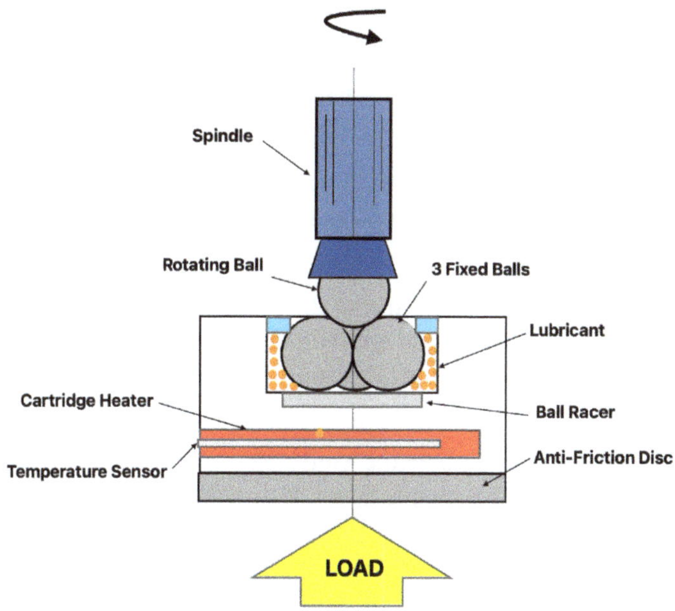

Figure F18 Four ball test setup. Source: Plant Lubrication for Industrial Facilities – Fourth Edition 2024 Kenneth E. Bannister.

The four ball method is crucial for formulating and selecting lubricants that will perform reliably in high-stress environments, ensuring the longevity and efficiency of mechanical components.

350. Fourier Transform Infrared Spectroscopy (FTIR)

Fourier transform infrared spectroscopy (FTIR) is an analytical technique used to identify organic, polymeric, and some inorganic materials. In the context of lubrication, FTIR is employed to analyze oil samples for contaminants, oxidation levels, additive depletion, and other chemical changes. It provides insights into the health of the lubricant and helps in predictive maintenance by identifying potential issues before they lead to equipment failure.

How FTIR Works:

FTIR measures the infrared absorption of a sample across a range of wavelengths. Each type of chemical bond absorbs infrared light at a specific frequency, creating a unique spectral fingerprint. By analyzing this fingerprint, FTIR can determine the chemical composition of a lubricant and detect any changes or contaminants present.

Key Applications in Lubrication:

1. **Oxidation monitoring**: FTIR detects oxidation products, which indicate lubricant degradation. This helps in assessing whether the oil needs to be replaced or if corrective actions are necessary.
2. **Contaminant identification**: It can identify the presence of water, glycol, soot, fuel dilution, and other contaminants that affect lubricant performance.
3. **Additive depletion**: FTIR tracks the depletion of important additives, such as antiwear agents and antioxidants, providing information on the remaining useful life of the lubricant.

Advantages of FTIR in Lubricant Analysis:

- **Nondestructive**: The sample is not destroyed during analysis, allowing for further testing if needed.
- **Rapid results**: FTIR provides fast and accurate results, making it suitable for routine oil analysis and monitoring
- **Comprehensive information**: Offers a broad analysis of chemical changes and contamination levels, helping to maintain equipment reliability and performance.

351. Free Air

Free air in the context of lubrication and fluid systems refers to the air present in its uncompressed state, as it exists under normal atmospheric conditions. In lubrication and hydraulic systems, managing free air is important to maintain system efficiency and prevent performance issues.

Key Points About Free Air in Lubrication Systems:

1. **Contamination risk**: The presence of free air in a lubrication system can lead to contamination, increasing the risk of oxidation and degradation of the lubricant.
2. **Reduced efficiency**: Free air can cause cavitation and decreased efficiency in hydraulic systems and pumps, potentially leading to damage or malfunction.
3. **Foam formation**: When free air mixes with lubricants, it can lead to the formation of foam, reducing the lubricant's ability to form a protective film between surfaces.

Proper system design and maintenance practices, such as the use of deaeration devices and antifoam additives, help to minimize the impact of free air and ensure optimal system performance.

352. FRL

See "Filter, Regulator, Lubricator."

353. Fretting

Fretting refers to a type of wear damage that occurs when two contacting surfaces experience small, repeated relative movements, typically under load. This form of wear is characterized by the generation of fine debris particles, often consisting of oxidized material, which can accelerate wear and cause surface damage over time.

Key Characteristics of Fretting:

- **Micro-motions**: The small, oscillatory movements between the two surfaces lead to material removal or surface degradation.

- **Oxidation**: The wear debris generated often oxidizes, forming abrasive particles that can exacerbate surface damage.
- **Surface damage**: Fretting typically results in pitting, micro-cracking, and a roughened surface, which can compromise the structural integrity of the material.

Common Occurrences:

- **Bearings and joints**: Fretting is often seen in bearings, splines, or joints where there is a small amplitude of movement between parts.
- **Bolted connections**: It can also occur in bolted or clamped assemblies subjected to vibration or fluctuating loads.
- **Contact points**: Fretting damage frequently appears where surfaces come into contact under pressure and vibration, such as in electrical connectors or shaft couplings.

Prevention Strategies:

- **Lubrication**: Applying the appropriate lubricant can help reduce friction and minimize fretting wear.
- **Surface treatments**: Techniques like surface hardening or coatings can increase resistance to fretting.
- **Design modifications**: Reducing relative motion between components through design changes can mitigate fretting.

354. Friction

Friction is the resistance to motion that occurs when two surfaces slide or attempt to slide across each other. It is a fundamental force that affects the efficiency of mechanical systems and is a key factor in lubrication science. Webster's dictionary describes friction as "the force, which opposes the movement of one surface sliding or rolling over another with which it is in contact." Simply put, friction is the resistive force that retards motion. Friction can be categorized into different types. See Figure F19.

Types of Friction:

1. **Static friction**: The force that keeps an object at rest and prevents it from moving until a certain threshold is overcome.
2. **Kinetic (sliding) friction**: The resistance encountered when one surface slides over another, typically lower than static friction.

3. **Rolling friction**: The force resisting the motion of a rolling object, such as a wheel or ball, across a surface.

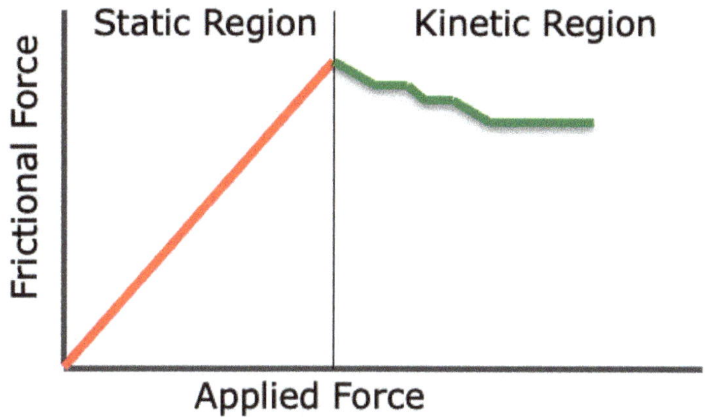

Figure F19 Static friction becomes kinetic friction as the object begins to move. Courtesy: ENGTECH Industries Inc.

Relevance to Lubrication:

- **Lubricants** are used to reduce friction between surfaces, minimizing wear and energy loss. By forming a protective film, lubricants decrease the direct contact between metal surfaces, thus lowering both friction and the resulting heat.
- **Coefficient of friction**: This is a dimensionless value that represents the ratio of the force of friction between two bodies to the normal force pressing them together. Different lubricants are chosen based on their ability to lower this coefficient under specific operating conditions.

Applications:

- **Mechanical systems**: Reducing friction improves energy efficiency and extends the lifespan of machinery.
- **Automotive**: In engines, minimizing friction between moving parts increases performance and reduces fuel consumption.

To understand friction, we must recognize that there are two unchanging fundamental laws that govern friction, these being:

- **Friction** varies directly with load
- **Friction** is independent of surface area

355. Friction Control

Friction control involves methods and strategies used to manage and minimize the resistance between two contacting surfaces to improve performance and reduce wear and energy loss in mechanical systems. Effective friction control is essential in industrial machinery, automotive applications, and various equipment to maintain efficiency and extend component life.

Methods of Friction Control:

1. **Lubrication**: Applying lubricants (like oils or greases) reduces direct contact between surfaces, forming a film that minimizes friction and prevents wear.
2. **Surface treatments**: Techniques such as coatings, surface texturing, or hardening are used to reduce friction or increase wear resistance.
3. **Material selection**: Using materials with low friction coefficients, such as PTFE (Teflon) or ceramics, can help control friction in specific applications.
4. **Design optimization**: Engineering components to have smoother surfaces, appropriate tolerances, and proper alignment can minimize unnecessary friction.

Importance of Friction Control:

- **Efficiency improvement**: Reducing friction conserves energy and enhances the efficiency of engines, gear systems, and mechanical operations.
- **Wear reduction**: Minimizing friction decreases wear, prolonging the lifespan of components and reducing the need for frequent maintenance or replacements.
- **Heat management**: Friction generates heat, which can damage components. Controlling friction helps in thermal management and prevents overheating.

356. Friction Modifier Additive

Friction modifiers are chemical additives used in lubricants to reduce friction between surfaces, particularly under conditions where boundary lubrication occurs. These additives improve the efficiency of engines and machinery by minimizing energy loss due to friction, thus enhancing fuel economy and performance while reducing wear.

How Friction Modifiers Work:
Friction modifiers create a low-friction film on metal surfaces, which reduces the direct contact between them. This film can be formed through physical adsorption or a chemical reaction with the metal surface. They are particularly effective in situations where the lubricant film is thin and metal-to-metal contact is likely.

Common Types of Friction Modifiers:
1. **Organic friction modifiers**: These include fatty acids, esters, and amides, which are often derived from natural sources like vegetable oils.
2. **Molybdenum compounds**: Molybdenum disulfide (MoS_2) and other molybdenum-based additives are used for their exceptional friction-reducing properties.
3. **Graphite and other solid lubricants**: These are used in greases and certain oils to provide long-lasting friction reduction.

Applications:
- **Industrial equipment**: Reduces energy consumption and wear in machinery operating under heavy loads or extreme conditions.
- **Automotive**: Used in engine oils, transmission fluids, and gear lubricants to improve fuel efficiency and performance.

357. EPA

See "Fourier Transform Infrared Spectroscopy."

358. Fuel Contamination

Fuel contamination refers to the presence of unwanted substances in fuel, which can degrade its quality and adversely affect the performance and efficiency of engines and fuel systems. Contaminants can include water, dirt, microbial growth, fuel degradation products, and other impurities.

Common Types of Fuel Contaminants:
1. **Water**: Water can enter fuel systems through condensation, leaks, or poor handling practices. It can cause corrosion, promote microbial growth, and reduce the fuel's efficiency.
2. **Particulates**: Dirt, dust, and metal particles can enter fuel systems during refueling or from degraded components. These contaminants can clog fuel filters and injectors, leading to engine damage.

3. **Microbial growth**: Bacteria and fungi can thrive in fuel, especially in the presence of water, creating sludge and acids that damage the fuel system.
4. **Gums and varnishes**: These are the result of fuel oxidation or degradation, which can leave sticky residues that clog fuel lines and filters.

Effects of Fuel Contamination:

- **Reduced engine performance**: Contaminants can impair fuel combustion, causing loss of power, rough running, or misfires.
- **Increased wear and corrosion**: Abrasive particles and corrosive substances can damage fuel injectors, pumps, and other engine components.
- **Fuel system blockages**: Clogged filters and injectors can restrict fuel flow, leading to engine stalling or failure.
- **Microbial contamination issues**: Microbes can produce sludge and acids, which can corrode fuel tanks and components.

Prevention and Management:

1. **Proper storage**: Store fuel in clean, sealed containers to prevent moisture and dirt from entering.
2. **Regular maintenance**: Inspect and clean fuel tanks, filters, and lines regularly.
3. **Water separators**: Use water separators to remove water from the fuel system.
4. **Biocides**: Apply biocides to prevent microbial growth, especially in diesel fuels.

Regular monitoring and quality control are essential to prevent fuel contamination and ensure the longevity and efficiency of engines and fuel systems.

359. Full Film Lubrication – Hydrodynamic (HDL)

Full film lubrication – hydrodynamic, also known as **hydrodynamic lubrication** or **HDL**, refers to a lubrication regime in which a complete, continuous film of lubricant separates two moving surfaces, preventing direct metal-to-metal contact. This film is thick enough to fully support the load applied to the surfaces, reducing friction and minimizing wear.

How a Full Film Lubrication Works:

- **Fluid pressure**: As one surface moves relative to the other, the lubricant is drawn into the converging gap between the surfaces, creating a pressure that supports the load.
- **Film thickness**: The thickness of the lubricant film is sufficient to accommodate surface roughness, ensuring no direct contact between the surfaces. See Figure F20.

Key Factors Affecting Full Film Lubrication:

1. **Viscosity**: The lubricant's viscosity must be appropriate to form a stable film. Higher viscosity provides better film formation but may increase resistance to motion.
2. **Speed**: Higher relative speeds between surfaces enhance the formation of a full lubricant film.
3. **Load**: The system must be designed so that the lubricant film can support the applied load without breaking down.

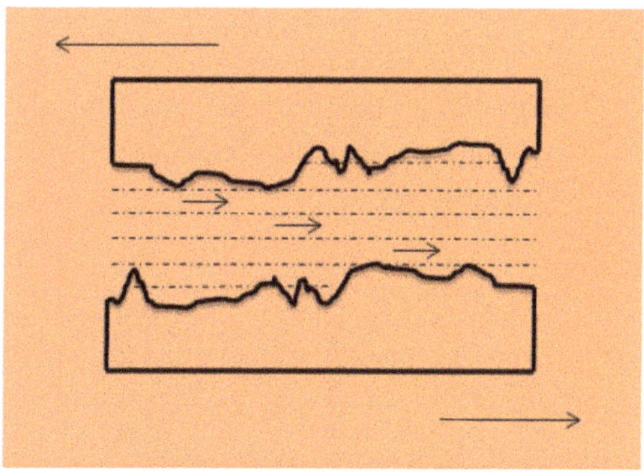

Figure F20 Full film lubrication with full separation under movement.

Benefits of Full Film Lubrication:

- **Reduced friction**: The lubricant film minimizes friction between surfaces, improving energy efficiency.
- **Minimized wear**: Prevents metal-to-metal contact, reducing the risk of surface wear and extending the lifespan of components.
- **Heat dissipation**: The lubricant helps to dissipate heat generated during operation, maintaining optimal temperature conditions.

To combat friction and wear successfully, a lubricant film must be present at all times between the mating bearing surfaces. The degree of protection and subsequent bearing surface life are directly relevant to the lubricant's working film thickness, load, speed, and lubricant viscosity.

360. Full Film Lubrication – Hydrostatic (HSL)

Full film lubrication – hydrostatic (**HSL**) refers to a type of lubrication regime where an external pressure source, such as a pump, is used to maintain a continuous lubricant film between two surfaces. Unlike **hydrodynamic lubrication**, where the lubricant film is generated by the relative motion of the surfaces, hydrostatic lubrication provides a film that is independent of motion. This means that a full fluid film can be maintained even when the surfaces are stationary or moving at very low speeds. See Figure F21.

How Hydrostatic Lubrication Works:

- **External pressure supply**: A pump delivers lubricant under high pressure into the space between the surfaces. This pressure creates a film thick enough to separate the surfaces, supporting the load and preventing metal-to-metal contact.
- **Load carrying capacity**: The pressure and film thickness are maintained regardless of the speed or load variations, as long as the pump continues to supply lubricant.

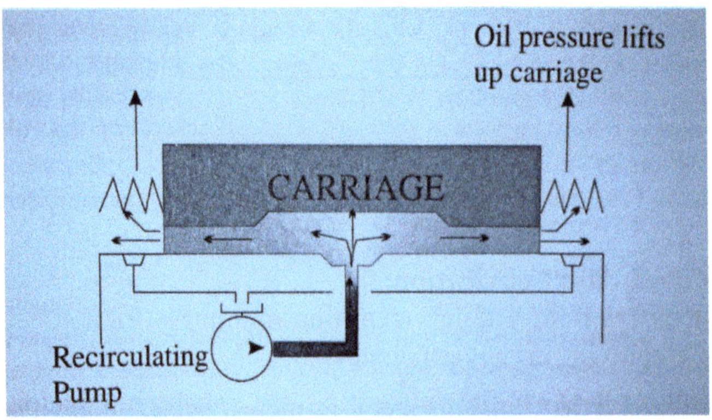

Figure F21 Full film lubrication – hydrostatic (HSL). Courtesy: ENGTECH Industries Inc.

Key Characteristics:

1. **No metal-to-metal contact**: The lubricant film fully supports the load, even when the surfaces are stationary, eliminating wear.

2. **Controlled film thickness**: The film thickness can be precisely controlled by adjusting the pressure of the lubricant.
3. **High stability**: Hydrostatic systems provide stable lubrication and are less affected by changes in speed or load compared to hydrodynamic systems.

Applications:

- **Machine tools**: Precision machines, such as CNC machines, use hydrostatic lubrication for guideways and spindles to maintain high accuracy.
- **Heavy machinery**: Used in applications requiring high load-carrying capacity, such as large thrust bearings and heavy-duty equipment.
- **Turbines and rotating equipment**: Hydrostatic bearings are often used in turbines and other critical rotating machinery where reliable lubrication is essential.

Advantages:

- **Reduced friction and wear**: Since there is no metal-to-metal contact, components experience minimal friction and wear.
- **High load capacity**: The system can support heavy loads even at low speeds or while stationary.
- **Consistent performance**: The lubrication performance is consistent, providing reliability and precision in applications requiring stability.

Considerations:

- **Complexity and cost**: Hydrostatic systems are more complex and costly due to the need for external pumps and control mechanisms.
- **Energy consumption**: The pump requires energy to maintain the pressure, which can increase operational costs.

Hydrostatic lubrication is essential for applications where precision, stability, and high load capacity are critical, making it ideal for use in advanced industrial and engineering systems.

361. Full-flow Filter

A **full-flow filter** is a type of filtration device used in lubrication systems to filter all of the oil or lubricant as it circulates through an engine or mechanical system. The full-flow filter is placed in the main oil circulation pathway, ensuring that the entire volume of oil passes through the filter before being delivered to critical engine or machinery components. See Figure F22.

Key Features of Full-flow Filters:

- **Primary filtration**: These filters are designed to capture and remove larger particles and contaminants from the oil to protect the system's vital components from wear and damage.
- **Low restriction**: Full flow filters must allow oil to flow through easily and quickly to ensure adequate lubrication and prevent pressure drops, which could harm the engine or system.
- **Bypass valve**: Many full-flow filters are equipped with a bypass valve that opens if the filter becomes clogged or if the oil is too thick (e.g., during cold starts), allowing oil to bypass the filter and continue circulating to prevent damage from oil starvation.

Applications:

- **Industrial machinery**: Used in hydraulic systems, compressors, and other machinery where continuous, clean lubrication is crucial for optimal performance and equipment longevity.
- **Automotive engines**: Full-flow oil filters are commonly used in car and truck engines to ensure that contaminants, such as metal particles, dirt, and sludge, are removed from the oil.

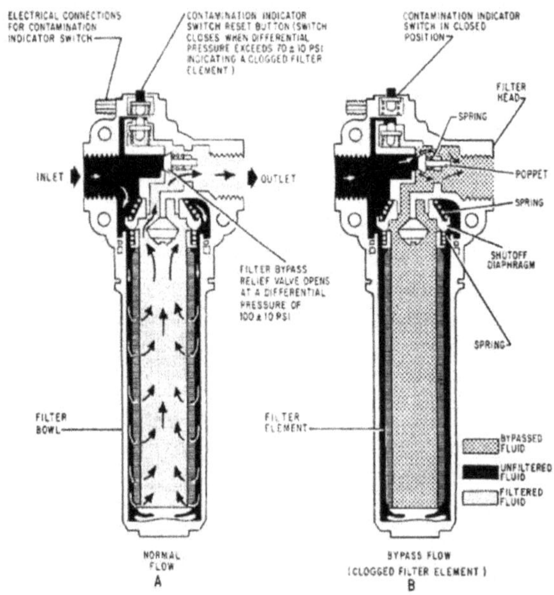

Figure F22 Full-flow filter cross-section in normal flow and bypass flow. Source: UFI Hydraulic Filters.

Advantages:

- **Continuous protection**: Since all the oil is filtered, full-flow filters provide continuous protection against contaminants.
- **Reliability**: Essential for the reliable operation of engines and machinery, helping to minimize wear and prolong the life of components.

Limitations:

- **Filtering efficiency**: Full-flow filters are generally designed to remove larger particles to avoid restricting oil flow, meaning they may not capture very fine contaminants. For more comprehensive filtration, a bypass or secondary filtration system may be used in conjunction with a full-flow filter.

362. Functions of a Lubricant

Lubricants play a critical role in mechanical systems by performing a variety of functions that ensure smooth operation, reduce wear, and enhance the efficiency of machinery. Below are the 10 **key functions of a lubricant**:

1. **Reduce friction**
   - Lubricants form a thin film between moving surfaces, minimizing direct contact and reducing friction.
   - This improves efficiency, reduces heat generation, and extends component life.

2. **Minimize wear and tear**
   - By reducing metal-to-metal contact, lubricants prevent surface damage, abrasion, and scuffing.
   - Antiwear and extreme-pressure additives in lubricants enhance this protective function under high-load conditions.

3. **Cool**
   - Lubricants carry away heat generated by friction or external sources, helping maintain optimal operating temperatures.
   - Effective heat dissipation prevents overheating and thermal damage to components.

4. **Prevent corrosion**
   - Lubricants protect metal surfaces by forming a barrier against moisture, air, and other corrosive substances.

- ○ Additives like rust inhibitors enhance this function in humid or wet environments.

5. **Remove contaminant**
   - ○ Lubricants carry contaminants such as dirt, wear particles, and combustion byproducts away from critical areas.
   - ○ Filters and separators in lubrication systems help remove these particles from the fluid.

6. **Seal**
   - ○ Lubricants help seal gaps between components, such as in pistons and cylinders, to prevent fluid or gas leakage.
   - ○ In engines, they improve compression and prevent contaminants from entering critical areas.

7. **Dampen vibration**
   - ○ Lubricants can absorb and dampen vibrations in machinery, reducing noise and wear caused by mechanical stress.

8. **Reduce energy consumption**
   - ○ By reducing friction and heat loss, lubricants improve the energy efficiency of machinery and engines.

9. **Clean**
   - ○ Lubricants, especially in automotive and industrial systems, help dissolve and remove sludge, varnish, and other deposits.

10. **Enhance equipment longevity**
   - ○ Proper lubrication minimizes component wear, overheating, and corrosion, significantly extending the lifespan of machinery.

Selecting the right lubricant for specific applications ensures that these functions are effectively fulfilled, optimizing performance and reliability across a wide range of mechanical systems.

363. FZG Four Square Gear Oil Test

FZG four square gear oil test is a standardized method used to evaluate the performance of gear oils under high-stress conditions. The test determines

the load-carrying capacity, wear protection, and scuffing resistance of a lubricant when used in gear systems.

The **FZG test** is widely recognized in the industry and provides essential data for assessing the suitability of lubricants for use in industrial gearboxes, automotive transmissions, and other high-load applications.

Test Procedure:

1. **Test setup**:
   - The test uses a "Four-square" gear test rig equipped with a pair of standardized helical gears.
   - The gears are immersed in the lubricant being tested.
2. **Operating conditions**:
   - The gears are operated at a specified speed and temperature while under a progressively increasing load.
   - Load is applied in steps, with each step lasting a fixed duration.
3. **Assessment**:
   - After each load step, the gear teeth are inspected for wear, scuffing, or damage.
   - The test continues until visible scuffing or wear exceeds the allowable limit.

Key Parameters Measured:

1. **Load stage**:
   - The test identifies the highest load stage (or step) the gear oil can handle without causing significant wear or scuffing.
2. **Wear protection**:
   - The test evaluates the lubricant's ability to prevent material loss from the gear teeth under increasing loads.
3. **Scuffing resistance**:
   - Determines the lubricant's capability to protect gears from scuffing, which occurs when metal-to-metal contact happens due to lubricant film breakdown.

Applications:

- **Industrial gearboxes**: Ensures the lubricant can handle high loads and resist wear in heavy-duty machinery.

- **Automotive transmissions**: Tests the oil's performance in automotive gear systems where high-speed and high-load conditions are common.
- **Lubricant development**: Helps manufacturers formulate and certify gear oils for specific applications.

Benefits:

- **Reliability**: Provides a standardized evaluation of gear oil performance, helping users select suitable lubricants.
- **Equipment protection**: Ensures the gear oil can withstand operating stresses, reducing the risk of gearbox failure.
- **Cost efficiency**: Helps identify lubricants that extend gear life and reduce maintenance costs.

G

364. Gage (Gauge)

Gage (or **gauge**) is a general term used to describe a device or instrument that measures and displays a specific property, such as pressure, temperature, flow, or level, in a system. In the context of lubrication systems, gages are crucial for monitoring critical parameters to ensure proper operation and maintenance.

Common Types of Gages in Lubrication Systems:

1. **Pressure gages**:
   - Measure the pressure of the lubricant within a system.
   - Ensure that the lubricant is being delivered at the correct pressure to prevent under or overlubrication.
2. **Temperature gages**:
   - Monitor the temperature of the lubricant to ensure it is within optimal operating limits.
   - High temperatures can cause lubricant degradation, while low temperatures can impact flow properties.
3. **Flow gages**:
   - Measure the flow rate of the lubricant to ensure it is reaching all critical components in the system.
   - Useful for detecting Go (green)–No GO (Red) situations such as blockages or system failures. See Figure G1.
4. **Level gages**:
   - Indicate the amount of lubricant in a reservoir or tank.
   - Ensure adequate lubricant levels are maintained to prevent equipment damage.

DOI: 10.1201/9788743806608-370

Figure G1 Tell-tale Go-No Go flow gauge indicating full oil flow through the filter.

Applications:

- **Industrial equipment**: Monitoring pressure and flow in centralized lubrication systems.
- **Automotive**: Measuring oil pressure and temperature in engines to ensure proper lubrication.
- **Hydraulic systems**: Monitoring fluid pressure and temperature to prevent system failure.

Importance of Gages in Lubrication:

- **Prevent equipment damage**: Ensure proper lubrication by monitoring critical parameters.
- **Improve efficiency**: Help maintain optimal system performance by detecting issues early.
- **Enhance safety**: Prevent system overpressure or overheating, which could lead to failure or accidents.

Gages are essential tools in lubrication management, providing real-time data to ensure systems operate reliably and efficiently. Proper maintenance and calibration of gages are critical for accurate readings and effective system monitoring.

365. Galling (Wear)

Galling is a severe form of adhesive wear that occurs when two metal surfaces in relative motion experience excessive friction, leading to material transfer, surface damage, and deformation. It typically happens when metal surfaces are poorly lubricated or subjected to high loads, causing localized welding and tearing at the contact points.

Galling occurs when two interacting surfaces come into contact under load causing a high-pressure zone that can create significant friction between the surfaces. The two surfaces can at this point bond at microscopic contact points due to the heat and pressure. When the surfaces slide, these bonds can break unevenly, pulling and transferring material from one surface to the other. Repeated bonding and tearing cause rough, irregular wear patterns, leading to bearing surface degradation or failure.

Figure G2 Image of typical galling and material transfer on an SS bolt. Source: Unknown.

Galling is most often found in threaded fasteners (see Figure G2), particularly stainless-steel fasteners, that are put together under great loads without lubrication. It is also found where metal components slide over one another in hydraulic systems, valves, and guideways.

Characteristics of Galling:

- **Surface scoring**: Visible scratches or grooves on the metal surfaces.
- **Material buildup**: Transferred material appears as raised areas or deposits.
- **Rapid progression**: Once galling starts, it can worsen quickly if not addressed.

The causes of galling are many and can include the following:

- **Insufficient lubrication**: Lack of a protective lubricant film increases direct metal-to-metal contact.

- **Material properties**: Metals with similar hardness or ductility, like stainless steels, are more prone to galling.
- **High loads and speeds**: Greater stress and friction exacerbate the bonding of surfaces.
- **Surface finish**: Rough or poorly machined surfaces can increase the likelihood of galling.

Prevention Strategies:

1. **Adequate lubrication**: Use appropriate lubricants or antiseize compounds to reduce friction and wear.
2. **Material selection**: Use dissimilar metals or coatings with lower adhesion tendencies.
3. **Surface treatments**: Polishing or applying surface coatings, like hard anodizing, reduces the risk of galling.
4. **Proper design**: Ensure appropriate tolerances and alignment to minimize excessive loads.

366. Gang Plate

Gang plate in the context of lubrication systems refers to a centralized mounting plate or manifold used to consolidate multiple lubrication points into a centralized location. Gang plates are designed to ensure efficient and organized distribution. Figure G3 shows two "ganged" sets of grease points on a column (notice the grease nipples have been incorrectly painted).

Key Features of a Gang Plate:

1. **Centralized design**: Combines multiple lubrication outlets or components in one compact, easily accessible unit.
2. **Ease of installation**: Simplifies the mounting of multiple lubrication elements, reducing installation time and effort.
3. **Durable construction**: Typically made of robust materials like aluminum or steel to withstand harsh industrial environments.
4. **Customizable layout**: Allows for flexibility in the number and arrangement of lubrication points to match system requirements.

Figure G3 Two "Ganged" sets of grease points mounted on a column. Courtesy: ENGTECH Industries Inc.

Applications:

- **Industrial machinery**: Used in both manual and automated lubrication systems for manufacturing and heavy machinery.
- **Vehicle and fleet maintenance**: Supports lubrication in complex systems like trucks or buses.

367. Gas (Gaseous) Lubrication

Gas lubrication, or **gaseous lubrication**, is a lubrication regime in which a gas – commonly air or an inert gas like nitrogen – is used as the lubricating medium between two surfaces. Unlike traditional liquid or solid lubricants, gas lubrication forms a thin film that separates the surfaces, reducing friction and wear while allowing for clean and low-resistance operation.

How Gas Lubrication Works:

A lubricating gas is introduced under pressure into the space between the surfaces, creating a thin film that prevents direct contact.

The gas film, though extremely thin, is capable of supporting loads by generating pressure through hydrodynamic or aerostatic principles.

The absence of liquid or solid lubricants results in extremely low friction coefficients.

Types of Gas Lubrication:

1. **Hydrodynamic gas lubrication**: Generated by relative motion between surfaces, which draws the gas into the lubricating gap, creating a load-bearing film.
2. **Aerostatic gas lubrication**: Gas is externally supplied under pressure to maintain a stable lubricating film, even at low or zero relative motion between surfaces.

Advantages:

Gas lubrication is ideal for applications where contamination by liquid or solid lubricants is unacceptable (e.g., in electronics or vacuum environments). Gas films offer exceptionally low resistance, reducing energy loss and remain effective at both high and low temperatures where liquid lubricants may degrade or freeze. Furthermore, lubricating gases- like air or nitrogen are safe as they are nontoxic and environmentally friendly.

Applications:

- **Precision equipment**: Used in high-precision instruments like gyroscopes and optical systems.
- **High-speed machinery**: Suitable for high-speed spindles, turbines, and compressors where liquid lubricants might cause drag.
- **Cleanroom environments**: Ideal for semiconductor manufacturing and other cleanroom operations where cleanliness is critical.
- **Aerospace**: Used in spacecraft and satellites, where traditional lubricants may evaporate in a vacuum.

Limitations:

1. **Lower load capacity**: Gas films have a lower load-carrying ability compared to liquid lubricants.
2. **Complexity**: Systems may require precise control of gas pressure and flow.
3. **Sensitive to surface conditions**: Requires smooth, well-machined surfaces for optimal performance.

Gas lubrication is a specialized technology suited for high-precision, clean, and low-resistance applications, offering unique advantages in environments where traditional lubricants are unsuitable.

368. Gasket

A **gasket** is a mechanical seal used to fill the space between two or more mating surfaces to prevent leakage of fluids or gases when those surfaces are under compression. Gaskets are commonly used in machinery, piping, and equipment where tight seals are required to maintain lubrication system integrity and performance.

Key Features of Gaskets:

1. **Compressibility**: Designed to deform under compression, creating a tight seal that compensates for surface irregularities.
2. **Material versatility**: Made from a variety of compressible materials that include rubber, silicone, metal, cork, felt, and PTFE, depending on the application.
3. **Chemical and temperature resistance**: Selected based on the type of fluid or gas being sealed and the operating environment.

Functions:

- **Leak prevention**: Prevents the escape of fluids, gases, or contaminants between joined surfaces.
- **Pressure maintenance**: Helps maintain system pressure by ensuring no leakage occurs.
- **Vibration dampening**: Reduces vibrations and noise in machinery by cushioning contact surface

369. Gear Lubricants

Gear lubricants are specially formulated lubricants designed to protect and enhance the performance of gear systems. They reduce friction, prevent wear, and ensure efficient power transmission under various operating conditions. Gear lubricants must withstand high loads, pressures, and temperatures while maintaining their protective properties.

Functions of Gear Lubricants:

1. **Friction reduction**: Minimizes metal-to-metal contact between gear teeth, reducing wear and energy loss.
2. **Wear protection**: Prevents damage like scuffing, pitting, or abrasion caused by high loads and surface interactions.
3. **Thermal regulation**: Dissipates heat generated during gear operation to maintain optimal system temperatures.
4. **Corrosion resistance**: Protects gears from moisture and contaminants that can cause rust and corrosion.
5. **Shock absorption**: Cushions impact loads in gears, especially in heavy-duty or high-speed systems.

Types of Gear Lubricants:

1. **Gear oils**:
   - **Mineral-based oils**: Cost-effective and suitable for standard applications.
   - **Synthetic oils**: Offer superior performance, thermal stability, and oxidation resistance, ideal for high-stress environments.
   - **Food-grade oils**: Used in food processing equipment to meet safety and hygiene standards.
2. **Greases**:
   - Thickened lubricants for gears exposed to moisture, dirt, or where oil retention is challenging.
   - Common in low-speed, high-load gear applications.
3. **Solid Lubricants**:
   - Materials, like graphite or molybdenum disulfide, are used in extreme conditions where liquid lubricants may fail.

Applications of Gear Lubricants:

As implied by their name, gear lubricants are designed for specific use for the lubrication of meshing gears that must manage both sliding friction and rolling friction during the meshing process in many types of operating environments that include:

1. **Industrial gearboxes**: Used in manufacturing, mining, and power plants to handle heavy loads and variable operating conditions.

2. **Automotive transmissions**: Protects manual, automatic, and differential gears in vehicles.
3. **Marine equipment**: Provides durability and resistance to moisture and salt in marine gear systems.
4. **Agricultural machinery**: Keeps gears operating efficiently in dusty and demanding environments.

370. Gear Lubrication

Gear lubrication refers to the application of lubricants to gears to reduce friction, prevent wear, and ensure smooth and efficient power transmission. Proper gear lubrication is essential for minimizing energy loss, reducing heat generation, protecting against corrosion, and extending the lifespan of gear systems.

Functions of Gear Lubrication:

1. **Friction reduction**: Forms a protective film between gear teeth to reduce metal-to-metal contact and minimize friction.
2. **Wear prevention**: Protects against pitting, scuffing, and abrasive wear caused by high loads and surface interactions.
3. **Heat dissipation**: Carries away heat generated during operation to prevent overheating and thermal degradation.
4. **Corrosion protection**: Provides a barrier against moisture and contaminants that could lead to corrosion and rust.
5. **Shock absorption**: Cushions impact loads in gears, especially in heavy-duty or high-speed applications.

Gear lubrication is critical for ensuring the efficiency, reliability, and longevity of gear systems in a wide range of applications. Proper selection, application, and maintenance of lubricants can significantly enhance the performance of gears under varying operating conditions.

371. General Purpose Grease

General purpose grease, also referred to as **multipurpose** grease, is a versatile lubricating grease designed for use in a broad range of applications where specialized performance characteristics are not required. It provides basic lubrication, wear protection, and corrosion resistance for bearings, bushings, hinges, and other mechanical components.

Key Features:

1. **Versatility**:
   - Suitable for a variety of **low to moderate**-speed and load applications.
   - Commonly used in automotive, industrial, and household equipment.
2. **Basic protection**:
   - Reduces friction and wear between moving surfaces.
   - Offers resistance to moisture and oxidation to prevent corrosion.
3. **Ease of use**:
   - Often formulated for easy application with manual or automated grease delivery systems.

Composition:

General-purpose grease typically consists of:

- **Base oil**: Mineral oil or synthetic oil to provide the primary lubricating properties.
- **Thickener**: Usually a soap-based thickener, such as lithium or calcium soap, to give the grease its semisolid structure.
- **Antioxidant additive**: Prevent grease degradation due to oxidation.
- **Rust and corrosion inhibitors**: Protect metal surfaces from moisture and contaminants.

Limitations:

Because it is a general-purpose lubricant, it will have a limited operating temperature range and may not perform well in extremely high or low temperatures, heavy load, or high-pressure applications and while it will offer basic moisture protection, it may not resist heavy water exposure or washout.

Advantages:

As it will operate in most applications a general-purpose grease formulation offers many advantages that include:

1. **Cost-effectiveness**: Affordable and readily available.
2. **Convenience**: Eliminates the need to stock multiple specialized greases for general use.

3. **Low maintenance**: Easy to apply and manage for routine lubrication.

General-purpose grease is ideal for general maintenance tasks and light-duty applications, offering a reliable and economical solution for basic lubrication needs. For demanding or specialized applications, however, a grease formulated for specific performance characteristics may be required.

372. Gill, August H.

August H. Gill (1885–1965) is an American lubrication pioneer born in Canton, Massachusetts who became a founding member of the ASTM committee for petroleum products and lubricants. Gill was a prominent figure in the field of lubrication engineering and tribology, known for his groundbreaking contributions to the understanding and science of lubrication science. His work laid the foundation for many modern lubrication reliability practices and engineering principles used today.

Key Contributions:

1. **Author of "The Oil Book" (1922)**:
   - Gill's influential text provided a comprehensive guide to lubrication technology, addressing both theoretical and practical aspects of lubricant selection, application, and system design.
   - The book remains a reference point in lubrication science and is considered a seminal work in the field.
2. **Advancements in lubrication theory**:
   - He contributed to the understanding of film formation, friction, and wear, offering insights that improved the efficiency and reliability of machinery.
   - His research helped bridge the gap between tribology theory and industrial application.
3. **Development of standards**:
   - Gill was involved in efforts to standardize lubricant testing and classification, ensuring consistency and reliability in product performance.
   - His work influenced the development of industry practices and standards for oil testing.

4. **Industry leadership**:
   - As an educator and researcher, Gill mentored many engineers and scientists in the emerging field of lubrication engineering.
   - He collaborated with major industrial organizations to apply his findings to practical machinery and systems.

Legacy:

- **Tribology pioneer**: Gill is regarded as one of the early pioneers of tribology, contributing to the systematic study of friction, lubrication, and wear.
- **Impact on industrial applications**: His work greatly influenced industries such as automotive, aerospace, and manufacturing, where efficient lubrication is critical.
- **Influence on education**: Gill's writings and teachings have been foundational in lubrication engineering curricula.

August H. Gill's contributions have had a lasting impact on the field of lubrication engineering, shaping both theoretical research and practical applications for decades.

373. Glycol Contamination

Glycol contamination refers to the presence of glycol-based coolants (typically ethylene glycol or propylene glycol) in lubricating oil, which can significantly impair the performance and integrity of both the lubricant and the machinery it serves. Glycol contamination often occurs due to leaks in engine cooling systems, seals, or gaskets, and it is particularly problematic in engines, compressors, and hydraulic systems.

Sources of Glycol Contamination:

1. **Coolant system leaks**:
   - Leaking head gaskets or cylinder liners in engines.
   - Cracks in cooling system components, such as radiators or hoses.
2. **Improper maintenance**: Cross-contamination during fluid handling or storage.
3. **Seal failures**: Failed seals between coolant and lubricant systems.

Effects of Glycol Contamination:

1. **Lubricant degradation**:
    - Glycol reacts with oil, forming sludge, varnish, or gel-like substances that clog oil passages and filters.
    - It reduces the lubricant's viscosity and film strength, impairing its protective properties.
2. **Corrosion**: Glycol contamination promotes the formation of acidic byproducts, leading to corrosion of metal components.
3. **Increased wear**: Glycol and water mix to create abrasive conditions, accelerating wear on bearings, gears, and other moving parts.
4. **Reduced thermal stability**: Glycol contamination decreases the oil's ability to dissipate heat effectively, increasing the risk of overheating.
5. **Engine damage**: Combustion-related issues, such as preignition or knocking, can result from glycol contamination in engines.

Detection:

1. **Oil analysis**:
    - **Spectroscopy**: Detects the presence of glycol or its byproducts.
    - **FTIR analysis**: Identifies glycol contamination through specific absorption peaks.
2. **Visual inspection**: Look for sludge, emulsification, or discoloration in the lubricant.
3. **Coolant tracer dyes**: Tracer dyes added to coolant systems can reveal leaks when mixed with the lubricant.

Prevention and Mitigation:

1. **Regular maintenance**:
    - Inspect and replace worn seals, gaskets, and cooling system components.
    - Monitor coolant levels and pressure to detect leaks early.
2. **Oil and coolant separation**: Use equipment designed to prevent cross-contamination between cooling and lubrication systems.
3. **Oil analysis programs**: Implement routine oil testing to detect contamination before it causes significant damage.

4. **Timely response**: If contamination is detected, immediately drain and flush the system, replacing both the coolant and lubricant.

Glycol contamination is a critical issue that can lead to costly repairs, increased downtime, and equipment failure if not addressed promptly.

374. Graphite

Graphite is a naturally occurring form of crystalline carbon (see Figure G4) known for its unique physical and chemical properties, making it a widely used material in lubrication and other industrial applications. Its layered structure, composed of graphene sheets, allows for excellent lubrication due to weak bonds between the layers, enabling them to slide over each other with minimal resistance.

Figure G4 Graphite – a natural crystalline carbon.

Key Properties of Graphite:
Graphite delivers many properties that make it well-suited as a solid lubricant or additive. These include:

1. **Excellent lubricity**:
   - Graphite acts as a solid lubricant, reducing friction and wear between contact surfaces.

- Its lubricating ability is enhanced in dry or high-temperature environments where liquid lubricants may fail.

2. **High thermal resistance**: Remains stable at extreme temperatures, making it suitable for high-heat applications like furnaces and turbines.
3. **Chemical stability**: Resistant to most chemicals, ensuring durability in harsh environments.
4. **Electrical conductivity**: Conducts electricity, allowing it to be used in specialized applications like electrical brushes.
5. **Self-lubricating**: Requires no additional lubricant, making it ideal for environments where traditional oils or greases cannot be used.

Graphite Applications:

1. **Dry lubrication**: Used as a powder or coating in high-temperature or vacuum conditions, such as aerospace or foundry applications.
2. **Additive in greases**: Added to greases to enhance their lubricating properties, especially for heavy-duty and extreme-pressure applications.
3. **Gear and chain lubrication**: Ideal for open gears and chains exposed to dust, dirt, or high temperatures.
4. **High-temperature bearings**: Used in applications like kiln or oven bearings where traditional lubricants would degrade.
5. **Mold release agent**: Prevents sticking of molten metals or plastics in molds.

Advantages/Disadvantages:

Graphite functions well in extreme conditions, including high temperatures and vacuums, it is nontoxic and environmentally safe, long-lasting, and resistant to degradation over time. Despite all of its many advantages, graphite's properties diminish when exposed to water or moisture and is not suitable for all environments where liquid lubrication is necessary. In addition, excessive use can cause wear on softer surfaces.

Graphite is a versatile and reliable lubricant for specialized applications, particularly in extreme operating conditions where liquid lubricants cannot be used. This ability to function where liquid or grease-based lubricants cannot, makes it indispensable in many industrial and high-performance environments.

375. Gravimetric Analysis

Gravimetric analysis is a quantitative analytical technique used to measure the amount of a specific substance in a sample by converting it into a stable compound, isolating it, and weighing it. In lubrication and oil analysis, gravimetric methods are often employed to determine the concentration of solids, contaminants, or additives in a lubricant.

Key Principles of Gravimetric Analysis:

1. **Sample conversion**: The substance of interest is chemically or physically transformed into a solid, stable form (e.g., precipitate or residue).
2. **Isolation**: The solid compound is separated from the rest of the sample, typically through filtration or evaporation.
3. **Weighing**: The isolated solid is dried and weighed to determine its mass, which is used to calculate the concentration of the target substance.

Applications in Lubrication:

1. **Contaminant detection**: Measure the concentration of insoluble contaminants, such as dirt, wear debris, or carbon deposits, in used oil.
2. **Additive measurement**: Determine amounts of specific additives in a lubricant, such as detergents or antiwear compounds.
3. **Filter performance testing**: Quantify the particles captured by filters to evaluate their efficiency.
4. **Sludge and residue analysis**: Assess the accumulation of sludge or varnish in lubricating systems by isolating and weighing these residues.

Gravimetric analysis is an essential tool in lubricant analysis, providing critical data on the condition of oils and the effectiveness of filtration systems. Its precision and reliability make it a standard method in maintenance and quality assurance practices.

376. Grease

Grease is a semisolid lubricant composed of a base oil, a thickener, and performance-enhancing additives. It is widely used to reduce friction and wear in machinery and equipment, particularly in applications where liquid lubricants would not stay in place or where relubrication is infrequent. Grease provides long-lasting lubrication, protects against contaminants, and helps seal out moisture.

Components of Grease:

1. **Base oil**:
   - Typically, a mineral or synthetic oil that provides the primary lubricating properties.
   - The choice of base oil determines the grease's operating temperature range and viscosity.
2. **Thickener**:
   - A soap or nonsoap material that gives grease its semisolid structure.
   - Common thickeners include lithium, calcium, aluminum, or complex soaps, as well as nonsoap thickeners like clay or polyurea.
3. **Additives**:
   - Enhance specific properties, such as:
   - **Antiwear**: Reduces friction and surface damage.
   - **Extreme pressure (EP)**: Prevents wear under high loads.
   - **Corrosion inhibitors**: Protects against rust and oxidation.
   - **Water resistance**: Improves performance in wet environments.

Key Properties of Grease:

1. **Consistency**:
   - Measured by the National Lubricating Grease Institute (NLGI) scale, ranging from 000 (fluid-like) to 6 (hard).
   - NLGI grade determines the suitability of grease for specific applications.
2. **Temperature stability**:
   - Grease must maintain its structure and performance across its operating temperature range.
3. **Water resistance**:
   - Essential for applications exposed to moisture or water spray.
4. **Load-carrying capacity**:
   - Ability to protect against wear under heavy loads or extreme pressure.

Advantages of Grease:

1. **Long-lasting**: Requires less frequent relubrication compared to oils.
2. **Sealing properties**: Protects against dirt, dust, and moisture.
3. **Adhesion**: Stays in place even under vibration or load.

Limitations:

1. **Heat sensitivity**: High temperatures can cause grease to soften and leak out.
2. **Flow characteristics**: Grease does not flow easily, making it unsuitable for some high-speed applications.
3. **Complex application**: Requires specific tools and techniques for precise application.

Grease is a versatile and reliable lubricant essential for maintaining the efficiency and longevity of machinery across a wide range of industries.

377. Grease Caddy

A **grease caddy** is a portable or fixed device/container designed to store or transport grease guns and/or grease cartridges for easy application in maintenance and lubrication tasks. Caddies are designed to keep grease and grease guns apart so they do not cross-contaminate in storage or transportation within the workplace. Figure G5 shows a fixed mount single grease gun wall storage device. These devices can often be found mounted on mobile grease carts.

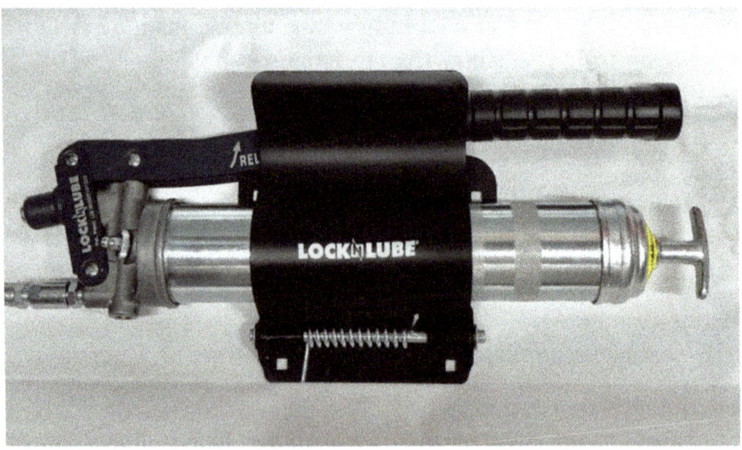

Figure G5 Wall-mounted single grease gun caddy device. Courtesy: **ENGTECH** Industries Inc. and Lock n' Lube Lubrication Corporation.

1. Grease Gun

A **grease gun** is a manual or powered tool designed to apply grease to specific lubrication points on machinery, equipment, or vehicles. It ensures precise delivery of grease to bearings, joints, and other components, helping reduce friction, prevent wear, and extend the lifespan of moving parts. Figure G6 shows a high-quality lever-operated grease gun.

Key Features of a Grease Gun:

1. **Delivery mechanism**: Uses pressure to dispense grease through a nozzle or coupler into lubrication fittings (e.g., Zerk fittings).
2. **Control**: Provides precise application, reducing waste and ensuring that the correct amount of grease reaches the lubrication point.
3. **Interchangeable nozzles**: Equipped with different tips, such as needle or flat nozzles, for compatibility with various fittings and applications.

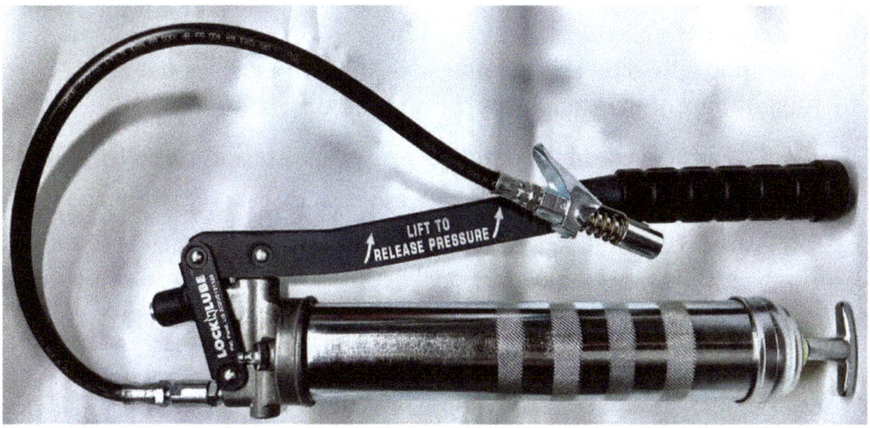

Figure G6 A high-quality lever-operated grease gun. Courtesy: **ENGTECH** Industries Inc. and Lock n Lube Corporation.

1. **Manual grease guns**:
   - **Lever-operated**: Requires pulling a lever to pump grease; suitable for moderate applications.
   - **Pistol-grip**: Operated with one hand, offering greater control and ease of use.

2. **Powered grease guns**:
   - **Battery-operated (cordless)**: Convenient for large-scale lubrication tasks, delivering consistent pressure.
   - **Pneumatic (air-powered)**: Uses compressed air to pump grease, ideal for industrial or heavy-duty applications.
3. **Hand pump grease guns**:
   - Compact and portable, used for small lubrication jobs or tight spaces.

Advantages:

1. **Precision**: Delivers grease directly to lubrication points, minimizing waste and ensuring effective lubrication.
2. **Portability**: Lightweight and easy to carry, making it suitable for field applications.
3. **Durability**: Made from robust materials to withstand frequent use in demanding environments.

Considerations:

1. **Compatibility**: Ensure the grease in the gun matches the grease type and fittings of the equipment being greased.
2. **Maintenance**: Clean the nozzle before every use and check for leaks to maintain performance.
3. **Pressure control**: Grease guns have the potential to reach pressures as high as **15,000 psi** which can easily damage a bearing seal if too much grease is dispensed into the bearing cavity. To avoid seal failure:
   - Calculate the displacement of a single grease stroke;
   - Identify the amount per shot on the gun;
   - Calculate how many shots are required for each bearing point and record on the work order master;
   - Dispense only the calculated amount into the bearing each and every time.
4. **Delivery volume**: The volume of grease delivered per stroke will vary considerably from gun to gun. Many work orders state manual grease delivery by “shot” where one shot equals one full cycle of the grease gun lever or trigger. If this method of measure is to work effectively, all

grease guns must be standardized within the plant. The "shot" size must be measured and divided into the evaluated bearing requirement to attain the correct measure every time for each bearing, regardless of who the operator is. Failure to standardize in this manner will put all bearings at risk of over or underlubrication every time they are lubricated.

A **grease gun**, when approached and used in an engineered and responsible manner is an inexpensive and indispensable tool for efficient and targeted lubrication, ensuring equipment reliability and reducing maintenance costs in various industries.

378. Grease Lubrication

Grease lubrication is the process of using lubricating grease to reduce friction, protect against wear, and enhance the performance of machinery and equipment. Grease, a semisolid lubricant, is particularly effective in applications where liquid lubricants would not stay in place, such as vertical or slow-moving components.

Key Characteristics of Grease Lubrication:

1. **Thickened structure**:
   - Grease consists of a base oil, a thickener (soap or nonsoap), and additives.
   - The thickener gives grease its consistency, allowing it to stay in place.
2. **Viscous film**: Forms a long-lasting lubricating film that adheres to surfaces, even in challenging conditions like vertical orientations or under vibration.
3. **Sealing properties**: Grease helps seal out contaminants such as dirt, moisture, and dust, protecting critical components.
4. **Temperature range**: Operates within specific temperature ranges, depending on the type of grease and its formulation.

Advantages of Grease Lubrication:

1. **Long-lasting**: Requires less frequent reapplication compared to oil.
2. **Adherence**: Stays in place even under vibration or high loads.
3. **Sealing capability**: Protects against contaminants and reduces corrosion risks.

4. **Versatility**: Suitable for various industrial, automotive, and marine applications.

379. Grease Monkey

A **grease monkey** is a derogatory colloquial term used for a person(s) involved in the machine lubrication process.

History

The "**Grease Monkey**" term is believed to have originated from the 19th-century Industrial Revolution cotton mills that required continual lubrication of the steam-engine driven line shaft drive systems. A network of overhead gantries was dangerously navigated by children who hand-applied tallow-based grease lubricant to the equipment from above while the machinery kept on running. Figure G7 depicts an actual enhanced and colorized photo of child mill workers taken in the early 1900s who would have been responsible for greasing operations.

Figure G7 English child mill workers in 1910. Source: BrightStyleGrant.

Their hunched-over scurried movements above the equipment were thought to emulate monkeys in tree canopies giving birth to the term "grease monkey." Unfortunately, many of these children were injured or killed performing these dangerous greasing operations.

Today

Today, the term carries a more light-hearted connotation for persons working with greasing devices such as machine operators, mechanics, millwrights, and lubricators.

380. Gullborg, Arthur

Arthur Gullborg was an innovative figure in lubrication technology. In 1916, working as a lubricator for his father's die-casting company in Chicago he decided to reduce his workload by inventing a revolutionary manual grease system. His invention included both a screw-type grease gun and a grease fitting. The "fitting" consisted of a braided metal hose with a special end connection. Gulborg named his system, "The Alemite® High-Pressure Lubricating System" after the Alemite® Die Casting and Manufacturing Company where he worked. See Figure G8.

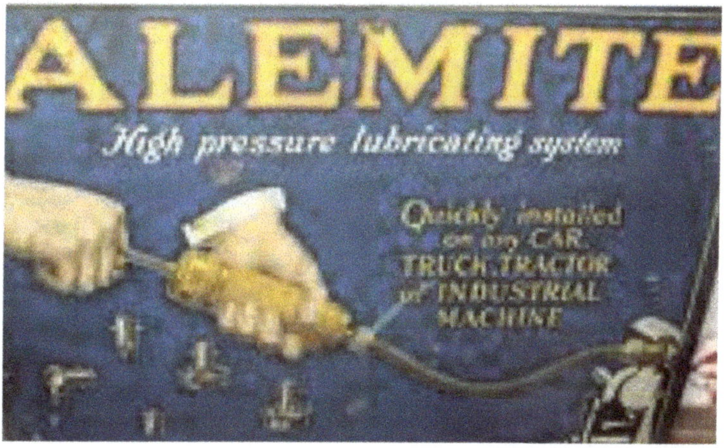

Figure G8 Early sales ad for Gullborg's invention.

Gullborg's invention revolutionized the way industry lubricated its bearings and his invention is the father of all manual greasing systems in use today.

381. Gum

See "Varnish."

382. Gunfil® Lubricating device

A **Gunfil®** device is a specialized auto-lubrication device designed to replace conventional grease fittings. Originally designed by John R Battle this auto lube device offers a more precise and controlled delivery of grease to bearings and other lubrication points. It can be installed horizontally, vertically, or at any angle, making it suitable for various mechanical systems. The device uses a spring-loaded grease applicator to deliver lubricant only while the bearing is in motion, preventing overlubrication and waste.

Figure G9 The modern version of the original Gunfil® lubricator developed a century before by John H. Battle. Source: LDI Industries Industrial Catalog.

Although the self-lubricating device ensures a continued supply of grease to the bearing it essentially relies on bearing back-pressure to regulate the grease delivery. In instances of large bearing clearances or worn bearings, back pressure is reduced and can result in an increased lubricant delivery. Care must be taken to monitor flow to know when to refill the unit via the grease nipple shown on the side of the reservoir in Figure G9.

H

383. Hand Packing

Hand packing refers to a manual method of applying grease to bearings that require lubrication and is often employed for low-volume tasks or single-bearing fills. This technique is typically used when automated systems or grease guns are unavailable, or for specific applications where precision and control over the grease application is critical.

Steps for Hand Packing Bearings:

1. **Prepare the area**:
   - Clean the workspace, bearings, and wash hands to prevent contamination.
   - Use gloves to maintain hygiene and reduce the risk of contaminants entering the bearing.
2. **Apply grease**:
   - Place a dollop of grease in your palm.
   - Hold the bearing in your opposite hand and press the bearing edge into the grease repeatedly.
3. **Force grease through the bearing**:
   - Continue pressing until grease emerges from the opposite side of the bearing, ensuring it fills all spaces between the rolling elements and raceways.
4. **Inspect and reapply**: Rotate the bearing and repeat the process until the entire bearing is evenly packed with grease.

Advantages:

1. **Cost-effective**: Requires minimal equipment.
2. **Precision**: Allows for controlled application and inspection of the lubrication process.
3. **Accessibility**: Suitable for hard-to-reach or delicate components.

DOI: 10.1201/9788743806608-390

Limitations:

1. **Time-consuming**: Less efficient than automated or mechanical lubrication methods.
2. **Risk of contamination**: Requires careful handling to avoid introducing dirt or debris into the bearing.
3. **Consistency**: Manual packing may not provide uniform grease distribution compared to automated methods.

Hand packing remains a reliable and versatile method for greasing bearings, especially in situations requiring precision or where automated systems are unavailable. Proper technique and **cleanliness** are essential to ensure optimal lubrication and equipment performance.

384. Hardness

Hardness refers to a material's resistance to deformation, indentation, or penetration when subjected to a force. In the context of lubrication, hardness is often considered in relation to the surfaces being lubricated, as well as in the evaluation of additives and contaminants that may influence wear and performance.

Key Types of Hardness:

1. **Surface hardness**:
   - The resistance of a material's surface to wear, abrasion, or indentation.
   - Relevant for evaluating the durability of bearings, gears, and other components subjected to sliding or rolling contact.
2. **Bulk hardness**: Refers to the overall hardness of a material, often assessed in metals or alloys.
3. **Lubricant additive hardness**: Hardness may apply to solid additives, such as molybdenum disulfide or graphite, which form protective layers under extreme pressure.
4. **Particle hardness**: Evaluates the abrasiveness of contaminants in lubricants that can lead to wear or damage.

Methods of Measuring Hardness:

There are numerous methods and apparatus used to measure surface hardness, these include:

1. **Brinell hardness test (BHN)**:
   - Uses a steel or tungsten carbide ball to measure the indentation left on the material.
   - Common for softer metals and alloys.
2. **Rockwell hardness test (HR)**:
   - Measures depth of penetration under a specific load, with various scales for different materials.
   - frequently used for harder metals.
3. **Vickers hardness test (VHN)**:
   - Employs a diamond-shaped indenter and measures the size of the indentation.
   - Suitable for thin materials or coatings.
4. **Mohs scale**:
   - Ranks materials based on their ability to scratch softer substances.
   - Useful for assessing mineral-based additives or contaminants.

Importance in Lubrication:
Knowing the surface hardness of the material to be lubricated allows the correct choice of lubricant to be made and will increase the life cycle by influencing the following:

1. **Wear prevention**: Matching lubricant properties to the hardness of mating surfaces minimizes wear and prolongs component life.
2. **Compatibility**: Ensures that lubricant additives do not damage softer materials or fail to protect harder surfaces.
3. **Contamination control**: Identifying hard particles in lubricants helps in mitigating abrasive wear and maintaining performance.
4. **Equipment design**: Hardness data is essential for designing components that withstand operational stresses.
5. **Lubricants**: Solid lubricants like graphite or boron nitride enhance performance by interacting with surface hardness under pressure.

385. Hazardous Disposal

Hazardous disposal refers to the proper handling, treatment, and removal of substances or materials classified as hazardous to human health, the environment, or both. These materials can include chemicals, oils, lubricants, industrial waste, or contaminated equipment that pose risks due to their toxicity, flammability, corrosiveness, or reactivity.

Key Elements of Hazardous Disposal:

1. **Identification**:
   - Materials must be classified as hazardous according to regulations like those from the Environmental Protection Agency (EPA), local state, and corporate environmental authorities.
   - Examples include mixed lubricants that cannot be recycled, used lubricants, contaminated rags, spent cleaning solvents, lubrication spill clean-up debris.
2. **Collection**: Hazardous materials are to be contained in approved, clearly labeled containers to prevent leaks or spills during storage and transport.
3. **Storage**: Temporary storage in compliant facilities designed to prevent contamination of soil, water, or air. Storage areas must be clearly labeled and identified using signage similar to that shown in Figure H1.
4. **Transport**: Handled by licensed transporters with proper documentation to ensure compliance with environmental laws.
5. **Disposal or treatment**: Includes methods such as:
   - **Incineration**: Burning hazardous waste at high temperatures to neutralize harmful components.
   - **Recycling**: Processing materials like used oil or metals for reuse.
   - **Landfill**: Disposal in specialized hazardous waste landfills designed with containment systems to prevent leaching.

Regulations Governing Hazardous Disposal:

Managing hazardous waste may require meeting standards and regulations under one or more standards that include, but are not limited to the following:

1. **EPA guidelines** (United States): Under the Resource Conservation and Recovery Act (RCRA), hazardous waste is regulated to ensure safe handling, treatment, and disposal.
2. **OSHA standards**: For worker safety during handling and transportation of hazardous materials.

3. **International standards**: The Basel Convention governs the trans-boundary movement of hazardous wastes.

Figure H1 Typical hazardous waste warning sign.

Examples in Lubrication and Maintenance:

1. **Used oil**: Disposed of through recycling or incineration due to contaminants from wear metals, additives, and fuel residues.
2. **Grease**: Disposed of as hazardous waste if it contains heavy metals or other toxic additives.
3. **Filters**: Oil and fuel filters may require special disposal as they can contain residual hazardous substances.

Importance of Proper Hazardous Disposal:

1. **Environmental protection**: Prevents contamination of water sources, soil, and air.
2. **Human health**: Reduces exposure to toxic substances that can cause illness or injury.
3. **Regulatory compliance**: Avoids legal penalties and fines for improper disposal.
4. **Sustainability**: Encourages recycling and reuse of materials, minimizing waste.

Conclusion:
Hazardous disposal is a critical process that ensures the safe and responsible management of materials that could otherwise harm people and the environment. It is vital for industries such as manufacturing, maintenance, and lubrication to adhere to proper disposal protocols and regulations.

386. HDL

See "Hydrodynamic Lubrication."

387. Heat Exchanger

A **heat exchanger** is a mechanical device designed to transfer heat between two or more fluids (liquid or gas) without mixing them. Heat exchangers are widely used in industrial processes, HVAC systems, automotive engines, and lubrication systems to regulate temperature, enhance efficiency, and protect equipment from thermal damage.

Key Features of Heat Exchangers:

1. **Thermal transfer**: Facilitates efficient heat transfer between fluids to maintain optimal operating temperatures.
2. **Separated fluids**: The fluids do not come into direct contact, ensuring the integrity of each fluid and preventing contamination.
3. **Energy efficiency**: By recycling heat, heat exchangers improve energy efficiency, reducing operational costs.

Applications in Lubrication Systems:

1. **Oil coolers**: Regulate the temperature of lubricating oils in engines, gearboxes, and compressors, preventing thermal breakdown and maintaining viscosity.
2. **Hydraulic systems**: Maintain optimal fluid temperatures to ensure consistent performance and prevent overheating.
3. **Bearing cooling**: Used to dissipate heat generated in high-speed or heavy-load bearings, reducing wear and extending service life.

Maintenance Considerations:

1. **Regular cleaning**: Prevents fouling or scaling, which can reduce efficiency.

2. **Inspection for leaks**: Ensures no cross-contamination between fluids.
3. **Corrosion monitoring**: Protects against material degradation, especially in aggressive environments.
4. **Proper fluid selection**: Ensures compatibility with the materials of construction to avoid damage.

In lubrication systems, heat exchangers play a critical role in maintaining fluid performance, protecting machinery, and enhancing system reliability.

388. Helical Coil Sampling

See "High Pressure (HP) Tap Sampling."

389. Hertzian Contact Area

The **Hertzian contact area** (see Figure H2) refers to the localized region of contact between two curved surfaces under load, typically seen in rolling or sliding components such as bearings, gears, or cam-followers. The concept is based on Hertzian contact theory, developed by Heinrich Hertz, which describes the stress distribution and deformation of contacting surfaces.

Key Characteristics:

1. **Contact shape**: The contact area is typically elliptical or circular, depending on the geometry of the surfaces in contact (e.g., spherical or cylindrical).
2. **Localized stress**: Contact stresses (Hertzian stresses) are highest at the center of the contact area and decrease toward the edges.
3. **Elastic deformation**: Under normal operating conditions, the deformation of the surfaces remains within the elastic limit, meaning the materials return to their original shape after the load is removed.
4. **Load distribution**: The size of the Hertzian contact area increases with higher loads, spreading the stress across a larger surface to prevent damage.

Applications in Lubrication and Tribology:

1. **Bearing design**: Hertzian contact theory helps engineers determine stress levels in ball and roller bearings, ensuring that the materials can withstand the loads without yielding or excessive wear.

2. **Gear teeth**: Used to evaluate stress in meshing gear teeth, ensuring durability and proper lubrication under load.
3. **Rolling elements**: In rollers and shafts, the theory calculates stresses to prevent surface fatigue, pitting, or cracking.
4. **Lubrication regime**: The Hertzian contact area determines the lubrication regime (e.g., boundary, mixed, or elastohydrodynamic) and influences the choice of lubricant and additives.

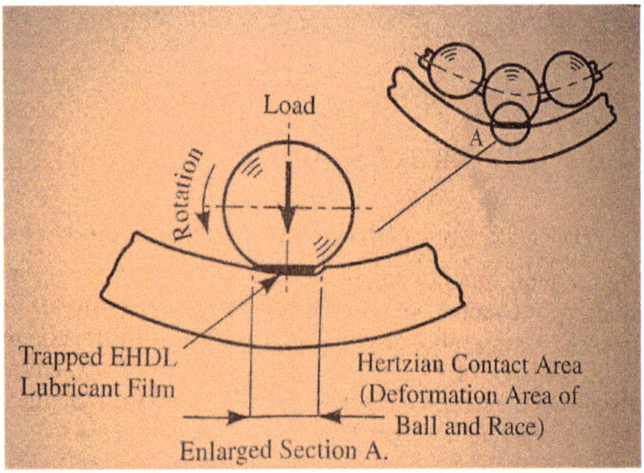

Figure H2 Hertzian contact areas in a ball bearing. Source: Machinery Handbook 27th Edition lubrication section written by Kenneth E. Bannister.

Importance in Maintenance and Lubrication:

1. **Stress management**: Proper analysis ensures stresses remain within acceptable limits to avoid surface fatigue or wear.
2. **Lubrication efficiency**: Ensuring the right lubrication film thickness in the Hertzian contact area minimizes friction and wear.
3. **Predictive maintenance**: Monitoring wear in the contact area helps predict potential failures and optimize maintenance schedules.

The Hertzian contact area is a fundamental concept in tribology and mechanical design, providing insights into stress distribution and deformation in loaded components. Understanding and managing the Hertzian contact area is essential for ensuring the durability, efficiency, and reliability of machinery, especially in applications involving rolling or sliding contact. Proper lubrication tailored to this area can significantly enhance equipment performance and longevity.

390. Hi-lo Indicator

A **hi-lo level indicator** is a visual or electronic device used to monitor the fluid level within a reservoir, tank, or lubrication system. It is designed to display both the maximum (Hi) and minimum (Lo) levels of the fluid, ensuring that the system operates within the optimal range. This indicator is essential for maintaining proper lubrication, hydraulic, or cooling system performance.

Key Features:

1. **High-level (Hi) indication**: Indicates the maximum fluid level to prevent overfilling, which could lead to overflow or system inefficiency.
2. **Low-level (Lo) indication**: Alerts operators when fluid levels drop below the minimum threshold, preventing potential issues like insufficient lubrication or pump cavitation.
3. **Visual or electronic display**:
   - **Visual indicators**: Sight glasses, float gauges, or graduated markers.
   - **Electronic indicators**: Sensors with alarms or digital readouts for remote monitoring.
4. **Durable design**: Constructed with materials resistant to the fluid being monitored, such as oils, coolants, or hydraulic fluids.

Applications:

1. **Lubrication systems**: Ensures sufficient oil or grease levels for bearings, gears, and moving components.
2. **Hydraulic systems**: Maintains optimal fluid levels for consistent pressure and performance.
3. **Industrial machinery**: Used in compressors, turbines, and pumps to monitor coolant or lubricant reservoirs.
4. **Automotive and marine systems**: Tracks fuel, oil, or coolant levels in vehicles, ships, and other equipment.

Types of Indicators:

1. **Sight glass**: Simple and cost-effective; allows operators to visually check fluid levels. See Figure H3.
2. **Float-style indicator**: Uses a floating element connected to a gauge or arm to show Hi and Lo levels.

3. **Electronic sensor**: Sends real-time fluid level data to a control system or triggers alarms when levels are out of range.
4. **Magnetic indicator**: Features a magnetic float and an external gauge for noncontact level monitor.

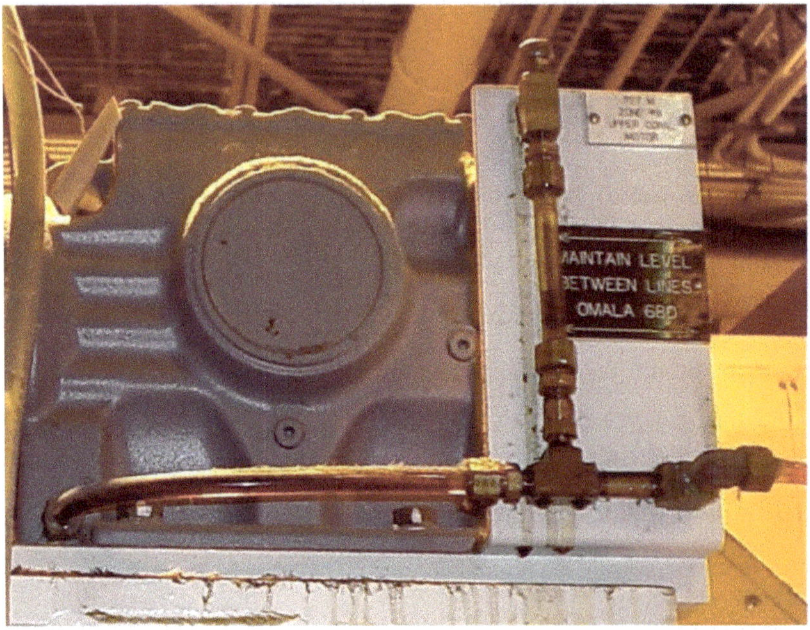

Figure H3 Typical column sight glass level indicator with a hi-lo marking in place. Courtesy: ENG**TECH** Industries Inc.

391. High-pressure (HP) Tap Sampling

High-pressure tap sampling refers to the process of extracting fluid samples, such as hydraulic oil, lubricants, or other operating fluids, directly from a pressurized system without shutting it down. This method is commonly used in industrial and mobile equipment to analyze the condition of fluids and assess the health of machinery through oil analysis.

Key Features of High-pressure Tap Sampling:

1. **System compatibility**: Designed for systems operating at high pressures, typically found in hydraulic, lubrication, or fuel systems.

2. **Inline sampling**: Allows fluid to be collected directly from the system without interrupting its operation, ensuring real-time sampling under working conditions.
3. **Controlled flow**: Sampling devices regulate flow and pressure, ensuring safe and accurate collection of fluid samples.
4. **Clean sampling**: Prevents contamination by using dedicated sampling ports and valves, preserving the integrity of the sample.

Advantages:

- **Minimizes downtime**: Eliminates the need to stop machinery, reducing disruptions to operations.
- **Accurate representation**: Captures fluid samples under actual working conditions, providing a more accurate representation of fluid and system health.
- **Safety**: High-pressure sampling ports are designed to handle the system's pressure, ensuring safe extraction.
- **Improved maintenance**: Facilitates proactive maintenance by enabling regular and precise fluid condition monitoring.

Components of a High-pressure Sampling System:

Components for sampling are illustrated in Figure H4 and include:

- **Sampling ball valve**: Installed on the system to provide a secure point for fluid extraction.
- **Sampling port adapter**: Connects the sampling equipment to the system, ensuring a tight seal.
- **Pressure-relief coil**: Controls pressure during sampling to avoid system damage or unsafe conditions.
- **Sample container**: A clean, sealed container to collect and store the fluid sample for analysis.

Applications:

This sampling method can be applied anywhere oil under high pressure needs to be sampled, for example:

- **Hydraulic systems**: Used to monitor oil cleanliness, viscosity, and contamination levels in high-pressure hydraulic circuits.
- **Pressured lubrication systems**: Collects lubricant samples from bearings, gearboxes, or compressors for wear analysis and condition monitoring.

- **Industrial machinery**: Essential for equipment with high-pressure operations, such as turbines, presses, and mining machinery.

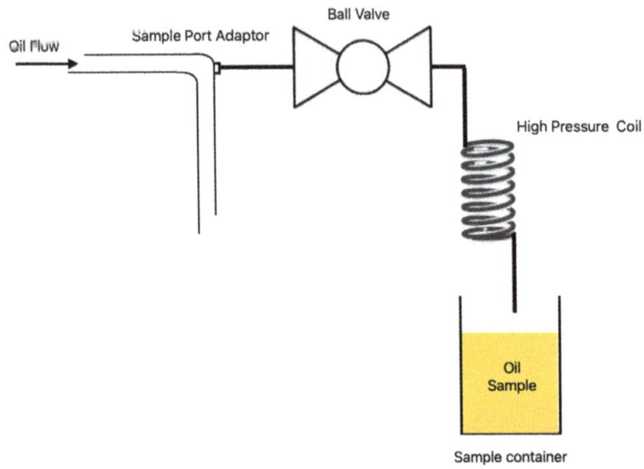

Figure H4 High-pressure coil tap sampling representation.

Best Practices:

- **Cleanliness**: Ensure all sampling tools and containers are clean to avoid contamination.
- **Sampling location**: Choose locations where fluid flow is consistent and representative of the system's overall condition (e.g., downstream of pumps or filters).
- **Regular scheduling**: Implement a sampling schedule to monitor trends and detect issues early.

392. High-temperature Grease

High-temperature grease is a specialized lubricating grease designed to perform under extreme heat conditions without breaking down, losing consistency, or compromising lubrication. It is formulated with base oils, thickeners, and additives that provide excellent thermal stability, oxidation resistance, and protection against wear in high-temperature environments.

Key Properties:

1. **Thermal stability**: Maintains its structure and lubrication properties at elevated temperatures, typically above 120 °C (248 °F) and up to 260 °C (500 °F) or more for specialized greases.

2. **Oxidation resistance**: Resists chemical breakdown caused by prolonged exposure to high temperatures and oxygen, preventing sludge or varnish formation.
3. **Load-carrying capacity**: Offers superior protection against wear and friction, even under heavy loads and high-temperature stress.
4. **Consistency retention**: Retains its consistency, ensuring it does not harden or become too fluid when exposed to heat.
5. **Adhesion**: Sticks well to metal surfaces to provide consistent lubrication and protection in harsh conditions.

Common Thickeners and Additives:

1. **Thickeners**:
   - **Lithium complex**: Popular for high-temperature performance and water resistance.
   - **Calcium sulfonate**: Offers excellent thermal stability and protection against corrosion.
   - **Polyurea**: Ideal for high-temperature, long-lasting lubrication.
2. **Additives**:
   - **Extreme pressure (EP) additives**: Enhance load-bearing capacity.
   - **Antiwear agents**: Minimize surface damage under stress.
   - **Antioxidants**: Prevent oxidation and extend grease life.

Common Applications:

- **Industrial equipment**: Used in furnaces, ovens, and machinery operating in foundries and steel mills.
- **Aerospace**: Applied in jet engines and components exposed to extreme heat.
- **Marine**: Suitable for machinery with exposure to both heat and moisture.
- **Power plants**: Ensures smooth operation of turbines, boilers, and generators.

High-temperature grease is essential for maintaining performance and durability in extreme heat conditions.

393. High Shear Rate Viscosity

High shear rate viscosity refers to the measurement of a lubricant's resistance to flow under conditions of high shear stress, typically occurring in machinery components like bearings, gears, and engine piston rings where high-speed relative motion exists. It is a critical property for understanding how a lubricant performs under intense operating conditions.

Key Characteristics:

1. **Shear stress**: High shear rate viscosity describes how the lubricant behaves when subjected to high shear forces caused by rapid motion between surfaces.
2. **Dynamic operating conditions**: Lubricants experience high shear rates in high-speed or heavily loaded components, such as crankshaft bearings and high-speed gearboxes.
3. **Temperature sensitivity**: Viscosity at high shear rates can change significantly with temperature, making it important to assess performance across a range of temperatures.
4. **Non-Newtonian behavior**: Some lubricants (e.g., multigrade oils) exhibit shear thinning, where the viscosity decreases as the shear rate increases. This behavior is critical in maintaining effective lubrication under dynamic conditions.

Importance in Lubrication:

- **Load carrying**: Ensures the lubricant maintains sufficient film strength to support loads and prevent metal-to-metal contact.
- **Energy efficiency**: Optimized viscosity under high shear rates reduces friction and energy losses in high-speed machinery.

Measurement Methods:

High shear rate viscosity is measured using specialized equipment, such as:

- **High shear viscometers**: Devices like tapered plugs or capillary viscometers simulate high shear conditions to determine lubricant viscosity.
- **ASTM testing standards**: Commonly tested using methods like ASTM D4683 or ASTM D4741 to measure high-temperature high-shear (HTHS) viscosity, typically at 150 °C.

Factors Affecting High Shear Rate Viscosity:

- **Viscosity index improvers**: Multigrade oils use additives to maintain viscosity stability across a wide range of shear rates and temperatures.
- **Base oil selection**: Synthetic oils often perform better under high shear rates compared to mineral oils due to their uniform molecular structure.
- **Additive performance**: Additives like antiwear agents and friction modifiers influence viscosity behavior under shear stress.

394. Homogenization

Homogenization is a mechanical process used to mix or blend two or more immiscible substances, such as oil and water, into a stable and uniform mixture. In lubrication and industrial applications, homogenization is often employed to ensure consistent distribution of additives in lubricants, emulsions, or other fluid formulations, preventing separation and enhancing performance.

Key Characteristics:

1. **Uniform mixture**: Produces a stable and consistent fluid by reducing particle or droplet size and evenly distributing them throughout the medium.
2. **Particle size reduction**: Breaks down particles or droplets into smaller sizes, often in the micron or submicron range, to improve stability and performance.
3. **Stabilization**: Prevents phase separation in emulsions or suspensions, ensuring the mixture remains homogenous over time.

Applications in Lubrication:

1. **Additive blending**:
    - Ensures that viscosity modifiers, detergents, and other additives are uniformly distributed within the base oil.
2. **Emulsion preparation**:
    - Used to create oil–water emulsions, such as soluble cutting fluids or fire-resistant hydraulic fluids, by stabilizing the mixture.
3. **Improving lubricant stability**:
    - Homogenized lubricants resist separation, ensuring consistent performance in machinery.

Homogenization Process:

1. **High-pressure homogenizers**:
   - Fluids are forced through a narrow valve at high pressure, breaking particles or droplets into smaller sizes and dispersing them evenly.
2. **Ultrasonic homogenizers**:
   - Use ultrasonic waves to create cavitation, which disrupts and disperses particles for a uniform mixture.
3. **Mechanical agitation**:
   - Employs shear forces through mixers or blenders to combine and stabilize components.

Benefits:

1. **Improved product consistency**:
   - Ensures uniform properties across batches, reducing variability and improving performance.
2. **Enhanced stability**:
   - Prevents separation, sedimentation, or coalescence, extending the shelf life of lubricants and emulsions.
3. **Optimized performance**:
   - Uniformly distributed additives ensure effective lubrication, anti-wear protection, and contaminant resistance.
4. **Reduced waste**:
   - Homogenized products minimize the need for remixing or reprocessing, lowering costs and resource usage.

Challenges:

1. **Equipment maintenance**:
   - High-pressure and ultrasonic homogenizers require regular upkeep to maintain efficiency and prevent wear.
2. **Energy-intensive**:
   - The process can consume significant energy, particularly in industrial-scale operations.
3. **Material compatibility**:
   - Proper material selection is essential to avoid contamination or chemical reactions during the process.

Applications Beyond Lubrication:

1. **Food industry**:
   - Used to create uniform emulsions in products like milk, sauces, and dressings.
2. **Pharmaceuticals**:
   - Ensures consistent distribution of active ingredients in suspensions and emulsions.
3. **Cosmetics**:
   - Produces stable creams, lotions, and gels by blending oil and water phases.

Homogenization is a critical process in lubrication, ensuring uniformity, stability, and optimal performance of fluids and lubricants.

395. Housing (Bearing)

A **bearing housing** is a mechanical component designed to support and secure bearings within machinery. It provides structural support, ensures proper alignment, and protects the bearing from contaminants and mechanical damage. Bearing housings are essential in various applications, including industrial machinery, automotive systems, and construction equipment.

Key Functions:

- **Support and alignment**: Bearing housings maintain the correct position and alignment of bearings, ensuring smooth rotational movement and load distribution.
- **Protection**: They shield bearings from external elements such as dust, dirt, moisture, and chemicals, which can adversely affect bearing performance and longevity.
- **Load distribution**: Bearing housings help distribute loads from the bearing to the surrounding structure, contributing to the overall stability and efficiency of the machinery.

Common Types:

- **Pillow block bearings**: These housings have a mounting base with a bearing insert, allowing for easy horizontal and vertical installation and maintenance. See Figure H5.

- **Flanged bearings**: Featuring a flange with two or four mounting holes, these housings are usually mounted on a vertical machine side frame allowing the shaft to be supported through the frame, suitable for applications where space constraints are a concern. See Figure H5.

Figure H5 Overlubricated two bolt housing flange bearing and an underslung pillow block housing bearing. Courtesy: ENGTECH Industries Inc.

- **Plummer blocks**: Designed for high-load applications, plummer blocks are robust housings that provide secure bearing support.

Materials:
Bearing housings are typically made from materials such as cast iron, steel, stainless steel, or thermoplastics, selected based on factors like load capacity, corrosion resistance, and operating environment.

396. HSL

See "Hydrostatic Lubrication."

397. Hydraulic

The term **hydraulic** refers to systems, mechanisms, or components that use pressurized fluid, typically oil or water, to generate, transmit, and control power. Hydraulic systems are widely used in industrial machinery, automotive systems, aerospace, construction equipment, and more due to their ability to transmit large amounts of power efficiently and with precise control.

Key Components of Hydraulic Systems:

1. **Hydraulic fluid**: Acts as the medium for transmitting power and provides lubrication, cooling, and contamination removal.
2. **Pump**: Converts mechanical energy into hydraulic energy by pressurizing the fluid.
3. **Hydraulic actuator**: Converts hydraulic energy into mechanical motion. Includes:
   - **Cylinders**: Produce linear motion.
   - **Motors**: Produce rotary motion.
4. **Valves**: Control the direction, pressure, and flow rate of the hydraulic fluid.
5. **Reservoir**: Stores and supplies hydraulic fluid to the system.
6. **Filters**: Remove contaminants to protect components and maintain fluid performance.
7. **Hoses and pipes**: Transport hydraulic fluid between components.
8. **Seals**: Prevent fluid leaks and maintain pressure.

Advantages of Hydraulic Systems:

1. **High power-to-weight ratio**: Can generate significant force with relatively small components.
2. **Precise control**: Smooth and accurate control of speed, direction, and force.
3. **Versatility**: Suitable for a wide range of applications, from low-force movements to high-power tasks.
4. **Durability**: Operates effectively in harsh environments.

Challenges and Considerations:

1. **Fluid maintenance**: Hydraulic fluid must be clean and properly maintained to prevent contamination and wear.

2. **Leakage**: Leaks can cause system inefficiencies and environmental concerns.

3. **Heat management**: Hydraulic systems generate heat and may require cooling mechanisms to maintain efficiency.

4. **Pressure management**: Overpressure can damage components, requiring the use of relief valves for safety.

398. Hydraulic Cavitation

See "Cavitation."

399. Hydraulic Fluid

Hydraulic fluid is the medium used to transmit power within a hydraulic system. It serves multiple roles beyond power transmission, including lubrication, heat dissipation, and contamination control. The performance of hydraulic systems heavily depends on the quality, properties, and maintenance of the hydraulic fluid.

Key Functions of Hydraulic Fluid:

1. **Power transmission**: Transmits energy from the pump to actuators, enabling motion and force application.

2. **Lubrication**: Reduces friction between moving components such as valves, pumps, and cylinders, preventing wear and tear.

3. **Cooling**: Dissipates heat generated by friction and compression within the hydraulic system.

4. **Contamination removal**: Suspends and transports contaminants to filters, helping to maintain system cleanliness and efficiency.

5. **Corrosion protection**: Provides a protective barrier to prevent oxidation and corrosion of system components.

Types of Hydraulic Fluids:

1. **Mineral oil-based**: Derived from petroleum; commonly used in general hydraulic systems.

2. **Synthetic fluids**: Engineered for extreme temperatures or environmental conditions, offering improved stability and longevity.

3. **Water-based**: Includes water-glycol and water–oil emulsions, offering fire resistance but requiring corrosion inhibitors.
4. **Biodegradable fluids**: Environmentally friendly options made from synthetic esters or vegetable oils, used in eco-sensitive applications.

Key Properties of Hydraulic Fluids:

1. **Viscosity**: Determines the fluid's resistance to flow, affecting power transmission and lubrication efficiency.
2. **Thermal stability**: Ensures consistent performance at high temperatures without breaking down.
3. **Oxidation resistance**: Prolongs fluid life by preventing sludge and deposit formation.
4. **Fire resistance**: Reduces the risk of ignition in high-temperature environments.
5. **Compressibility**: Low compressibility ensures efficient energy transfer and minimal system lag.

Maintenance Considerations:

1. **Regular monitoring**: Check for contamination, viscosity changes, or degradation through oil analysis.
2. **Filtration**: Use filters to remove contaminants and maintain fluid cleanliness.
3. **Temperature control**: Prevent overheating to avoid oxidation and breakdown of the fluid.
4. **Proper handling**: Store in sealed containers to prevent moisture ingress and contamination.

Conclusion:

Hydraulic fluid is a critical component of hydraulic systems, influencing efficiency, reliability, and longevity. Selecting the appropriate fluid type, monitoring its condition, and following proper maintenance practices are essential for optimal system performance.

400. Hydro Cracking Refining Process

The **hydrocracking process** is an advanced petroleum refining technique used to convert heavy crude oil fractions into lighter, more valuable products such as gasoline, diesel, jet fuel, and high-quality base oils. It combines catalytic cracking and hydrogenation under high temperature and pressure, breaking down larger hydrocarbon molecules into smaller, more desirable ones while removing impurities like sulfur and nitrogen.

Key Steps in the Hydrocracking Process:

1. **Feedstock preparation**: Heavy oil fractions, such as vacuum gas oils or residual oils, are pretreated to remove contaminants like metals and salts that can poison the catalyst.
2. **Hydrogenation**: Hydrogen is introduced to saturate hydrocarbons, remove impurities (e.g., sulfur, nitrogen, and oxygen), and stabilize the molecules.
3. **Catalytic cracking**: The treated feedstock passes over a catalyst at high temperature (400–450 °C) and pressure (up to 200 bar), breaking large hydrocarbon molecules into smaller ones.
4. **Separation**: The cracked products are cooled and sent to fractionation towers to separate them into various lighter fractions, such as naphtha, kerosene, and diesel.
5. **Recycling**: Unconverted heavier fractions are recycled back into the hydrocracking unit for further processing.

Key Features of Hydrocracking:

- **Hydrogen addition**: Unlike traditional cracking processes, hydrocracking adds hydrogen to the molecules, enhancing stability and improving the quality of the products.
- **Catalyst use**: Catalysts containing metals like platinum or nickel are used to facilitate the cracking and hydrogenation reactions.
- **High yield**: Produces a higher yield of lighter and cleaner fuels compared to conventional catalytic cracking.
- **Flexibility**: Can be adjusted to produce different product slates based on market demand (e.g., more diesel or jet fuel).

Advantages of Hydrocracking:

- **High-quality products**: Produces fuels and lubricants with low sulfur content, meeting strict environmental regulations.

- **Enhanced base oils**: Generates Group II and Group III base oils with high viscosity index and oxidation stability, essential for modern lubricants.
- **Impurity removal**: Effectively removes sulfur, nitrogen, and aromatic compounds, making products cleaner and more stable.
- **Versatility**: Converts a wide range of feedstocks, including heavy crude oils and residues, into valuable products.

Applications in Lubrication:

1. **Base oil production**:
   - Hydrocracking is a key process for producing high-quality base oils used in synthetic and semisynthetic lubricants.
   - These base oils exhibit superior performance in terms of thermal stability, oxidation resistance, and low-temperature fluidity.
2. **Cleaner fuels**: Fuels produced through hydrocracking help reduce engine deposits, enhancing the performance and longevity of lubricated components.

Challenges:

1. **High operating costs**: Requires significant investment in equipment, catalysts, and hydrogen production.
2. **Catalyst sensitivity**: Catalysts are prone to deactivation by contaminants like metals and asphaltenes, necessitating stringent feedstock pretreatment.
3. **Hydrogen dependency**: The process relies on a steady supply of hydrogen, which can increase costs and environmental impact depending on the hydrogen source.

The hydrocracking refining process plays a pivotal role in modern petroleum refining, enabling the production of cleaner, higher-quality fuels and base oils. Its versatility, efficiency, and ability to meet environmental standards make it a cornerstone of the oil and lubrication industries.

401. Hydrocarbon

In the context of lubrication, a hydrocarbon refers to a class of organic compounds made up solely of carbon and hydrogen atoms that serve as the foundational molecules for many base oils and lubricants. These hydrocarbons determine the chemical properties and performance characteristics of lubricating oils.

Types of Hydrocarbons in Lubrication:

1. **Paraffinic hydrocarbons**:
   - Linear or branched chains of carbon atoms.
   - Common in mineral oils.
   - Advantages: High thermal and oxidative stability, good viscosity index.
   - Example: Used in automotive and industrial lubricants.
2. **Naphthenic hydrocarbons**:
   - Saturated cyclic structures (rings).
   - Known for low pour points and good solvency for additives.
   - Common in transformer oils and specialty lubricants.
3. **Aromatic hydrocarbons**:
   - Contain conjugated ring structures with double bonds.
   - Excellent solvent properties but less stable under oxidation.
   - Used in certain industrial lubricant formulations.
4. **Synthetic hydrocarbons**:
   - Man-made hydrocarbon compounds, such as **polyalphaolefins (PAOs)**.
   - Offer superior thermal stability, low-temperature performance, and extended service life.
   - Commonly used in high-performance and extreme environment applications.

Importance of Hydrocarbons in Lubrication:

- **Base oils**: Hydrocarbons form the primary component of many lubricants, influencing viscosity, volatility, and oxidation resistance.
- **Additive compatibility**: Hydrocarbon structure affects the solubility and effectiveness of additives like antioxidants and detergents.
- **Performance characteristics**: Tailored hydrocarbon chains provide specific properties such as wear protection, film strength, and temperature stability.

Understanding hydrocarbon structures and their behavior in lubrication systems is critical for designing and selecting lubricants that meet specific industrial and automotive requirements.

402. Hydrodynamic Lubrication (HDL)

Hydrodynamic lubrication (HDL) is a lubrication regime in which a full fluid film completely separates two moving surfaces under load, preventing direct metal-to-metal contact. This regime relies on the relative motion of the surfaces to generate a pressure within the lubricant film, sufficient to support the applied load and minimize friction and wear. Any friction present in this state is due to fluid friction within the lubricant, which will commence to increase once the film thickness (See Film Thickness Ratio – Lambda) becomes greater than Lambda 4. Figure H6 demonstrates hydrodynamic full film lubrication (HDL) at work in a journal bearing

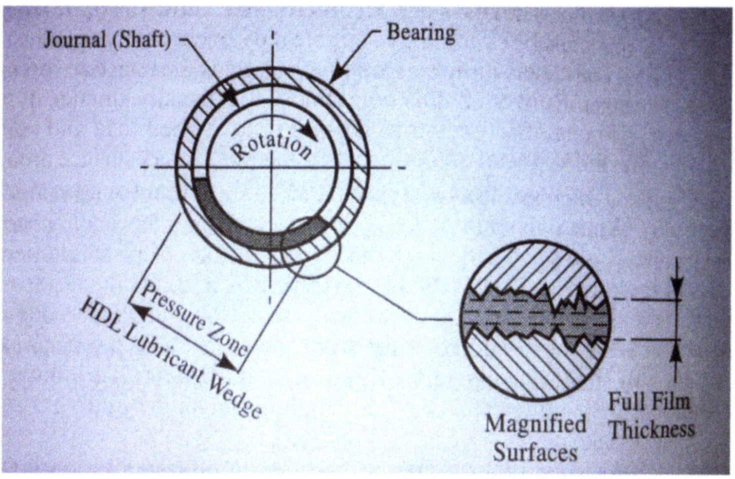

Figure H6 Hydrodynamic lubrication with full surface separation. Source: Machinery Handbook 27th Edition, Kenneth E Bannister.

Key Features of HDL

- **Full fluid film separation**:
  A continuous layer of lubricant prevents metal-to-metal contact between the moving surfaces. This fluid film eliminates surface asperity interactions, minimizing wear.

- **Load support through fluid pressure**:
  The lubricant film pressure is generated by the motion and shape of the surfaces, typically forming a wedge-shaped zone allowing the fluid film to bear the applied load, protecting surfaces from damage.

- **Dynamic conditions**:
  HDL requires relative motion between the surfaces to maintain the lubricant film. Stationary conditions or insufficient speed can lead to boundary or mixed lubrication regimes.

- **Low friction**:
  The absence of direct surface contact results in minimal friction, dominated by the viscous shear of the lubricant.

Conditions for HDL

- **Surface design**:
  Shapes that promote fluid film generation, such as tapered or wedge profiles, are essential.

- **Relative speed**:
  Adequate relative motion between surfaces is necessary to sustain the hydrodynamic film.

- **Lubricant properties**:
  Proper viscosity and flow characteristics are critical for film formation and load-bearing capacity.

Common Applications

- **Journal bearings**: The rotating shaft generates a hydrodynamic film within the bearing clearance, supporting the shaft load and reducing friction.

- **Thrust bearings**: Flat or tapered surfaces in relative motion develop a lubricant film to support axial loads.

- **Sliding surfaces**: Mechanisms like pistons or guides use HDL to minimize wear and ensure smooth operation.

Hydrodynamic lubrication plays a crucial role in reducing wear, increasing efficiency, and extending the life of components in systems where relative sliding motion occurs. It is fundamental to the reliable operation of many industrial and mechanical systems.

403. Hydrogeneration

Hydrogenation is a chemical process that involves adding hydrogen molecules to unsaturated compounds, such as alkenes or alkynes, and converting them into saturated compounds. In the context of lubricants, hydrogenation is employed to enhance the stability and performance of base oils used in lubricant formulations.

Process Overview:
Hydrogenation of base oils involves reacting unsaturated hydrocarbons with hydrogen gas in the presence of a catalyst, typically at elevated temperatures and pressures. This reaction saturates the carbon–carbon double bonds, resulting in more stable and less reactive hydrocarbons.

Benefits:

- **Improved oxidative stability**: Saturated hydrocarbons are less prone to oxidation, leading to longer oil life and reduced formation of harmful acids and sludge.
- **Enhanced viscosity index**: Hydrogenated base oils exhibit a more consistent viscosity across a wide temperature range, ensuring effective lubrication under varying operating conditions.
- **Reduced aromatic content**: The process decreases the concentration of aromatic compounds, which can be detrimental to engine components and the environment.

Applications:

- **Automotive lubricants**: Hydrogenated base oils commonly used in engine oils provide superior performance and longevity.
- **Industrial lubricants**: development of machinery lubricants requiring high thermal stability and resistance to oxidative degradation.

404. Hydrolysis

Hydrolysis in lubrication refers to the chemical breakdown of lubricant molecules when exposed to water. This reaction can lead to the degradation of both the base oil and its additives, resulting in the formation of acids and other by-products that compromise the lubricant's performance.

Key Characteristics:

- **Chemical decomposition**: Hydrolysis involves the splitting of chemical bonds in the lubricant molecules due to the presence of water, leading to the formation of acids and other degradation products.
- **Impact on additives**: Additives within the lubricant, such as antiwear agents, can also undergo hydrolysis, resulting in the formation of acids that further degrade the lubricant's performance.

Effects on Lubricant Performance:

- **Acid formation**: Hydrolysis can produce carboxylic acids, which can lower the pH of the lubricant, leading to corrosion and increased wear on metal surfaces.
- **Viscosity reduction**: The breakdown of lubricant molecules can lead to a decrease in viscosity, impairing the lubricant's ability to form a protective film and increasing the risk of metal-to-metal contact.
- **Additive depletion**: Hydrolysis can deplete essential additives, such as antiwear agents, reducing the lubricant's effectiveness and potentially leading to increased wear and component failure.

Mitigation Strategies:

- **Water contamination control**: Implementing measures to prevent water ingress into the lubrication system is crucial. This includes sealing systems properly and monitoring for leaks.
- **Regular monitoring**: Routine analysis of the lubricant can help detect early signs of hydrolysis, such as changes in acidity or the presence of degradation products.
- **Selecting appropriate lubricants**: Choosing lubricants with high hydrolytic stability, especially in applications prone to water exposure, can reduce the risk of hydrolysis-related degradation.

Understanding and managing hydrolysis is essential for maintaining the performance and longevity of lubricants, thereby ensuring the reliability and efficiency of machinery and equipment.

405. Hydrolytic Stability

The **ASTM D2619-09** standard test method, commonly known as the "Coke bottle test," is used to assess a lubricant's hydrolytic stability. This test involves mixing the lubricant with water, adding a copper strip, and subjecting the mixture to heat and rotation. The results indicate the lubricant's resistance to acid formation and degradation.

Understanding and ensuring hydrolytic stability is essential for selecting and maintaining lubricants that perform reliably in environments where water contamination is a concern.

406. Hydrometer

A **hydrometer** is an instrument used to measure the density or specific gravity of liquids, including lubricants. In the context of lubricants, hydrometers are employed to assess the density of oil and petroleum products, which is crucial for quality control and ensuring the proper formulation of lubricants.

It typically consists of a sealed, hollow glass tube with a weighted bulb at one end to ensure it floats upright. The tube is graduated with scales that correspond to various properties, such as specific gravity, allowing users to determine the liquid's density by observing the point where the liquid's surface intersects the scale. See Figure H7.

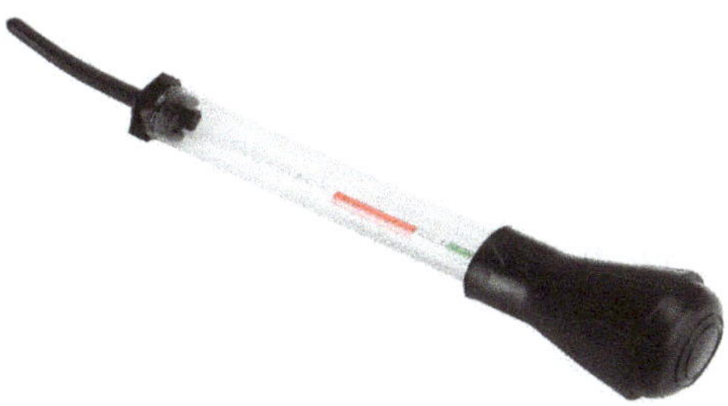

Figure H7 Typical hydrometer.

Key Features:

- **Design**: Hydrometers consist of a sealed, hollow glass tube with a weighted bulb at one end to ensure it floats upright. The tube is graduated with scales that correspond to various properties, such as specific gravity, allowing users to determine the liquid's density by observing the point where the liquid's surface intersects the scale.
- **Calibration**: They are calibrated with specific scales, such as specific gravity, to provide accurate measurements. Some hydrometers may include multiple scales for different applications.

Applications in Lubricant Analysis:

- **Quality control**: Measuring the density of lubricants helps ensure they meet specified standards, which is essential for maintaining the performance and longevity of machinery.
- **Formulation verification**: Hydrometers assist in verifying that lubricant formulations contain the correct proportions of base oils and additives, ensuring optimal performance characteristics.
- **Batch consistency**: Regular density measurements can detect variations between batches, allowing for adjustments to maintain consistent product quality.

By providing a simple yet effective means of measuring liquid density, hydrometers are essential tools in the lubricant industry, ensuring product quality and consistency.

407. Hydrophilic Lubrication

Hydrophilic lubrication involves utilizing water or water-based solutions as lubricants to reduce friction and wear between surfaces. This approach leverages the natural affinity of hydrophilic (water-attracting) substances to form stable, lubricating films in aqueous environments.

Key Characteristics:

- **Water-based lubricants**: Hydrophilic lubrication employs water or water-based solutions, such as hydrogels or aqueous emulsions, to create lubricating films between surfaces.
- **Surface interactions**: Hydrophilic surfaces attract water molecules, facilitating the formation of a lubricating layer that reduces direct contact between surfaces, thereby minimizing friction and wear.

Applications:

- **Biomedical devices**: Hydrophilic lubrication is utilized in medical devices, such as catheters and implants, to reduce friction and enhance biocompatibility. For instance, a study developed a biomimetic bi-layered hydrophilic lubrication coating on silicone elastomers to improve their lubricity and wear resistance.
- **Cartilage replication**: Research has explored hydrophilic lubrication mechanisms inspired by natural cartilage to develop artificial joints with enhanced lubricating properties. This approach aims to mimic the

lamellar-repulsive mechanism of natural lubrication, where hydrophilic surfaces interact to reduce friction.

- **Nanotribology**: At the nanoscale, hydrophilic lubrication is studied to understand and control frictional forces. For example, investigations into the slippery behavior of lubricating fluid-coated smooth hydrophilic surfaces have provided insights into nanoscale friction and lubrication mechanisms.

Hydrophilic lubrication offers advantages in specific applications where water-based solutions are preferable due to their biocompatibility, environmental friendliness, and cost-effectiveness. However, the effectiveness of hydrophilic lubrication depends on factors such as surface properties, operating conditions, and the specific requirements of the application.

408. Hydrophobic Lubrication

Hydrophobic lubrication involves utilizing hydrophobic (water-repelling) substances to reduce friction and wear between surfaces. These substances form lubricating films that minimize direct contact between surfaces, thereby enhancing performance and longevity.

Key Characteristics:

- **Water repellency**: Hydrophobic lubricants repel water, preventing corrosion and degradation of components.

- **Lubricant film formation**: Hydrophobic substances create stable films that separate surfaces, reducing friction and wear.

Applications:

- **Pharmaceutical industry**: Hydrophobic lubricants, such as magnesium stearate, are commonly used in tablet manufacturing to improve processing and product quality.

- **Biomedical devices**: Hydrophobic lubrication mechanisms, inspired by natural joints, are studied to develop artificial joints with enhanced lubricating properties.

- **Nanotribology**: Research into hydrophobic lubrication at the nanoscale aims to understand and control frictional forces in micro- and nano-devices.

By leveraging the properties of hydrophobic substances, hydrophobic lubrication offers advantages in various applications, including corrosion resistance, enhanced performance, and increased component lifespan.

409. Hydrostatic Lubrication (HSL)

Hydrostatic lubrication (HSL) is a lubrication method where an external pressure source, such as a pump, supplies lubricant to the interface between two surfaces to hydraulically separate and "float" one surface over another (see Figure H8). This pressurized lubricant forms a film that separates the surfaces, preventing direct contact and thereby reducing friction and wear.

This film regime is typically utilized on precision machines and machine tools such as a plunge grinder in which the grindstone carriage is "floated" into the workpiece to perform precision grinding on gears, etc. HSL is similar to HDL in that it also provides full film separation.

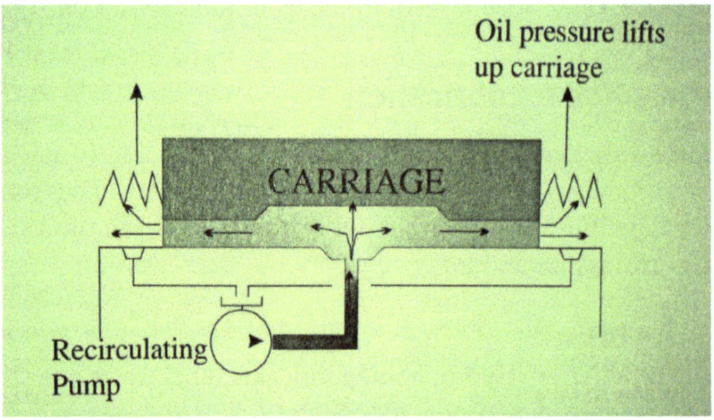

Figure H8 Hydrostatic machine tool lift system. Source: Machinery Handbook 27th Edition, Kenneth E Bannister.

Key Characteristics:

- **External pressure application**: Unlike hydrodynamic lubrication, which relies on the relative motion between surfaces to generate pressure, hydrostatic lubrication uses an external pump to pressurize the lubricant.
- **Stable lubricant film**: The pressurized lubricant maintains a consistent film between surfaces, ensuring separation even under varying loads and speeds.
- **High load capacity**: This method is particularly effective in applications requiring high load-bearing capacity, as the external pressure can support substantial loads without direct surface contact.

Applications:

- **Machine tool slides and bearings**: Hydrostatic lubrication is commonly used in precision machinery to ensure smooth movement and high load capacity.
- **High-precision bearings**: In applications where minimal wear and high precision are critical, such as in aerospace and semiconductor manufacturing, hydrostatic bearings provide the necessary support.
- **Hydraulic systems**: Hydrostatic lubrication is utilized in hydraulic systems to maintain pressure and ensure the proper functioning of components.

410. Hydrostatic Side-feed Lubricator

A **hydrostatic side-feed lubricator** is a type of lubricator commonly used in steam locomotives to supply lubricating oil to the cylinders and other moving parts. It operates on the principle of hydrostatic pressure, where steam enters a condensing chamber, and the resulting condensate creates pressure that forces oil into the lubricator's delivery system. This design ensures a consistent and controlled flow of oil to the engine components, thereby reducing friction and wear. See Figure H9.

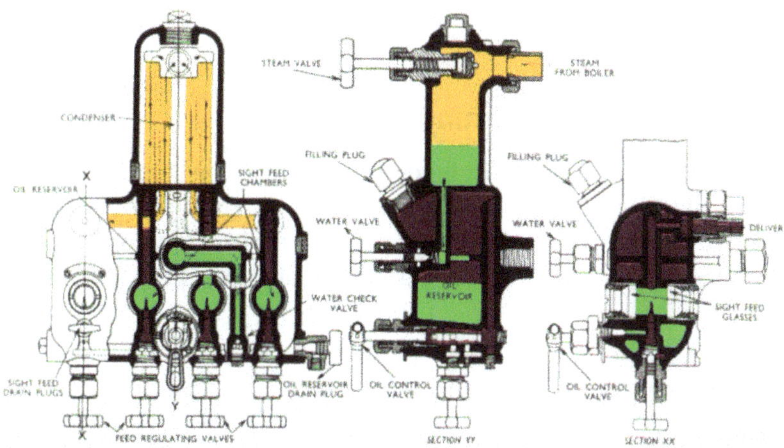

Figure H9 Cross-section of a typical steam engine hydrostatic side feed oiler.

Key Features:

- **Hydrostatic operation**: Utilizes steam pressure to generate the necessary force for oil delivery, ensuring a steady supply of lubricant to critical engine parts.
- **Side-feed design**: Features a side-mounted oil reservoir, allowing for easier access and maintenance.
- **Multiple feeds**: Equipped with multiple delivery lines, enabling the distribution of oil to various parts of the locomotive, such as the cylinders, air pumps, and valve gear.

Applications:

- **Steam locomotives**: Widely used in steam-powered trains to maintain engine efficiency and longevity by providing necessary lubrication to moving parts.
- **Industrial steam engines**: Employed in stationary steam engines where consistent lubrication is essential for optimal performance.

The hydrostatic side-feed lubricator's design ensures that steam engines receive the necessary lubrication to operate smoothly, thereby enhancing performance and extending the lifespan of engine components.

411. Hydrotreated Base Stock

See "Base Oil Group" (Group II)

412. Hydrotreating Refining Process

Hydrotreating is a catalytic refining process that enhances the quality of lubricant base oils by removing impurities such as sulfur, nitrogen, oxygen, and metals. This process involves reacting the feedstock with hydrogen gas in the presence of a catalyst under high temperature and pressure conditions.

Key Characteristics:

- **Improved oxidation stability**: By removing sulfur and other reactive compounds, hydrotreating enhances the base oil's resistance to oxidation, thereby extending the service life of lubricants.
- **Enhanced viscosity index (VI)**: The process can increase the VI of base oils, ensuring consistent performance across a wide range of temperatures.

- **Reduced aromatic content**: Hydrotreating reduces the aromatic hydrocarbons in base oils, which can improve lubricity and reduce the formation of sludge and deposits.

Typical Hydrotreating Three-step Process:

1. **Feed preparation**: The feedstock, often a heavier distillate, is mixed with hydrogen gas and preheated to the desired temperature.
2. **Reaction**: The mixture enters a fixed-bed reactor containing a catalyst, where hydrogenation reactions occur, converting impurities into hydrogen sulfide, ammonia, and other by-products.
3. **Separation**: The treated oil is separated from the by-products, and the hydrogen sulfide is removed, often through a Claus process to recover elemental sulfur.

By effectively removing impurities, hydrotreating ensures that lubricant base oils meet stringent performance standards, contributing to the longevity and efficiency of engines and machinery.

413. Hydroxystereate

Hydroxystearate refers to a class of chemical compounds derived from stearic acid, a saturated fatty acid. These compounds are characterized by the addition of a hydroxyl group (-OH) to the stearic acid molecule, resulting in various derivatives with distinct properties and applications.

Common Hydroxystearate Compounds:

- **Lithium 12-hydroxystearate**: This lithium salt of 12-hydroxystearic acid is widely used in the production of lithium-based greases. These greases are valued for their water resistance, oxidative stability, and mechanical performance, making them suitable for applications in vehicles, aircraft, and heavy machinery.
- **Glyceryl hydroxystearate**: An ester of glycerol and hydroxystearic acid, glyceryl hydroxystearate is utilized in cosmetic formulations as an emulsifying agent, emollient, and bodying agent. It contributes to the stability and texture of products like moisturizers, shampoos, and lotions.
- **Isostearyl hydroxystearate**: This ester, derived from isostearyl alcohol and hydroxystearic acid, serves as an occlusive emollient in skincare

products. It helps reduce transdermal moisture loss, enhancing skin hydration and providing a soft, smooth feel.

Applications:

- **Lubricants**: Hydroxystearates, particularly lithium 12-hydroxystearate, are integral components in the formulation of lubricating greases. Their inclusion enhances the grease's performance by improving water resistance and oxidative stability.
- **Cosmetics and personal care**: Compounds like glyceryl hydroxystearate and isostearyl hydroxystearate are employed in various cosmetic products to improve texture, stability, and skin feel. They function as emulsifiers, emollients, and bodying agents, contributing to the overall efficacy and sensory attributes of the products.

Hydroxystearates are versatile compounds with applications spanning from industrial lubricants to cosmetic formulations, valued for their ability to enhance product performance and stability.

414. Hygrometric

The term **hygrometric** pertains to the measurement of humidity, specifically the moisture content in the air. In the context of lubrication, understanding and controlling humidity is crucial, as excessive moisture can adversely affect lubricant performance and the longevity of machinery components.

Impact of Humidity on Lubrication:

- **Oxidation**: Elevated humidity levels can accelerate the oxidation of lubricants, leading to the formation of acids and sludge. This degradation diminishes the lubricant's effectiveness and can cause corrosion of metal surfaces.
- **Contamination**: Moisture can introduce contaminants into the lubrication system, compromising the lubricant's integrity and potentially causing abrasive wear.
- **Viscosity changes**: Water contamination can alter the viscosity of lubricants, affecting their ability to form protective films and increasing the risk of metal-to-metal contact.

Managing Humidity in Lubrication Systems:

- **Sealing and containment**: Implementing effective sealing mechanisms prevents the ingress of moisture into lubrication systems.

- **Desiccant breathers**: Utilizing desiccant breathers can absorb moisture from the air, maintaining optimal humidity levels within a lubricant reservoir or storage container (see Desiccant Breather).
- **Regular monitoring**: Employing hygrometric instruments to monitor humidity levels ensures that the lubrication environment remains within the desired parameters.

By maintaining appropriate humidity levels, the performance and lifespan of lubricants and machinery components are significantly enhanced, leading to more efficient and reliable operations.

415. Hygroscopic

Hygroscopic refers to a substance's ability to absorb moisture from its environment. In the context of lubrication, this property is significant because certain lubricants, especially synthetic oils like polyol esters (POE) and polyalkylene glycols (PAG), are hygroscopic. This means they can absorb water from the air, leading to potential issues such as hydrolysis, which can degrade the lubricant and cause corrosion in machinery components.

Implications of Hygroscopic Lubricants:

- **Hydrolysis risk**: When these lubricants absorb moisture, they may undergo chemical reactions that break down the lubricant, producing acids and other by-products that can damage equipment.
- **Corrosion potential**: The presence of water in the lubricant can lead to corrosion of metal parts, compromising the integrity and performance of machinery.

Managing Hygroscopic Lubricants:

- **Storage**:
  - Store lubricants in sealed containers to prevent moisture absorption.
  - Utilize desiccant-style breathers to control moisture in the reservoir headspace.
- **Handling**: Minimize exposure to air during maintenance or refilling to reduce the risk of water contamination.
- **Monitoring**: Regularly check for signs of water contamination and replace the lubricant if necessary.

Understanding the hygroscopic nature of certain lubricants is crucial for maintaining equipment longevity and ensuring optimal performance.

416. Hypoid Gear Oil

Hypoid gear oil is a specialized lubricant formulated to meet the unique demands of hypoid gears (see Figure H10), commonly found in automotive differentials and certain industrial machinery. These gears are designed to transmit power between shafts that are not parallel, creating a sliding and rolling motion that generates high levels of friction and wear. To withstand these extreme conditions, hypoid gear oil is formulated with unique additives and properties to protect and extend the life of these gears.

Figure H10 Typical hypoid gear and pinion set.

Key Characteristics:

- **Extreme pressure (EP) additives**: Hypoid gear oils contain high levels of EP additives, which are designed to withstand the high contact pressures and sliding friction generated by hypoid gears. These additives form a protective film on the gear surfaces, preventing metal-to-metal contact and reducing wear.
- **High viscosity**: These oils have a higher viscosity compared to standard gear oils, ensuring a consistent lubricating film under the high-pressure conditions typical of hypoid gear operation.
- **Oxidation resistance**: Formulated to resist oxidation, hypoid gear oils maintain their performance over extended periods, even under high-temperature conditions.

- **Rust and corrosion protection**: They provide protection against rust and corrosion, safeguarding metal components from environmental factors.

Applications:
- **Automotive differentials**: Hypoid gear oils are essential for the proper functioning of automotive differentials, ensuring smooth operation and longevity of the gear components.
- **Heavy-duty trucks**: These oils are used in heavy-duty truck differentials, where high torque and load handling are needed.
- **Industrial machinery**: Certain industrial applications that utilize hypoid gears also require these specialized lubricants to maintain optimal performance.

Using hypoid gear oil for hypoid gears is crucial for maintaining the health and longevity of the gears.

I

417. ICML

See the "International Council for Machinery Lubrication."

418. ILSAC

See "International Lubricant Standardization and Approval Committee."

419. Immiscible

In lubrication, **immiscible** refers to the inability of two liquids to mix or form a homogeneous solution. When two immiscible liquids are introduced into a lubrication system, they tend to separate into distinct layers due to differences in polarity and molecular structure. For example, water and mineral oil are immiscible because of their differing polarities; water molecules are highly polar, while mineral oil consists of nonpolar hydrocarbon molecules.

Implications for Lubrication Systems:

- **Layer formation**: In systems where immiscible liquids are present, such as water and oil, the liquids will form separate layers as shown in Figure I1. This separation can affect the distribution and effectiveness of the lubricant, potentially leading to inadequate lubrication and increased wear.
- **Compatibility issues**: Introducing an immiscible liquid into a lubrication system can lead to compatibility problems, as the two liquids may not function synergistically. This incompatibility can result in performance degradation and potential damage to components.
- **Maintenance challenges**: The presence of immiscible liquids can complicate maintenance procedures, as special care is needed to manage the separation and ensure that each liquid is properly handled and disposed of.

DOI: 10.1201/9788743806608-425

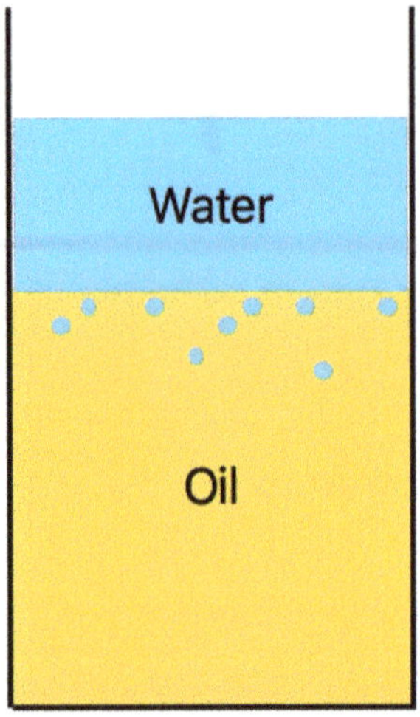

Figure I1 Oil and water are immissicible liquids.

To maintain optimal performance and longevity of machinery, it is crucial to ensure that the lubricants used are compatible, and if necessary, miscible with other fluids present in the system.

420. Indicator (Lubrication KPI's)

In the context of lubrication, an **indicator** refers to a measurable parameter or tool used to assess the effectiveness and efficiency of lubrication practices. These indicators provide insights into the management and performance of lubrication programs and systems, helping to identify areas for improvement and ensuring optimal machinery operation.

Key Performance Indicators (KPIs) in Lubrication:
KPIs are specific metrics that evaluate various aspects of lubrication management. They are typically categorized into leading and lagging indicators:

- **Leading indicators**: These KPI metrics predict future performance by monitoring activities and processes. Examples include:
  - **Scheduled lubrication compliance**: The percentage of lubrication tasks completed as per the maintenance schedule.
  - **Lubricant contamination levels**: Monitoring the cleanliness of lubricants to prevent contamination-related failures.

- **Lagging Indicators**: These metrics reflect past performance and outcomes. Examples include:
  - **Mean time between lubrication failures (MTBLF)**: The average time between lubrication caused equipment failures, indicating the effectiveness of the lubrication program in preventing wear.
  - **Lubrication maintenance costs**: The total expenses related to lubrication maintenance, including labor and materials. These are usually compared as a % of overall maintenance costs.

By analyzing these indicators, organizations can make informed decisions to enhance lubrication practices, leading to improved equipment reliability and reduced downtime.

Additional Indicators:
Beyond KPIs, other indicators can be used to assess lubrication effectiveness:

- **Oil analysis results**: Regular testing of lubricant samples to detect contaminants, degradation, and wear particles.
- **Lubricant consumption rates**: Monitoring the amount of lubricant used over time to identify leaks or excessive consumption.

Implementing a comprehensive set of indicators allows for proactive maintenance strategies, ensuring that lubrication systems function optimally and contribute to the overall efficiency of operations.

421. Induction Period

In lubrication, the **induction period** refers to the initial phase during which a lubricant will resist oxidation thanks to the presence of antioxidants and other stabilizing additives. During this period, the lubricant maintains its performance characteristics, effectively protecting machine components. Once antioxidants are depleted, the lubricant will begin to oxidize, leading to an increase in viscosity and the commencement of acid, sludge, and deposit build-up that will cause rapid deterioration in the effectiveness of the lubricant.

Key Points:

- **Oxidation resistance**: The induction period is characterized by the lubricant's ability to resist oxidation, preserving its protective qualities.

- **Additive depletion**: The duration of the induction period depends on the concentration and effectiveness of antioxidants and other additives in the lubricant.
- **Performance decline**: After the induction period, the lubricant's performance deteriorates, leading to potential engine damage if not replaced or replenished.

Understanding the induction period is crucial for determining appropriate oil change intervals and ensuring optimal machine and engine performance.

422. Industrial Fluid Handling (IFH)

Industrial fluid handling (IFH) refers to the **storage**, **transfer**, **filtration**, **and dispensing of industrial lubricants**, hydraulic fluids, and other process oils used in manufacturing, maintenance, and heavy machinery operations. Proper fluid handling ensures contamination control, fluid integrity, and optimized lubricant performance across industrial applications.

Key Functions of Industrial Fluid Handling (IFH)

- **Ensures proper lubricant storage and dispensing**: Prevents cross-contamination and degradation by using dedicated storage and transfer systems.
- **Improves contamination control**: Reduces ingress of dirt, moisture, and airborne particles that can degrade lubricants and damage machinery.
- **Optimizes lubricant transfer and application**: Ensures precise fluid dispensing and delivery, reducing waste and improving efficiency.
- **Extends equipment and lubricant life**: Proper handling minimizes fluid oxidation, varnish formation, and premature breakdown.

Components of Industrial Fluid Handling Systems

- **Fluid storage tanks and containers**: Used for bulk storage of lubricants, hydraulic fluids, and coolants to maintain fluid integrity.
- **Filtration and contamination control systems**: Inline filters, desiccant breathers, and kidney-loop filtration units remove particles, water, and contaminants.

- **Lubricant dispensing equipment**: Pumps, metering valves, and grease guns ensure accurate lubricant delivery.
- **Fluid transfer systems**: Hoses, pipelines, and dedicated transfer pumps prevent cross-contamination between lubricant types.
- **Fluid monitoring and analysis tools**: Sensors and on-site testing equipment track fluid condition, viscosity, and contamination levels.

423. Inert

In lubrication, **inert** refers to substances that are chemically stable and unreactive under specific operating conditions. Inert lubricants do not readily undergo chemical reactions, making them ideal for applications where chemical stability is crucial.

Key Characteristics of Inert Lubricants:

- **Chemical stability**: Most lubricants are designed to resist chemical reactions, such as oxidation or degradation, ensuring consistent performance over time.
- **Nonreactivity**: Lubricants do not react with other substances, including metals and gases, preventing unwanted chemical interactions that could compromise machine integrity.
- **High thermal stability**: Inert lubricants can withstand extreme temperatures without breaking down, making them suitable for high-temperature applications.

Applications:

- **Aerospace industry**: Inert lubricants are used in aerospace applications where chemical stability is essential for the reliability of components exposed to harsh environments.
- **Semiconductor manufacturing**: Inert lubricants are employed in semiconductor manufacturing to prevent chemical reactions that could damage sensitive equipment.
- **High-temperature machinery**: Inert lubricants are ideal for machinery operating at elevated temperatures, as they maintain their properties without degradation.

By utilizing inert lubricants, industries can enhance the longevity and reliability of their equipment, ensuring optimal performance in demanding conditions.

424. Infra-red Thermography

Infrared thermography is a noninvasive diagnostic technique that utilizes infrared cameras to detect and visualize heat patterns emitted by objects. In the context of lubrication, this method is particularly valuable for identifying thermal anomalies in machinery components, which can indicate lubrication-related issues.

Applications in Lubrication:

- **Bearing inspections**: Elevated temperatures in bearings often signal inadequate lubrication or contamination. Infrared thermography can detect these temperature anomalies, allowing for timely maintenance. When motors bearings are lubricated incorrectly, lubricating grease can find its way into the winding causing it to overheat and eventually fail. Thermography can easily detect overheated motors as depicted in Figure I2.

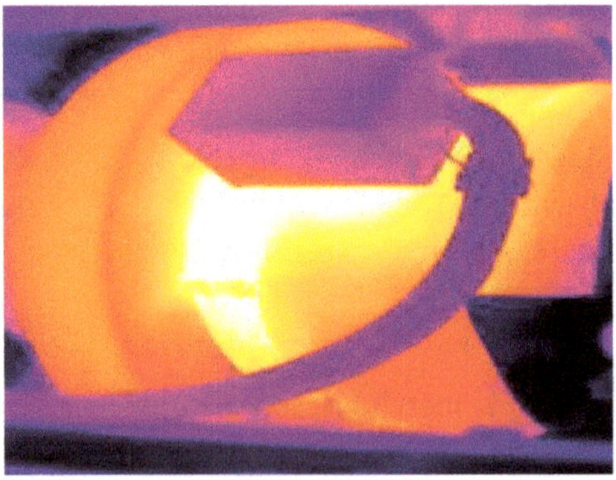

Figure I2 Typical "hot" electric motor requiring further internal inspection.

- **Lubricant delivery system balancing**: Individual delivery points can be calibrated using bearing temperature signatures.
- **Misalignment detection**: Misaligned components generate excessive heat due to increased friction. Thermal imaging can reveal these hot spots, prompting corrective actions to prevent further damage.

- **Fluid flow assessment**: In systems where lubricants are circulated, thermal imaging can identify areas of restricted or uneven fluid flow, for example, heat exchanger or distribution blocks, where restriction may lead to overheating and premature component wear or failure.
- **Lubricant pollution**: Lubricant loss on land and in water can be identified using thermal imaging as the lubricant fluorescence emits radiation at a different rate than its surroundings identifying the presence of lubricant loss/pollution. See Figure I3.

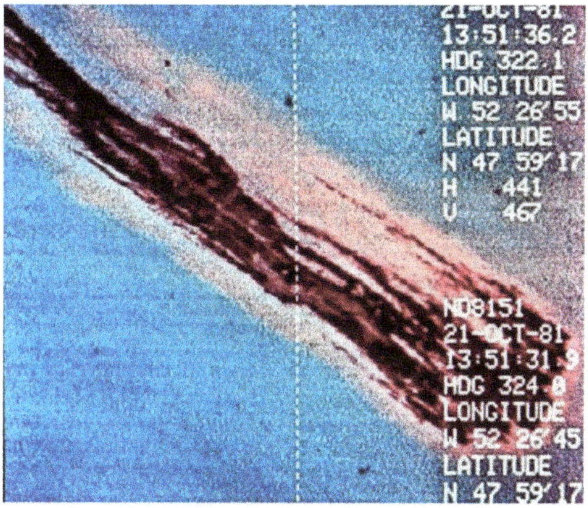

Figure I3 Overhead view depicting an oil slick presence in a water body. Source: Environment Canada.

By integrating infrared thermography into maintenance routines, industries can proactively address lubrication-related issues, enhancing equipment reliability and extending service life.

425. Inhibitor

In lubrication, an **inhibitor** is a chemical additive incorporated into a lubricant to enhance its performance and protect machinery components. These additives serve various functions, including preventing corrosion, reducing wear, and improving the lubricant's overall stability.

Types of Inhibitors in Lubricants:

a. **Corrosion inhibitor**: These additives form protective films on metal surfaces, shielding them from corrosive elements such as moisture and oxygen. They are essential in preventing rust and oxidation, thereby extending the lifespan of equipment.

b. **Antioxidants**: Antioxidant additives slow down the oxidation process of the lubricant, maintaining its effectiveness over time. This is particularly important in high-temperature applications where oxidation can lead to the formation of acids and sludge.

c. **Antiwear agents**: These compounds create protective layers on metal surfaces, reducing friction and wear. They are crucial in high-pressure environments where metal-to-metal contact can lead to significant damage.

d. **Metal deactivators**: These additives neutralize metal ions that can catalyze the degradation of the lubricant, thereby enhancing its stability and performance.

e. **Dispersants**: Dispersant additives keep contaminants, such as soot and sludge, suspended in the lubricant, preventing them from settling and causing damage to engine components.

Benefits of Using Inhibitors in Lubricants:

- **Enhanced equipment longevity**: By preventing corrosion and wear, inhibitors help extend the lifespan of machinery components.
- **Improved lubricant stability**: Inhibitors maintain the lubricant's effectiveness over time, reducing the need for frequent oil changes.
- **Reduced maintenance costs**: Protecting equipment from corrosion and wear leads to fewer repairs and downtime, resulting in cost savings.

426. Inorganic Compound

In the context of lubricants, **inorganic compounds** are chemical substances that do not contain carbon–hydrogen bonds. These compounds are utilized in various forms to enhance the performance and longevity of lubricants.

Types of Inorganic Compounds in Lubricants:
Solid lubricants: These are inorganic materials that function as lubricants in their solid state, reducing friction and wear between moving parts. Common examples include:

- **Graphite**: Known for its layered structure, graphite provides excellent lubrication properties, especially in high-temperature applications.
- **Molybdenum disulfide (MoS_2)**: This compound offers low friction and high load-carrying capacity, making it suitable for extreme pressure conditions.

Inorganic additives: These compounds are incorporated into lubricants to impart specific properties:

- **Extreme pressure (EP) additives**: Inorganic EP additives, such as molybdenum disulfide, form protective films on metal surfaces under high-pressure conditions, preventing wear and damage.
- **Corrosion inhibitors**: Inorganic compounds can act as corrosion inhibitors, protecting metal surfaces from rust and degradation.

Benefits of Inorganic Compounds in Lubricants:

- **Enhanced load-carrying capacity**: Inorganic solid lubricants like molybdenum disulfide provide high load-carrying capacity, making them ideal for heavy-duty applications.
- **High-temperature stability**: Inorganic compounds often exhibit superior thermal stability, allowing lubricants to perform effectively at elevated temperatures.
- **Corrosion resistance**: Inorganic additives can form protective layers on metal surfaces, enhancing the lubricant's ability to prevent corrosion.

Incorporating inorganic compounds into lubricants is essential for performance optimization, extending equipment lifespan, and ensuring reliability across various industrial applications.

427. Inorganic Thickener

In lubricants, particularly greases, **inorganic thickeners** are substances that help deliver a semisolid consistency to the lubricant without relying on metallic soaps. These thickeners are typically derived from inorganic compounds such as silica, clay, or calcium sulfonate. They are often used in specialized applications where traditional soap-based thickeners may not be suitable.

Common Types of Inorganic Thickeners:

- **Calcium sulfonate**: Known for its excellent shear stability, high-temperature resilience, and resistance to corrosive environments, calcium sulfonate greases are favored in demanding applications.
- **Bentonite clay**: Bentonite grease, formulated with bentonite clay, is known as a nonmelt lubricant because of its indefinite dropping point. It resists temperature changes, and offers superior wear protection, exceptional water tolerance, good mechanical stability, and strong adhesion.
- **Fumed silica**: Fumed silica is an inorganic nano thickener that does not melt, providing efficient lubricating performance at high temperatures. It has a high dropping point and offers good thickening and thixotropic performance.

Advantages of Inorganic Thickeners:

- **High-temperature stability**: Inorganic thickeners can withstand elevated temperatures without degrading, making them suitable for high-heat applications.
- **Water resistance**: Certain inorganic thickeners, like calcium sulfonate, offer excellent water resistance, ensuring the lubricant remains effective in wet conditions.
- **Corrosion resistance**: Inorganic thickeners can enhance the lubricant's ability to protect metal surfaces from corrosion, extending equipment lifespan.

Considerations:

While inorganic thickeners offer several benefits, they may have limitations in certain applications. For example, some inorganic thickeners may not provide the same level of shear stability as soap-based thickeners. Therefore, selecting the appropriate thickener depends on the specific requirements of the application, including operating temperature, load conditions, and environmental factors.

In summary, inorganic thickeners play a crucial role in formulating lubricants that meet the demands of specialized applications, offering advantages such as high-temperature stability, water resistance, and corrosion protection.

428. Insolubles

In the context of lubricants, **insolubles** refer to solid or liquid materials that do not dissolve in the base oil. These materials can originate from various sources, including:

- **Oxidation products**: As lubricants degrade over time, they can oxidize, leading to the formation of solid particles suspended within the oil.
- **Additive degradation**: The breakdown of chemical additives in the lubricant can produce insoluble by-products.
- **Contaminants**: External particles, such as dust, soot, or metal wear debris, can enter the lubrication system and become suspended in the oil.

Significance of Measuring Insolubles:
Monitoring the level of insolubles in lubricants is crucial for several reasons:

- **Oxidization**: An increase in insolubles can signal that the oil is oxidizing more rapidly which may necessitate an oil change to maintain optimal lubrication.
- **Increased viscosity**: As insolubles accumulate, they can cause the lubricant to thicken, resulting in higher viscosity. This thickening can impede oil flow and reduce its ability to circulate effectively. This has the potential to increase friction and wear of the lubricated parts.
- **Formation of sludge and varnish**: Insolubles can combine with other degradation products to form sludge and varnish. These substances can further increase viscosity and impair machine performance.
- **Assessment of filtration efficiency**: High levels of insolubles might indicate that the oil filtration system is not effectively removing contaminants, potentially leading to increased wear and reduced equipment lifespan.

Testing for Insolubles:
The determination of insolubles typically involves:

- **Centrifugation**: A sample of the used oil is mixed with a solvent and then centrifuged. The insoluble materials collect at the bottom of the test tube, allowing for quantification.

- **Filtration**: Alternatively, the oil can be filtered through a preweighed filter paper. The increase in weight after filtration indicates the amount of insolubles present.

Regular oil changes and oil analysis monitoring of insolubles help in maintaining the health of the lubrication system, ensuring efficient operation, and preventing potential equipment failures.

429. Instantaneous Lubrication Failure (ILF)

Instantaneous lubrication failure (ILF) occurs when a lubricant film breaks down suddenly, leading to immediate metal-to-metal contact, excessive wear, and potentially catastrophic equipment failure. This failure mode is often triggered by extreme operating conditions, lubricant contamination, or sudden loss of lubrication flow, resulting in rapid component damage.

Key Causes of Instantaneous Lubrication Failure

- **Sudden oil or grease starvation**: Occurs when the lubricant supply is interrupted due to pump failure, blocked lines, or reservoir depletion.
- **Extreme temperature spikes**: High temperatures degrade lubricant viscosity, causing the film to thin and fail under load.
- **Overloading and excessive surface pressure**: Increased stress can exceed the lubricant's film strength, leading to immediate surface welding and wear.
- **Contamination (water, dirt, or metal particles)**: Foreign particles disrupt lubricant film formation, accelerating wear.
- **Incorrect lubricant selection**: A lubricant with inadequate viscosity, poor film strength, or insufficient extreme pressure (EP) additives may fail under high loads.
- **High-speed and shock loads**: Rapid acceleration or sudden impact loads can disrupt lubrication film formation, leading to scuffing or seizure.

Effects of Instantaneous Lubrication Failure

- **Severe friction and wear**: Causes adhesive wear, galling, and scuffing of metal surfaces.

- **Rapid temperature rise**: Excessive heat due to friction can cause thermal damage to bearings, gears, and other components.
- **Component seizure and catastrophic failure**: Bearings, gears, and other moving parts can weld together or fracture due to lubrication loss.
- **Increased maintenance costs and downtime**: Requires emergency repairs or full component replacement, leading to unplanned downtime.

430. Insulating Oil

Insulating oil is a **high-purity dielectric fluid** used in electrical equipment to provide **insulation, cooling, and arc suppression**. It prevents electrical discharge, dissipates heat, and protects components from oxidation and contamination in transformers, circuit breakers, and switchgear.

Key Functions of Insulating Oil

- **Electrical insulation**: Prevents short circuits and dielectric breakdown by providing high-voltage resistance.
- **Heat dissipation and cooling**: Transfers heat away from windings, cores, and contacts, preventing overheating.
- **Arc quenching and fire suppression**: Minimizes electrical arcing in high-voltage applications, reducing the risk of fire and system failure.
- **Oxidation and moisture protection**: Forms a protective barrier that prevents metal corrosion and water absorption, maintaining system integrity.

Types of Insulating Oil

- **Mineral oil-based insulating oils** – Most commonly used in transformers and circuit breakers, offering good dielectric properties and oxidation resistance.
- **Synthetic insulating oils** – Includes silicone-based and ester-based fluids, known for higher thermal stability and fire resistance.
- **Biodegradable and natural ester oils** – Used in environmentally sensitive applications, **providing low toxicity and renewable alternatives**.
- **Silicone insulating oils** – High-temperature resistant, ideal for sealed electrical systems and high-voltage applications.

Applications of Insulating Oil

- **Power transformers and distribution transformers**: Prevents dielectric failure and enhances heat dissipation.
- **High-voltage circuit breakers**: Acts as an arc suppression medium to prevent flashovers and electrical faults.
- **Capacitors and switchgear**: Provides insulation and cooling to extend equipment lifespan.
- **High-voltage cables and bushings**: Protects against dielectric loss and partial discharge.

431. Interfacial Tension (IFT)

Interfacial tension (IFT) is a lubricant property that measures the **force per unit length at the interface between two immiscible fluids**, such as oil and water. It indicates the lubricant's ability to resist contamination, oxidation, and degradation, making it a crucial parameter in transformer oils, hydraulic fluids, and industrial lubricants. A high IFT value suggests clean oil with minimal contamination, while a low IFT value indicates oxidation, water contamination, or additive depletion.

Key Functions of Interfacial Tension in Lubrication

- **Detects oil degradation and contamination**: A decreasing IFT value signals oxidation, water contamination, or detergent additive depletion.
- **Prevents emulsification and phase separation issues**: Ensures that oil repels water, reducing the risk of sludge, corrosion, and viscosity breakdown.
- **Indicates lubricant cleanliness and aging**: Fresh lubricants typically have higher interfacial tension, while used or degraded oils exhibit lower IFT values.
- **Improves lubricant stability**: A lubricant with proper IFT ensures consistent film strength and separation from water.

Factors Affecting Interfacial Tension

- **Oxidation and acid formation**: As a lubricant oxidizes, it forms polar degradation byproducts, reducing IFT.

- **Water contamination**: The presence of dissolved or free water lowers IFT, leading to emulsification and corrosion.
- **Detergent and dispersant additives**: Some additives alter IFT to improve oil–water separation or emulsification resistance.
- **Contaminants and insolubles**: Dirt, wear debris, and external contaminants can decrease IFT, affecting lubricant performance.

432. Interlock Control

In industrial systems, **interlock control** refers to a safety mechanism that ensures the proper sequence of operations by linking the functions of different devices or processes. This interdependence prevents conflicting actions, thereby safeguarding equipment and personnel.

Types of Interlocks:

1. **Hardwired interlocks**: These are physical connections that directly link devices, ensuring that one cannot operate unless the other is in a specific state. They are often used as primary safety devices and cannot be bypassed by control logic.
2. **Safety interlocks**: Implemented through control logic, safety interlocks prevent actions that could lead to unsafe conditions. They can be changed or bypassed only by modifying the control logic.
3. **Process interlocks**: These interlocks automate the control of a process or device, ensuring that operations occur in a predefined sequence. They are always active during the run mode.
4. **Permissive interlocks**: A specific type of interlock that prevents actions from taking place until certain predefined conditions are met. For example, a pump may not start until the suction valve is open.

Applications in Lubrication Systems:

In lubrication systems, interlock controls are crucial for ensuring that components operate within safe parameters. For instance, in centrifugal compressors, interlocks can prevent the compressor from starting if conditions such as low seal gas pressure, high vibration, or **low lubrication oil pressure** are detected. This prevents potential damage to the compressor and ensures its longevity.

By implementing interlock controls, industries can enhance the safety, reliability, and efficiency of their operations, ensuring that processes occur in a controlled and predictable manner.

433. International Council of Machinery Lubrication – ICML

The International Council for Machinery Lubrication (ICML) is a vendor-neutral, not-for-profit organization founded in 2001 to facilitate the growth and development of machine lubrication as a technical field of endeavor. Among its various activities, ICML offers global skill certification testing for individuals in the fields of machine condition monitoring, lubrication, and oil analysis. ICML currently provides certification for the Machine Lubricant Analyst (MLA), Machine Lubrication Technician (MLT), and Laboratory Lubricant Analyst (LLA). The MLA and LLA ICML certifications have been pioneered into ISO 18436-4 Standards. ICML is also the creator of the ICML 55 world lubrication standard.

434. International Council of Machinery Lubrication – 55 Standard

The **ICML 55® Standard** is a comprehensive framework developed by the International Council for Machinery Lubrication (ICML) to guide organizations in establishing, implementing, and maintaining an effective lubrication management system (LMS). The standard aligns with ISO 55000 asset management principles, providing a structured approach to managing lubricated mechanical assets.

Structure of the ICML 55® Standard:

1. **ICML 55.0**: **Optimized lubrication of mechanical physical assets overview**
   This document introduces the importance of a robust LMS and its alignment with broader asset management goals. It serves as an executive summary, helping stakeholders understand the relevance of lubrication management within organizational contexts.

2. **ICML 55.1®**: **Requirements for the optimized lubrication of mechanical physical assets**
   55.1® is the part of the ICML standard an organization is audited against by accredited third-party auditors. This document outlines the

twelve interrelated areas essential for a sustainable lubrication program that auditors use to assess the level of a company's lubrication management program. These areas include:

- **Skills**: Job tasks, training, and competency
- **Machine**: Lubrication and condition monitoring readiness
- **Lubricant**: System design and selection
- **Lubrication**: Planned and corrective maintenance tasks
- **Tools**: Lubrication support facilities and tools
- **Inspection**: Machine and lubricant inspection
- **Lubricant analysis**: Condition monitoring and analysis
- **Troubleshoot**: Fault/failure troubleshooting and root cause analysis
- **Waste**: Lubricant waste handling and management
- **Energy**: Energy conservation and environmental impact
- **Reclaim**: Oil reclamation and system decontamination
- **Management**: Program management and metrics

These areas collectively provide a blueprint for developing and sustaining an effective LMS.

3. **ICML 55.2**: **Guideline for the optimized lubrication of mechanical physical assets**
   This document offers practical guidance on implementing the requirements outlined in ICML 55.1. It serves as a detailed manual for organizations to apply best practices in lubrication management.

4. **ICML 55.3**: **Guideline for the internal assessment and audit of lubricated mechanical assets**
   This document offers practical guidance for internal auditors readying their organization for an ICML 55.1® and for accredited auditor training.

Purpose and Benefits:
The ICML 55 Standard aims to:

- Enhance asset reliability and performance through optimized lubrication practices.

- Align lubrication management with broader asset management strategies.
- Provide a structured approach to developing and maintaining an effective LMS.
- Offer a framework for organizations to achieve certification, demonstrating a commitment to best practices in lubrication management.

By adhering to the ICML 55 Standard, organizations can establish a systematic approach to lubrication management, leading to improved machinery reliability, reduced downtime, and optimized operational efficiency.

435. Inductively Coupled Plasma Atomic Emission Spectroscopy (ICP-AES)

See "Atomic Emission Spectroscopy."

436. International Lubricant Standardization and Approval Committee (ILSAC)

The **International Lubricant Standardization and Approval Committee (ILSAC)** is a collaborative organization formed in 1992 by the American Automobile Manufacturers Association (AAMA) and the Japan Automobile Manufacturers Association (JAMA). Its primary objective is to establish and maintain standardized performance specifications for automotive lubricants, ensuring that engine oils meet the evolving needs of modern vehicles.

Key Functions of ILSAC:

- **Development of lubricant specifications**: ILSAC defines the requirements for engine oils, focusing on aspects such as fuel economy, engine protection, and compatibility with emission control systems. These specifications are designated as ILSAC GF-x, with "x" representing the version number.
- **Collaboration with industry standards**: ILSAC works alongside other organizations, including the American Petroleum Institute (API), the Society of Automotive Engineers (SAE), and the American Society for Testing and Materials (ASTM), to create a comprehensive Engine Oil Licensing and Certification System (EOLCS). This system allows engine oil marketers who meet specified requirements to use the API Engine Oil Quality Marks, aiding consumers in identifying quality engine oils for their vehicles.

ILSAC GF Specifications:

ILSAC's specifications are periodically updated to address advancements in engine technology and environmental considerations. For example, the ILSAC GF-6 standard, introduced in 2019, includes two subspecifications:

- **GF-6A**: Fully backward-compatible with ILSAC GF-5, offering improved fuel economy, enhanced engine protection, and better performance.
- **GF-6B**: Specifically designed for SAE 0W-16 viscosity oils, providing similar benefits as GF-6A but tailored for vehicles requiring this lower viscosity grade.

These standards ensure that engine oils meet the stringent requirements of modern engines, promoting efficiency and longevity.

In summary, ILSAC plays a crucial role in the automotive industry by setting and updating lubricant standards that align with technological advancements and environmental goals, thereby supporting the development of high-performance, efficient, and environmentally friendly engine oils.

437. Interval-based Lubrication

Interval-based lubrication refers to a mature lubrication management strategy in which lubricants are applied to machinery components at predetermined time intervals, regardless of the equipment's actual operating condition. This approach is often based on manufacturer recommendations or historical data, specifying lubrication schedules in terms of operating hours, calendar dates, or production cycles.

Advantages of Interval-based Lubrication:

- **Simplicity**: Easy to implement and manage, as it relies on fixed schedules.
- **Predictability**: Maintenance activities are planned in advance, facilitating resource allocation.

Disadvantages of Interval-based Lubrication:

- **Potential for Overlubrication**: Applying lubricant more frequently than necessary can lead to excess heat generation and pressure, causing seals to fail. Conversely, underlubrication results in insufficient protection, accelerating wear.
- **Inefficiency**: May result in unnecessary lubricant usage and increased maintenance costs if intervals are not optimized.

In contrast, **condition-based lubrication** involves monitoring the actual condition of the lubricant and machinery components and adjusting lubrication schedules based on real-time data. This method aims to optimize lubricant usage, reduce waste, and enhance equipment reliability.

Transitioning from interval-based to condition-based lubrication can lead to more efficient maintenance practices, potentially extending equipment lifespan and reducing operational costs.

438. IP

The Institute of Petroleum (United Kingdom), has now merged with the Energy Institute (EI). See "Energy Institute."

439. ISO

The **International Organization for Standardization (ISO)** is an independent, nongovernmental international organization that develops and publishes standards across a wide range of industries and sectors. Established in 1947 and headquartered in Geneva, Switzerland, ISO comprises national standards bodies from over 170 countries, each representing their nation's interests in standardization.

Purpose and Functions:

- **Standard development**: ISO brings together experts from various fields to create international standards that ensure products and services are safe, reliable, and of high quality. These standards cover everything from manufacturing processes to environmental management.
- **Facilitating trade**: By establishing common standards, ISO helps reduce barriers to international trade, ensuring that products and services meet consistent quality benchmarks globally.
- **Promoting innovation**: ISO standards support innovation by providing frameworks that guide the development of new technologies and processes, ensuring they meet international quality and safety requirements.

Relevant ISO Standards:

- **ISO 9001**: This standard specifies the criteria for a quality management system and is the most widely recognized standard for quality management worldwide.
- **ISO 14001**: Focused on environmental management, this standard provides a framework for organizations to improve their environmental

performance through more efficient use of resources and reduction of waste.

- **ISO 27001**: This standard outlines the requirements for an information security management system, helping organizations keep information assets secure.
- **ISO 18436**: used as a guideline for the standardization of condition monitoring and diagnostics training and certification programs. Lubrication training falls under part four of the standard.
- **ISO 55001**: This standard focuses on the management of all corporate assets that add value to an organization. This standard is used for the certification of maintenance and lubrication operations.
- **ISO 4406-99**: A lubricant cleanliness code standard (see Fluid Target Cleanliness Levels)

Structure and Membership:
ISO operates through a network of technical committees, each focusing on specific areas of standardization. The organization is governed by a council and managed by a central secretariat. Membership is open to national standards bodies, with each country typically having one member.

Through its extensive network and comprehensive standards, ISO plays a crucial role in ensuring quality, safety, and efficiency across various industries worldwide.

440. ISO VG

ISO VG stands for **International Standards Organization Viscosity Grade**, a classification system used to denote the viscosity of industrial lubricants. This system assigns a numerical value to lubricants based on their kinematic viscosity measured at 40 °C (104 °F). The higher the ISO VG number, the thicker the oil, indicating a higher viscosity.

Understanding ISO VG:

- **Viscosity measurement**: Viscosity refers to a fluid's resistance to flow. In the ISO VG system, lubricants are categorized by their kinematic viscosity at 40 °C, expressed in centistokes (cSt). For example, an ISO VG 100 lubricant has a kinematic viscosity between 90 and 110 cSt at 40 °C.
- **Classification**: The ISO VG scale ranges from 2 to 1500, with each grade representing a specific viscosity range. For instance, ISO VG

32 corresponds to a kinematic viscosity between 29.8 and 35.2 cSt at 40 °C.

Importance of ISO VG:

- **Equipment compatibility**: Selecting the appropriate ISO VG grade ensures that the lubricant's viscosity aligns with the operational requirements of machinery, promoting efficient performance and longevity.
- **Standardization**: The ISO VG system provides a standardized method for classifying lubricants, facilitating clear communication and consistency across the industry.

ISO VG versus SAE:

While ISO VG is primarily used for industrial lubricants, the Society of Automotive Engineers (SAE) classification system is commonly applied to automotive oils. SAE grades are determined based on viscosity measurements at 100°C, whereas ISO VG grades are based on measurements at 40 °C. This difference can lead to variations in viscosity characteristics between the two systems.

Understanding ISO VG is essential for selecting the correct lubricant for industrial applications, ensuring optimal equipment performance and longevity.

441. Isothermal Lubrication

Isothermal lubrication refers to a lubrication condition where the temperature of the lubricant remains constant throughout the lubrication process, despite friction and mechanical stresses. This is an idealized concept in tribology used in fluid film lubrication modeling and analysis, ensuring consistent viscosity, film thickness, and lubrication performance.

Key Functions of Isothermal Lubrication

- **Maintains lubricant film thickness**: A constant temperature ensures stable viscosity, preventing excessive thinning or thickening of the lubricant.
- **Prevents thermal degradation**: Minimizing temperature fluctuations helps reduce oxidation, varnish formation, and additive breakdown.
- **Ensures predictable lubrication behavior**: In isothermal conditions, lubricants maintain consistent flow and film strength, reducing wear and energy loss.

- **Enhances load-carrying capacity**: By preventing excessive temperature rise, the lubricant retains its ability to handle high loads without viscosity loss.

Factors Affecting Isothermal Lubrication

- **Heat dissipation efficiency**: Effective cooling through heat exchangers, circulation systems, or convection helps maintain isothermal conditions.
- **Lubricant thermal stability**: High-quality synthetic lubricants with low volatility and oxidation resistance improve isothermal lubrication.
- **Operating load and speed**: High-speed and high-load applications generate more frictional heat, making isothermal conditions harder to maintain.
- **External temperature control**: Industrial systems use controlled cooling loops or thermal regulation to stabilize lubricant temperature.

J

442. Jost, Peter H.

Sir Peter Jost (January 25, 1921–June 7, 2016) was a distinguished British mechanical engineer renowned for founding the discipline of Tribology, the study of friction, wear, and lubrication in interacting surfaces. His pioneering work has had a profound impact on engineering, leading to significant advancements in machinery efficiency and longevity.

Early Life and Education:
Born in Berlin, Germany, Jost fled Nazi persecution and settled in the United Kingdom. He pursued his education at Liverpool Technical College and Manchester College of Technology, laying the foundation for his future contributions to engineering.

Career and Contributions:
In 1966, Jost authored the influential "Jost Report," commissioned by the UK Department of Education and Science. This report highlighted the substantial economic costs associated with friction, wear, and corrosion, emphasizing the need for dedicated research and education in lubrication and tribology. It was within this report that he introduced the term "tribology," coining a term that would become central to mechanical engineering.

Beyond his research, Jost was instrumental in establishing the International Tribology Council, an organization dedicated to promoting tribology worldwide. His advocacy and leadership have been recognized through numerous honors, including the Order of the Rising Sun from Japan and the Austrian Cross of Honour for Science and Art.

Legacy:
Jost's legacy endures through the Peter Jost Charitable Foundation, established in 1986, which supports education and research in tribology. His work has laid the groundwork for modern tribology, influencing various industries

DOI: 10.1201/9788743806608-451

by enhancing the understanding and management of friction and wear in mechanical systems.

Peter Jost's dedication to advancing tribology has left an indelible mark on engineering, underscoring the critical role of understanding material interactions in the development of efficient and durable machinery.

443. Jost Report

The **Jost Report**, formally titled "**Lubrication (Tribology)**: **Education and Research**: **A Report on the Present Position and Industry's Needs**," was published in 1966 under the leadership of Dr. H. Peter Jost. Commissioned by the UK Department of Education and Science, the report aimed to assess the state of lubrication education and research, and to identify the needs of industry in this area.

Key Contributions:

- **Introduction of "Tribology"**: The report introduced the term "tribology," derived from the Greek word "tribos," meaning rubbing or attrition. This term encapsulates the science and technology of interacting surfaces in relative motion, encompassing friction, wear, lubrication, and bearing design.
- **Economic impact assessment**: It highlighted the significant economic costs associated with friction, wear, and corrosion, estimating that these factors accounted for 1.1%–1.4% of the UK's GDP at the time.
- **Recommendations for research and education**: The report emphasized the need for dedicated research and education in tribology to address these economic challenges. It proposed the establishment of specialized institutes and the integration of tribology into engineering curricula.

Impact and Legacy:

The Jost Report was instrumental in elevating the importance of tribology within engineering disciplines. It led to the creation of national centers focused on tribological research and education, fostering interdisciplinary collaboration among mechanical engineers, materials scientists, physicists, and chemists.

In recognition of his contributions, Dr. Jost received numerous honors, including the Order of the Rising Sun from Japan in 2011 and the Austrian Cross of Honour for Science and Art in 2001.

The Jost Report remains a seminal document in the field of tribology, underscoring the critical role of understanding and managing friction and wear in engineering applications.

444. Journal Bearing

A **journal bearing** is a simple plain bearing designed to support a rotating shaft, known as the journal, within a stationary housing. Unlike rolling element bearings, journal bearings operate without rolling elements; instead, they rely on a thin film of lubricant to separate the shaft from the bearing surface, thereby reducing friction and wear.

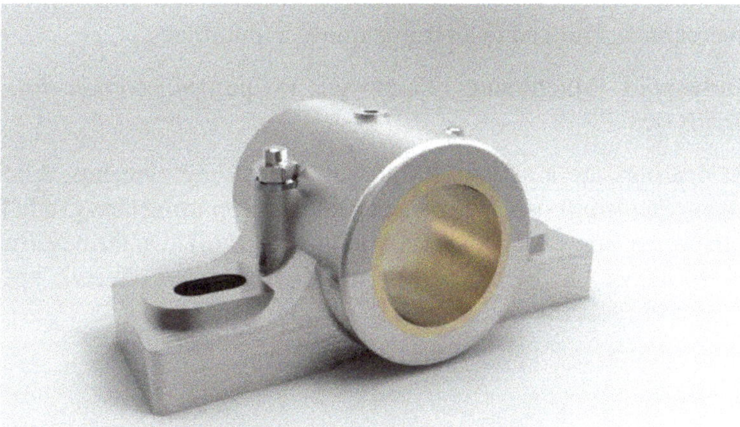

Figure J1 Typical bronze journal bearing. Image: Ahmed Shawki.

Key Features:

- **Design**: Consists of a soft metal cylindrical sleeve or lined (babbit) shell that encases the shaft, with a clearance gap to accommodate the lubricant film. See Figure J1.
- **Lubrication**: The lubricant, often oil, is introduced into the bearing to create a hydrodynamic film that supports the shaft and minimizes direct contact between the shaft and bearing surfaces.
- **Load support**: Primarily designed to handle radial loads, which are forces acting perpendicular to the shaft's axis.

Types of Journal Bearings:

- **Plain journal bearings**: The simplest form, consisting of a smooth, cylindrical surface, sometimes referred to as plain bearings or radial bearings.
- **Grooved journal bearings**: Feature grooves on the bearing surface to enhance lubricant distribution and improve load-carrying capacity.

- **Tilting pad bearings**: Comprise multiple pads that can tilt to accommodate shaft misalignment and dynamic loads.

Applications:
Journal bearings are widely used in various industries, including:

- **Automotive**: Supporting crankshafts and camshafts in internal combustion engines.
- **Aerospace**: Utilized in aircraft engines and turbines.
- **Industrial Machinery**: Employed in pumps, compressors, and gearboxes.

Journal bearings are a simple design with no rolling elements, leading to lower manufacturing costs. They are capable of supporting heavy radial loads and with proper lubrication, they offer long service life. On the negative side, their performance can be affected by misalignment and effective operation relies on proper lubrication.

K

445. Karl Fischer

See Fischer, Karl.

446. Karl Fischer Titration Test

A Karl Fischer titration test is a precise analytical method used to determine the water content in various substances, including lubricants, fuels, chemicals, and pharmaceuticals. This test is highly regarded for its accuracy in measuring both free water and chemically bound water, making it an essential tool for quality control and contamination analysis.

Karl Fischer Titration Test Method:

1. **Principle**: The Karl Fischer method is based on the chemical reaction between water and iodine in the presence of sulfur dioxide and an alcohol, typically methanol. The titration continues until all the water has reacted with the iodine, and the endpoint is detected either visually or electrochemically.

2. **Types of Karl Fischer Titration**:
   - **Volumetric titration**: Suitable for samples with higher water content (typically 1% to 100%).
   - **Coulometric titration**: Used for samples with very low water content (ppm levels), where iodine is generated electrochemically.

Applications in Lubrication:

- **Oil and lubricant testing**: Moisture can be highly detrimental to the performance of lubricants and oils, leading to corrosion, oxidation, and reduced efficiency. The Karl Fischer method provides a reliable way to measure water content and ensure product quality.

DOI: 10.1201/9788743806608-455

- **Quality control**: Used in manufacturing environments to test the moisture content in raw materials, finished products, and storage conditions to prevent contamination and degradation.

Advantages:

- **Accuracy and precision**: Capable of detecting minute quantities of water, making it highly effective for quality control in sensitive applications.
- **Versatility**: Applicable to a wide range of substances, including oils, solvents, and pharmaceuticals.

The Karl Fischer Titration remains a critical method in laboratories worldwide for its effectiveness in moisture analysis, contributing to advancements in various scientific and industrial fields.

447. Key Performance Indicator (KPI)

Key performance indicators (KPIs) are essential metrics used to evaluate the effectiveness and efficiency of lubrication practices within industrial operations. By monitoring these indicators, organizations can identify areas for improvement, enhance machinery performance, and reduce lubrication and maintenance costs.

Types of Lubrication KPIs:

1. **Leading indicators**: These metrics provide early insights into potential issues, allowing for proactive measures. Examples include:
   - **Compliance with lubricant sampling schedules**: Adherence to planned lubricant sampling ensures timely analysis, facilitating early detection of contamination or degradation.
   - **Percentage of planned lubrication tasks completed on time**: This measures adherence to scheduled lubrication activities, reflecting the effectiveness of maintenance planning.
2. **Lagging indicators**: These metrics assess outcomes after events have occurred, helping to evaluate the success of past actions. Examples include:
   - **Mean time between lubrication failures (MTBLF)**: This metric calculates the average operational time between lubrication-related failures, indicating the reliability of lubrication practices.

- **Lubrication-related maintenance costs as a percentage of total maintenance costs**: This KPI assesses the financial impact of lubrication-related maintenance activities on overall maintenance expenditures.

Implementing effective lubrication KPIs:

- **Data collection**: Regular and accurate data gathering is crucial for KPI analysis. Utilizing automated systems can enhance data accuracy and timeliness.
- **Benchmarking**: Establishing baseline values for each KPI allows for performance comparison over time, aiding in identifying trends and areas needing attention.
- **Continuous improvement**: Regularly reviewing and adjusting KPIs ensures they remain aligned with organizational goals and industry best practices.

By effectively utilizing these KPIs, organizations can optimize their lubrication management programs, leading to improved equipment reliability, reduced downtime, and lower maintenance costs.

448. Kinematic Viscosity

Kinematic viscosity is a fundamental property of fluids that quantifies their internal resistance to flow under the influence of gravity. It is defined as the ratio of dynamic viscosity to the fluid's density:

$$\nu = \frac{\mu}{\rho}$$

where:

- ν = Kinematic viscosity (m^2/s)
- μ = Dynamic viscosity (Pa·s or N·s/m^2)
- ρ = Density (kg/m^3)

This ratio provides insight into how a fluid flows under the influence of gravity, considering both its internal friction (dynamic viscosity) and its density.

Units of Measurement:

- **SI unit**: Square meters per second (m^2/s)

- **CGS unit**: Stokes (St), where 1 St = 10^{-4} m²/s
- **Common subunit**: Centistokes (cSt), where 1 cSt = 10^{-6} m²/s

For practical purposes, kinematic viscosity is often expressed in centistokes (cSt), especially in industries like lubrication and hydraulics.

Applications:
Kinematic viscosity is crucial in various fields:

- **Lubrication engineering**: Determines the flow characteristics of lubricants, influencing their performance in reducing friction and wear.
- **Fluid dynamics**: Assists in analyzing and predicting fluid flow behavior in pipes, channels, and natural systems.
- **Meteorology**: Helps in understanding the movement of atmospheric fluids, such as air currents.

Kinematic viscosity plays a crucial role in lubrication systems, as it determines how easily a lubricant flows at specific temperatures. Proper selection ensures that the lubricant provides sufficient film strength to prevent metal-to-metal contact while maintaining smooth flow.

449. Kinematic Friction

Kinetic friction, also known as dynamic or sliding friction, is the force that resists the relative motion between two surfaces in contact when one or both are moving. See Figure D11. This force acts opposite to the direction of motion, aiming to slow down or stop the movement.

Examples:
- **Sliding a box**: When you push a box across the floor, the force you apply must overcome the kinetic friction between the box and the floor to initiate and maintain movement.
- **Braking in vehicles**: When a vehicle brakes, the tires experience kinetic friction with the road surface, which slows down the vehicle.

Applications:
- **Engineering design**: Understanding kinetic friction is crucial in designing mechanical systems where moving parts interact, such as in bearings, gears, and conveyor belts.

- **Safety considerations**: In automotive engineering, knowledge of kinetic friction helps in designing braking systems and tires to ensure effective stopping power.
- **Material selection**: Selecting materials with appropriate coefficients of kinetic friction is essential in applications like manufacturing processes, where precise control of movement is required.

By comprehending and applying the principles of kinetic friction, engineers and designers can optimize performance, enhance safety, and improve the durability of mechanical systems.

450. KPI

See "Key Performance Indicator."

L

451. Labyrinth Seal

A **labyrinth seal** is a mechanical sealing device designed to prevent the leakage of fluids or gases by creating a complex, tortuous path that the fluid must navigate. This intricate pathway effectively restricts the flow of contaminants into sensitive areas, such as bearings or machinery housings, thereby enhancing the reliability and longevity of mechanical systems.

Design and Functionality:

- **Structure**: Labyrinth seals consist of interlocking components, often featuring grooves or teeth that mesh together. These interlocking parts form a series of chambers or passages, creating a maze-like path for the fluid. The design ensures that the fluid encounters multiple changes in direction and pressure, significantly reducing the likelihood of leakage. See Figure L1.
- **Noncontact operation**: Operating without direct contact between moving and stationary parts, labyrinth seals minimize friction and wear. This noncontact nature contributes to their durability and effectiveness over extended periods.

Applications:

Labyrinth seals are utilized across various industries, including:

- **Aerospace**: In gas turbine engines, labyrinth seals are employed to prevent the leakage of oil and maintain pressure integrity within the engine components.
- **Mining and mineral processing**: SKF's noncontact labyrinth seals are used in vibrating screen bearings to keep contaminants out and lubricants in, thereby extending service life and enhancing reliability.
- **Oil and gas industry**: Labyrinth seals are applied in pumps, compressors, and turbines to prevent leakage of fluids and gases, ensuring the integrity and safety of equipment.

DOI: 10.1201/9788743806608-462

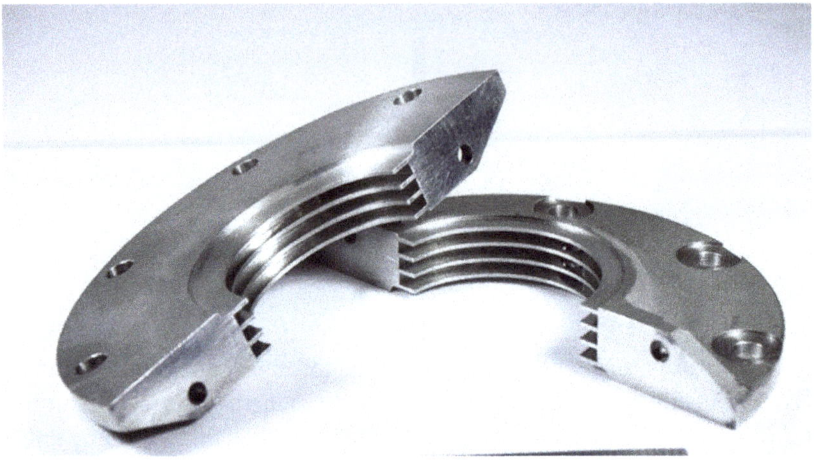

Figure L1 Split labarynth sea.

Advantages:

- **Enhanced durability**: The noncontact design reduces wear and tear, leading to a longer service life of the seal and associated components.
- **High-speed operation**: Labyrinth seals can function effectively at high rotational speeds, making them suitable for applications like gas turbines.
- **Resistance to contaminants**: The intricate design effectively keeps contaminants out, which is particularly beneficial in harsh operating environments.

While labyrinth seals offer numerous benefits, their performance is highly dependent on precise manufacturing and installation. Proper alignment and clearance are crucial to ensure the seal's effectiveness and to prevent potential issues such as increased leakage or mechanical failure.

In summary, labyrinth seals are vital components in mechanical systems, providing effective sealing solutions through their intricate, noncontact design. Their ability to prevent leakage and resist contaminants makes them indispensable in various high-performance applications.

452. Lacquer

See "Varnish."

453. Lambda Thickness Ratio

See "Film Thickness Ratio – Lambda."

454. Laminar Particles

In lubrication, **laminar particles** are thin, flat wear debris generated during rolling contact fatigue. These particles are typically elongated with high aspect ratios, often ranging from 20:1 to 50:1, and may exhibit holes and folds along their length. They are produced when surface fatigue leads to the formation of pits and spalls under high load and low-contact surface area conditions. As these particles are forced out of their original setting, they acquire their distinctive shape.

Formation Mechanism:
Laminar particles originate from the rolling contact between surfaces, such as in bearings or gears. Under high load and low-contact surface area, surface fatigue occurs, leading to the formation of pits and spalls. These defects are then expelled from the contact area, resulting in the creation of laminar particles.

Figure L2 Laminar wear particles. Source: Fluidlife.

Characteristics:

- **Shape**: Elongated with high aspect ratios, often between 20:1 and 50:1.
- **Surface features**: May have holes and folds along their length. See Figure L2.
- **Size**: Can vary, with some particles being several millimeters in size, especially in large bearings.

Significance in Lubrication:
The presence of laminar particles in a lubricant can indicate ongoing rolling contact fatigue within the system. Monitoring these particles through wear particle analysis helps in assessing the health of the machinery and can guide maintenance decisions to prevent further damage.

Understanding the formation and characteristics of laminar particles is crucial for developing effective lubrication strategies and ensuring the longevity and reliability of mechanical systems.

455. LCAT I

The **ISO LCAT-1 designation** refers to the **Lubrication Analyst ISO Category I** certification, a globally recognized qualification that validates an individual's proficiency in lubrication management and analysis. This certification is essential for professionals involved in the maintenance and reliability of machinery, as it ensures adherence to international standards and best practices in lubrication.

Key Aspects of ISO LCAT-I Certification:

- **International standards compliance**: The LCAT-1 certification aligns with ISO standards, including ISO 18436-1, ISO 18436-3, and ISO 18436-4, which pertain to condition monitoring and diagnostics of machines.
- **Training and examination**: The certification process typically involves a structured training program followed by an examination.
- **Competency development**: The program focuses on developing competencies in areas such as dispensing and handling lubricants, maintaining lubrication devices, lubrication theory, oil sampling, and operation of portable lubrication analysis tools.
- **Global recognition**: Achieving the LCAT-1 certification signifies a commitment to professional excellence and is recognized across various industries worldwide.

456. LCAT II

The **ISO LCAT II designation** refers to the **Lubrication Analyst ISO Category I** certification, a globally recognized qualification that validates an individual's proficiency in lubrication management and analysis. This certification is essential for professionals involved in the maintenance and reliability of machinery, as it ensures adherence to international standards and best practices in lubrication.

Key Aspects of ISO LCAT-II Certification:

- **International standards compliance**: The LCAT-II certification aligns with ISO standards, including ISO 18436-1, ISO 18436-3, and ISO 18436-4, which pertain to condition monitoring and diagnostics of machines.
- **Training and examination**: The certification process typically involves a structured training program followed by an examination. Entrance to an LCAT II program requires successful Completion of the LCAT I program.
- **Competency development**: The program focuses on developing maintenance and reliability competencies in areas that include areas such as lubrication theory, oil sampling and handling, wear debris monitoring, and contamination measurement and control.

457. Leak Detection

Leak detection in lubrication involves identifying and locating unintended releases of lubricants from machinery systems. Prompt detection is crucial to prevent equipment damage, maintain operational efficiency, and minimize environmental impact.

Common Methods for Leak Detection:

1. **Visual inspection**: Regularly examining equipment for visible signs of leaks, such as oil stains or puddles, can help identify external leaks. This method is straightforward but may not detect internal leaks.
2. **Infra-red inspection**: I/R camera system can easily detect high light oil leakage effluence. See Figure L3.
3. **Ultrasonic testing**: Utilizing ultrasonic equipment during normal operation can detect leaks by identifying the high-frequency sounds produced by escaping fluids. This method is effective for pinpointing the source of the leak.
4. **Dye penetrant testing**: Applying fluorescent dyes to the lubrication system can make leaks visible under ultraviolet light, aiding in locating small or hard-to-see leaks.

5. **Pressure testing**: Monitoring pressure levels within the lubrication system can reveal inconsistencies indicative of leaks. This method is particularly useful for detecting internal leaks.

6. **Acoustic emission monitoring**: Detecting the sounds produced by leaks can help identify their location. This method is effective for locating leaks in pressurized systems.

Figure L3 I/R image effluence from a vehicle oil leak. Courtesy: Flir Corporation.

458. Line Monitor Protection

Line monitor protection in lubrication systems involves the use of monitoring devices to ensure the integrity and functionality of lubrication lines, thereby preventing potential failures and enhancing equipment reliability. These devices detect issues such as blockages, leaks, or pressure anomalies within the lubrication lines, allowing for timely maintenance actions.

Key Components and Functions:

- **Monitoring devices**: These include sensors and indicators that continuously assess the condition of lubrication lines. For example, pressure, flow, and temperature within the lubrication system. Figure L4 shows numerous high-pressure monitor valves mounted on a progressive divider lube block, designed to automatically open and relieve line pressure in the case of a pinched line or blocked bearing hydraulic lockup.

Figure L4 Progressive divider block line high pressure monitor valves. Courtesy: ENGTECH Industries Inc.

Flow monitoring: Devices such as the SKF Flowline Monitor measure and control the flow rate in oil circulation lubrication systems, ensuring consistent lubricant delivery to critical components.

- **Pressure monitoring**: Pressure switches and sensors monitor the pressure within lubrication lines, alerting operators to potential issues like blockages or leaks. SKF's DSD pressure switches are designed for use in oil and fluid grease lubrication systems with pressures up to 45 bar (652 psi).
- **Leak detection**: Devices such as the hose connection control (HCC) monitor high-pressure lubrication hoses for line breakage, alerting operators immediately if a fault occurs.

459. Lithium Grease

Lithium grease is a widely used, multipurpose lubricant known for its versatility and effectiveness across various applications. It is formulated by combining lithium soap – typically lithium hydroxide or lithium 12-hydroxystearate – with mineral or synthetic oils, resulting in a smooth, buttery texture that facilitates easy application.

Key Characteristics:

- **High melting point**: Lithium grease has a high dropping point, typically around 190–220 °C (374–428 °F), allowing it to perform well in high-temperature environments.
- **Water resistance**: It offers excellent resistance to water washout, making it ideal for applications exposed to moisture, rain, or humidity.
- **Oxidation stability**: Lithium grease maintains its lubricating properties over time, even in harsh conditions, which helps extend the life of the machinery or components.
- **Adhesive properties**: It sticks well to metal surfaces, ensuring long-lasting lubrication and reducing the need for frequent reapplication.

Common Applications:

- **Automotive industry**: Used in wheel bearings, chassis components, and constant-velocity joints due to its high-temperature resistance and water resistance.
- **Industrial machinery**: Applied in bearings, gears, and other moving parts to reduce friction and wear, thereby enhancing equipment longevity.
- **Household appliances**: Employed in garage doors, window tracks, and other household items to ensure smooth operation and prevent rust.

Types of Lithium Grease:

- **Standard lithium grease**: Suitable for general-purpose lubrication tasks, including automotive, marine, and household applications.
- **Moly lithium grease**: Contains molybdenum disulfide (moly), providing additional protection against extreme pressure and heavy loads. Ideal for applications involving high stress, such as automotive chassis, bearings, and construction equipment.

In summary, lithium grease is a versatile and reliable lubricant that offers excellent performance across a wide range of applications, making it a preferred choice in various industries.

460. Live Zone Sampling

Live zone sampling is a method of collecting oil analysis lubricant samples from areas within a machine's oil circuit where the fluid is actively flowing and well-mixed. This approach ensures that the sample accurately represents

the lubricant's condition as it circulates through the system, providing valuable insights into the health of both the lubricant and the machinery.

Key Aspects of Live Zone Sampling:

- **Optimal sampling locations**: To obtain a representative oil sample, it is essential to collect oil from turbulent zones within the system. Ideal locations include areas downstream of critical components such as pumps, bearings, gears, and actuators, as well as sections with turbulent flow, like elbows or T-junctions in the piping. Sampling from these areas helps capture the true condition of the lubricant as it interacts with various machine parts. See Figure L5.

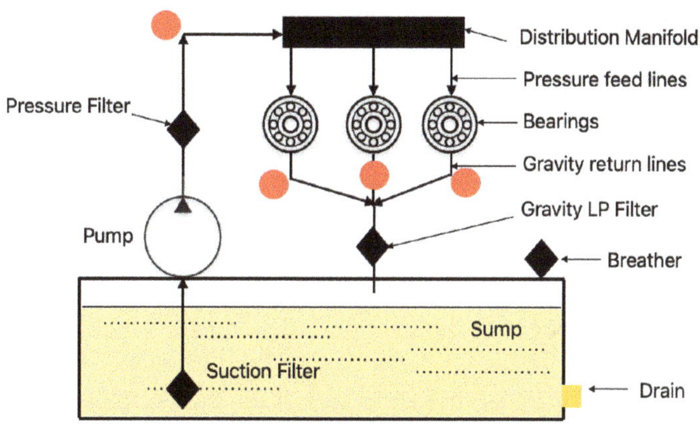

Figure L5 Typical live zone sampling points.

- **Avoiding contaminated areas**: Sampling from stagnant or "dead" zones, such as dead pipe legs or hoses, should be avoided, as these areas may contain settled contaminants that do not reflect the lubricant's current state. Additionally, sampling after filters or from sumps can lead to inaccurate assessments due to the removal of particles and the accumulation of debris over time.

- **Consistent sampling practices**: For effective monitoring, it is crucial to collect samples from the same location and at the same depth each time. This consistency allows for accurate trending of lubricant conditions and early detection of potential issues.

Benefits of Live Zone Sampling:

- **Accurate lubricant assessment**: By capturing the lubricant's condition in its active state, live zone sampling provides a true representation of its performance, enabling more precise analysis and maintenance decisions.

- **Early detection of issues**: Regular sampling from live zones facilitates the early identification of contaminants, wear particles, and other anomalies, allowing for timely interventions that can prevent equipment damage and extend machinery lifespan.
- **Enhanced maintenance strategies**: The data obtained from live zone sampling supports predictive maintenance approaches, helping to optimize maintenance schedules and reduce unplanned downtime.

Incorporating live zone sampling into a comprehensive oil analysis program is essential for maintaining equipment reliability and ensuring the longevity of machinery. By focusing on obtaining quality samples from active flow areas, maintenance teams can gain a deeper understanding of lubricant performance and make informed decisions to enhance operational efficiency.

461. LLA I – Laboratory Lubricant Analyst – Level I

The Level I **Laboratory Lubricant Analyst (LLA I)** certification designation is awarded through the International Council for Machine Lubrication (see ICML).

The purpose of the Level I Laboratory Lubricant Analyst (LLA I) certification is to verify that laboratory personnel from both commercial oil analysis labs and private in-house labs practicing in the field of lube oil analysis as it is applied to machinery condition monitoring and maintenance, are qualified to perform the following tasks: receive, handle, and prepare samples for lube oil analysis to ensure the integrity of the sample and the accuracy of subsequent tests data.

Fully trained personnel at the LLA I level can be expected to:

- Select appropriate tests for diagnosing common machine lubrication problem conditions as they relate to machine wear, lubricant condition, and lubricant contamination.
- Conduct and run appropriate lube oil analysis tests according to appropriate standards (ISO, ASTM, etc.).
- Interpret oil analysis test data to identify and determine abnormal machine wear conditions.

- Interpret oil analysis test data to identify and determine conditions of wrong or degraded lubricant.
- Interpret oil analysis test data to identify and determine contaminated lubricant conditions.

Common job titles for the individual who would become Level I LLA certified include Lab Technician, Lube Analyst, etc. Generally, this individual is regularly involved with receiving lube oil samples for lab analysis, running lube oil analysis tests, and interpreting oil analysis test data. The individual is likely to be directly involved in all aspects of running lab analysis as well as providing basic advice and recommendations to end-user practitioners of oil analysis. As such, a basic understanding of machinery lubrication, contamination control, and machinery maintenance are required.

462. LLA II – Laboratory Lubricant Analyst – Level II

The Level II **Laboratory Lubricant Analyst (LLA II)** certification designation is awarded through the International Council for Machine Lubrication (see ICML). These individuals must demonstrate skills in the performance of oil analysis activities typically required of a laboratory technician working in the used lubricant analysis field.

Level II is directed toward advanced diagnostics, troubleshooting instrument errors, and managing lab processes, including testing and service quality.

463. LM

See "Lubrication Management."

464. LMS

See "Lubrication Management System."

465. Load-carrying Capacity

Load-carrying capacity in lubrication refers to the ability of a lubricant to support applied loads without causing damage to the contacting surfaces. This capacity is crucial for preventing wear, reducing friction, and extending the lifespan of mechanical components.

Factors Influencing Load-carrying Capacity:

1. **Lubricant viscosity**: Higher viscosity lubricants form thicker films, enhancing load support. However, excessively high viscosity can lead to increased friction and energy loss.
2. **Additives**: Extreme pressure (EP) additives, such as sulfur–phosphorus compounds, chemically react under high pressure to form protective films, thereby increasing load-carrying capacity. For instance, the four ball extreme pressure tester evaluates the load-carrying capacity and weld point of lubricants without EP additives.
3. **Lubrication regime**: The lubrication regime – hydrodynamic, mixed, or boundary lubrication – affects load support. In hydrodynamic lubrication, the load is carried by the fluid pressure generated by the relative motion between the journal and the bearing. In boundary lubrication, the load is carried by direct contact between the surfaces. The load-carrying capacity increases with increasing Sommerfeld number, indicating a transition from boundary to hydrodynamic lubrication.
4. **Surface texture and material**: Smooth surfaces and materials with high load-bearing properties, such as Frelon (a PTFE-based material), can significantly enhance load-carrying capacity. Frelon-lined bearings, for example, can support loads four to eight times greater than comparable ball bearings.

Measuring Load-carrying Capacity:
The load-carrying capacity of lubricants is often assessed using standardized tests:

- **Four ball test**: Determines the weld point and load-carrying capacity by applying increasing pressure until the lubricant fails.
- **Four ball wear test**: Evaluates the wear scar diameter to assess the wear protection provided by the lubricant under specific load conditions. See Figure F17.

Understanding and optimizing the load-carrying capacity of lubricants is essential for designing efficient and durable mechanical systems, as it directly influences the performance and longevity of components under load.

466. Load Wear Index

The **load wear index (LWI)** is a parameter used to assess a lubricant's ability to minimize wear under applied loads. It is determined through the **four-ball**

extreme pressure (EP) test, specifically ASTM D2596, which evaluates the lubricant's performance under increasing pressure until a seizure occurs.

Calculation of Load Wear Index:
The LWI is calculated based on the wear scars produced on the three stationary balls during the test. The procedure involves:

1. **Conducting the four-ball EP test**: Apply incremental loads to the lubricant until a seizure occurs.
2. **Measuring wear scars**: After the test, measure the diameter of the wear scars on the three stationary balls.
3. **Calculating LWI**: The LWI is determined using the wear scar diameters from the ten highest nonseizure loads. This calculation provides an index of the lubricant's ability to minimize wear at applied loads.

Significance of Load Wear Index:
A higher LWI indicates better performance in reducing wear under load, which is crucial for:

1. **Enhancing equipment longevity**: Minimizing wear extends the lifespan of machinery components.
2. **Improving operational efficiency**: Reduced wear leads to smoother operation and less downtime.
3. **Optimizing maintenance schedules**: Understanding wear characteristics aids in planning maintenance activities effectively.

For detailed procedures and calculations, refer to **ASTM D2596**, which outlines the standard test method for measuring extreme-pressure properties of lubricating grease using the four-ball method.

467. Low-pressure (LP) Sampling Valve

Low-pressure valve sampling involves using specialized spring-loaded, push-button ball valves to extract fluid samples from process lines operating at low pressures. See Figure L6. These valves are designed to safely and efficiently collect representative samples without disrupting the ongoing process.

Key Features of Low-pressure Ball Valve Sampling:

1. **Cavity-free design**: These valves are engineered to minimize dead space, preventing the accumulation of residual products that could

contaminate future samples. This design ensures that each sample is representative of the current process conditions.

2. **Low sump area**: The valves feature a low sump area, reducing the volume of product that can accumulate between sampling events. This design improvement enhances product mixing and prevents the formation of residues that could affect sample integrity.

3. **Spring-loaded sealing system**: A spring-loaded sealing mechanism maintains consistent torque across varying temperatures, ensuring reliable operation and prolonging the valve's lifespan.

4. **Versatility**: Suitable for a wide range of applications, including chemical, pharmaceutical, food, and cosmetic industries, these valves can handle various media such as gases, liquids, emulsions, syrups, suspensions, granulates, and powders.

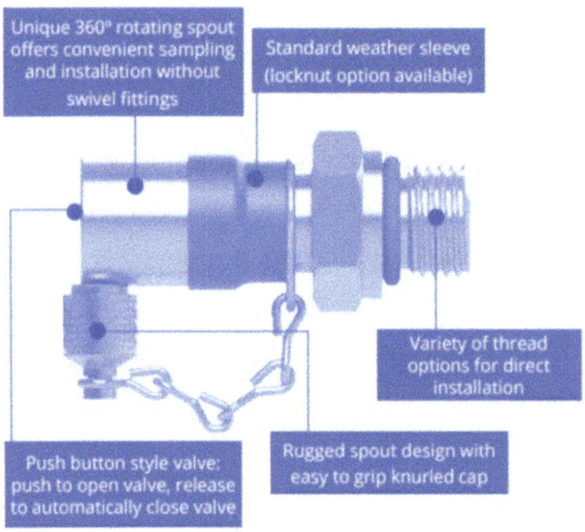

Figure L6 Typical low-pressure KP-style sampling valve for oil.

Benefits:

1. **Enhanced safety**: By enabling safe and controlled sampling, these valves reduce the risk of exposure to hazardous materials during the sampling process.

2. **Improved sample integrity**: The cavity-free and low-sump designs ensure that each sample accurately represents the process fluid, leading to more reliable analytical results.

3. **Operational efficiency**: These valves facilitate easy cleaning and maintenance, minimizing downtime and ensuring continuous, uninterrupted sampling operations.

468. Lube Cart

See "Filter Cart."

469. Lube Point

A **lube point** refers to a specific location on machinery or equipment where lubrication is applied to reduce friction, prevent wear, and ensure the smooth operation of moving parts.

Key Aspects of Lube Points:

1. **Identification**: Lube points are typically marked or indicated on equipment to ensure they are easily located during maintenance procedures.
2. **Lubrication methods**: Depending on the equipment design, lubrication can be applied manually, automatically, or through centralized lubrication systems.
3. **Maintenance schedule**: Regular inspection and lubrication of these points are crucial to prevent component failure and maintain optimal performance.

For example, in industrial machinery, lube points can be found on bearings, gears, and conveyor systems.

470. Lube for Life

The term "**lube for life**" commonly refers to components, such as bearings or gearboxes designed to operate without the need for regular lubrication changes throughout their service life. Manufacturers achieve this by using sealed bearings or incorporating lubricants that are intended to last the lifetime of the component. However, it is important to note that "lube for life" does not mean indefinite maintenance-free operation. Factors like operating conditions, load, and speed can influence the actual lifespan of these components. Therefore, while these components may not require routine lubrication, they are not immune to wear and may still require maintenance or replacement over time.

471. Lubricant

A **lubricant** is a substance – such as oil, grease, or specialized fluids – applied between surfaces in relative motion to reduce friction, minimize wear, and dissipate heat. This application enhances the efficiency and longevity of mechanical systems by facilitating smoother movement between components.

Key Functions of Lubricants:

1. **Friction reduction**: Lubricants form a film between moving parts, decreasing direct contact and thereby reducing friction.
2. **Heat dissipation**: By absorbing and transferring heat away from contact surfaces, lubricants help maintain optimal operating temperatures.
3. **Wear prevention**: Lubricants protect surfaces from wear and tear by minimizing direct contact and reducing the likelihood of material degradation.
4. **Corrosion resistance**: Certain lubricants provide a protective barrier against moisture and contaminants, preventing rust and corrosion.

Common Types of Lubricants:

1. **Oils**: Widely used in engines and machinery, oils can be mineral-based or synthetic, offering various viscosities and additives tailored to specific applications.
2. **Greases**: Composed of oil thickened with soap or other thickening agent, greases are ideal for applications requiring lubrication over extended periods or under heavy loads.
3. **Solid lubricants**: Materials like graphite or molybdenum disulfide are used in high-temperature or high-pressure environments where liquid lubricants may not be effective.

472. Lubricant Consolidation Program

A **lubricant consolidation program** is a strategic approach designed to reduce the number of different lubricants used within an organization or facility, streamlining the lubrication process for improved efficiency, cost savings, and inventory management. This program focuses on selecting multifunctional lubricants that can serve multiple applications, thereby minimizing the variety of products required.

Key Objectives:

1. **Simplified inventory**: Reduces the number of lubricant types, making storage, handling, and inventory management more efficient.
2. **Cost savings**: Lowers procurement and storage costs by purchasing fewer, but more versatile, lubricants in bulk.
3. **Improved maintenance practices**: Simplifies training for maintenance personnel by reducing the complexity of choosing and applying different lubricants.
4. **Reduced risk of misapplication**: Minimizes the chances of cross-contamination and the use of incorrect lubricants that could lead to equipment damage or reduced performance.

Implementation Steps:

1. **Assessment of current lubricants**: Conduct an audit to identify all types and volumes of lubricants currently stored and used within the plant and determine any redundancies.
2. **Application analysis**: Evaluate equipment needs to identify opportunities for consolidation without compromising performance or protection.
3. **Selection and documentation of versatile lubricants**: Choose high-performance, multiapplication lubricants that can meet the requirements of different equipment types and minimize the number of lubricants carried in the plant.
4. **Changeover**: Remove all unwanted lubricants from the plant and update all work orders requiring updated or new lubricant references. All old lubricant references must be removed and replaced with new information on machinery, manuals, and stock/purchasing references. Finally, replace old Safety Data Sheets with updated lubricant SDSs.
5. **Training and education**: Ensure maintenance teams are trained on the use of the consolidated lubricants and understand their applications.
6. **Ongoing monitoring**: Continuously review the program to make adjustments and maintain optimal lubrication practices.

Benefits:

- **Streamlined operations**: A more straightforward lubrication process reduces errors and improves maintenance efficiency.
- **Enhanced equipment reliability**: Consistent use of the correct lubricants helps extend the life of machinery and reduce unscheduled downtime.
- **Environmental impact**: Less variety in lubricant types can lead to reduced waste and more sustainable practices.
- **Cost reduction**: Less lubricants reduce stock management costs and cross-contamination downtime incidents

473. Lubricant Delivery

A **lubricant delivery cycle** refers to the interval between successive applications of lubricant to machinery components, ensuring they receive the necessary lubrication to function optimally. This cycle is crucial for maintaining equipment performance, reducing wear, and extending the lifespan of components.

Key Aspects of Lubricant Delivery Cycles:

- **Frequency of lubrication**: Determining the appropriate interval between lubrication applications is essential. This interval depends on factors such as equipment type, operating conditions, and manufacturer recommendations.
- **Lubricant quantity**: Applying the correct amount of lubricant during each cycle is vital. This is termed optimum full film lubrication. Overlubrication can lead to fluid friction and energy consumption, while underlubrication will place the bearing into a boundary state that may result in inadequate protection and premature wear. See Figure L7.
- **Lubricant type**: Selecting the appropriate lubricant – considering factors like viscosity, additives, and compatibility with materials – is crucial for effective lubrication.
- **Lubrication method**: The method of applying lubricant, whether manual or automated, should be chosen based on the specific requirements of the equipment and operational environment.

- **Monitoring and adjustment**: Regular monitoring of lubrication effectiveness and adjusting the delivery cycle as needed can help optimize performance and prevent issues related to lubrication.

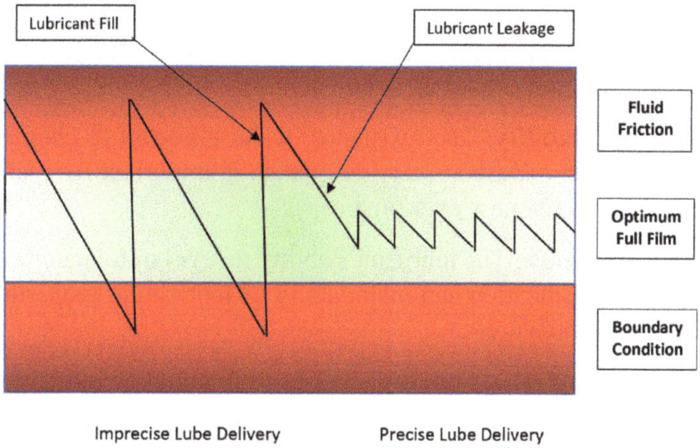

Figure L7 Lubricant delivery cycle chart. Source: Plant Lubrication for Industrial Facilities – Kenneth E. Bannister, 2024.

474. Lubricant Design Parameter

Lubricant design parameters are the specific characteristics and requirements that guide the selection and formulation of lubricants to ensure optimal performance in various applications. These parameters are crucial for achieving desired outcomes in machinery operation, such as reducing friction, minimizing wear, and enhancing efficiency.

Key Lubricant Design Parameters:

- **Viscosity**: The thickness or flow resistance of a lubricant, which affects its ability to form a protective film between moving parts. Selecting the appropriate viscosity is essential for ensuring effective lubrication under specific operating conditions.
- **Additive composition**: The inclusion of various chemical additives, such as antiwear agents, corrosion inhibitors, and extreme pressure additives, tailored to meet the specific demands of the application. These additives enhance the lubricant's performance and longevity.

- **Operating temperature range**: The range of temperatures within which the lubricant maintains its desired properties, ensuring consistent performance under varying thermal conditions. This parameter is vital for applications exposed to high or low temperatures.
- **Load-carrying capacity**: The ability of the lubricant to support applied loads without causing damage to the contacting surfaces, thereby preventing wear and extending component life.
- **Oxidation stability**: The resistance of the lubricant to chemical degradation when exposed to oxygen, which is important for maintaining lubricant effectiveness over time.
- **Water resistance**: The lubricant's ability to repel or tolerate water, preventing emulsification and maintaining its lubricating properties in the presence of moisture.
- **Compatibility**: The lubricant's ability to function effectively with the materials and components it contacts, including seals, gaskets, and metal surfaces, to prevent degradation or failure.

By carefully considering these design parameters, lubrication engineers and specialists can select or formulate lubricants that meet the specific requirements of their application.

475. Lubricant Dispensing System

A **lubricant dispensing system** is an A-to-Z setup designed to store, manage, and deliver lubricants to machine-bearing surfaces in the appropriate amount, in the right place, and at the right time, thereby enhancing equipment performance and extending component life.

Key Components of a Lubricant Dispensing System:

1. **Storage containers**: These are used to hold lubricants in bulk quantities, often made from materials like stainless steel or polyethylene to prevent contamination and degradation. Modern storage solutions are designed to minimize the risk of spills and contamination, featuring color-coded systems and spill containment kits.
2. **Pumps and hoses**: Pumps transfer lubricants from storage containers to machinery dispensing unit reservoirs. The choice of pump and hose depends on the type of lubricant and the required flow rate.

3. **Machine dispensing units**: These lubricating units control the amount of lubricant dispensed, ensuring precision and preventing over or under-lubrication. They can be manual or automated, with automated systems offering features like metered dispensing and remote monitoring.
4. **Filtration systems**: Filters remove contaminants from lubricants before they are applied to machinery, maintaining lubricant quality and protecting equipment from damage.
5. **Control and monitoring systems**: These paper-based or computerized systems track lubricant levels, usage, and dispensing intervals, providing data for maintenance schedules and inventory management. Advanced systems may include touch-screen controls and status monitoring for remaining lubricant volume and filter conditions.

Benefits of an engineered Lubricant Dispensing System:

- **Improved equipment longevity**: Consistent and accurate lubrication reduces wear and tear on machinery components.
- **Enhanced operational efficiency**: Automated dispensing minimizes downtime and labor costs associated with manual lubrication.
- **Reduced contamination risk**: Proper storage and handling prevent lubricant contamination, which can lead to equipment failure.
- **Environmental compliance**: Well-designed systems help manage spills and leaks, ensuring adherence to environmental regulations.

476. Lubricant Film Regime

A **Lubricant film regime** refers to the specific operating conditions under which a lubricant can form a film between two contacting surfaces, thereby reducing friction and wear. The nature of this film depends on factors such as load, speed, temperature, and surface roughness, and it determines the lubrication regime governing the interaction between the surfaces.

Key Lubrication Regimes:

- **Boundary lubrication**: Occurs under high load and low-speed conditions where the lubricant film is insufficient to separate the surfaces completely. In this regime, asperities (microscopic high points) on the surfaces come into direct contact, and damage is prevented by protective additives that promote sliding rather than welding surface asperities.

- **Mixed lubrication**: Represents an intermediary condition between boundary and hydrodynamic/elastohydrodynamic lubrication regimes. In this regime, the load is supported by both asperity contact and hydrodynamic pressures present in the lubricant film.
- **Hydrodynamic (HD) lubrication**: Occurs when a full film of lubricant supports and separates the surfaces, creating a working clearance. This regime is characterized by a thick lubricant film that completely separates the contacting surfaces, preventing direct contact and minimizing wear.
- **Elastohydrodynamic (EHD) lubrication**: Occurs in rolling contacts, such as ball bearings or roller bearings, where the softer material makes up the rolling element, which deforms to enlarge the contact area between mating surfaces. At full EHD lubrication, the generated lubricant film completely separates the surfaces.

Understanding the lubricant film regime is crucial for selecting appropriate lubricants and designing components that operate efficiently under specific condition.

477. Lubricant Film Replenishment Rate

Lubricant replenishment rate refers to the amount of lubricant supplied to a system over a specific period to maintain optimal performance and prevent component wear. Proper replenishment ensures that moving parts are adequately lubricated, reducing friction and extending the lifespan of machinery. See Figure L8.

Key Factors Influencing Lubricant Replenishment Rate:

1. **Operating conditions**: High-speed operations or elevated temperatures can accelerate lubricant degradation, necessitating more frequent replenishment.
2. **Lubricant type and viscosity**: The choice of lubricant and its viscosity affect how quickly it depletes. For instance, high-viscosity lubricants may replenish more slowly due to increased drag.
3. **Component design**: The geometry and surface finish of components influence lubricant distribution and retention, impacting the replenishment rate.

4. **Sealing integrity**: Effective seals prevent lubricant leakage, maintaining the necessary levels for proper lubrication.

Calculating Lubricant Replenishment Rate:
To determine the appropriate replenishment rate, consider the following approach:

- **Determine film thickness**: Calculate the required lubricant film thickness based on component specifications and operating conditions.
- **Calculate surface area**: Measure the contact area between moving parts.
- **Compute volume**: Multiply the film thickness by the surface area to find the lubricant volume needed.
- **Establish frequency**: Divide the total volume by the desired replenishment interval to determine the replenishment rate.

Lubricant and Delivery Method	Film Thickness	Time
Manual Application – Terminating Grease	.002" / .050mm Film	4 Hours
Automatic Application – Terminating Grease	.001" / .025mm Film	4 Hours
Automatic Application – Terminating Oil	.001" / .025mm Film	1 Hour
Automatic Application – Circulating Oil	.001" / .025mm Film	1 Minute

Figure L8 Typical lubricant replenishment rate. Source: Plant Lubrication for Industrial Facilities – Kenneth E. Bannister 2024.

478. Lubricant Receiving Procedure

A **lubricant receiving procedure** outlines the systematic steps for accepting, inspecting, and storing lubricants upon delivery to ensure their quality and suitability for use. Proper execution of this procedure is vital to prevent contamination, maintain lubricant integrity, and protect machinery.

Key Steps in the Lubricant Receiving Procedure:

1. **Predelivery preparation**:
   - **Documentation**: Ensure that purchase orders, delivery schedules, and Certificates of Analysis (COAs) are prepared and accessible.

- **Storage readiness**: Verify that storage areas are clean, organized, and equipped with the necessary tools for unloading and inspection.

2. **Receiving the Delivery**:
   - **Inspection of containers**: Examine all lubricant containers for signs of damage, such as dents, cracks, or leaks, which could indicate contamination or product loss.
   - **Verification of product details**: Confirm that the delivered lubricants match the ordered specifications, including type, quantity, and packaging.
3. **Quality Control Checks**:
   - **Review of Certificates of Analysis (COAs)**: Assess COAs to ensure the lubricant meets required lubricant cleanliness standards and specifications.
   - **Sampling and testing**: If necessary, collect samples for in-house testing to verify quality before storage.
4. **Documentation and Record-keeping**:
   - **Delivery records**: Document delivery details, including date, time, supplier information, and any discrepancies noted during the inspection.
   - **Inventory management**: Update inventory systems to reflect new stock levels and track lubricant usage.
5. **Storage**:
   - **Proper placement**: Store lubricants in designated areas, ensuring containers are upright, tightly sealed, and protected from environmental factors like temperature extremes and moisture.
   - **Labeling**: Clearly label storage containers with product information, including type, batch number, and expiration date, to facilitate easy identification and rotation.
6. **Handling and Dispensing**:
   - **Clean equipment**: Use clean tools and equipment when handling lubricants to prevent contamination.
   - **Controlled dispensing**: Implement procedures to dispense lubricants without introducing contaminants, such as using dedicated pumps and hoses for each lubricant type.
   - **Stock rotation**: Ensure lubricants are always fresh through the utilization of FIFO methods. See First In, First Out method.

By adhering to a structured lubricant-receiving procedure, organizations can maintain lubricant quality, prevent equipment damage, and ensure operational efficiency. Regular training and audits are recommended to uphold these standards.

479. Lubricant Storage Vessel

A **lubricant storage vessel** is a container or tank designed to safely store lubricants – such as oils and greases – before they are used in machinery or equipment. Proper storage is essential to maintain the lubricant's quality, prevent contamination, and ensure optimal performance when applied.

Key Considerations for Lubricant Storage Vessels:

1. **Material compatibility**: The vessel should be constructed from materials compatible with the specific type of lubricant to prevent chemical reactions or degradation. Common materials include stainless steel, mild steel, anodized aluminum, and high-density polyethylene (HDPE).
2. **Design Features**:
   - **Breathers and ventilators**: These components allow for air exchange while filtering out contaminants, and maintaining the integrity of the stored lubricant.
   - **Sight gauges**: Installed to monitor the lubricant level, ensuring timely replenishment and preventing overfilling.
   - **Drain valves**: Facilitate the removal of old or contaminated lubricant, aiding in maintenance and system cleanliness.
3. **Storage Conditions**:
   - **Temperature control**: Maintain storage temperatures between 60 and 100 °F (15 to 38 °C) to prevent degradation and maintain lubricant properties.
   - **Protection from contaminants**: Ensure vessels are sealed and stored in clean, dry environments to prevent the ingress of water, dust, or other contaminants.
4. **Labeling and identification**: Each vessel should be clearly labeled with the lubricant type, grade, and any other pertinent information to prevent cross-contamination and ensure correct usage.
5. **Maintenance and Inspection**:
   - **Regular inspections**: Conduct routine checks for signs of leaks, corrosion, or contamination.

- **Inventory management**: Implement a first-in, first-out (FIFO) system to ensure older lubricants are used first, reducing the risk of using expired or degraded products.

By adhering to these guidelines, organizations can ensure that their lubricant storage vessels effectively preserve lubricant quality.

480. Lubricant Transfer Vessel

A **lubricant transfer vessel** is a specialized closed container or system designed to safely and efficiently transfer lubricants between different locations, equipment, or storage units. These vessels are required to ensure the lubricant is protected from airborne contamination during the transfer process.

Key Features of Lubricant Transfer Vessels:

1. **Material compatibility**: Constructed from materials resistant to corrosion and degradation, such as stainless steel, or high-density polyethylene (HDPE) plastics to maintain lubricant integrity during transfer as depicted in Figure L9.
2. **Design and functionality**:
   - **Pressure management**: Equipped with mechanisms to control and monitor pressure, ensuring safe transfer operations.
   - **Isolation valves**: Incorporate valves to control the flow and prevent backflow, enhancing safety and efficiency.
   - **Transparent components**: Some designs include transparent sections to allow visual monitoring of fluid levels and flow.
3. **Capacity and dimensions**: Available in various sizes to accommodate different volumes of lubricant, tailored to specific operational requirements.
4. **Safety Features**:
   - **Pressure relief systems**: Designed to prevent overpressurization, reducing the risk of equipment failure.
   - **Emergency shut-off mechanisms**: Include features to quickly halt the transfer process in case of emergencies.
5. **Regulatory compliance**: Manufactured to meet industry standards and regulations, ensuring safe and environmentally responsible operation.

Figure L9 Molded HDPE plastic lubricant transfer vessels with color-coded caps and tags. Courtesy: Fluid Defense Systems.

Applications:
Manufacturing: Facilitate the movement of lubricants from storage vessels to machine reservoirs within production facilities.

481. Lubricate

Lubricate is a verb that means to apply a lubricant, such as oil or grease, to a surface to reduce friction, facilitate movement, and prevent wear. This process is essential in for basic machinery maintenance. See Figure L10.

Definitions:

- **Cambridge dictionary**: To use a substance such as oil to make a machine or device operate more easily, or to prevent something from sticking or rubbing.
- **Merriam-Webster**: To make smooth or slippery; to apply a lubricant to.

In essence, to lubricate means to make a surface slippery or smooth by applying a suitable substance, thereby reducing friction and facilitating easier movement or operation.

Figure L10 Grease lubricating a bearing. Courtesy: LocknLube Corporation.

482. Lubricating Grease

See "Grease."

483. Lubrication

Lubrication is the process and science of introducing a lubricant – commonly oil, grease, or other specialized fluids – between moving surfaces to reduce friction, prevent wear, and facilitate smooth operation. This practice is essential in various applications, including machinery maintenance, automotive engines, and industrial equipment, to enhance performance and extend the lifespan of components.

Key Functions of Lubrication:

- **Reduce friction**: By forming a thin film between surfaces, lubrication minimizes direct contact, thereby decreasing friction and associated heat generation.

- **Prevent wear**: Lubricants act as a protective barrier, reducing surface wear and tear, which is crucial for maintaining the integrity of moving parts.

- **Protect against corrosion**: Certain lubricants contain additives that form protective layers on metal surfaces, shielding them from moisture and corrosive elements.

- **Transmit power**: In hydraulic systems, lubricants serve as the medium for power transmission, enabling the movement of machinery components.

- **Cool components**: Lubricants absorb and dissipate heat generated by moving parts, helping to maintain optimal operating temperatures.

484. Lubrication Cycle

A **lubrication cycle** is the interval frequency at which lubricant is charged into the bearing surface area; frequency can be measured in time, machine cycle operations, or measurable machine conditions. See Figure L11.

Automated lubrication system frequency cycles are generally much shorter in duration than manual application cycles operating on the principle of applying a small amount of lubricant on a more frequent basis (minutes/hours compared to days as in manual lubrication) to ensure the bearing always stays in an optimum full film condition.

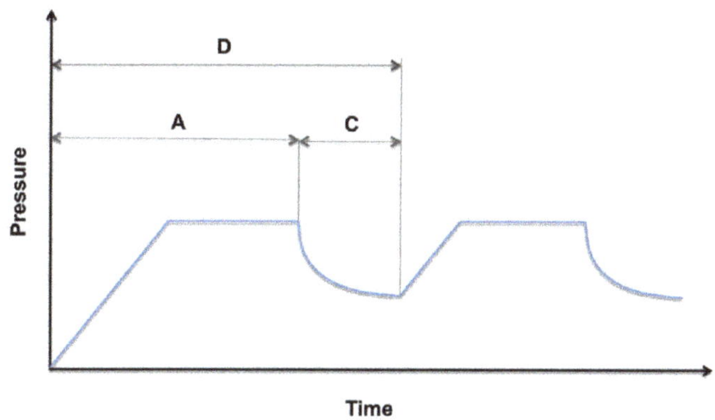

A = Lubrication cycle time
C = Interval time
D = Operating cycle time
A + C = D

Figure L11 Typical series progressive divider system lubrication cycle. Source: Plant Lubrication for Industrial Facilities, Fourth Edition, Kenneth E. Bannister, 2024.

485. Lubrication Failure

See "Failure Codes."

486. Lubrication Film

A **lubrication film** is a thin layer of lubricant – such as oil or grease – formed between two moving surfaces to reduce friction, prevent wear, and facilitate smooth operation. This film acts as a barrier, ensuring that the surfaces do not come into direct contact, thereby minimizing heat generation and extending the lifespan of the components.

Key Characteristics of Lubrication Films:

- **Thickness**: The thickness of the lubrication film varies depending on the lubrication regime:
  - **Hydrodynamic lubrication**: In this regime, the lubricant forms a thick film, typically around 7000 Å (angstroms) in thickness, fully separating the moving surfaces.
  - **Elastohydrodynamic lubrication**: Occurs under high-pressure conditions, where the lubricant film thickness is reduced due to elastic deformation of the surfaces.
- **Boundary lubrication**: In this regime, the lubricant film is insufficient to separate the surfaces completely, leading to direct contact at the asperities (microscopic high points) of the surfaces.
- **Viscosity**: The viscosity of the lubricant influences the formation and stability of the lubrication film. Higher viscosity lubricants can maintain a thicker film under varying operating conditions.
- **Additives**: Lubricants often contain additives that enhance film strength, reduce wear, and improve performance under extreme conditions.

Functions of Lubrication Films:

- **Friction reduction**: By separating moving surfaces, the lubrication film reduces direct contact, thereby decreasing friction and associated heat generation.
- **Wear prevention**: The film acts as a protective barrier, preventing metal-to-metal contact and reducing wear on components.
- **Heat dissipation**: Lubrication films help in dissipating heat generated during operation, maintaining optimal operating temperatures.
- **Contamination removal**: The movement of the lubrication film can help in carrying away contaminants and wear particles from the contact area.

487. Lubrication Interval

See "Lubricant Delivery" and "Lubrication Cycle."

488. Lubrication Dunning–Kruger Effect

The **Dunning–Kruger effect** is a cognitive bias where individuals with low ability in a specific area overestimate their competence, while those with higher ability may underestimate their skills. This phenomenon occurs because individuals lacking expertise are often unable to recognize their own lack of knowledge, leading them to believe they are more capable than they are. See Figure L12 for the Dunning–Kruger effect in lubrication.

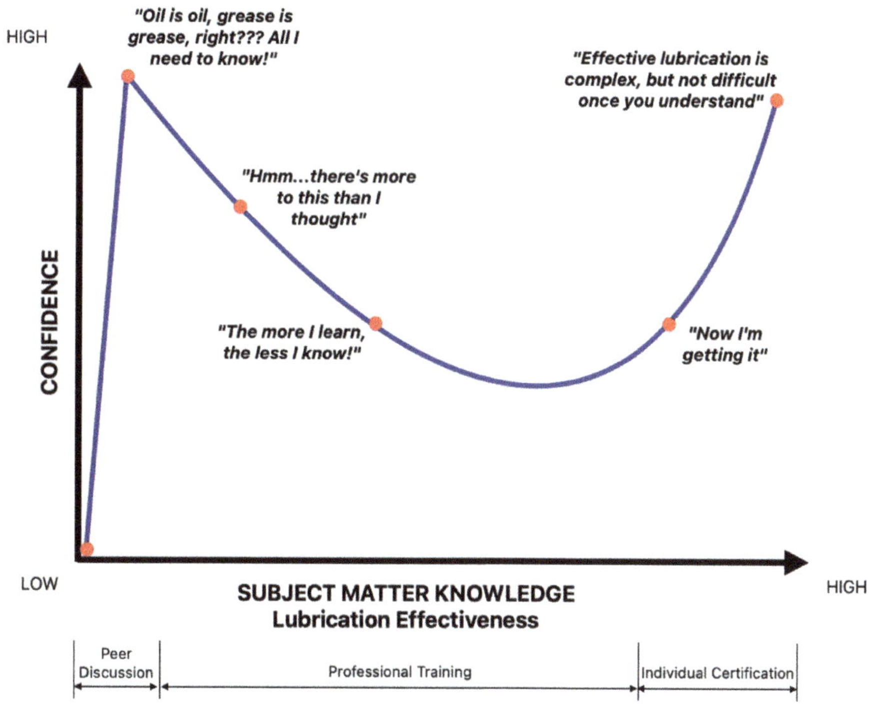

Figure L12 The lubrication Dunning–Kruger effect that prevails in the industry.

Key Aspects of the Dunning–Kruger Effect:

- **Overestimation of ability**: Individuals with limited knowledge or skills in a particular domain may rate their abilities as above average, not realizing their deficiencies.
- **Underestimation by experts**: Conversely, those with substantial expertise might rate their abilities as average or below, not recognizing their proficiency.

- **Lack of metacognitive skills**: The effect is linked to a deficiency in metacognition – the ability to assess one's own knowledge and performance accurately.

Implications:

- **Decision-making**: Overconfident individuals may make poor decisions due to their inflated self-assessments.
- **Learning and development**: Recognizing this bias is crucial for personal growth, as it encourages individuals to seek feedback and engage in continuous learning.

Understanding the Dunning–Kruger Effect highlights the importance of self-awareness and the value of feedback in personal and professional development.

489. Lubrication Management System – LMS

Lubrication management system involves the systematic organization and application of lubricants to machinery and equipment to enhance performance, reliability, and longevity. By ensuring that components receive the appropriate type and amount of lubricant at optimal intervals, organizations can reduce friction, prevent wear, and minimize downtime.

Key Components of Lubrication Management:

- **Lubricant selection**: Choosing the appropriate lubricant type, grade, and additives based on equipment specifications and operating conditions.
- **Lubricant application**: Determining the correct quantity and frequency of lubricant application to each component.
- **Monitoring and analysis**: Regularly assessing lubricant condition and performance through oil analysis and equipment inspections to identify potential issues before they lead to failures.
- **Training and documentation**: Educating personnel on best practices and maintaining detailed records of lubrication activities to ensure consistency and traceability.

Benefits of Effective Lubrication Management:

- **Enhanced equipment reliability**: Proper lubrication reduces friction and wear, leading to fewer breakdowns and extended equipment lifespan.

- **Cost savings**: By preventing premature component failure and reducing maintenance needs, organizations can achieve significant cost reductions.
- **Energy efficiency**: Well-lubricated machinery operates more efficiently, consuming less energy and contributing to sustainability goals.
- **Regulatory compliance**: Adhering to lubrication best practices helps meet industry standards and environmental regulations.

Implementing a structured lubrication management program is essential for organizations aiming to optimize equipment performance, reduce operational costs, and enhance overall productivity.

490. Lubrication Mapping

Lubrication mapping is a **structured process used to identify, document, and optimize lubrication points, schedules, and requirements for machinery and equipment**. It involves creating a detailed visual representation of all lubrication points within a facility, specifying lubricant types, application methods, and maintenance intervals. Lubrication mapping is essential for preventive maintenance programs, ensuring proper lubricant application, reducing failures, and improving equipment reliability.

Key Functions of Lubrication Mapping

- **Identifies and documents lubrication points**: Provides a complete visual reference of all greasing and oiling locations.
- **Optimizes lubricant selection and application methods**: Ensures each component receives the correct lubricant type and delivery method.
- **Establishes a structured lubrication schedule**: Defines application frequency to maintain consistent lubrication and prevent over or underlubrication.
- **Improves maintenance efficiency and standardization**: Helps technicians follow a clear lubrication plan, reducing errors and increasing productivity.

Components of a Lubrication Map

- **Lubrication point identification**: Labels **bearings, gears, chains, motors, hydraulic systems, and other key lubrication points**
- **Lubricant type and specification**: Specifies **oil or grease type, viscosity, and additive requirements for each point**.

- **Application method**: Defines **manual greasing, automated lubrication, drip, or circulating systems.**
- **Lubrication frequency and volume**: Set **intervals based on OEM recommendations, operating conditions, and environmental factors.**
- **Equipment criticality ranking**: Prioritizes **critical assets that require closer monitoring and frequent lubrication.**

491. Lubrication Operation Effectiveness Review – LOER

A **lubrication operation effectiveness review (LOER)** is a comprehensive assessment conducted by an accredited lubrication consultant to evaluate and enhance an organization's lubrication practices. The primary goal of a LOER is to identify improvement opportunities across various aspects of lubrication management, including:

- **Application**: Ensuring that the right lubricant is applied in the correct amount, at the appropriate location, and at the optimal time.
- **Purchasing**: Streamlining the procurement process to acquire lubricants that meet quality standards and are cost-effective.
- **Storage**: Implementing proper storage conditions to maintain lubricant integrity and prevent contamination.
- **Transfer and dispensing**: Utilizing dedicated equipment to transfer and dispense lubricants, thereby reducing the risk of cross-contamination and ensuring accurate application.
- **Disposal**: Establishing environmentally responsible methods for disposing of used lubricants.

By conducting a LOER, organizations can develop a structured approach to delivering the right lubricant in the right amount, at the right place, and at the right time, whether using manual methods or fully automated systems.

Implementing the recommendations from a LOER can lead to several benefits:

- **Cost reduction**: Optimizing lubricant usage and minimizing waste can result in significant cost savings.
- **Enhanced equipment reliability**: Proper lubrication practices reduce wear and extend the lifespan of machinery.

- **Improved energy efficiency**: Well-lubricated components operate more efficiently, contributing to energy savings.
- **Environmental compliance**: Adhering to best practices in lubricant management helps meet environmental regulations and reduces the ecological footprint

492. Lubrication Room

A **lubrication room**, often referred to as a **lube room or facility**, is a designated area within a facility where lubricants – such as oils and greases – are stored, managed, and dispensed for use in machinery and equipment and can double as a lubrication work center where lubrication repair work can take place. Proper design and maintenance of a lube room are crucial for ensuring the quality and effectiveness of lubricants, thereby enhancing equipment reliability and extending component lifespan. See Figure L13 for a poor versus excellent example of a lube room.

Key Components of a Lubrication Room:

- **Storage facilities**: Adequate space and shelving to store lubricant containers, including drums, totes, and smaller packages, in an organized manner.
- **Dispensing equipment**: Tools such as pumps, taps, and hoses designed for the safe and efficient transfer of lubricants from storage to application points.
- **Contamination control measures**: Systems to prevent contamination, including air filtration, moisture control, and cleanliness protocols for equipment and storage containers.
- **Safety features**: Proper ventilation, fire suppression systems, and spill containment measures to ensure a safe working environment.
- **Inventory management systems**: Methods for tracking lubricant usage, monitoring stock levels, and scheduling reorders to maintain optimal inventory.

Best Practices for Lubrication Room Design:

- **Organization**: Implement a logical layout with clearly labeled storage areas and designated workspaces to facilitate efficient operations.

- **Cleanliness**: Maintain a clean environment to prevent contamination of lubricants, which can lead to equipment damage and increased maintenance costs.
- **Accessibility**: Ensure that all necessary tools and lubricants are easily accessible to authorized personnel, reducing downtime and improving workflow.
- **Training**: Provide regular training for staff on proper lubrication practices, safety protocols, and equipment handling to ensure consistency and reliability.

Implementing a well-designed lubrication room is a fundamental aspect of a comprehensive lubrication management program, contributing to improved equipment performance, reduced maintenance costs, and enhanced operational efficiency.

Figure L13 Lubrication room – before and after. Courtesy: ENG**TECH** Industries Inc. and Fluid Defense Systems (Whitmore Manufacturing).

493. Lubrication Route

A **lubrication route** is a structured plan that outlines the sequence and schedule for performing lubrication tasks across various machines or equipment within a facility. These routes are designed to ensure that each component receives the appropriate type and amount of lubricant at specified intervals, thereby enhancing equipment reliability and extending component lifespan.

Key Elements of a Lubrication Route:

- **Identification of lubrication points**: Cataloging all machines and components that require lubrication, including bearings, gears, pumps, and other moving parts.
- **Lubricant specifications**: Determining the correct type, grade, and quantity of lubricant for each component based on manufacturer recommendations and operating conditions.
- **Scheduling**: Establishing a routine for lubrication tasks, which can be time-based (e.g., daily and weekly) or condition-based (e.g., after a certain number of operating hours).
- **Standard operating procedures (SOPs)**: Developing detailed, step-by-step instructions for performing lubrication tasks to ensure consistency and quality.
- **Documentation and tracking**: Maintaining records of completed lubrication tasks, including dates, personnel involved, and any observations or anomalies.

Benefits of Implementing a Lubrication Route:

- **Consistency**: Standardized procedures ensure that lubrication tasks are performed uniformly, regardless of the technician.
- **Efficiency**: A well-planned route minimizes downtime and optimizes the use of maintenance personnel.
- **Proactive maintenance**: Regular lubrication helps in the early detection of potential issues, reducing the risk of unexpected equipment failures.
- **Cost savings**: Proper lubrication reduces wear and tear, leading to longer equipment lifespan and lower maintenance costs.

494. Lubrication Schedule

A **lubrication schedule** is a structured plan that outlines the timing and procedures for applying lubricants to machinery and equipment. This schedule ensures that each component receives the appropriate type and amount of lubricant at specified intervals, thereby enhancing equipment reliability and extending component lifespan.

Key Components of a Lubrication Schedule:

1. **Identification of lubrication points**: Cataloging all machines and components that require lubrication, including bearings, gears, pumps, and other moving parts.
2. **Lubricant specifications**: Determining the correct type, grade, and quantity of lubricant for each component based on manufacturer recommendations and operating conditions.
3. **Scheduling**: Establishing a routine for lubrication tasks, which can be time-based (e.g., daily and weekly) or condition-based (e.g., after a certain number of operating hours).
4. **Standard operating procedures (SOPs)**: Developing detailed, step-by-step instructions for performing lubrication tasks to ensure consistency and quality.
5. **Documentation and tracking**: Maintaining records of completed lubrication tasks, including dates, personnel involved, and any observations or anomalies.

495. Lubrication Specialist Technician

See "CLS Designation" and "MLT Designation."

496. Lubrication Work Center

See "Lubrication Room."

497. Lubricity

Lubricity refers to the ability of a substance to reduce friction between surfaces in contact, thereby minimizing wear and enhancing the efficiency of mechanical systems. In the context of lubricants, lubricity is a critical property that determines how effectively a lubricant can form a protective film between moving parts, reducing direct contact and associated friction.

Lubricity is not a material property and as such, cannot be measured directly. Instead, it is assessed through tests that quantify a lubricant's performance in specific systems, often by determining the extent of wear caused to a surface under defined conditions. For example, lubricants with higher lubricity result in smaller wear scars on test surfaces.

498. Lubripedia – An Encyclopedic Illustrated Glossary of 1001 Lubrication Terms and References

Lubripedia refers to a comprehensive and organized collection of knowledge, definitions, and best practices related to lubrication technology, maintenance, and management. It serves as a centralized reference for professionals in industries such as manufacturing, automotive, aerospace, and energy, where effective lubrication is critical to equipment reliability and efficiency.

Purpose of Lubripedia:

1. **Education**:
    - Provides detailed explanations of lubrication terms, concepts, and techniques to help users understand and apply best practices.
2. **Standardization**:
    - Promotes consistency in lubrication practices by offering clear definitions and guidelines based on industry standards.
3. **Problem-solving**:
    - Offers insights into troubleshooting lubrication-related issues, such as wear, contamination, or system inefficiencies.
4. **Innovation**:
    - Shares updates on new technologies, trends, and advancements in lubrication to keep users informed and competitive.

Key Features of Lubripedia:

1. **Terminology database**:
    - 1001 Definitions of lubrication-related terms like viscosity, film strength, NLGI grades, and additive types.
2. **Maintenance guidelines**:
    - Instructions on lubrication schedules, selection, and application methods.
3. **Case studies**:
    - Real-world examples illustrating the impact of proper (or improper) lubrication.
4. **Tools and techniques**:
    - Information on lubrication equipment, such as grease guns, filtration systems, and monitoring tools.

5. **Compliance and safety**:
   - Details on regulations and safety protocols for handling lubricants in industrial settings.

Audience:

1. **Maintenance engineers**:
   - For planning and optimizing lubrication strategies.
2. **Operators and technicians**:
   - For on-the-job reference and troubleshooting.
3. **Industry trainers**:
   - As a resource for educating teams about lubrication best practices.
4. **Students and researchers**:
   - As a knowledge base for studying lubrication science and technology.

499. Luster (Grease)

In the context of lubricants, luster refers to the visual appearance of a grease lubricant, particularly its sheen or glossiness. This characteristic is assessed through visual inspection and is part of the broader evaluation of a lubricant's appearance, which also includes color, texture, and consistency. These visual attributes can provide insights into the lubricant's quality and suitability for specific applications.

For instance, greases are often evaluated for their luster to ensure they meet the desired standards for performance and aesthetics. A uniform and appropriate luster can indicate proper formulation and consistency, which are essential for effective lubrication.

It is important to note that while luster is a useful visual indicator, it should be considered alongside other properties such as viscosity, chemical composition, and performance characteristics to fully assess a lubricant's suitability for a given application.

500. LWI

See "Load Wear Index."

501. L10/L10a Rating

The L10 rating is a standard used to estimate the lifespan of rolling element bearings, such as ball and roller bearings. It represents the number of revolutions or operating hours at a specific speed that 90% of a group of identical bearings are expected to achieve or exceed before experiencing fatigue failure. In other words, the L10 rating provides a 90% reliability level, meaning there is a 10% chance that a bearing will fail before reaching the L10 life

Key Factors Influencing the L10 Rating:

- **Lubrication quality**: The type and condition of the lubricant used can significantly impact bearing life. Proper lubrication reduces friction and wear, leading to a longer lifespan.
- **Operating temperature**: Elevated temperatures can accelerate lubricant degradation and bearing wear. Operating within recommended temperature ranges helps maintain bearing integrity.
- **Contamination levels**: Presence of contaminants such as dirt, water, or chemicals can lead to premature bearing failure. Implementing effective sealing and filtration systems is crucial.
- **Load conditions**: Variations in load magnitude and direction can affect bearing performance. Ensuring loads are within specified limits helps in achieving the predicted L10a life.
- **Speed variations**: Operating speeds outside the recommended range can influence bearing fatigue life. Maintaining consistent and appropriate speeds is essential.

The L10a rating is an extended bearing life estimate that accounts for the use of better quality bearing steels manufactured in electric arc furnaces.

It is important to note that both the L10 and L10a life are statistical estimates based on ideal conditions. Actual bearing lifespan can be influenced by the key factors listed above.

M

502. Machine Condition Factors

Machine condition factors are factors employed by lubrication engineers to describe the operating and environmental conditions used to determine suitable machine lubricants and develop an appropriate lubrication schedule.

Key Operating Conditions Affecting Lubricant Choice and Lubrication Schedule:

- **Temperature extremes**:
  - **High temperatures***:* Elevated operating temperatures can accelerate lubricant degradation, necessitating more frequent lubrication intervals to maintain effective protection.
  - **Low temperatures***:* Cold environments can cause lubricants to thicken, impairing their ability to flow and lubricate effectively. This may require adjustments in lubricant selection and application frequency.
- **Load variations**:
  - Heavy or fluctuating loads increase stress on machine components, leading to higher friction and wear. In such conditions, more frequent lubrication is necessary to ensure adequate protection.
- **Environmental Factors**:
  - **Dust and contamination***:* Operating in dusty or dirty environments can introduce contaminants into the lubrication system, compromising lubricant integrity and necessitating more regular maintenance.
  - **Moisture exposure***:* Exposure to water or high humidity can lead to corrosion and lubricant degradation, requiring adjustments in lubrication schedules and lubricant types.

DOI: 10.1201/9788743806608-514

- **Vibration and shock loads**:
  - Machines subjected to high levels of vibration or shock loads may experience accelerated lubricant breakdown, necessitating more frequent lubrication intervals to maintain component protection.
- **Operational speeds**:
  - High-speed operations can cause increased heat generation, leading to faster lubricant degradation. This may require more frequent lubrication to ensure effective performance.

Best Practices:

- **Condition-based lubrication**: Implementing lubrication schedules based on real-time equipment conditions, such as temperature and load, can optimize lubricant usage and extend equipment life.
- **Regular monitoring**: Utilizing condition monitoring tools like vibration analysis and oil analysis can help detect early signs of underlubrication or overlubrication, allowing for timely adjustments to lubrication schedules.
- **Customized lubricant selection**: Choosing lubricants formulated for specific operating conditions ensures optimal performance and protection. For example, high-temperature applications might require synthetic lubricants, while standard conditions may be suitable for mineral oils.

By considering these operating conditions and implementing tailored lubrication strategies, organizations can enhance equipment reliability, reduce downtime, and extend the lifespan of machinery.

503. Macpherson Contamination Effect

The **Macpherson contamination effect curve** illustrates the relationship between lubricant cleanliness and bearing fatigue life. Developed by researchers including P.B. Macpherson, this curve demonstrates how varying levels of particulate contamination in lubricants influence the lifespan of rolling element bearings See Figure M1.

Key Insights from the Macpherson Curve:

- **Particle size impact**: The curve indicates that particles larger than approximately 10 micrometers (μm) have minimal effect on bearing

life. However, particles around 5 μm are particularly detrimental, as they are of similar size to the clearances between rolling elements and races, leading to increased wear and reduced bearing lifespan.

- **Filtration efficiency**: Implementing filtration systems that effectively remove particles in the 1–5 μm range can significantly enhance bearing life. For instance, improving filtration to 3 μm can increase bearing life by up to sevenfold compared to systems with 25 μm filtration.
- **Optimal cleanliness levels**: The curve suggests that while removing particles larger than 10 μm is beneficial, achieving filtration below 3 μm yields diminishing returns. Therefore, targeting the 1–5 μm particle size range is most effective for prolonging bearing life.

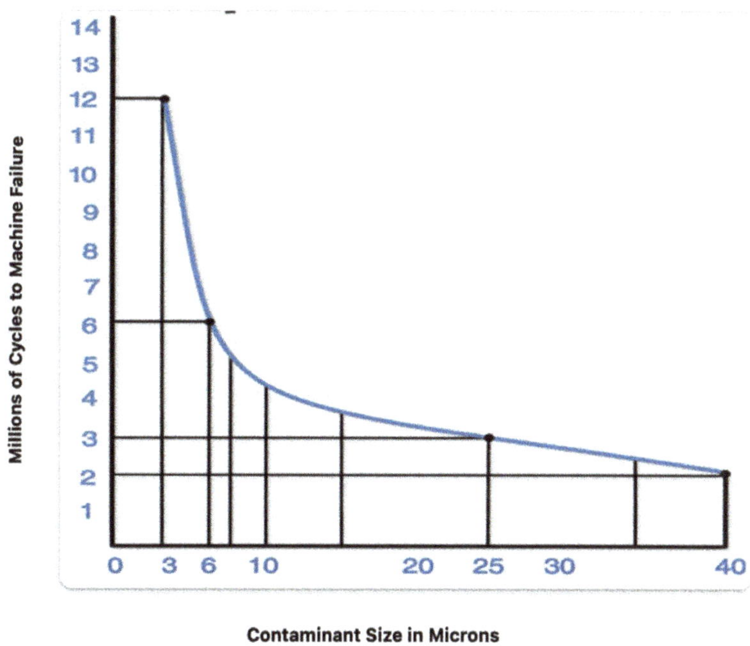

Figure M1 Macpherson contamination curve. Source: Plant Lubrication for Industrial Facilities, Fourth Edition, 2024 Kenneth E. Bannister

Understanding and applying the Macpherson Contamination Effect Curve enables industries to optimize lubrication practices, select appropriate filtration systems, and maintain lubricant cleanliness, thereby enhancing equipment reliability and reducing maintenance costs.

504. Magnetic Filter

A magnetic filter is a device that utilizes magnetic fields to remove ferrous (iron-containing) particles from liquids, gases, or powders. The magnetic filter is a first-stage filtration device that is most often screwed into a threaded hole or bulkhead fitting located on the bottom of the lubrication system oil reservoir. If the reservoir has a cleanout window in the reservoir, a simple magnet can be placed on the reservoir bottom provided it is made of ferrous material.

By attracting and capturing metallic particles from recirculated lubricant contained in the reservoir, the magnetic filter helps maintain the purity of the lubricant and protects downstream equipment from wear and damage.

How Magnetic Filters Work:
Magnetic filters consist of high-strength magnets, often made from materials like neodymium. As the contaminated fluid passes across the magnet, ferrous particles are attracted to and retained by the magnetic surfaces. This process effectively removes contaminants without the need for consumable filter media, allowing for continuous operation and easy cleaning. The magnet filter must be removed and cleaned of metal debris whenever the lubricant is changed out.

505. Maintain

When referencing lubrication, the verb **"Maintain"** means:

- **To keep in an existing state**: Preserving something in its current condition without allowing it to decline or deteriorate.
- **To keep in good condition**: Regularly servicing or repairing to ensure proper functioning.

506. Manual Centralized Grease Delivery System

A **manual centralized grease delivery system** is a lubrication setup where a single pump distributes grease to multiple lubrication points via a series progressive divider metering block across machinery or equipment. Operated manually, this system ensures that each lubrication point receives the necessary amount of grease, promoting efficient operation and extending the lifespan of the equipment.

The same grease delivery system can be operated without a pump by placing a grease nipple in the top of the divider block and pumping grease using a manual grease gun until the block cycle pin indicator shows all pistons have cycled in the block and delivered an engineered amount of grease to the bearing. See Figure M2.

The benefit of this type of system is that it can be easily converted to a fully automated pump system by connecting the pump to the block where the grease nipple or manual pump is connected at any point in the future.

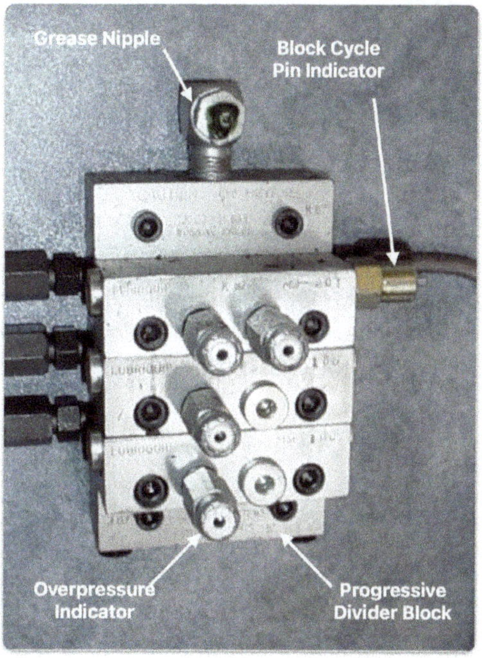

Figure M2 Manual series progressive divider block. Courtesy: **ENGTECH** Industries Inc.

Key Components:

1. **Reservoir**: Holds the grease, keeping it clean and readily available for distribution.

2. **Pump**: Manually operated to pressurize and deliver grease from the reservoir to various lubrication points.

3. **Distribution lines**: Pipes or hoses that transport grease from the pump to the lubrication points.

4. **Metering block**: Control the amount of grease delivered to each lubrication point, ensuring precise application.

5. **Lubrication points**: Specific areas on the machinery where grease is applied, such as bearings, gears, and joints.

Advantages:

- **Cost-effective**: Manual systems are generally less expensive to install and maintain compared to automatic systems.
- **Simplicity**: Fewer components make manual systems straightforward to operate and troubleshoot.
- **Flexibility**: Easily adaptable to various machinery types and sizes

507. Manual Centralized Oil Delivery System

A **manual centralized oil delivery system** is a lubrication setup where a single pump distributes oil to multiple lubrication points across machinery or equipment. See Figure M3. Operated manually, this system ensures that each component receives the necessary amount of oil, promoting efficient operation and extending the lifespan of the equipment.

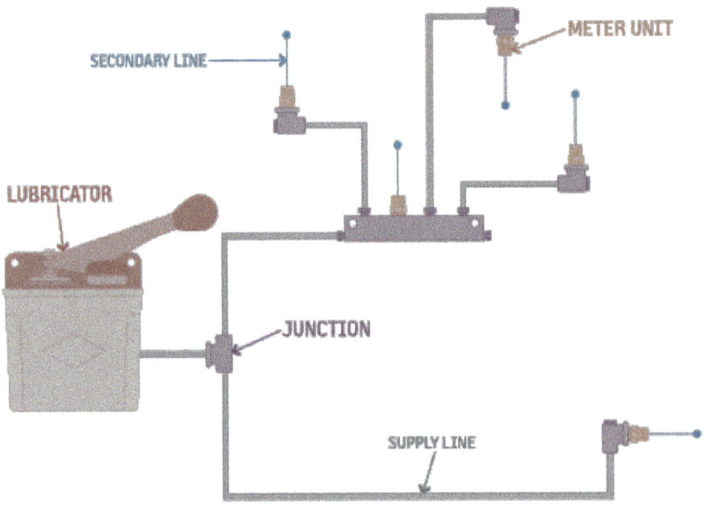

Figure M3 Manual centralized oil delivery system. Courtesy: Bijur–Delimon Corporation.

Key Components:

1. **Lubricator**: Manually operated to pressurize and deliver oil from the reservoir to various lubrication points. Holds the oil, keeping it clean and readily available for distribution.

2. **Distribution lines**: Pipes or hoses that transport oil from the pump to the lubrication points.
3. **Manifold**: Centralizes the lubricant feed and splits into secondary feed lines to the lube points
4. **Metering units**: Control the amount of oil delivered to each lubrication point, ensuring precise application.
5. **Lubrication points**: Specific areas on the machinery where oil is applied, such as bearings, gears, and joints.

Advantages:

6. **Cost-effective**: Manual systems are generally less expensive to install and maintain compared to automatic systems.
7. **Simplicity**: Fewer components make manual systems straightforward to operate and troubleshoot.
8. **Flexibility**: Easily adaptable to various machinery types and sizes.

508. Manufacturing Maintenance Readiness (MMR)

Manufacturing maintenance readiness (**MMR**) is a structured methodology designed by the Author Kenneth E. Bannister in conjunction with the NRC (National Research Council) Canada to ensure that maintenance and lubrication programs are fully integrated into the manufacturing process from the early design stages through to full operational implementation. MMR is designed to help organizations maximize asset reliability, optimize maintenance practices, and reduce downtime by aligning maintenance and lubrication strategies with manufacturing objectives.

Key Functions of MMR in Lubrication and Maintenance

- **Ensures equipment is designed for maintainability**: Integrates lubrication and maintenance accessibility into equipment design and selection.
- **Aligns maintenance and lubrication production goals**: Balances production efficiency with proactive maintenance strategies to avoid unexpected failures.
- **Optimizes lubrication programs**: Ensures proper selection of lubricants, application methods, and monitoring practices for improved equipment performance.

- **Facilitates standardized maintenance procedures**: Develops consistent lubrication and maintenance workflows based on industry best practices.
- **Improves predictive and preventive maintenance integration**: Uses condition monitoring tools, oil analysis, and performance tracking to anticipate failures before they occur.

Key Components of Manufacturing Maintenance Readiness

- **Asset criticality assessment (ACA)**: Identifies mission-critical equipment requiring prioritized maintenance and lubrication strategies.
- **Lubrication management and contamination control**: Establishes proper storage, handling, and application of lubricants to reduce failure risks.
- **Work order and maintenance planning**: Ensures maintenance teams follow structured lubrication schedules and work procedures.
- **Condition-based maintenance (CBM)**: Uses vibration analysis, oil analysis, and infrared thermography to detect early signs of wear.
- **Training and workforce readiness**: Equips maintenance teams with skills and knowledge in lubrication best practices, failure analysis, and asset reliability.
- **Lifecycle cost analysis (LCA)**: Evaluates long-term maintenance and lubrication costs to optimize asset performance and operational efficiency.
- **Spare parts and lubrication inventory management**: Ensures the right lubricants, filters, and spare parts are available to prevent downtime.

Applications of MMR in Industrial Lubrication and Maintenance

- **Any industrial facility** wanting to prepare for an **ISO 55001** Asset Management or **ICML 55®** Lubrication Certification Program.
- **Manufacturing plants and production facilities**: Ensures machines and lubrication systems operate efficiently without unexpected failures.
- **Heavy equipment and mining operations**: Implements preventive maintenance programs to reduce equipment breakdowns.
- **Power generation and turbine systems**: Integrates lubrication monitoring systems for long-term performance tracking.

- **Food and beverage processing**: Ensures compliance with food-grade lubrication standards (NSF H1, ISO 21469) while maintaining high reliability.
- **Aerospace and automotive manufacturing**: Enhances precision lubrication practices for improved part quality and machine performance.

Benefits of Manufacturing Maintenance Readiness (MMR)
Performing MMR brings together the Engineering, Production, Maintenance, and Lubrication teams to work together as a cohesive unit to do what is best for the assets under their care so as to:

✓ Improve overall equipment reliability and uptime

✓ Reduce maintenance costs by preventing unexpected failures

✓ Optimize lubrication and maintenance workflow

✓ Extend asset life by ensuring proper lubrication management

✓ Enhance workforce efficiency with structured maintenance programs

Conclusion: Manufacturing Maintenance Readiness (MMR) is a proactive approach that integrates lubrication and maintenance best practices into the manufacturing process, ensuring optimized equipment performance, reduced downtime, and improved operational efficiency. By implementing structured lubrication management, predictive maintenance techniques, and condition monitoring, organizations can enhance asset reliability and maximize manufacturing productivity.

509. Material Safety Data Sheet

See "Safety Data Sheet."

510. McCoy, Elijah

Elijah McCoy (1844–1929) was a lubrication pioneer with over 50 lubrication patents to his credit. Elijah, now a member of the National Inventors Hall of Fame was born in Canada, schooled as a mechanical engineer in Scotland, and worked as a Foreman on the Michigan state railroad steam trains performing manual lubrication duties. In 1872 he patented his famous steam-actuated single-point McCoy lubricator that became known as "the real McCoy," as all others were thought to be feeble copies in comparison.

Elijah McCoy's contributions to engineering, particularly through the development of the McCoy Lubricator, have left a lasting legacy in industrial machinery lubrication.

511. McCoy Lubricator

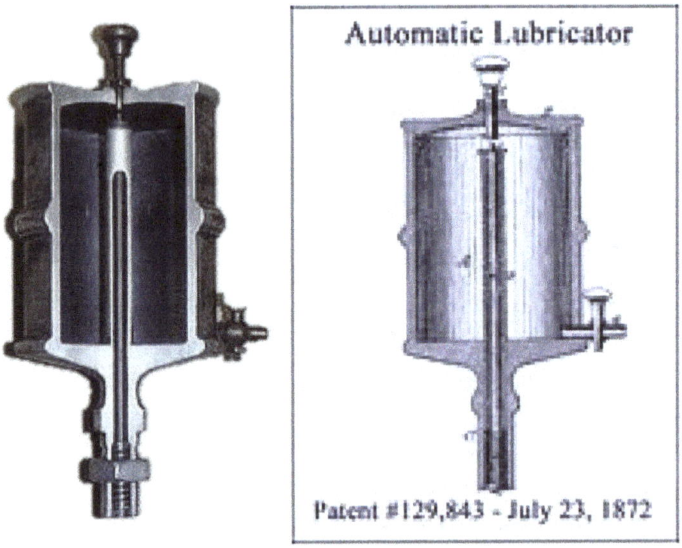

Figure M4 Actual cutaway of an early McCoy lubricator beside the patent application drawing. Courtesy: ENG**TECH** Industries Inc.

The **McCoy lubricator** is an automatic lubrication device invented by Canadian–American engineer Elijah McCoy in 1872. This innovation revolutionized the maintenance of steam engines by enabling continuous lubrication of moving parts during operation, thereby enhancing efficiency and reducing downtime. See Figure M4.

Key Features:

- **Automatic lubrication**: The McCoy lubricator delivered oil directly to the engine's moving components, ensuring consistent lubrication without manual intervention.
- **Sight-feed mechanism**: Later models incorporated a glass tube, allowing operators to monitor and adjust the oil flow rate, ensuring optimal lubrication.

Impact and Legacy:

- **Industrial adoption**: The McCoy lubricator became standard equipment on steam engines across North America, significantly improving operational efficiency.
- **Cultural significance**: The phrase "the real McCoy" emerged, reflecting the high quality and reliability of McCoy's invention.

512. Mean Time Between Lubrication Failure

Mean time between lubrication failures (MTBLF) is a metric used to assess the reliability of lubrication systems in machinery. It represents the average operational time between instances where the lubrication system fails to provide adequate lubrication, leading to potential equipment damage or failure.

Understanding MTBLF:

- **Definition**: MTBLF quantifies the average duration a machine operates before experiencing a lubrication failure. A higher MTBLF indicates more reliable lubrication, while a lower MTBLF suggests frequent lubrication issues.
- **Calculation**: MTBLF is determined by dividing the total operational time by the number of lubrication failures during that period:

$$\text{MTBLF} = \frac{\text{Total Operational Time}}{\text{Number of Lubrication Failures}}$$

For example, if a machine operates for **1000** hours and experiences **five** lubrication failures, the MTBLF would be **200** hours.

Importance of MTBLF:

- **Predictive maintenance**: Monitoring MTBLF helps in scheduling maintenance activities, such as lubrication system inspections and component replacements, to prevent unexpected failures.
- **Reliability assessment**: A low MTBLF may indicate issues like inadequate lubrication, contamination, or equipment wear, prompting a review of lubrication practices and system integrity.
- **Operational efficiency**: Ensuring a high MTBLF contributes to smoother operations, reduced downtime, and extended equipment lifespan.

Improving MTBLF:

- **Regular inspections**: Conduct routine checks of lubrication systems to identify and address potential issues before they lead to failures.
- **Quality lubricants**: Use lubricants that meet the manufacturer's specifications to ensure optimal performance and compatibility with equipment.
- **Training**: Educate maintenance personnel on proper lubrication techniques and the importance of maintaining clean and effective lubrication systems.

By focusing on MTBLF, organizations can enhance equipment reliability, reduce maintenance costs, and improve overall operational performance.

513. Measurement Standards

Measurement standards in lubrication are **established guidelines and testing methods used to evaluate lubricant properties**, performance, and contamination levels. These standards ensure consistency, reliability, and compatibility across industries, helping manufacturers, engineers, and maintenance professionals make data-driven decisions regarding lubricant selection, monitoring, and maintenance.

Key Functions of Measurement Standards in Lubrication

- **Ensures lubricant quality and performance**: Defines minimum requirements for viscosity, oxidation stability, and contamination control.
- **Facilitates industry compliance and safety**: Helps meet ISO, ASTM, SAE, and OEM specifications for various lubricants.
- **Standardizes lubricant testing methods**: Provides uniform procedures for evaluating lubricant properties across different applications.
- **Aids in predictive maintenance and condition monitoring**: Helps detect early signs of oil degradation, contamination, and wear.

Common Measurement Standards and Their Applications

1. **Viscosity and flow properties**
    - **ASTM D445 (kinematic viscosity test)**: Determines the fluidity of lubricants under controlled conditions.
    - **Saybolt universal viscosity (ASTM D88)**: Measures viscosity at specific temperatures for oil classification.

- **Brookfield viscosity (ASTM D2983)**: Evaluates low-temperature viscosity in hydraulic and gear oils.

2. **Contamination and cleanliness standards**
   - **ISO 4406 (particle count classification)**: Assesses solid contamination in lubricating oils and hydraulic fluids.
   - **NAS 1638 (cleanliness standard)**: Used in aerospace and hydraulic systems for fluid contamination control.
   - **ASTM D6304 (Karl Fischer water content test)**: Measures water contamination in lubricants.
3. **Oxidation and degradation testing**
   - **ASTM D943 (turbine oil oxidation stability test – TOST)**: Evaluates oil oxidation resistance over time.
   - **ASTM D2272 (rotating pressure vessel oxidation test – RPVOT)**: Determines oxidation stability under high-pressure conditions.
   - **FTIR spectroscopy (ASTM E2412)**: Detects oxidation, additive depletion, and contamination.
4. **Wear and extreme pressure (EP) properties**
   - **Four-ball wear test (ASTM D4172)**: Measures antiwear performance in lubricants.
   - **Timken OK load test (ASTM D2509)**: Evaluates extreme pressure (EP) lubricant capabilities.
   - **FZG gear wear test (ISO 14635)**: Determines lubricant performance under high-load conditions.
5. **Lubricant compatibility and chemical composition**
   - **ASTM D1401 (water separation test)**: Measures oil's ability to separate from water.
   - **ASTM D892 (foaming tendency test)**: Evaluates foam formation in hydraulic and gear oils.
   - **ASTM D471 (seal compatibility test)**: Determines lubricant compatibility with rubber seals and elastomers.
6. **Industry-specific lubrication standards**
   - **ISO 6743/ISO 12924**: International classification of lubricants based on application and performance.

- **SAE J300 (engine oil viscosity classification)**: Defines automotive engine oil viscosity grades.
- **DIN 51517 (industrial gear oil standard)**: Regulates gear oil properties and performance.
- **AGMA 9005 (gear lubricant standard)**: Used for industrial gear lubrication requirements.
- **FDA and NSF H1/H2**: Defines food-grade lubricants for food processing equipment.

514. Mechanical Stability (Grease)

Mechanical stability refers to a grease's ability to maintain consistent consistency and structure under mechanical stress, such as shearing forces encountered in bearings and gears. Greases with high mechanical stability resist breakdown and maintain their properties during operation, ensuring continuous protection and performance.

Key Aspects of Mechanical Stability:

- **Consistency retention**: A grease's ability to retain its consistency when worked is its shear stability or mechanical stability.
- **Resistance to deformation**: Greases with high mechanical stability resist breakdown and maintain their properties during operation, ensuring continuous protection and performance.
- **Impact on performance**: Greases that soften as they are worked are called thixotropic, while those that harden when worked are called rheopectic.

Understanding and selecting greases with appropriate mechanical stability is crucial for ensuring optimal performance and longevity of machinery components.

515. Melting Point (Grease)

The **melting point** of grease, often referred to as its **dropping point**, is simply the temperature at which grease transitions from solid to liquid state. This property is crucial for determining the upper-temperature limit at which grease can effectively lubricate machinery components.

Understanding Dropping Point:

- **Definition**: The dropping point is the temperature at which a grease sample, when heated, will begin to flow through an opening according to standardized testing methods.

- **Significance**: While the dropping point indicates the temperature at which grease retains its structure, it is not the maximum temperature at which grease may be used. The maximum operating temperature is typically lower than the dropping point to ensure optimal performance and prevent degradation.

Typical Dropping Points for Common Greases:

- **Lithium-based greases**: These greases generally have a dropping point between 190 and 220 °C (374 to 428 °F), making them suitable for a wide range of applications.
- **Calcium-based greases**: Calcium-based greases typically have a lower dropping point, around 120 °C (248 °F), limiting their use in high-temperature environments.
- **Clay-thickened greases**: Some clay-thickened (bentonite) greases may have very high dropping points, with dropping points noted on the product data sheets as 500 °C or greater. For these nonmelting products, the lubricating oil burns off at high temperatures, leaving behind hydrocarbon and thickener residues.

516. Metal Deactivator

Metal deactivators are specialized additives incorporated into lubricants to mitigate the catalytic effects of metal ions, particularly from nonferrous metals like copper and its alloys. These metal ions can accelerate oxidation reactions within the lubricant, leading to premature degradation and potential equipment damage.

Functions of Metal Deactivators:

- **Formation of protective films**: Metal deactivators adsorb onto metal surfaces, creating a protective barrier that prevents metal ions from interacting with the lubricant. This adsorption involves the polar part of the metal deactivator molecule interacting with the metal surface, while the lipophilic part remains in contact with the lubricant.
- **Sequestration of metal ions**: They bind to dissolved metal ions, such as copper, effectively neutralizing their catalytic activity and reducing the rate of oxidation reactions.

Common Metal Deactivators:

- **Benzotriazole and derivatives**: These compounds are widely used to protect against copper-induced oxidation in lubricants.

- **Tolyltriazole**: Effective in preventing corrosion and oxidation in systems containing copper and its alloys.

Applications in Lubricants:

- **Engine oils**: Metal deactivators are added to engine oils to prevent metal-induced oxidation and corrosion, extending the life of the lubricant and equipment.
- **Industrial lubricants**: In hydraulic fluids, gear oils, and other industrial lubricants, these additives protect metal surfaces from corrosion and degradation

517. Metal Wetting

Metal wetting refers to the ability of a liquid to spread over and adhere to a metal surface, forming a uniform film. This phenomenon is crucial in various industrial processes, including lubrication, soldering, and metal forming.

Key Factors Influencing Metal Wetting:

1. **Surface energy of the metal**: Metals with higher surface energy tend to promote better wetting by liquids. Surface treatments or coatings can modify a metal's surface energy to enhance or reduce wetting properties.
2. **Liquid properties**: The viscosity, surface tension, and chemical composition of the liquid significantly affect its wetting behavior on metal surfaces.
3. **Surface contaminants**: Oxide layers, oils, or other contaminants on metal surfaces can hinder wetting by creating barriers between the liquid and the metal. Proper cleaning and surface preparation are essential to achieve optimal wetting.

518. Metalworking Lubricant

Metalworking lubricants are essential fluids used in machining and metal-forming processes to reduce friction, dissipate heat, and prevent corrosion, thereby enhancing tool life and improving the quality of the finished product.

Types of Metalworking Lubricants:
There are four basic types of metalworking lubricants:

1. **Cutting fluids**: These are used during machining operations like turning, milling, and drilling to cool and lubricate the cutting tool and workpiece. They help in heat dissipation, chip removal, and reducing tool wear. See Figure M5.

Figure M5 Metalworking process using a metal working lubricant. Source: Unknown.

2. **Forming fluids**: Applied in processes such as stamping, drawing, and forging, forming fluids provide lubrication to reduce friction between the tool and workpiece, facilitating the shaping of metal without cracks or defects.

3. **Quenching fluids**: Used in heat treatment processes, quenching fluids cool heated metal parts rapidly to achieve desired hardness and mechanical properties. They also help in preventing oxidation and scaling during cooling.

4. **Rust preventatives**: These lubricants form protective films on metal surfaces to prevent corrosion during storage and transportation.

Functions of Metalworking Lubricants:

- **Cooling**: They absorb and dissipate heat generated during metalworking processes, preventing overheating of tools and workpieces.

- **Lubrication**: By reducing friction, lubricants minimize tool wear and improve surface finishes.
- **Corrosion protection**: They form protective barriers against moisture and chemicals, preventing rust and corrosion.
- **Chip removal**: Lubricants help in flushing away chips and debris from the cutting area, maintaining clear visibility and preventing recutting.

Considerations for Selection:

- **Compatibility**: Choose lubricants compatible with the materials being processed and the equipment used.
- **Environmental impact**: Choose lubricants that are biodegradable and nontoxic to reduce environmental harm.
- **Health and safety**: Select lubricants that are safe for operators, minimizing exposure to harmful chemicals.
- **Performance requirements**: Ensure the lubricant meets the specific demands of the machining process, such as extreme pressure resistance or high-temperature stability.

519. Meter Unit

A **meter unit** is a cyclic **metering valve** utilized in a single-line resistance (SLR) centralized lubrication system. Designed by Joseph Bijur and pioneered by the original Bijur Lubricating Corporation in 1924, this control unit is responsible for **restricting and metering** the delivery of **lubricating oil** in a cyclic delivery lubrication system.

How an SLR Meter Unit Works

Meter units can be positioned at each individual bearing point or in an upstream manifold within the delivery system. The lubricant flow is controlled by a cyclic pump and once activated, the pump delivers a fixed amount of oil into the delivery system. As the oil is forced under pressure into the system, the fluid pressure opens the 5 lb spring check valve inside each meter unit simultaneously allowing lubricant to flow into the bearing point. Flow is apportioned by the diameter of the meter pin held in the center of the meter unit – see Figure M6.

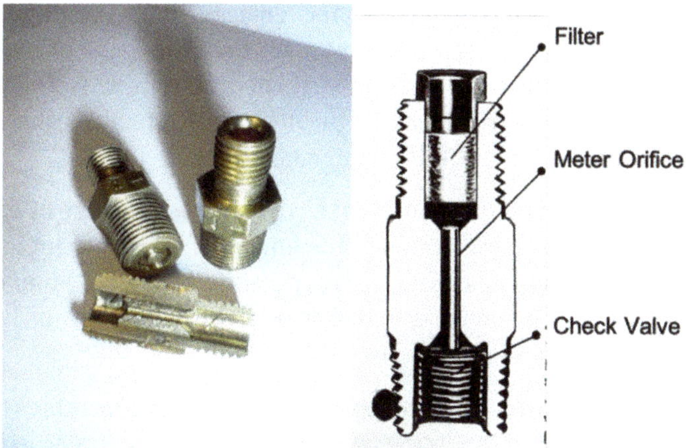

Figure M6 Single-line resistance meter unit. Images Courtesy: ENG**TECH** Industries Inc and BIjur–Delimon Corporation.

As an apportioning device, the meter unit is numbered based on the diameter of the meter pin. The higher the number, the less restriction and the higher the flow. For example, an FJB-2 meter unit will allow twice the flow of the number FJB-1 sized meter unit and visa-versa. The stamped arrow on the meter unit designates the flow direction as meter units are unidirectional.

As the lubricant charge flows into the bearings, the system pressure will eventually drop. Once the pressure drops below 5 psi, all of the meter unit check valves will commence closing and the system is then fully closed until the next lubrication cycle begins.

520. Metrics – Lubrication

Lubrication metrics are quantitative measures used to evaluate the effectiveness of lubrication practices within an organization. By monitoring these metrics, companies can identify areas for improvement, reduce maintenance costs, and enhance equipment reliability.

Key Lubrication Metrics:

1. **Mean time between lubrication failures (MTBLF)**: This metric calculates the average operational time between instances where lubrication issues lead to equipment failure. A decreasing MTBLF may indicate inadequate lubrication practices, such as incorrect lubricant selection, improper application, or failure to adhere to lubrication schedules.

2. **Lubrication-related maintenance cost as a percentage of total maintenance cost**: This metric assesses the proportion of maintenance expenses attributed to lubrication-related activities. A high percentage may suggest inefficiencies in lubrication practices that could be addressed to reduce overall maintenance costs.
3. **Lubrication-related maintenance cost as a percentage of total operating cost**: This metric evaluates the impact of lubrication-related maintenance on the organization's total operating expenses. Monitoring this metric helps in understanding the financial implications of lubrication practices on overall operations.
4. **Lubrication work-order compliance**: This metric measures the adherence to scheduled lubrication tasks. High compliance rates indicate that lubrication schedules are being followed, contributing to equipment reliability and performance.
5. **Overall lubrication effectiveness (OLE)**: OLE evaluates the timeliness of lubrication preventive maintenance (PMs), contamination control, and lubricant health to ensure optimal machine reliability and performance. It drives proactive behaviors on the plant floor to ensure that the lubricant and lubrication process are effectively protecting machine components.
6. **Oil analysis metrics**: Regular oil analysis provides insights into lubricant condition, contamination levels, and wear particles. Key metrics include viscosity, acid number, and particle count, which help in assessing lubricant health and determining optimal change intervals

521. MF Lubrication

See "Mixed Film Lubrication."

522. Micro-pitting

Micro-pitting is a type of surface fatigue wear that occurs in gears, bearings, and rolling contact surfaces due to high loads, inadequate lubrication, and cyclic stress. It manifests as tiny surface cracks and depressions, which can progress into severe material degradation, reducing component lifespan, efficiency, and reliability.

Key Causes of Micro-pitting

- **Thin or insufficient lubricant film**: Inadequate elastohydrodynamic lubrication (EHL) results in metal-to-metal contact, leading to micro-cracks.
- **High contact stress and load cycling**: Repeated loading and unloading create localized fatigue stress, weakening the surface layer.
- **Surface roughness and asperity contact**: Poor surface finish increases stress concentration, accelerating crack formation.
- **Contaminants and debris in lubricant**: Particles can act as abrasives, damaging the oil film and initiating micro-cracks.
- **Oil viscosity and additive insufficiency**: Low-viscosity lubricants or degraded additives reduce load-carrying capacity, increasing fatigue wear.

523. Micron

A **micron**, also known as a **micrometer**, is a unit of length equal to one-millionth of a meter (10^{-6} meters). It is commonly used to measure extremely small objects or distances. See Figure M7. In the field of lubrication, this includes lubricant film thickness, contaminants, and filter ratings.

Conversion:

- 1 micron = 0.001 millimeters (mm)
- 1 micron = 0.00003937 inches

Examples of Objects Measured in Microns:

- **Hydrodynamic lubricant film thickness**: 4 microns
- **Contaminants**: Approximately 1–10 microns in length
- **Average human hair**: 80 microns in diameter
- **Red blood cells**: About 10 microns in diameter
- **Dust particles**: Typically 0.5–5 microns in diameter

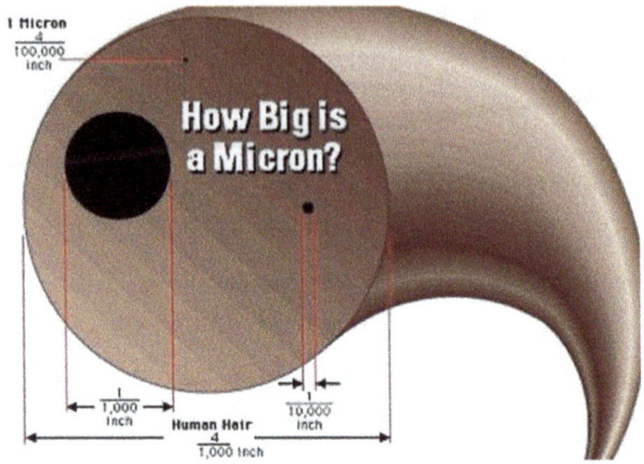

Figure M7 How big is a micron?

524. Microscopy

In lubrication, **microscopy** involves using advanced microscopic analysis techniques to examine the microstructure and morphology of lubricants, such as greases and oils, as well as wear particles and surfaces. This analysis provides valuable insights into lubricant performance, wear mechanisms, and contamination levels, facilitating the development of more effective lubrication strategies.

Key Microscopy Techniques in Lubrication:

1. **Scanning electron microscopy (SEM)**: SEM offers high-resolution imaging of lubricant microstructures and wear particles, enabling the identification of particle size, shape, and distribution. This information is crucial for understanding wear mechanisms and optimizing lubricant formulations.

2. **Atomic force microscopy (AFM)**: AFM provides detailed topographical images of lubricant surfaces at the nanoscale, allowing for the assessment of surface roughness and the study of lubricant film thickness. This technique is essential for evaluating the effectiveness of lubricants in reducing friction and wear.

3. **Transmission electron microscopy (TEM)**: TEM enables the examination of lubricant microstructures at atomic resolution, offering insights into the arrangement of molecules and particles within the lubricant matrix. This information is vital for understanding

the fundamental properties of lubricants and their interactions with surfaces.

4. **Optical microscopy**: Optical microscopy is used to observe the dispersion and stability of nanoparticles in lubricants, providing a cost-effective method for evaluating the uniformity of nanoparticle distribution. This technique is particularly useful in the development of nano-lubricants.

Applications of Microscopic Analysis in Lubrication:

- **Wear particle analysis**: Microscopic examination of wear particles extracted from lubricants helps in identifying the type and severity of wear occurring in machinery components. This analysis aids in diagnosing issues such as abrasive wear, adhesive wear, and fatigue.
- **Lubricant development**: Understanding the microstructure of lubricants through microscopy allows for the design of formulations with enhanced properties, such as improved load-carrying capacity, better thermal stability, and reduced friction.
- **Contamination assessment**: Microscopic techniques can detect and characterize contaminants in lubricants, such as particles, water droplets, and degradation products. Identifying these contaminants is crucial for maintaining lubricant integrity and preventing equipment damage.

In summary, microscopic analysis is a powerful tool in lubrication science, providing detailed insights into lubricant behavior, wear mechanisms, and contamination levels.

525. Mineral Base Oil

Mineral base oil is a fundamental component in the formulation of lubricants, serving as the primary fluid that facilitates the reduction of friction and wear in machinery. Derived from crude oil through refining processes, mineral base oils are categorized into different groups based on their refining methods and properties.

Refining Process:

Mineral base oils are petroleum-based oils produced by refining crude oil to remove impurities and unwanted components. The refining process involves various steps, including distillation, solvent extraction, and purification, resulting in a clear, stable, and low-viscosity oil with lubricating properties.

Categories of Mineral Base Oils:
The American Petroleum Institute (API) classifies mineral base oils into three main groups:

- **Group I**: These oils are solvent-refined and typically have a viscosity index (VI) between 80 and 120. They contain less than 90% saturates and more than 0.03% sulfur.
- **Group II**: Produced by hydrocracking, these oils have a higher VI, ranging from 80 to 120, and contain greater than 90% saturates with less than 0.03% sulfur.
- **Group III**: These highly refined oils have a VI over 120 and contain greater than 90% saturates with less than 0.03% sulfur. They are often considered synthetic due to their superior performance characteristics.

Properties and Applications:
Mineral base oils are characterized by their viscosity, pour point, and oxidative stability, which influence the performance of the finished lubricant. They are commonly used in automotive and industrial lubricants, such as engine oils, gear oils, and hydraulic fluids. The specific properties of each type vary depending on its source and refining process, impacting their suitability for different applications.

526. Miscibility

In industrial lubrication, **miscibility** refers to the ability of different lubricants or additives to mix uniformly without phase separation. Understanding miscibility is crucial for ensuring optimal performance and longevity of machinery.

Key Considerations:

- **Compatibility assessment**: Before mixing lubricants or adding additives, it is essential to evaluate their compatibility to prevent issues like phase separation, which can lead to equipment malfunction. This assessment often involves consulting compatibility charts or conducting compatibility tests.
- **Additive interactions**: Different additives can interact in ways that affect the lubricant's performance. For instance, combining certain additives may result in precipitation or degradation, compromising the lubricant's effectiveness. Therefore, understanding the chemical interactions between additives is vital.

- **Lubricant changeovers**: When transitioning between different lubricants, especially those with varying base oils or additive packages, it is important to consider their miscibility. Incompatibility can lead to issues like sludge formation or filter plugging. Proper flushing and cleaning procedures are recommended to ensure system compatibility.

527. Mist Lubrication System

An oil **mist lubrication system** is a specialized method used to lubricate machinery components, particularly bearings, by delivering a fine mist of oil particles suspended in the air directly to the lubrication points. When set up correctly, this approach can ensure consistent and efficient lubrication, enhancing equipment reliability and extending component lifespan.

Key Components (see Figure M8):

- **Oil mist generator**: The core of the system, where oil is atomized into a fine mist using compressed air. This process typically involves a venturi or vortex chamber to produce the oil–air mixture.

- **Distribution network**: A network of piping that transports the oil mist from the generator to various lubrication points throughout the machinery.

- **Metering units**: Reclassifiers that control the mist flow to applicable bearing points, ensuring the correct particle size and distribution for effective lubrication.

- **Air supply system**: Provides clean, dry air free from contaminants to maintain the quality of the oil mist.

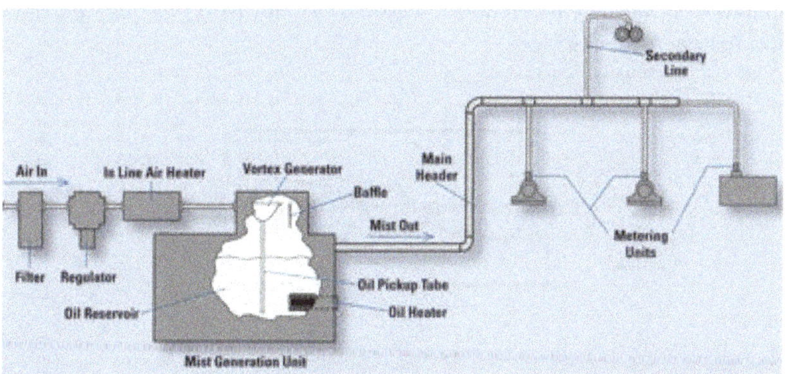

Figure M8 Typical mist lubrication system. Source: Anatomy of a Centralized Lubrication System – Oil Mist Systems. Lubrication Management and Technology Magazine – Ken Bannister.

Advantages:

- **Enhanced bearing life**: By providing a continuous supply of fresh oil, oil mist lubrication reduces friction and wear, leading to longer bearing life.
- **Reduced maintenance**: The system's design minimizes the need for manual lubrication, reducing maintenance efforts and associated costs.
- **Environmental benefits**: Oil mist systems can be more environmentally friendly compared to traditional lubrication methods, as they reduce oil consumption and minimize spillage but proper design is crucial to ensure the correct oil mist concentration and distribution. Open vent designs can create oily, environmentally unfriendly workplaces if not carefully monitored and controlled.

Considerations:

- **Oil selection**: Choosing the appropriate oil with suitable viscosity and additive properties is essential for optimal performance.
- **Monitoring and maintenance**: Regular monitoring of the system's performance and periodic maintenance are necessary to ensure continued efficiency and to prevent potential issues.

528. Mixed Film Lubrication (MF)

Mixed film lubrication is an intermediate lubrication regime where the load is supported by both lubricant and partial contact between surface asperities. This occurs when the lubricant film is sufficiently thick to separate the surfaces but not thick enough to prevent all asperity contact as depicted in Figure M9. In this regime, the coefficient of friction decreases significantly compared to boundary lubrication, as the lubricant film provides partial separation between the surfaces.

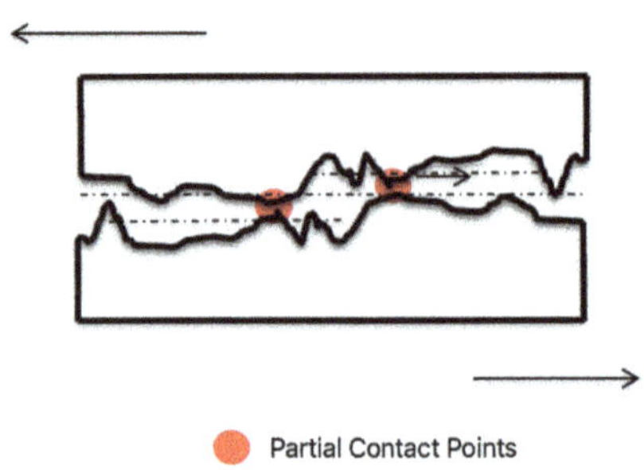

Figure M9 Mixed film lubrication. Courtesy: ENG**TECH** Industries Inc.

Key Characteristics:

- **Partial surface separation**: The lubricant film is thick enough to separate the surfaces to some extent, but asperity contact still occurs.
- **Reduced friction**: The presence of the lubricant film reduces friction compared to boundary lubrication, leading to improved performance.
- **Transition zone**: Mixed lubrication serves as a transitional phase between boundary lubrication (where surfaces are in direct contact) and hydrodynamic lubrication (where surfaces are fully separated by a lubricant film).

Mixed film is also thought of as an "unstable" regime, generally caused by insufficient lubricant, heavy loads at rest, or the use of a lubricant with too low of a viscosity.

529. Mixibility

See "Miscibility."

530. MLA I (ICML Machine Lubricant Analyst – Level I)

MLA I Machine Lubricant Analyst (MLA) – Level 1 is a certification designation awarded through the International Council for Machine Lubrication (see ICML). The purpose of the Level I Machine Lubricant Analyst certification is to verify that technicians practicing in the field of lube oil analysis, as it is applied to machinery condition monitoring and maintenance, are qualified to perform the following tasks:

- Manage lubricant delivery, storage, and dispensation.
- Manage a route for machinery relubrication and/or inspection.
- Properly change and/or top the oil in mechanical equipment found in common industrial sites.
- Use simple techniques to select lubricants with the proper base oil and additive system for machinery commonly found in industrial settings.
- Use simple techniques to select grease lubricants appropriate for machines commonly found in industrial settings.
- Use simple techniques to select grease application methods (including automated delivery) that are least intrusive and most effective for machines commonly found in industrial settings.

- Use simple techniques to estimate regrease volume and interval for machines commonly operated in industrial settings.
- Properly maintain automatic lubrication systems (auto-grease, mist systems, etc.).
- Employ basic oil analysis techniques to identify and troubleshoot abnormal lubricant degradation conditions, and use simple techniques to adjust the lubricant specification accordingly.

Common job titles for the individual who would become Level I MLA certified include Lubrication Technician, PM Technician, Millwright, Mechanic, etc. Generally, this individual has regular contact with the machine and has routine influence over the condition of lubricants and hydraulic fluids in use. The individual is likely to be directly involved in the machine lubrication process.

531. MLA II (ICML – Machine Lubricant Analyst – Level II)

MLA II Machine Lubricant Analyst (MLA) – Level II is a certification designation awarded through the International Council for Machine Lubrication (see ICML). This certification aligns with ISO 18436-4 standards and focuses on enhancing skills in oil sampling, lubricant health monitoring, and contamination control. The purpose of the Level II Machine Lubricant Analyst certification is to verify that technicians practicing in the field of lube oil analysis, as it is applied to machinery condition monitoring and maintenance, are qualified to perform the following tasks:

- Install sampling hardware such that a representative sample may be efficiently and repeatedly extracted from mechanical systems commonly found in industrial and fleet applications.
- Effectively extract representative samples from typical industrial machinery, and prepare them for analysis.
- Employ basic oil analysis techniques to identify and troubleshoot conditions of wrong or degraded lubricant, and devise and implement corrective actions.
- Employ basic oil analysis to identify and troubleshoot contaminated lubricant conditions, and devise and implement corrective actions.
- Employ basic oil analysis techniques to identify and troubleshoot abnormal machine wear conditions, and devise and implement corrective actions.

- Inspect and receive new oil deliveries.

Common job titles for the individual who would become Level II MLA certified include Oiler, Lube Technician, Lubricator, PM or PDM Technician, Millwright, Mechanic, Reliability Technician, etc. Generally, this individual has regular contact with the machine and has routine influence over the condition of lubricants and hydraulic fluids in use. The individual is likely to be directly involved in oil sampling and troubleshooting routine problems identified by lab test results and field lubricant inspections. As such a basic understanding of lubrication, oil analysis tests, oil analysis terminology, and data interpretation are required.

532. MLA III (ICML – Machine Lubricant Analyst – Level III)

MLA III Machine Lubricant Analyst (MLA) – Level III certification designation is awarded through the International Council for Machine Lubrication (see ICML). The purpose of the Level III Machine Lubricant Analyst (MLA III) certification is to verify that technicians practicing in the field of lube oil analysis, as it is applied to machinery condition monitoring and maintenance, are qualified to perform the following tasks:

- Design an oil analysis program, including selecting tests, setting targets, alarms, and limits, determining test frequency, and assigning machine-specific sampling methods.
- Diagnose abnormal machine and lubricant conditions to make or advise appropriate decisions and oversee implementation.
- Understand and communicate to managers, supervisors, and lube technicians, the role of lubricant analysis in a reliability-centered maintenance (RCM) program.
- Deploy proactive lubrication monitoring to reduce component failure rate and extend machine life.
- Manage lubricant analysis information and associated software programs.
- Evaluate the performance of outside oil analysis services and make corrective changes where required.
- Use oil analysis to assure lubricant supplier quality conformance.
- Develop, deploy, and manage procedures for lubricant sampling and analysis.

- Select on-site oil analysis instruments as required.
- Work within a multitechnology machine condition-monitoring program and facilitate synergistic integration.
- Financially rationalize the lubricant analysis program.
- Provide training, guidance, and support to the organization and to those individuals who are seeking MLA certification.
- Document the contributions oil analysis makes to the organization.

Common job titles for the individual who would become Level III MLA certified include Lubrication Technician or Engineer, PM or PDM Technician, Engineer or Manager, Reliability Technician Engineer or Manager, Maintenance Technician or Engineer, etc. Generally, this individual manages the lubricant analysis program and/or serves as a senior diagnostician. The individual is likely to be involved with troubleshooting tough problems identified by routine analysis and inspections. The Level III MLA-certified individual is likely to supervise and/or support Level I and II MLA-certified individuals.

533. MLE (ICML – Machine Lubrication Engineer)

MLE Machinery Lubrication Engineer (MLE) – is the highest certification level awarded by the International Council for Machine Lubrication (see ICML). The purpose of the Machinery Lubrication Engineer (MLE), an engineering-grade certification, is to evaluate an individual's knowledge and understanding of and ability to apply the lubrication engineering skills required to effectively provide engineering support to a typical industrial plant, mill, or site as a staff, contract or consulting lubrication engineer. **Note: The certification does not address all aspects of lubrication and/or the tribological sciences. The MLE certification is a peer-reviewed certification of the candidate's lubrication engineering skills. It does not represent licensure or registration to practice engineering as a professional or chartered engineer, which is governed by applicable laws and regulations.**

The certification is designed to verify that an MLE-certified professional is qualified to:

- Employ analytical methods applicable to the science and application of machinery lubrication so as to optimize lubricant selection, perform lubrication practices, and manage and monitor lubrication programs to a high degree of effectiveness

- Systematically analyze abnormal events where lubrication is the cause or a contributing cause
- Reach diagnostic and prognostic conclusions and implement any associated action plans, including but not limited to changes in machinery design, lubricant selection, lubricant application method, lubrication maintenance plans, and analytical methods prescribed.
- Assess potential machine/system failure due to ineffective lubrication practices and develop a proactive reliability-centered approach based on failure consequences for new and existing lubrication requirements.
- Manage a lubrication team, including lubrication technicians and lubricant analysts.
- Manage key relations with lubricant suppliers, lubrication accessory suppliers, lubrication service providers, and oil analysis laboratories.

Common job titles for the individual who would become MLE certified include Lubrication Engineer, Plant Engineer, PdM Engineer, Manufacturing Engineer, Process Engineer, Reliability Engineer or Manager, Maintenance Engineer, etc. Generally, this individual oversees the development, implementation, and management of the lubrication program. The individual is likely to be involved with troubleshooting tough problems identified by routine analysis and inspections. The MLE-certified individual is likely to supervise and/or support Level I MLT, Level II MLT, Level I MLA, Level II MLA, and Level III MLA-certified individuals.

534. MLT I (ICML – Machine Lubrication Technician Level I)

MLT I Machine Lubrication Technician (MLT) – Level I certification designation is awarded through the International Council for Machine Lubrication (see ICML). The purpose of the Level I Machine Lubrication Technician certification is to verify that technicians practicing in the field of machinery lubrication, as it is applied to machinery condition monitoring and maintenance, are qualified to perform the following tasks:

- Manage lubricant delivery, storage, and dispensation.

- Manage a route for machinery relubrication and/or inspection.
- Properly change and/or top the oil in mechanical equipment found in common industrial sites.
- Use simple techniques to select lubricants with the proper base oil and additive system for machinery commonly found in industrial settings.
- Use simple techniques to select grease lubricants appropriate for machines commonly found in industrial settings.
- Use simple techniques to select grease application methods (including automated delivery) that are least intrusive and most effective for machines commonly found in industrial settings.
- Use simple techniques to estimate regrease volume and interval for machines commonly operated in industrial settings.
- Properly maintain automatic lubrication systems (auto-grease, mist systems, etc.).
- Employ basic oil analysis techniques to identify and troubleshoot abnormal lubricant degradation conditions, and use simple techniques to adjust the lubricant specification accordingly.

Common job titles for the individual who would become Level I MLT certified include Lubrication Technician, PM Technician, Millwright, Mechanic, etc. Generally, this individual has regular contact with the machine and has routine influence over the condition of lubricants and hydraulic fluids in use. The individual is likely to be directly involved in the machine lubrication process.

535. MLT II (ICML – Machine Lubrication Technician Level II)

MLT II Machine Lubrication Technician (MLT) – Level II certification designation is awarded through the International Council for Machine Lubrication (see ICML). The purpose of the Level II Machine Lubrication Technician (MLT) certification is to verify that technicians practicing in the field of machinery lubrication, as it is applied to machinery condition monitoring and maintenance, are qualified to perform the following tasks:

- Manage selection, application, and periodic consolidation of lubrication oils and grease lubricants, appropriate for machines commonly found in industrial settings. This should include regreasing intervals and volume, taking into consideration environmental, health, and safety issues.

- Design and manage a route for machinery relubrication and/or inspection.
- Create and manage lubrication PMs.
- Manage change and/or top of oil in mechanical equipment found in common industrial sites.
- Manage lubricant delivery, storage and dispensation, storage room design, and storage time calculation, taking into consideration environmental, health, and safety issues.
- Manage lubricant testing, and quality assurance, employing proper sampling procedures to ensure representative oil samples are taken consistently.
- Employ and interpret advanced oil analysis techniques and results to identify and troubleshoot lubricant degradation and/or abnormal conditions, and recommend cost-effective corrective action(s).
- Manage proper maintenance of lubrication equipment.

Common job titles for the individual who would become Level II MLT certified include Lubrication Technician or Specialist, PM Technician or Specialist, Machinery Reliability Technician or Specialist, etc. Generally, this individual has regular contact with plant machinery operators, mechanics, and engineering and has significant influence over the selection, troubleshooting, and monitoring of the lubricants and hydraulic fluids in use. The individual is likely to be directly involved in all aspects of the machinery lubrication process.

536. Moh Hardness Scale

The Mohs Hardness Scale is a classification scale developed in 1812 by German geologist Friedrich Mohs to measure the scratch resistance of various minerals through the ability of harder material to scratch softer ones., and compare against the relative hardness of ten minerals using a scale range from 1 to 10.

In order from softest to hardest, the scale includes the following material hardness ratings: (1) talc; (2) gypsum; (3) calcite; (4) fluorite; (5) apatite; (5) orthoclase; (7) quartz; (8) topaz; (9) corundum; and (10) diamond.

In the context of lubrication, understanding the hardness of materials is crucial for assessing the potential damage caused by particle contamination. Particles with higher hardness and angularity can cause significant damage to machinery components. For instance, particles with a Mohs hardness of 7 or higher, such as aluminum oxide, can be particularly damaging due to their sharp edges and high compressive strength. To mitigate such damage, it is essential to monitor and control particle contamination in lubricants through lubricant cleanliness and good filtration practices.

537. Moisture Analysis

Moisture analysis in lubricants is essential for maintaining equipment performance and longevity. Excessive water content can lead to corrosion, reduced lubrication efficiency, and accelerated wear.

Methods for Measuring Moisture in Lubricants:

1. **Karl Fischer titration**: This method quantifies water content by chemically reacting water with iodine in the presence of sulfur dioxide and a base. It is highly accurate and capable of detecting low levels of moisture. Certain additives in lubricants, such as detergents and extreme-pressure agents, can interfere with the accuracy of the test.

2. **Fourier transform infrared (FTIR) spectroscopy**: FTIR measures the absorption of infrared light by the lubricant, identifying water content based on specific absorption bands. Although a fast nondestructive analysis, it is less sensitive than Karl Fischer titration and may require baseline measurements for accurate results.

3. **Relative humidity sensor technology**: Utilizes sensors to detect moisture levels by measuring changes in relative humidity within the lubricant that provide quick, real-time measurements without the need for chemical reagents. This may be less precise than other methods and can be influenced by the presence of other volatile compounds.

By understanding and managing moisture levels in lubricants, industries can enhance equipment reliability,

538. Moisture Contamination

Moisture contamination in lubricants is a critical concern that can significantly impact equipment performance and longevity. Water can enter lubricants through various sources, including environmental humidity, condensation, leaks, and improper storage. Once present, moisture can exist in lubricants in three forms:

1. **Dissolved water**: Water molecules are uniformly dispersed throughout the lubricant. While this form is less harmful, it can still lead to chemical reactions that degrade the lubricant and promote corrosion.
2. **Emulsified water**: Water is suspended in the lubricant in tiny droplets, forming an emulsion. This state can disrupt the lubricant's protective film, leading to increased wear and potential bearing damage.
3. **Free water**: Water accumulates as a separate phase, often settling at the bottom of reservoirs. This form is particularly detrimental, as it can cause rusting, promote microbial growth, and lead to severe corrosion.

Effects of Moisture Contamination:

- **Corrosion**: Water can cause rusting and corrosion of metal components, leading to equipment failure.
- **Reduced lubrication efficiency**: Moisture can disrupt the lubricating film between moving parts, causing increased friction and wear.
- **Oil degradation**: Water contamination can accelerate the degradation of lubricating oils, reducing their effectiveness and leading to the formation of harmful byproducts.
- **Foaming**: Water can cause lubricants to foam, leading to air entrainment and reduced lubrication efficiency.
- **Microbial growth**: Moist environments can promote the growth of microbes, which can lead to the formation of sludge and deposits that impair equipment performance.

539. Moisture Target Limits

Moisture contamination in lubricants can lead to significant equipment issues, including corrosion, reduced lubrication efficiency, and accelerated wear. Establishing target moisture limits is essential for maintaining optimal lubricant performance and extending equipment lifespan.

General Guidelines for Acceptable Moisture Levels:

- **Engine oils**: Water content exceeding 300 parts per million (ppm) (0.03%) should be closely monitored. Levels reaching 1500–2000 ppm (0.15%–0.2%) are cause for concern.

- **Hydraulic fluids**: Water content above 200 ppm can lead to operational issues.
- **Turbine oils**: Moisture levels exceeding 100 ppm may affect performance.

540. Mold-release Lubricant

Mold-release lubricants are used in various manufacturing processes to facilitate the easy removal of molded items from molds without damaging the material or the mold itself. These lubricants create a barrier between the mold and the molded material to prevent sticking and ensure smooth demolding.

Three Types of Mold Release Lubricants:

1. **Silicone-based lubricants**: Silicon lubricants provide excellent release properties and are widely used due to their effectiveness and versatility.
2. **Water-based lubricants**: Water-based lubricants are environment-friendly alternatives that are easy to apply and provide effective release properties.
3. **Dry film lubricants**: These lubricants leave a dry, nonstaining film on the mold surface, reducing friction and preventing sticking.

Considerations When Choosing a Mold Release Lubricant:

- **Material compatibility**: Ensure the lubricant is compatible with the material being molded to prevent adverse reactions.
- **Environmental impact**: Opt for environmentally friendly options, such as water-based lubricants, to reduce ecological footprint.
- **Regulatory compliance**: For applications in food processing or medical devices, select lubricants that meet relevant regulatory standards.
- **Application method**: Choose a lubricant that aligns with your application process, whether it is spray, brush, or dip.

541. Molecular Film Lubrication

Molecular film lubrication refers to a lubrication regime where a **single molecular layer** of lubricant separates two surfaces, reducing friction and wear. This occurs in extreme conditions where a **continuous fluid film cannot be maintained**, and lubrication relies on **adsorbed or chemically bonded lubricant molecules**.

Key Characteristics of Molecular Film Lubrication

- **Occurs in extreme boundary conditions**: Found in boundary lubrication regimes where fluid film thickness is minimal.
- **Involves chemisorption and physisorption**: Lubricant molecules either chemically react with the surface (chemisorption) or adhere via physical forces (physisorption).
- **Utilizes polar additives and surface-active agents**: Special lubricant additives form protective molecular layers on metal surfaces.
- **Maintains lubrication under high load and low speed**: Provides a thin protective film where conventional hydrodynamic lubrication is ineffective.

How Molecular Film Lubrication Works

1. **Lubricant molecules attach to surface**: Polar lubricant molecules or solid lubricants adsorb onto the metal surface, forming a thin protective layer.
2. **Film prevents direct contact**: The molecular layer acts as a boundary film, reducing adhesion and surface wear.
3. **Self-healing mechanism**: As surfaces move, new molecules replace worn-out layers, maintaining lubrication.

542. Molybdenum Disulfide

Molybdenum disulfide (MoS_2) is an inorganic compound renowned for its exceptional lubricating properties. Its layered, hexagonal crystal structure, similar to graphite, allows for easy sliding between layers, resulting in a low coefficient of friction. See Figure M10.

Key Characteristics of MoS2 as a Lubricant:

- **Low coefficient of friction**: MoS_2 exhibits a coefficient of friction between 0.03 and 0.06, facilitating smooth movement between surfaces.
- **High load-bearing capacity**: It can withstand extreme pressures, making it suitable for high-load applications.
- **Thermal stability**: MoS_2 maintains its lubricating properties at temperatures up to approximately 350 °C in air and up to 1200 °C in inert or vacuum conditions.

- **Chemical inertness**: It is relatively unreactive, being unaffected by dilute acids and oxygen.

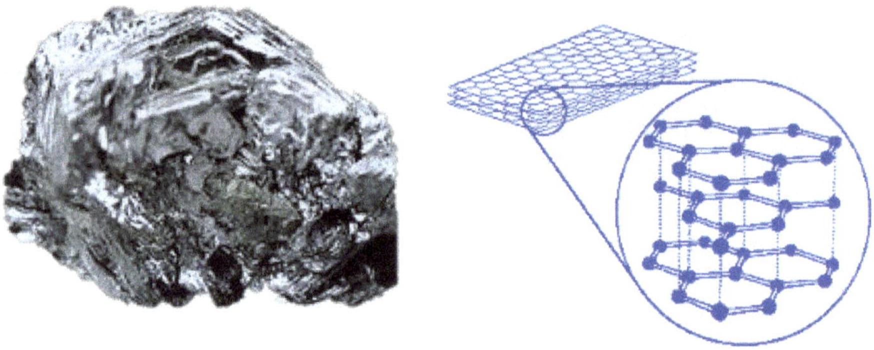

Figure M10 Crystaline structure of molybdenum disulfide.

Applications in Lubrication:

- **Solid lubricant**: MoS_2 is used as a dry lubricant in various applications, including aerospace, automotive, and industrial machinery, to reduce friction and wear.
- **Additive in lubricants**: It is incorporated into oils and greases to enhance their performance under extreme conditions.
- **Coatings**: MoS_2 coatings are applied to metal surfaces to provide lubrication in high-pressure environments.

In summary, molybdenum disulfide is a versatile and effective lubricant, particularly valuable in high-pressure, high-temperature, and chemically demanding environments.

543. MON

The **motor octane number (MON)** is a measure of a fuel's resistance to knocking or pinging during combustion in spark-ignition engines. Knocking occurs when the air-fuel mixture ignites prematurely, leading to inefficient engine performance and potential damage. A higher MON indicates greater resistance to knocking.

While MON primarily pertains to fuel characteristics, it has an effect on engine lubrication as engines designed for high-MON fuels are more likely to require lubricants designed to withstand the increased thermal and mechanical stresses associated with higher-performance MON fuels.

544. Morphological Analysis

In lubrication, **morphological analysis** involves examining the shape, size, and distribution of particles within lubricants to assess their condition and performance. This analysis is crucial for identifying wear patterns, contamination levels, and potential issues in machinery.

Key Aspects of Morphological Analysis in Lubrication:

- **Wear debris analysis**: By studying the morphology of particles in used lubricants, such as their shape, size, and color, analysts can identify various wear modes of rotating elements. This helps in determining the type and severity of wear occurring within the machinery.
- **Contamination detection**: Identifying foreign particles and assessing their morphology aids in detecting contamination sources, which can lead to equipment damage if not addressed.
- **Lubricant degradation assessment**: Changes in the morphology of particles can indicate chemical breakdown or degradation of the lubricant, signaling the need for maintenance or replacement.

Common Techniques for Morphological Analysis:

- **Scanning electron microscopy (SEM)**: Provides high-resolution images of particles, allowing for detailed morphological analysis.
- **Ferrography**: Involves creating a ferrogram slide from a used oil sample to examine particle morphology under magnification, aiding in identifying wear patterns.
- **X-ray fluorescence (XRF)**: Used to determine the elemental composition of particles, which, when combined with morphological analysis, helps in identifying wear modes and sources.

545. MoS_2

See "Molybdenum Disulfide."

546. Motor Oil

Motor oil is a specifically designed lubricant for internal combustion engines. Ensuring optimal engine performance and reliable operation requires the selection of the appropriate motor oil type.

Types of Motor Oil:

1. **Conventional oil**: Derived from crude oil, conventional oils are suitable for light-duty vehicles and are cost-effective.
2. **Synthetic oil**: Engineered through chemical processes, synthetic oils offer superior performance, especially in extreme temperatures, and are ideal for high-performance engines.
3. **Synthetic blend oil**: A mixture of synthetic and conventional oils, providing enhanced protection at a lower cost than full synthetics.
4. **High-mileage oil**: Formulated for vehicles with over 75,000 miles, these oils contain additives that help reduce oil consumption and leaks.

Selecting the Right Motor Oil:

- **Viscosity**: Indicated by numbers like 5W-30, where "W" stands for winter. The first number represents the oil's flow at 0 °F (−18 °C), and the second number indicates its flow at 100 °C. Choosing the correct viscosity ensures proper lubrication across various temperatures. The engine manufacturer will always advise the correct viscosity to use in its engines and failure to do so will often void the warranty.
- **API rating**: The American Petroleum Institute (API) rating, such as "API SP," signifies the oil's performance level. Always refer to your vehicle's manufacturer recommendations for the appropriate API rating. See API rating.

547. MRV Viscometer

The **Mini-rotary viscometer (MRV)** is a specialized instrument designed to assess the low-temperature pumpability of engine oils. It measures the viscosity and yield stress of lubricants after they have been cooled at controlled rates, typically over a period exceeding 45 hours, to final test temperatures between −10 and −40 °C.

Key Aspects of MRV Viscosity:

- **Yield stress measurement**: The MRV determines the yield stress of engine oils, which is the minimum stress required to initiate flow. This measurement is crucial for understanding how the oil will behave during cold starts and its ability to circulate effectively at low temperatures.
- **Viscosity profile**: By analyzing the viscosity at various temperatures, the MRV provides insights into the oil's performance under different operating conditions, particularly in cold environments.

548. MSDS

The **MSDS**, or Material Safety Data Sheet, is an old term that has been replaced with SDS, or Safety Data Sheet. See "SDS"

549. MTBLF

See "Mean Time Between Lubrication Failure."

550. Multigrade Oil

Multigrade oils are engine lubricants formulated to perform effectively across a wide range of temperatures, providing optimal engine protection and performance in both cold starts and high operating temperatures.

Understanding Multigrade Oils:

- **Viscosity ratings**: Multigrade oils are identified by a two-part viscosity rating, such as 5W-30.
  - The number before the "W" (e.g., 5) indicates the oil's viscosity at low temperatures, with lower numbers signifying better flow in cold conditions.
  - The number after the "W" (e.g., 30) represents the oil's viscosity at high operating temperatures, with higher numbers indicating thicker oil.
- **Additives**: These oils contain viscosity index improvers – polymer additives that expand at higher temperatures to thicken the oil, ensuring it remains effective across various temperature

Common Multigrade Oil Grades:

- **5W-30**: Offers excellent low-temperature performance and is suitable for many modern engines.
- **10W-40**: Provides a balance between cold start performance and high-temperature stability.
- **15W-40**: Ideal for warmer climates and older engines requiring thicker oil

551. Multipurpose Grease

Multipurpose grease is a versatile lubricant formulated to provide effective lubrication across a wide range of applications, including automotive, industrial, and household uses. It combines the properties of specialized greases, making it suitable for various components such as bearings, joints, and chassis.

Key Features of Multipurpose Grease:

- **Versatility**: Designed to perform well in diverse conditions, reducing the need for multiple specialized lubricants.
- **Protection**: Offers protection against wear, corrosion, and water wash-out, ensuring longevity and reliability of components.
- **Temperature range**: Typically effective within a broad temperature range, accommodating various operating environments.

Considerations When Choosing Multipurpose Grease:

- **Compatibility**: Ensure the grease is compatible with the materials and components in your application.
- **Operating conditions**: Select a grease that meets the specific temperature, pressure, and environmental conditions of your equipment.
- **Manufacturer recommendations**: Always refer to the equipment manufacturer's guidelines for lubricant specifications to ensure optimal performance and longevity.

N

552. Nano-lubricants

Nano-lubricants are lubricating fluids, greases, or additives that contain **nanometer-sized particles** (typically <100 nm in diameter) designed to improve lubrication performance.

These nanoparticles — which can be made of metals, metal oxides, carbon materials (like graphene or carbon nanotubes), or other compounds — are dispersed in a base lubricant (oil or grease).

The main purposes of nano-lubricants are to:

- **Reduce friction** by forming ultra-thin protective films between moving surfaces.
- **Decrease wear** by filling in microscopic surface irregularities.
- **Enhance thermal conductivity** to dissipate heat more effectively.
- **Increase load-bearing capacity** under high-pressure conditions.
- **Provide self-repairing effects**, as some nanoparticles can deposit into worn areas.

Typically, they are used in areas like automotive engines, industrial machinery, wind turbines, and even biomedical devices — anywhere extremely efficient, long-lasting lubrication is needed.

Nano-lubricants are designed work at the molecular and surface-interaction scale and offer five major advantages over regular lubricants. These include:

1. **Deliver a Size Advantage**
    - Nanoparticles are **1–100 nanometers** in size — small enough to enter the microscopic valleys (asperities) on metal surfaces.
    - This means they can **fill gaps and smooth out roughness**, creating a more continuous contact layer.
    - A smoother interface = less mechanical interlocking = **lower friction**.

DOI: 10.1201/9788743806608-565

2. **Form a Protective Tribofilm**
   - Many nanoparticles **react chemically or physically adhere** to the metal surface during sliding.
   - This reaction forms an ultra-thin **tribofilm** — a protective boundary layer a few nanometers thick.
   - This tribofilm:
     - Acts like a shield against direct metal-to-metal contact.
     - Reduces abrasive wear and scuffing.
     - Can **self-replenish** as more nanoparticles deposit.

3. **Introduce a Rolling Effect**
   - Certain nanoparticles (e.g., spherical tungsten disulfide, WS_2) act like **nano ball bearings**.
   - Instead of sliding, some of the motion is converted to a more efficient rolling, thereby reducing shear resistance and heat generation.

4. **Deliver a Thermal Conductivity Boost**
   - Many nanomaterials (like copper, graphene, or carbon nanotubes) conduct heat extremely well.
   - When dispersed in lubricants, they help **dissipate heat away from the contact zone**.
   - Lower operating temperature = less thermal degradation of the oil and surfaces.

5. **Enhance Extreme Pressure (EP) Performance**
   - Under high loads, nanoparticles can:
     - Deform plastically to absorb pressure.
     - Form high-strength protective layers that **prevent welding or pitting**.
   - This makes them excellent for heavy-duty gear systems and high-load applications.

Example in Action:
In an automotive engine, a nano-lubricant with **MoS_2 nanoparticles** can:

- Reduce boundary friction during cold starts.
- Lower wear over thousands of cycles.
- Extend oil change intervals.
- Improve fuel efficiency (less energy lost to friction).

553. Napthenic Crude Oil

Naphthenic crude oils are generally found in the coastal US regions, and in the South Americas, and are favored for the manufacture of compressor oils, refrigerant oils, and locomotive oils. This crude oil type is characterized by its higher content of naphthene, or cycloalkanes – hydrocarbons with a cyclic or ring-like structure that distinguishes it from paraffinic crude oils, which predominantly contain straight-chain alkanes.

Key Characteristics of Naphthenic Crude Oil:

- **Hydrocarbon composition**: Naphthenic crude oil contains a significant proportion of naphthenes, contributing to its unique chemical properties.
- **Physical properties**: It generally exhibits lower viscosity compared to paraffinic crude oil, allowing it to flow more easily at a given temperature. Additionally, naphthenic crude oil has a higher density, which affects its buoyancy and separation from other components during refining processes.
- **Refining characteristics**: The chemical composition and physical properties of naphthenic crude oil significantly impact the refining process. For instance, its lower viscosity and higher density influence the ease of separation and processing during refining.

Napthenic crude oils only show traces of wax giving it an improved (lower) pour point, but also a lower flash point than its paraffinic sibling, and a less desirable viscosity index in the range of 30 to 70.

554. National Lubricating Grease Institute

NLGI (National Lubricating Grease Institute) is an international organization that sets industry standards for lubricating greases, most notably through its NLGI consistency classification system.

Formed initially in 1933 as the "National Association of Lubricating Grease Manufacturers, Inc." to establish a code for the National Recovery Act demands it changed its name in 1937 to the now National Lubricating Grease Institute (NLGI).

The **NLGI** is a nonprofit organization established in 1933 to serve the lubricating grease industry. It plays a pivotal role in promoting research, development, and standardization within the grease sector.

Key Functions of NLGI:

- **Standardization**: NLGI develops and publishes industry standards for lubricating greases, ensuring consistency and quality across products.

- **Education**: The institute offers educational programs, publications, and events to disseminate knowledge about grease technology and applications.
- **Certification**: NLGI administers certification programs, such as the Certified Lubricating Grease Specialist (CLGS), to recognize expertise in grease technology.

555. Needle Bearing

A **needle roller bearing** is a type of roller bearing that utilizes long, slender cylindrical rollers resembling needles. These rollers are typically at least four times longer than their diameter, allowing the bearing to support high radial loads while maintaining a compact design (see Figure N1).

Key Features:

- **High load capacity**: The elongated rollers distribute the load over a larger contact area, enabling needle roller bearings to handle substantial radial loads effectively.
- **Compact design**: Their slim profile makes them ideal for applications with limited radial space.
- **Versatility**: Available in various designs, including drawn cup, machined ring, and caged configurations, needle roller bearings can be tailored to meet specific application requirements.

Figure N1 Cross-section of a typical needle bearing.

Common Applications:

Needle roller bearings are widely used in automotive, industrial, and aerospace applications, such as:

- **Automotive components**: They are employed in rocker arm pivots, compressors, and transmissions.
- **Industrial machinery**: Utilized in conveyors, pumps, and compressors to reduce friction and support heavy loads.
- **Aerospace systems**: Integrated into actuators, landing gear, and other critical components to ensure reliable performance under demanding conditions.

Advantages:

- **Space efficiency**: Their compact size allows for efficient use of space in design-constrained environments.
- **High stiffness**: The large number of small-diameter rollers contributes to high stiffness, enhancing the bearing's performance under load.

556. Neutralization Number

See "Total Acid Number (TAN)."

557. Newton, Isaac

Sir Isaac Newton (1643–1727) was one of the earliest lubrication pioneers. Recognized as a key figure in the development of classical mechanics, he contributed to the understanding of the forces that govern fluid dynamics that play a crucial role in lubrication theory.

Newton's Key Contributions to Lubrication Theory:

1. **Viscosity and fluid dynamics**:
   - In his groundbreaking work **Philosophiæ Naturalis Principia Mathematica** (1687), Newton formulates the laws of motion and universal gravitation. His work on the laws of motion provides a theoretical framework that would later be applied to the study of lubrication.

- Newton also contributed to the understanding of **viscosity**, a fundamental property of lubricants. Viscosity is the internal resistance of a fluid to flow, which influences the ability of lubricants to reduce friction between surfaces. Although Newton did not specifically study lubrication, his work on fluid behavior and viscosity is foundational.

2. **Newtonian fluids**:
   Newton's exploration of the behavior of fluids led to the distinction between **Newtonian** and **non-Newtonian fluids**. A **Newtonian fluid** has a constant viscosity regardless of the applied shear rate (the rate at which one layer of fluid moves over another), which is a critical characteristic for many lubricants used in engineering. Modern lubricants are often designed to behave as Newtonian fluids, making them predictable and effective in reducing friction.

3. **Laws of motion**:
   Newton's laws of motion, particularly the **second law of motion** (Force = Mass × Acceleration), apply directly to understanding how lubricants interact with moving surfaces. The ability of a lubricant to support loads, reduce friction, and maintain a fluid film between surfaces depends on understanding these basic principles of motion and force.

Newton's contributions to the understanding of fluid dynamics and viscosity paved the way for more advanced studies into lubrication and tribology (the study of friction, wear, and lubrication)at a much later date. His theories allowed engineers to develop more efficient lubrication systems by understanding how fluids behave under various conditions, especially in reducing friction in machines like bearings, gears, and engines. While Newton did not directly influence lubrication systems in the way modern engineers do, his work fundamentally shaped the principles that engineers use to design lubricants and lubrication systems today.

558. Newtonian Flow

Newtonian flow refers to the behavior of **Newtonian fluids**, which are fluids whose viscosity remains constant regardless of the applied shear rate. In other words, the relationship between shear stress and shear rate is linear, and the fluid's resistance to flow does not change with the rate at which it is deformed.

Key Characteristics of Newtonian Fluids:

- **Constant viscosity**: The viscosity of Newtonian fluids remains unchanged under varying shear rates.
- **Linear relationship**: There is a direct proportionality between shear stress and shear rate.
- **Predictable flow behavior**: Due to their constant viscosity, Newtonian fluids exhibit consistent and predictable flow characteristics.

Examples of Newtonian Fluids:

- **Water**: Exhibits constant viscosity under normal conditions.
- **Air**: Behaves as a Newtonian fluid in most practical applications.
- **Mineral oil**: Maintains constant viscosity under standard conditions.

Applications in Lubrication:

Understanding Newtonian flow is crucial in lubrication, as it allows for the prediction of fluid behavior between moving surfaces. The constant viscosity of Newtonian fluids ensures a stable lubricant film, reducing friction and wear in mechanical systems.

559. Nitration

In lubrication, **nitration** refers to the chemical degradation of engine oil caused by the formation of nitrogen-containing compounds during combustion. This process is particularly prevalent in natural gas-fueled engines, where high temperatures and pressures facilitate the reaction of atmospheric nitrogen with oxygen to form nitrogen oxides (NOx). These NOx compounds then interact with the lubricant, leading to the formation of organic nitrates and nitro compounds.

Key Aspects of Nitration in Lubrication:

- **Formation of nitrogen compounds**: During combustion, nitrogen reacts with oxygen to form NOx. These reactive compounds can then combine with the lubricant, resulting in the creation of organic nitrates and nitro compounds.
- **Impact on oil properties**: The accumulation of these nitrogen compounds can lead to several detrimental effects on the lubricant:

- **Increased viscosity**: Nitration can cause the oil to thicken, impairing its ability to flow and lubricate engine components effectively.
- **Formation of sludge and varnish**: The presence of nitration products can lead to the formation of sludge and varnish deposits, which can clog oil passages and reduce engine efficiency.
- **Corrosion**: Nitration products can increase the acidity of the oil, leading to corrosion of engine parts.
- **Contributing factors**: Several factors can accelerate nitration in engine oils:
- **Low oil sump temperatures**: Operating at lower oil temperatures can promote nitration.
- **Inefficient combustion**: Poor combustion efficiency, often due to improper air/fuel ratios or inadequate exhaust scavenging, can increase the formation of NOx, thereby enhancing nitration.
- **Engine load and speed**: Higher engine loads and lower operating speeds can contribute to increased nitration rates.

Monitoring and Mitigation:
Regular monitoring of oil chemistry is essential to detect and manage nitration. Techniques such as Fourier transform infrared (FTIR) spectroscopy can identify nitration products in the oil. To mitigate nitration, it is crucial to maintain optimal engine operating conditions, including proper air/fuel ratios, adequate exhaust scavenging, and appropriate oil temperatures. Additionally, selecting lubricants formulated with base oils and additives resistant to nitration can help reduce the risk of oil degradation

560. Nitrogen-purged Lubricant Storage

Nitrogen-purged lubricant storage is a contamination control method where **a nitrogen gas blanket is applied to lubricant storage tanks and containers** to prevent oxidation, moisture absorption, and degradation. This technique is particularly useful for high-performance lubricants, hydraulic fluids, and transformer oils, ensuring they maintain their chemical integrity and long-term performance.

Key Functions of Nitrogen-purged Lubricant Storage

- **Prevents oxidation and degradation**: Oxygen exposure accelerates oil oxidation, leading to sludge formation, viscosity changes, and additive

depletion. Nitrogen purging eliminates oxygen contact, extending lubricant life.

- **Reduces moisture contamination**: Many lubricants are hygroscopic (absorb moisture from the air), which can cause corrosion, emulsification, and reduced lubricant effectiveness. A nitrogen blanket creates a dry, moisture-free environment.
- **Enhances lubricant shelf life**: By maintaining a stable chemical environment, nitrogen purging preserves lubricant properties for long-term storage.
- **Minimizes volatility loss and evaporation**: Some lubricants, especially light synthetic oils, can lose critical components through evaporation. A nitrogen blanket reduces vaporization and maintains lubricant consistency.

How Nitrogen Purging Works in Lubricant Storage

1. **Sealed storage tank design**: Lubricants are stored in sealed tanks or containers equipped with nitrogen inlets.
2. **Continuous or on-demand nitrogen injection**: Nitrogen is introduced either continuously or when the pressure drops, maintaining a low-oxygen, moisture-free environment.
3. **Pressure regulation**: The nitrogen blanket maintains a slightly positive pressure to prevent air infiltration.
4. **Lubricant dispensing without air exposure**: As the lubricant is withdrawn, nitrogen automatically replaces the displaced volume, preventing outside air from entering.

561. NLGI

See "National Lubricating Grease Institute."

562. NLGI Standardized Grease Classification System

The **National Lubricating Grease Institute (NLGI)** has established a standardized classification system to define the consistency of lubricating greases. This system assigns a numerical grade to greases based on their hardness or softness, facilitating the selection of appropriate lubricants for various applications.

563. NLGI Grease Classification System

The **NLGI Grades** range from 000 to 6, (see Table N1) with each grade corresponding to a specific range of ASTM worked penetration values at 25 °C (77 °F). These values are measured in tenths of a millimeter, indicating the depth a standard cone penetrates into the grease under specified conditions (see Cone Test).

Applications of NLGI Grades:

- **Grades 000 to 1**: These lower consistency greases are suitable for applications requiring low viscosity, such as enclosed gear drives operating at low speeds and open gearing. Grades 0, 1, and 2 are used in highly loaded gearing, while grades 1 through 4 are often used in rolling contact bearings.
- **Grades 2 to 3**: These grades are commonly used in automotive, industrial, marine, and agricultural applications. They are excellent for antifriction bearings, wheel bearings, factory-sealed ball bearings, and some flexible couplings. Grade #3 greases have good sealing properties to protect the equipment and surfaces against pollution, corrosion, and environmental impact.
- **Grades 4 to 6**: Higher consistency greases are firmer and tend to stay in place, making them a good choice when leakage is a concern. They are suitable for applications requiring higher stiffness, such as water pumps and other high-speed, lightly loaded applications.

Table N1 NLGI appearance table.

NLGI Grade	ASTM Worked Penetration (25°C)	Appearance	Common Analogy
000	445-475	Fluid	Cooking oil
00	400-430	Semi-fluid	Apple sauce
0	355-385	Very soft	Brown mustard
1	310-340	Soft	Tomato paste
2	265-295	Normal	Peanut butter
3	220-250	Firm	Vegetable shortening
4	175-205	Very firm	Frozen yogurt
5	130-160	Hard	Smooth pâté
6	85-115	Very hard	Cheddar cheese

564. No-lead Grease

No-lead grease is a grease formulation free from lead-based additives, designed to provide extreme pressure (EP) protection, wear resistance, and lubrication performance while adhering to modern environmental and health regulations. Traditionally, lead was used in greases for its load-carrying capabilities and antiwear properties, but due to its toxicity and environmental hazards, it has been replaced with safer alternatives that offer comparable or superior performance.

Key Functions of No-lead Grease

- **Provides extreme pressure (EP) protection**: Uses alternative EP additives (such as molybdenum disulfide, sulfur–phosphorus compounds, or boron compounds) instead of lead to withstand high loads and shock conditions.
- **Reduces friction and wear**: Forms a durable lubricating film that prevents metal-to-metal contact in high-pressure applications.
- **Complies with environmental and health regulations**: Eliminates hazardous lead compounds, meeting EPA, OSHA, and REACH standards for safer industrial and commercial use.
- **Prevents corrosion and oxidation**: Contains corrosion inhibitors to protect surfaces from rust, oxidation, and degradation over time.
- **Enhances water resistance and contaminant protection**: Provides superior water washout resistance, making it ideal for marine, industrial, and outdoor applications.

Common Additives in No-lead Grease

- **Molybdenum disulfide (MoS_2)**: A solid lubricant that reduces friction, wear, and galling, often used in heavy-duty and high-load applications.
- **Sulfur–phosphorus compounds**: Serve as effective EP additives, replacing lead while ensuring strong film formation.
- **Boron compounds**: Improve load-carrying capacity and antiwear performance in extreme pressure environments.
- **Calcium sulfonate complex**: Provides excellent corrosion resistance, thermal stability, and water resistance.

565. Nondrip Lubricant

Nondrip lubricant refers to a specialized lubricant formulated to **adhere to surfaces without running, dripping, or flowing** away from the application point. These lubricants are designed for vertical surfaces, overhead components, chains, and open gears, where traditional lubricants may migrate or be displaced due to gravity, centrifugal force, or vibration.

Key Functions of Nondrip Lubricants

- **Prevents lubricant loss and waste**: Reduces the likelihood of lubricant runoff, evaporation, or excessive application, ensuring longer-lasting protection.
- **Enhances adhesion to surfaces**: Designed with tackifiers and viscosity modifiers to provide strong surface adhesion.
- **Reduces contamination and clean-up needs**: Keeps lubrication confined to critical points, preventing spills that could attract dirt and dust.
- **Improves lubrication retention in high-motion parts**: Ensures that lubricants stay in place on chains, gears, and vertical sliding components.

Common Additives and Formulation Properties

- **Tackifiers and adhesion agents**: Improve stickiness, preventing the lubricant from running off or being thrown off in motion.
- **High-viscosity base oils and thickeners**: Formulated with high-viscosity mineral or synthetic base oils, often combined with gel-like or solid components to resist flow.
- **Extreme pressure (EP) additives**: Provide additional wear protection under high loads in gearboxes, open gears, and chain systems.
- **Antisling and antisplash properties**: Reduces lubricant loss in high-speed rotating components.

566. Nonpressurized System

A **nonpressurized lubrication system** relies on gravity and splash mechanisms to distribute lubricant to gearing and engine components, eliminating the need for a pressurized pump. This approach is commonly found in industrial gearboxes and small engines.

Key Features:

- **Splash lubrication**: In this system, moving engine parts, like the crankshaft or a large gear, dip into the oil reservoir, picking up oil and splashing it onto other components, ensuring lubrication without pressurization.
- **Simplicity**: The absence of a pressurized pump reduces system complexity and maintenance requirements.
- **Cost-effectiveness**: Without the need for a pressurization mechanism, manufacturing costs are lower.

567. Non-Newtonian Fluid

Non-Newtonian lubricants are fluids whose viscosity changes under different shear rates or stresses, deviating from the constant viscosity behavior of Newtonian fluids. This variability allows for tailored lubrication properties in various applications.

Types of Non-Newtonian Lubricants:

1. **Shear-thinning (pseudoplastic) lubricants**:
   - **Behavior**: Viscosity decreases with increasing shear rate.
   - **Example**: Certain polymer-based lubricants that become less viscous under high shear conditions, improving flow and reducing friction.
2. **Shear-thickening (dilatant) lubricants**:
   - **Behavior**: Viscosity increases with increasing shear rate.
   - **Example**: Suspensions with high concentrations of solid particles that become more viscous under stress, providing enhanced load-bearing capacity.
3. **Bingham plastic lubricants**:
   - **Behavior**: Require a minimum yield stress to flow; once this threshold is surpassed, they flow like a fluid.
   - **Example**: Certain grease formulations that remain stable until a specific force is applied.

Applications in Lubrication:

- **Enhanced load-bearing**: Shear-thickening lubricants can increase viscosity under high-pressure conditions, improving load-carrying capacity.

- **Reduced friction**: Shear-thinning lubricants decrease viscosity under high shear rates, facilitating smoother movement and reducing friction.
- **Adaptive performance**: Bingham plastic lubricants provide stability under normal conditions and flow under specific stresses, offering versatility in various mechanical systems.

568. Nonpositive Seal

In lubrication systems, **nonpositive seals** are designed to allow a controlled, minimal amount of internal leakage. This slight leakage is intentional, as it facilitates the self-lubrication of moving parts, such as pistons and valves, by providing a continuous supply of lubricant to these components.

Key Characteristics of Nonpositive Seals:

- **Controlled leakage**: Unlike positive seals, which prevent any leakage, nonpositive seals permit a small, regulated amount of fluid to pass through.
- **Self-lubrication**: The intentional leakage ensures that moving parts receive adequate lubrication, reducing friction and wear.
- **Application specificity**: These seals are particularly useful in systems where internal lubrication is necessary for optimal performance and longevity of components.

Common Applications:

- **Hydraulic systems**: In hydraulic circuits, nonpositive seals are used to allow minute oil leakage, ensuring that spools and moving parts of valves are properly lubricated.
- **Pneumatic systems**: Similar to hydraulic systems, pneumatic systems may utilize nonpositive seals to maintain internal lubrication, enhancing the efficiency and lifespan of pneumatic actuators and valves.
- **Automotive transmissions**: In automatic transmissions, nonpositive seals allow a small amount of fluid to pass, aiding in the lubrication of moving parts within the transmission system.
- **Wiper seals and labyrinth seals**: Practical examples of nonpositive seals.

569. Nonsoap Thickener

See "Organic Thickener."

570. Nonwoven Media

Nonwoven media are integral to lubrication systems, serving primarily as filtration components to maintain oil cleanliness and enhance system performance. Utilizing fibers bonded together without weaving, the resulting fabric is built up from a random arrangement of filaments or fibers.

Applications in Lubrication:

- **Filtration**: Nonwoven filter media effectively capture contaminants such as particles, water, and sludge from lubricating oils, thereby preventing wear and corrosion in machinery. Their structure allows for high dirt-holding capacity and efficient filtration.
- **Oil mist collection**: In industrial settings, nonwoven media are utilized in oil mist collectors to trap airborne oil particles, ensuring a cleaner environment and reducing maintenance needs.

Advantages:

- **High dirt-holding capacity**: The porous structure of nonwoven media enables them to retain a significant amount of contaminants before requiring replacement.
- **Cost-effectiveness**: Manufacturing nonwoven media is generally more economical compared to other filtration materials, making them a cost-effective choice for various applications.
- **Versatility**: These media can be tailored to meet specific filtration requirements by adjusting fiber composition, density, and thickness.

571. NSF International

NSF International previously known as the National Sanitation Foundation is a globally recognized organization that develops public health standards and provides testing, auditing, and certification services across various sectors, including food, water, consumer products, and the environment, and has done so since 1944. In the context of lubricants, NSF plays a crucial role in ensuring that products used in food processing and other sensitive industries meet stringent safety and quality standards.

NSF lubricant classifications:
NSF classifies lubricants based on their potential for incidental food contact, assigning specific codes to guide their appropriate use:

- **NSF H1**: Lubricants registered under this category are suitable for use in food processing environments where incidental contact with food may occur. These lubricants must meet the US FDA's Code of Federal Regulations Title 21 Section 178.3570, ensuring they are odorless, colorless, and tasteless, with a limit of 10 parts per million (ppm) of lubricant base oils in food if incidental contact occurs.
- **NSF H2**: These lubricants are intended for use in food processing areas where there is no possibility of food contact. They are not suitable for applications where incidental food contact is possible.
- **NSF H3**: Soluble oils used to prevent rust on hooks, trolleys, and similar equipment in food processing environments. Equipment treated with H3 lubricants must be cleaned of the oils before production resumes.
- **NSF 3H**: Lubricants approved for direct contact with food products, typically used as release agents to prevent food from adhering to surfaces during processing.

ISO 21469 certification:
In addition to NSF H1 registration, NSF offers **ISO 21469 certification**, an internationally recognized standard that specifies hygiene requirements for the formulation, manufacture, and use of lubricants that may come into contact with food products during processing. This certification ensures that lubricants meet comprehensive safety and quality standards, providing an added layer of assurance for food processors.

O

572. O-ring Seal

An O-ring seal is a mechanical gasket designed to prevent the passage of fluids or gases in various applications. When compressed between two mating surfaces, the O-ring deforms under pressure creating a positive mechanical seal that resists pressure and prevents leakage.

Key Features:

- **Material composition**: O-rings are made from a variety of elastomers, including nitrile rubber (NBR), fluorocarbon (Viton®), silicone, and others, each selected based on compatibility with the operating environment and the substances being sealed. See Figure O1.
- **Design and function**: The circular cross-section of an O-ring allows it to deform under pressure, filling the gap between surfaces and creating a reliable seal.

Figure O1 Typical O-rings

DOI: 10.1201/9788743806608-586

Applications:

- **Automotive Industry**: Used in engines, transmissions, and braking systems to prevent fluid leaks.
- **Aerospace**: Employed in fuel systems, hydraulic systems, and cabin pressurization systems.
- **Industrial Equipment**: Found in pumps, valves, and hydraulic cylinders to ensure leak-proof operation.

Environmental Conditions: Factors such as exposure to ozone, UV light, and weathering can affect material performance.

Maintenance Tips:

- **Proper installation**: Ensure O-rings are installed without twisting or stretching to prevent damage.
- **Lubrication**: Applying an appropriate lubricant can reduce friction and wear during operation.

Key Features:

- **Material composition**: O-rings are made from a variety of elastomers, including nitrile rubber (NBR), fluorocarbon (Viton®), silicone, and others, each selected based on compatibility with the operating environment and the substances being sealed.
- **Design and function**: The circular cross-section of an O-ring allows it to deform under pressure, filling the gap between surfaces and creating a reliable seal.

573. Offline Oil Filtration System

Offline oil filtration systems are essential for maintaining the cleanliness and longevity of lubricants in industrial machinery. By operating independently of the main lubrication system, these units continuously remove contaminants such as solid particles and water, ensuring optimal fluid quality. See Figure O2.

Key Benefits of Offline Filtration:

- **Enhanced fluid cleanliness**: By operating independently of the main lubrication system, offline filtration units continuously remove contaminants such as solid particles and water, ensuring optimal fluid quality.

- **Extended equipment lifespan**: Maintaining clean lubricants reduces wear and tear on machinery components, leading to increased operational efficiency and reduced downtime.
- **Cost savings**: Improved fluid cleanliness can extend oil life, reducing the frequency of oil changes and associated disposal costs.

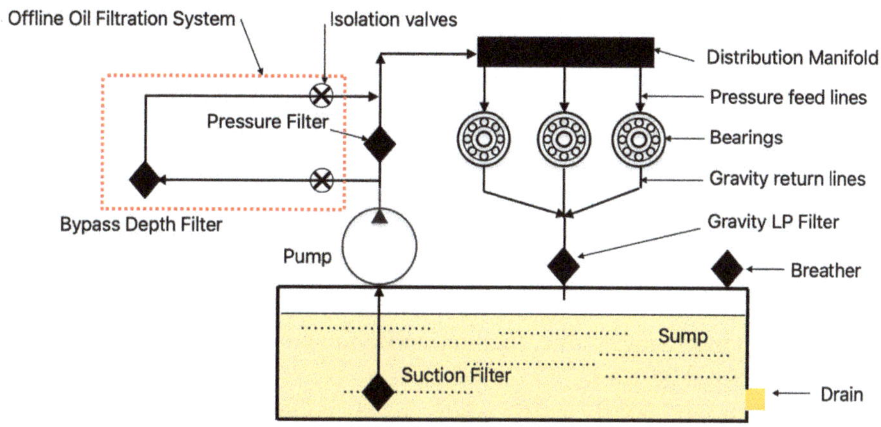

Figure O2 Offline oil filtration system.

Types of Offline Filtration Systems:

- **Portable units**: These mobile systems can be moved between different machines or areas, providing flexibility for maintenance activities.
- **Permanent installations**: Integrated into the machinery, these systems offer continuous filtration without the need for manual intervention.

574. Offline Oil Sampling

Offline oil sampling is a critical practice in predictive maintenance, enabling the assessment of lubricant conditions and the detection of potential issues before they lead to equipment failure. By analyzing oil samples, valuable insights can be gained into the condition of your machinery, allowing for timely interventions that enhance performance and extend equipment lifespan. See Figure O3.

Best Practices for Offline Oil Sampling:

1. **Strategic sampling points**: Collect samples from locations that provide a representative analysis of the oil's condition throughout the system. Primary sampling ports should be situated close to the oil reservoir, on the return line immediately before the gravity return line filter and reservoir, to capture oil that has lubricated all components. Secondary ports can be placed on the entrance to any off-line secondary/auxiliary bypass depth cleaning circuit and before and after specific components to identify localized issues.
2. **Consistent sampling procedures**: Use standardized methods to ensure sample integrity. Employ clean, contamination-free containers and avoid using drain plugs at the bottom of the tank, as they may collect sediments and contaminants that do not represent the oil circulating through the system.
3. **Timely analysis**: Promptly send collected samples to a certified laboratory for analysis to obtain accurate and actionable data. Delays can result in missed opportunities for early intervention.
4. **Regular monitoring**: Establish a routine sampling schedule to monitor oil conditions over time, facilitating trend analysis and early detection of potential issues.

Implementing these best practices ensures that offline oil sampling effectively contributes to machinery reliability and operational efficiency.

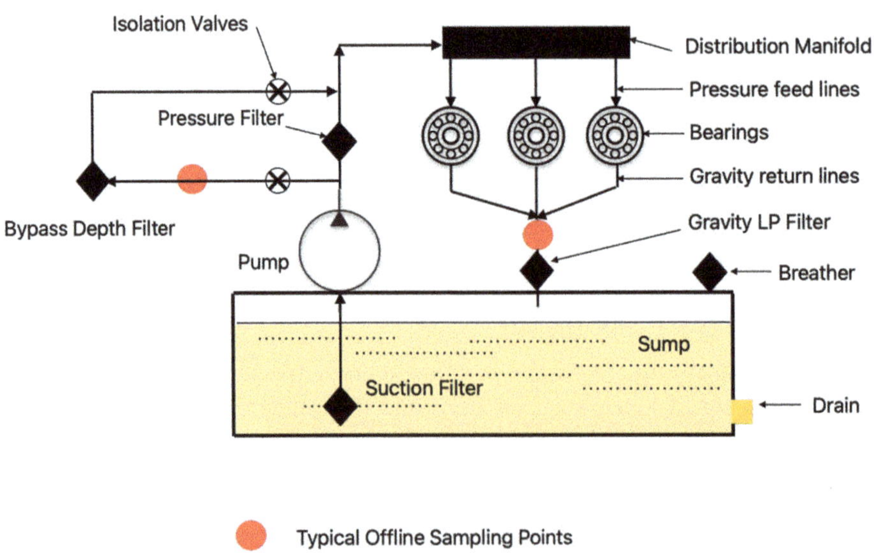

Figure O3 Primary offline sampling points.

575. Oil

Oil is a viscous, nonpolar liquid primarily composed of hydrocarbons. It is hydrophobic, meaning it does not mix with water and is flammable. Oils can be derived from various sources, including plants, animals, minerals, and synthetically produced compounds. They have a wide range of applications across different industries.

Types of Oil:

1. **Petroleum oil**: Commonly known as crude oil, it is extracted from underground reservoirs and refined into products like gasoline, diesel, and jet fuel.
2. **Vegetable oils**: Extracted from plants, these oils are used in cooking, cosmetics, and as biofuels.
3. **Animal fats**: Derived from animals, these oils are utilized in food products, soaps, and industrial applications.
4. **Essential oils**: Concentrated plant extracts used in aromatherapy, perfumes, and flavorings.
5. **Synthetic oils**: Man-made oils designed for specific industrial applications, offering enhanced performance characteristics.

Common Uses of Oil:

- **Lubrication**: Reduces friction between moving parts in machinery, extending their lifespan.

576. Oil Analysis

Oil analysis is a primary form of practice in predictive maintenance, involving the examination of lubricating oils to assess their condition, detect contaminants, and identify wear particles. This proactive approach enables early detection of potential issues, allowing for timely interventions that can prevent equipment failures and extend machinery lifespan.

Key Components of Oil Analysis:

1. **Fluid properties assessment**: Evaluating the oil's physical and chemical characteristics, such as viscosity, acidity (acid number), and oxidation levels, to determine its suitability for continued use.

2. **Contamination detection**: Identifying the presence of foreign substances like water, fuel, or dirt, which can compromise lubrication effectiveness and lead to equipment damage.

3. **Wear debris analysis**: Examining particles within the oil to detect signs of component wear, corrosion, or other forms of degradation, providing insights into the health of machine parts.

Benefits of Regular Oil Analysis:

- **Optimized maintenance schedules**: By understanding the oil's condition, maintenance activities can be performed based on actual needs rather than fixed intervals, reducing unnecessary downtime and costs.

- **Enhanced equipment reliability**: Early detection of potential issues allows for prompt corrective actions, minimizing the risk of unexpected failures and extending the operational life of machinery.

- **Cost savings**: Monitoring oil health can lead to extended oil change intervals and prevent costly repairs or replacements by addressing problems before they escalate.

Implementing an Oil Analysis Program (See Figure O4):

1. **Establish sampling procedures**: Develop standardized methods for collecting oil samples to ensure consistency and representativeness.

2. **Select appropriate tests**: Choose analysis tests that align with the specific equipment and operational conditions to obtain relevant data.

3. **Interpret results**: Analyze the data to identify trends, anomalies, and potential issues, facilitating informed decision-making.

4. **Take corrective actions**: Based on analysis findings, implement necessary maintenance or operational adjustments to address identified concerns.

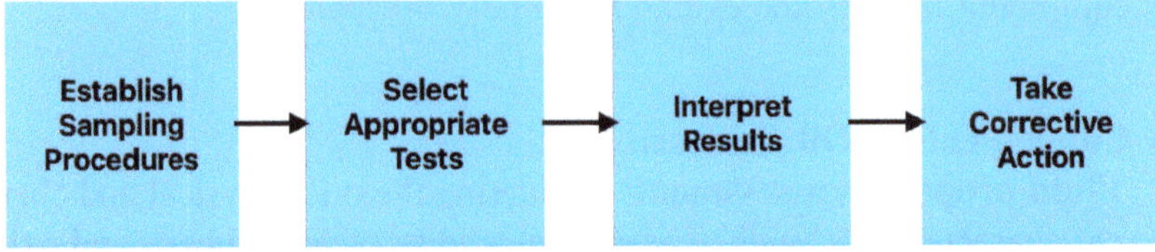

Figure O4 Four-stage oil analysis implementation program.

By integrating oil analysis into maintenance strategies, organizations can achieve more efficient operations, reduce unplanned downtime, and enhance the longevity of their equipment.

577. Oil Analysis Report

An **oil analysis report** is a comprehensive document that evaluates the condition of lubricating oil used in machinery. It provides insights into the oil's properties, identifies contaminants, and detects wear particles, enabling maintenance professionals to assess equipment health and predict potential failures.

Key Components of an Oil Analysis Report:

1. **Oil properties assessment**: This section examines the oil's physical and chemical characteristics, such as viscosity, acidity (acid number), and oxidation levels, to determine its suitability for continued use.

2. **Contamination detection**: Identifies the presence of foreign substances like water, fuel, or dirt, which can compromise lubrication effectiveness and lead to equipment damage.

3. **Wear debris analysis**: Examines particles within the oil to detect signs of component wear, corrosion, or other forms of degradation, providing insights into the health of machine parts. See Table O1 for all elements checked in for an oil analysis sample and breaks them down into typical collectives that are compared with a virgin oil sample to help identify what should and should not be.

Table O1 Wear debris elements checked for in oil analysis.

Element	Additive	Contaminant	Wear Metal	Coolant
Aluminum			X	
Barium	X			
Boron	X			X
Calcium	X			
Chromium			X	
Copper			X	
Iron			X	
Lead			X	
Magnesium	X			
Molybdenum	X			
Nickel			X	
Phosphorous	X			
Silicon	X	X		
Silver			X	
Sodium	X			X
Tin			X	
Vanadium	X	X		
Zinc	X			

Interpreting the Report:

- **Compare with baseline data**: Assess current results against historical data to identify trends and deviations.
- **Identify anomalies**: Look for unexpected increases in wear metals or contaminants, which may indicate emerging issues.
- **Consult with experts**: Engage with laboratory analysts or lubrication specialists to gain a deeper understanding of the findings and recommended actions.

578. Oil Analyzer

An **oil analyzer** is a device or system designed to assess the condition of lubricating oils used in machinery. By evaluating various parameters, oil analyzers provide critical insights into equipment health, enabling proactive maintenance and extending machinery lifespan.

Types of Oil Analyzers:

- **Laboratory-based oil analyzers**: These systems are typically used in specialized and controlled laboratory environments to conduct detailed analyses of oil samples. Typical laboratory analyzers can include an atomic emission spectroscope, a ferrogram analyzer, and a four-ball wear test machine.
- **Portable oil analyzers**: These suitcase kit devices are smaller portable versions of the laboratory test units that allow for accurate on-site analysis, providing immediate results and facilitating quick decision-making. See Figure O5.
- **Hand-held analyzer**: Small pocket-size hand analyzers that analyze the contamination levels in a sample oil drop and produce a traffic light condition code analysis (red/amber/green) within seconds. Primarily used as a quick "in-between oil change analysis" tool in harsh operating environments
- **Patch test analyzer**: Simple manual analyzing method that employs a portable microscopic to perform analysis of an oil drop placed on a special filter patch. Primarily used as a relatively inexpensive rapid condition test to determine if a full laboratory test is required. See Patch Test.

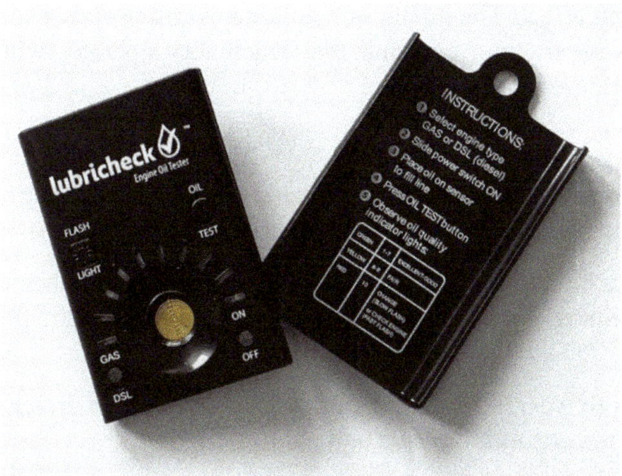

Figure O5 "Lubricheck" hand-held engine oil condition analyzer.

579. Oil Balls

The term **"oil balls"** is a colloquial term often used to describe contaminants or particles found within lubricating oil used in ball bearings. These particles can originate from various sources, including degraded lubricant, wear debris, or external contaminants. The presence of such particles can compromise bearing performance, leading to increased friction, wear, and potential failure. Regular oil analysis is essential to monitor and manage these contaminants, ensuring the longevity and efficiency of the bearings.

580. Oil Bath Lubrication

Oil bath lubrication is a method that is a self-contained arrangement in which a mechanical component, such as a gear, bearing, or chain, is partially or fully submerged in oil to provide continuous lubrication. This system ensures that the moving parts remain adequately lubricated, minimizing friction and wear.

Key Features of Oil Bath Lubrication:

- **Continuous lubrication**: The system allows for continuous oil circulation around the component, providing constant lubrication and cooling. This helps maintain smooth operation and reduce the risk of overheating.

- **Cooling effect**: The oil not only lubricates but also dissipates heat generated by friction, ensuring that machinery operates within optimal temperature ranges.

Common Applications:

- **Automotive systems**: Used in engine crankcases and gearboxes where parts are submerged in oil to ensure smooth operation.
- **Industrial equipment**: Gearboxes, bearings, and other components that require consistent lubrication and cooling benefit from oil bath lubrication.
- **Chain drives**: In machines like conveyors or chain drives, an oil bath keeps the chains lubricated and reduces friction.

581. Oil Bleed Rate

Oil bleed rate refers to the rate at which oil separates from grease under specific conditions. This separation is a natural property of greases, essential for effective lubrication. When grease is applied to components like bearings, the oil within the grease is released to form a lubricating film, reducing friction and wear.

Figure O6 Static bleed of a grease in a pail stored outdoors. Courtesy: ENGTECH Industries Inc.

Understanding Oil Bleed:

- **Static bleed**: Occurs when grease is stored or remains stationary, leading to oil separation due to gravity or temperature fluctuations. This is commonly observed as a thin layer of oil on the surface of grease stored outdoors. See Figure O6.
- **Dynamic bleed**: Happens during the operation of machinery, where mechanical stresses, temperature changes, or contamination cause oil to separate from the grease matrix. This ensures that the moving parts receive adequate lubrication.

Factors Influencing Oil Bleed Rate:

- **Grease composition**: The balance between the thickener system and base oil content affects oil separation. A higher proportion of thickener can reduce oil bleed, but some oil release is necessary for lubrication.
- **Base oil viscosity**: Lower-viscosity oils are more prone to separation. Greases with higher base oil viscosity typically exhibit reduced oil bleed rates.
- **Operating temperature**: Elevated temperatures can increase oil separation. Proper storage and operating conditions are crucial to minimize excessive oil bleed.
- **Mechanical stresses**: Vibrations, centrifugal forces, and high-speed operations can accelerate oil separation from grease.

Managing Oil Bleed:

- **Storage conditions**: Store greases in cool, dry environments to minimize static oil bleed. Avoid exposure to extreme temperatures and vibrations during storage.
- **Grease selection**: Choose greases with appropriate thickener systems and base oil viscosities suited to the specific application and operating conditions.
- **Regular monitoring**: Implement routine inspections to detect and address excessive oil bleed, ensuring optimal lubrication performance.

582. Oil Clearance

Oil clearance is the **gap between two mating parts**, typically between a **shaft** and a **bearing** (e.g., journal bearing or bushing), that allows a thin film of oil to form and lubricate the surfaces.

- It ensures **hydrodynamic lubrication**, where parts don't touch directly.
- The clearance must be **large enough** to allow oil to flow but **small enough** to maintain a pressure-supported oil film.

How Viscosity Affects Oil Clearance
Viscosity is the oil's resistance to flow. It directly impacts the oil film's thickness and behavior in the clearance gap.

High Viscosity Oil:
- Flows **slowly** and forms a **thicker** oil film.
- Can maintain separation better under high load.
- **Too thick** for small clearances — may cause **excess heat, drag, or oil starvation** at start-up.

Low Viscosity Oil:
- Flows **easily** and forms a **thinner** oil film.
- Better for **tight clearances** and cold starts.
- May not maintain an adequate oil film under high load or heat — risk of **metal-to-metal contact**.

The oil's viscosity must be matched to the **designed oil clearance**:
- **Tighter clearances** require **Lower viscosity lubricants**
- **Looser clearances** require **Higher viscosity lubricants**

583. Oil Consumption Rate

Oil consumption rate is a simple metric referring to the amount of oil used by a specific asset, machine, process line, or plant over a specific time period, typically measured in gallons or liters.

584. Oil Consumption Ratio

This metric is employed to understand a number of scenarios that include:

- **The total amount of oil (by oil type or all oil)** used in a given time period (annually, quarterly, etc.) divided by the total oil reservoir capacity of an individual or collective group of assets (e.g., by machine, machine type, area, production line, and plant)
- **Projected lubricant usage versus actual lubricant** usage (to determine if oil changes are actually taking place as per the PM system requirement.

- **Consumption increase rate over a longer period of** time to assess equipment wear over a period of time
- **Usage comparison in similar lubricated assets**
- **Usage comparison based on a trial of two or more viscosities** of the same lubricant in the same asset
- **Consumption rate versus the used oil rate**

585. Oil Cushioning

Oil cushioning refers to the use of **lubricating oil to absorb shocks, dampen vibrations, and reduce impact forces** in mechanical systems. This property is particularly important in hydraulic systems, dampers, shock absorbers, and heavy machinery, where controlled fluid motion helps minimize mechanical stress, wear, and energy loss.

Key Functions of Oil Cushioning

- **Absorbs shock loads and impact forces**: Forms a fluid barrier that softens sudden mechanical loads, preventing damage to components.
- **Dampens vibrations**: Reduces harmful oscillations in moving parts, improving stability and performance.
- **Controls motion and deceleration**: Ensures smooth braking and controlled movement in hydraulic actuators and shock absorbers.
- **Minimizes metal-to-metal contact**: Prevents excessive wear and mechanical stress by maintaining a protective oil film.

586. Oil Film Thickness

See "Lambda Thickness Ratio."

587. Oil Film Strength

Oil film strength refers to a lubricant's ability to maintain a protective film between two moving surfaces under load and extreme conditions. It is a critical factor in preventing metal-to-metal contact, reducing friction, and minimizing wear in mechanical systems such as bearings, gears, and engine components. A lubricant with high oil film strength can withstand heavy loads, high speeds, and temperature variations without breaking down.

Key Functions of Oil Film Strength

- **Prevents metal-to-metal contact**: Forms a protective barrier that reduces friction and prevents wear in high-stress areas.
- **Enhances load-carrying capacity**: Ensures stable lubrication under extreme pressure, reducing component damage.
- **Improves energy efficiency**: A strong oil film minimizes energy loss due to friction, improving mechanical performance.
- **Reduces wear and extends equipment life**: Maintains lubrication integrity even under high loads and elevated temperatures.

588. Oil Filter

An **oil filter** is a crucial component in any oil distribution system, for example, industrial bearing lubrication (total loss and recyclic oil systems). Hydraulic systems, internal combustion engines, etc., are designed to remove contaminants from oils. By filtering out particles such as dirt, metal shavings, and combustion by-products, oil filters ensure that only clean oil circulates through the lubricated bearing surfaces, thereby protecting them from wear and maintaining optimal performance.

Key Functions of Oil Filters:

- **Contaminant removal**: Oil filters capture harmful particles and debris present in the oil, preventing them from causing abrasion and damage to bearing surfaces.
- **Maintaining oil quality**: By removing contaminants, oil filters help maintain the oil's lubricating properties, ensuring efficient lubrication operation.

Types of Oil Filters:

- **Mechanical filters**: These filters use materials like pleated paper or synthetic fibers to trap contaminants as oil flows through them. They require periodic replacement to maintain effectiveness.
- **Magnetic filters**: Utilizing magnets, these filters attract and hold ferromagnetic particles, such as metal shavings, preventing them from circulating in the oil.

- **Centrifugal filters**: These filters use centrifugal force to separate contaminants from the oil, effectively removing particles without the need for filter media.

Components of a Typical Oil Filter (See Figure O7):

- **Filter element**: The core component that captures contaminants; it can be made from paper, synthetic fibers, or a combination of materials.

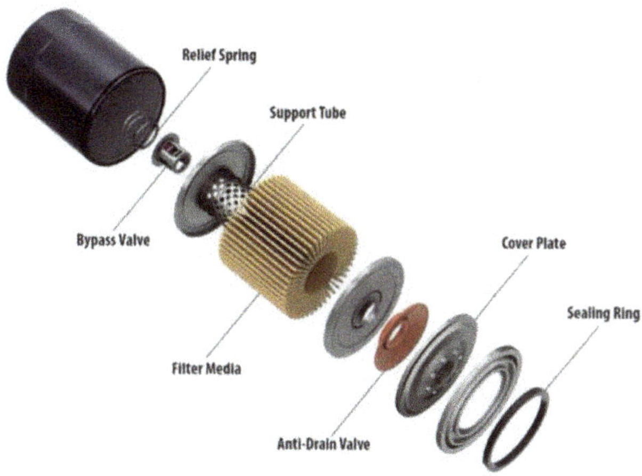

Figure O7 Typical mechanical filter components.

- **Canister**: The outer casing that houses the filter element, typically made of metal or durable plastic.
- **Antidrainback valve**: Prevents oil from draining out of the filter when the engine is off, ensuring immediate oil circulation upon startup.
- **Bypass valve**: Allows oil to bypass the filter if it becomes clogged, ensuring continuous oil flow to the engine.

Maintenance Considerations:
Regular replacement of oil filters is essential to maintain lubricant and bearing health. Neglecting to change the oil filter can lead to increased component wear, reduced performance, and potential damage leading to bearing failure.

589. Oil Groove

An **oil groove** is a channel or recess machined or molded into a surface—usually inside a **bearing, bushing,** or **gear housing**—to help **distribute lubricating oil** evenly across moving parts.

Purpose of an Oil Groove:

- To evenly distribute lubricant across the bearing surface to ensure consistent lubrication
- Promote oil circulation on open ended grooves

Common Types of Oil Grooves:

- **Circular (or circumferential)** – goes around the inner surface
- **Axial** – runs lengthwise along the component
- **Spiral** – combines axial and circular paths
- **Figure-eight or diamond** – complex patterns for more uniform coverage

Typical oil groove locations:

- Bearings (especially plain or journal bearings)
- Bushings
- Gearboxes
- Hydraulic and pneumatic components
- Linear oil path grooves on flat bearing surfaces

590. Oil Life Extension

Oil life extension refers to **strategies and technologies used to prolong the service life of lubricants**, reduce the frequency of oil changes, minimize waste, and improve overall equipment efficiency. By maintaining oil cleanliness, optimizing operating conditions, and using high-performance lubricants, industries can reduce costs, enhance machine reliability, and promote sustainability.

Key Methods for Oil Life Extension

- **Contamination control**: Prevents dirt, water, and particulate contamination through filtration, desiccant breathers, and proper storage.

- **Condition monitoring and oil analysis**: Regularly testing **for viscosity changes, oxidation, and additive depletion** allows for timely corrective actions.
- **Optimized operating conditions**: Reducing excessive heat, load, and pressure fluctuations helps slow down lubricant degradation.
- **High-performance lubricants and additives**: Using synthetic oils, antioxidant additives, and viscosity stabilizers enhances oil longevity.
- **Filtration and reconditioning**: Kidney-loop filtration, centrifugation, and vacuum dehydration remove contaminants, restoring oil performance.

Factors Affecting Oil Life

- **Oxidation stability**: Higher resistance to oxidation prevents sludge, varnish, and acidity buildup.
- **Thermal degradation** Resistance: Oils with high thermal stability resist breakdown in high-temperature environments.
- **Contaminant exposure**: Water, dirt, and wear debris accelerate oil degradation, reducing its service life.
- **Shear stability**: High-quality lubricants maintain viscosity under mechanical stress, extending usability.
- **Additive depletion**: Antioxidants, detergents, and dispersants must be maintained for optimal lubricant protection.

591. Oil Mist Lubrication

See "Mist Lubrication System."

592. Oil Processing Lifecycle Journey

The **oil processing lifecycle** journey encompasses the series of stages that crude oil undergoes from its initial extraction, and consumption, to disposal or rerefining for extended use. See Figure O8.

This journey involves five key phases:

1. **Exploration and appraisal**: Geologists and geophysicists conduct seismic surveys and geological studies to identify potential oil reserves. If promising, exploratory drilling is undertaken to confirm the presence

and viability of oil deposits. Upon confirming viable reserves, infrastructure is established, including drilling wells, constructing platforms or facilities, and setting up pipelines for transportation. This phase prepares the site for production activities. Oil is extracted from the reservoir through drilling. The extracted crude oil is then transported via pipelines, tankers, or rail to refineries for processing.

2. **Refining**: At refineries, crude oil undergoes various processes such as distillation, cracking, and reforming to separate and convert it into valuable products like gasoline, diesel, jet fuel, and petrochemical feedstocks.
3. **Distribution and storage**: Refined products are stored in tanks and distributed through pipelines, trucks, or ships to various bulk storage facilities, including distribution terminals and retail outlets.
4. **Consumption**: End-users, including industries and transportation sectors, utilize the refined products for energy, manufacturing, and other applications.
5. **Disposal or recommission**: Once oil products reach the end of their productive life, they can be reclaimed and reused if not mixed or contaminated. Contaminated oil products are disposed of in toxic disposal facilities, while noncontaminated used oils are recommissioned in specialized refineries and resold to customers.

Throughout this lifecycle, various stakeholders, including exploration companies, drilling contractors, refineries, and distribution networks, collaborate to ensure the efficient and safe processing of oil from extraction to end-use.

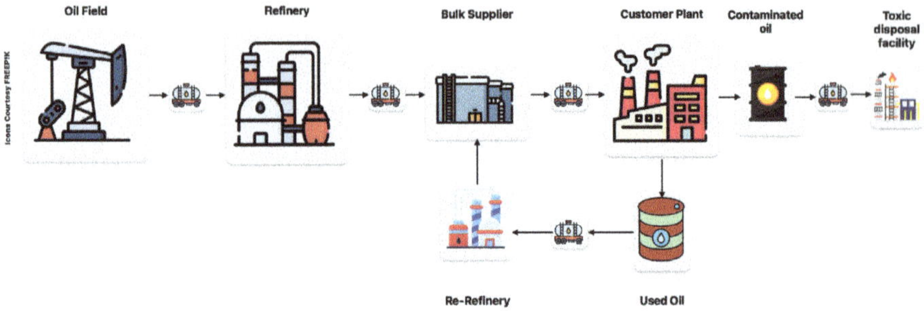

Figure O8 Typical oil processing journey. Source: ENGTECH Industries Inc. and Freepic Icons.

593. Oil Seal

An **oil seal**, also known as a shaft seal or grease seal, is a mechanical component designed to prevent the leakage of lubricants and the ingress of contaminants in machinery. Typically installed at the interface between a rotating shaft and a stationary housing, oil seals maintain the integrity of lubricated systems by ensuring that lubricants stay within the machinery and harmful substances are kept out.

Key Functions of Oil Seals:

- **Prevent lubricant leakage**: Oil seals create a barrier that retains lubricants within machinery components, ensuring optimal performance and reducing the need for frequent relubrication.
- **Protect against contaminants**: They prevent dirt, dust, and other harmful particles from entering machinery, safeguarding sensitive parts from damage and wear.
- **Maintain pressure**: Oil seals help maintain internal pressure within machinery, which is essential for the proper functioning of certain systems.

Common Types of Oil Seals:

1. **Radial oil seals**: The most commonly used type, consisting of an outer metal casing, an inner sealing lip made of rubber or elastomer materials, and a tension spring that provides radial force to maintain contact between the sealing lip and shaft surface. See Radial Lip Seal and Figure O9.

Figure O9 Typical radial lip seal.

2. **Cassette seals**: Advanced seals with multiple sealing lips, designed to provide superior protection against heavy contamination and extreme operating conditions.

3. **V-seals**: Also known as axial or V-seals, these are elastomeric, V-shaped rings that act as a lip to prevent the ingress of contaminants and retain lubricants.

Materials Used in Oil Seals:

- **Nitrile rubber (NBR)**: Commonly used for its good heat resistance properties and resistance to oils, hydraulic oils, and gasoline.
- **Fluorocarbon rubber (FKM)**: Offers excellent resistance to chemicals and high temperatures, making it suitable for demanding applications.
- **Polytetrafluoroethylene (PTFE)**: Known for its low friction properties and resistance to a wide range of chemicals.

594. Oil Slinger

An **oil slinger** is a mechanical component designed to direct or "sling" oil away from specific areas within an engine or machinery to prevent excessive oil accumulation and potential leaks. Typically made of metal, it rotates with the shaft and uses centrifugal force to distribute oil to various parts, ensuring proper lubrication and reducing the risk of oil leakage.

Key Functions of Oil Slingers:

- **Prevent oil leakage**: By directing oil away from seals and gaskets, oil slingers help prevent oil from reaching areas where it could cause leaks.
- **Enhance lubrication**: They ensure that moving parts receive adequate lubrication by distributing oil effectively, thereby reducing friction and wear.
- **Protect seals**: Oil slingers shield seals from direct exposure to oil flow, extending their lifespan.

See also "Oil Wiper."

595. Oil Wedge

An **oil wedge** is a fundamental concept in hydrodynamic lubrication, where a thin layer of oil forms between two surfaces in relative motion, such as a shaft and its bearing. As the shaft rotates, it draws oil into the converging gap between the surfaces, creating a wedge-shaped film. This oil wedge generates pressure that lifts the shaft away from the bearing surface, effectively preventing direct metal-to-metal contact and thereby reducing friction and wear.

Key Aspects of the Oil Wedge:

- **Formation**: The rotating shaft's motion draws oil into the bearing clearance, forming a wedge-shaped film.
- **Pressure generation**: The converging geometry of the gap causes the oil to build pressure, which supports the load and separates the surfaces. See Figure O10.
- **Film thickness**: The thickness of the oil film typically ranges from 2 to 100 microns, depending on factors like load, speed, and oil viscosity.
- **Lubrication regime**: The presence of the oil wedge signifies hydrodynamic lubrication, where the load is entirely supported by the oil film, minimizing direct contact between surfaces.

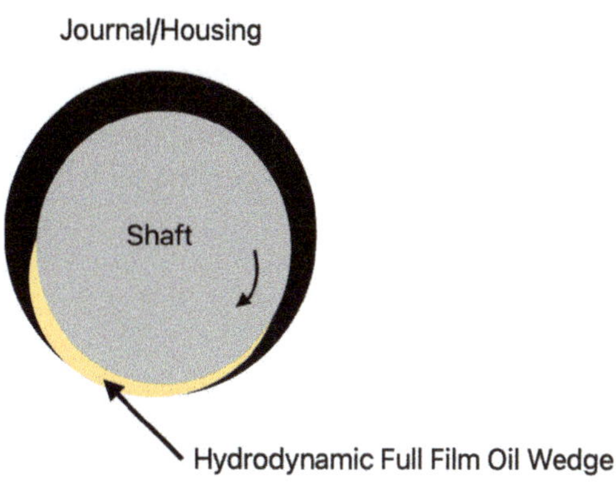

Figure O10 Hydrodynamic full film oil wedge.

596. Oil Whirl / Oil Whip

Oil whirl and **oil whip** are phenomena observed in rotating machinery, particularly in journal bearings, where the interaction between the rotating shaft and the lubricating oil leads to specific vibrational instabilities.

Oil Whirl:

Oil whirl is a self-excited fluid-induced instability where the oil film drives the shaft ahead of it in a forward circular motion, causing the shaft to follow a whirling path around the bearing clearance. This condition is characterized by subsynchronous vibrations, typically occurring at frequencies between 0.38 to 0.48 times the machine's running speed. The instability arises when the tangential force from the oil film becomes strong enough to disrupt the rotor's equilibrium, leading to a whirling motion.

Oil Whip:
Oil whip is a more severe form of instability that occurs when the frequency of oil whirl coincides with the system's natural frequency, often the rotor's first critical speed. This resonance condition leads to large amplitude vibrations, particularly at the rotor's mid-span, and can result in significant damage if not addressed promptly. Unlike oil whirl, which is a subsynchronous vibration, oil whip involves vibrations at the system's natural frequency, making it more destructive.

Key Differences:

- **Frequency**: Oil whirl occurs at subsynchronous frequencies (0.38 to 0.48 times the running speed), while oil whip occurs at the system's natural frequency.
- **Severity**: Oil whip is generally more destructive due to higher vibration amplitudes and potential for rotor damage.
- **Origin**: Oil whirl is driven by the tangential force of the oil film, whereas oil whip results from resonance between the oil whirl frequency and the system's natural frequency.

597. Oil Wiper

Oil wiper is a component used in machinery to manage the flow of oil and prevent leaks or excess oil buildup similar in manner to an oil slinger, but both function differently in terms of their mechanisms and roles in lubrication systems.

An **oil wiper** is designed to remove excess oil from a moving part, such as a piston rod or shaft. It scrapes the oil off as the component moves, ensuring that only the required amount of oil remains for proper lubrication. Commonly used in **reciprocating compressors** and **piston systems**, where oil needs to be removed from the moving parts after it performs its lubrication function.

As the moving part (e.g., piston rod) moves within the system, the oil wiper scrapes off the excess oil from the surface, directing it away from sensitive parts. Oil wipers are typically made from durable materials like rubber, PTFE, or other flexible, abrasion-resistant materials.

An **oil slinger** differs in that it is typically found in **rotating machinery** and uses centrifugal force to throw or "sling" oil from a rotating shaft to various parts of a machine, increasing lubrication to all moving parts and minimizing the chances of oil leaks.

Key Differences between Oil Wiper and Oil Slingers Include: :

1. **Function**:
    - An **oil wiper** removes excess oil, ensuring controlled lubrication and preventing leakage.
    - An **oil slinger** spreads oil using centrifugal force, providing lubrication to different parts of the system.
2. **Mechanism**:
    - **Oil wipers** scrape or wipe oil off moving parts to maintain a balanced oil level.
    - **Oil slingers** sling or spread oil to various parts of the machine, often using the rotation of the shaft.
3. **Applications**:
    - **Oil wipers** are used to remove oil from specific parts after lubrication, particularly in systems like compressors and piston rods.
    - **Oil slingers** are used in rotating machinery to distribute oil effectively to ensure the entire system remains lubricated.
4. **Maintenance**:
    - **Oil wipers** need to be maintained for flexibility and durability to continue scraping effectively.
    - **Oil slingers** need to be checked for wear, especially if metal-to-metal contact occurs, as slinging can lead to degradation of the slinger itself.

598. Oiler

An **oiler** is an old term used to describe a worker whose primary responsibility was/is to apply lubricants to machinery to ensure smooth operation and prevent wear. Historically prevalent across various industries – including maritime, railroads, steelmaking, and mining – the role of the oiler has typically revolved around the following key responsibilities.

Key Responsibilities:

- **Lubrication maintenance**: Regularly applying oil and grease to machinery components to reduce friction and prevent wear.
- **Equipment monitoring**: Observing machinery performance, including checking pressure and temperature gauges, to identify potential issues.
- **Assisting machine and plant engineers**: Supporting engineers in the servicing and repair of engine room machinery and equipment.
- **Record keeping**: Maintaining logs of engine performance, maintenance schedules, and oil usage.
- **Safety compliance**: Ensuring adherence to safety standards and environmental regulations concerning waste oil and lubricants.

With technological advancements, many oiler positions have been elevated to certified Lubrication Technicians who still perform many of the old oiler tasks but are now more involved in the management of lubricants and more sophisticated lubrication delivery systems performing many of the lubrication engineering tasks alongside the plant engineering group

599. Oiliness Agent

An **oiliness agent** is a chemical additive incorporated into lubricants to enhance their ability to adhere to metal surfaces, thereby reducing friction and wear under boundary lubrication conditions. These agents form a thin, adsorbed film on metal surfaces, providing a protective layer that minimizes direct metal-to-metal contact.

Key Functions of Oiliness Agents:

- **Enhance lubricant adhesion**: Oiliness agents improve the lubricant's capacity to stick to metal surfaces under pressure or load, ensuring continuous lubrication even in high-stress conditions.
- **Reduce friction**: By forming a protective film, these agents lower the coefficient of friction between moving parts, leading to smoother operation and reduced energy consumption.
- **Prevent wear**: The protective film created by oiliness agents acts as a barrier, reducing direct contact between metal surfaces and thereby minimizing wear and extending the lifespan of machinery components.

Common Oiliness Agents:

- **Fatty acids**: Organic acids derived from animal or vegetable fats that enhance the oil's ability to adhere to metal surfaces.
- **Fatty alcohols**: Alcohols with long hydrocarbon chains that improve the lubricating properties of oils.
- **Esters**: Compounds formed by the reaction of acids and alcohols, contributing to the oil's oiliness characteristics.
- **Sulfurized olefins**: Compounds formed by the reaction of olefins with sulfur, providing extreme pressure (EP) properties to the lubricant.

600. OMA – Level I

Oil Monitoring Analyst (OMA) – Level I is a designation awarded by the Society of Tribologists and Lubrication Engineers (see **STLE**) for persons.

The Oil Monitoring Analyst – Level I (OMA I) exam is designed to test and document the knowledge of an individual who conducts first echelon oil monitoring and analysis, as it applies to lubrication analysis and machine condition monitoring. The emphasis of this exam is the routine operation of an oil analysis program.

The OMA I designation is designed for individuals who have the responsibility of :

- Performing on-site oil sampling, ensuring the correct tests are performed
- Reviewing all basic oil analysis reports
- Making recommendations based on findings re: maintenance actions

These individuals are likely to be shop or plant personnel required to oversee the oil analysis program on a daily basis. Generally, this applies to individuals such as mechanics, engineers, operators, tradesmen, chemical managers, or on-site laboratory personnel.

601. OMA – Level II

Oil Monitoring Analyst (OMA) – Level II is a designation awarded the Society of Tribologists and Lubrication Engineers (see **STLE**).

The Oil Monitoring Analyst II (OMA II) exam is designed to test and document the knowledge of an individual who conducts second echelon oil monitoring and analysis, as it applies to lubrication analysis and machine condition monitoring. The emphasis of this level is the design, implementation, quality control and management of an oil analysis program.

The OMA II designation is designed for those individuals who have the responsibility of the overall management and efficiency of the oil analysis program. These individuals are normally at the supervisory level and have the authority to effectively control the overall program. Those individuals involved in these activities generally have the job function of a reliability engineer, lubrication engineer, tribologist, maintenance supervisor, or laboratory person (on-site and commercial).

602. ON

See "Octane Number."

603. Optical Particle Counter

An **optical particle counter (OPC)** is an instrument designed to detect, count, and measure the size of particles suspended in a fluid – such as air, liquids, or gases – by utilizing the principles of light scattering. When particles pass through a laser beam within the device, they scatter light; the intensity and pattern of this scattered light are then analyzed to determine the particle's size and concentration.

Key Components and Functionality:

- **Light source**: Typically, a laser that emits a coherent light beam, providing consistent illumination for particle detection.
- **Flow system**: A controlled flow mechanism directs the sample fluid through the detection chamber, ensuring particles pass through the laser beam.
- **Photodetector**: Captures the scattered light from particles and converts it into electrical signals.
- **Signal processing**: Analyzes the electrical signals to determine particle size and concentration, often categorizing particles into size bins for detailed analysis.

Applications:

- **Industrial processes**: Evaluating the cleanliness of lubricants in manufacturing settings, such as circulating oil and hydraulic systems, to prevent equipment damage.

604. Organic Compound

An **organic compound** is a chemical substance that contains carbon atoms covalently bonded to other elements, most commonly hydrogen, oxygen, and nitrogen.

Relationship to Lubrication:
In lubrication, organic compounds play a pivotal role as base oils and additives:

- **Base oils**: Many lubricants are derived from organic compounds, such as mineral oils and synthetic hydrocarbons. These base oils provide the fundamental lubricating properties necessary for reducing friction and wear in machinery.
- **Additives**: Organic compounds are also utilized as additives to enhance lubricant performance. For example, antioxidants, which are organic compounds containing elements like sulfur, nitrogen, or phosphorus, are incorporated to prevent oxidation and extend the lubricant's lifespan.

605. Organic Thickener

An **organic thickener** is a substance used to increase the viscosity of lubricants, such as greases, by forming a semisolid structure that holds the base oil in place. These thickeners are typically organic compounds that, when combined with base oils, create a gel-like consistency, allowing the lubricant to adhere to surfaces and provide effective lubrication.

Types of Organic Thickeners:

- **Simple soap thickeners**: Formed by the reaction of an organic acid (such as fatty acids) with an alkali metal to produce an organic salt. Common examples include lithium 12-hydroxystearate and calcium stearate.
- **Complex soap thickeners**: Created by combining a simple soap with a complexing agent, such as a short chain difunctional carboxylic acid,

boric acid, or an aromatic acid. This combination enhances the grease's high-temperature properties and shear stability.

- **Nonsoap thickeners**: These include polyurea, organophilic clay, fumed silica, and fluoropolymers. For instance, polyurea thickeners are formed by reacting a di-isocyanate with mono and/or diamines, resulting in a grease that is ashless and tends to be oxidatively stable.

Function in Lubricants:
Organic thickeners impart the desired consistency to lubricants, enabling them to:

- **Adhere to surfaces**: Maintain a protective film on moving parts, reducing friction and wear.
- **Provide stability**: Ensure the lubricant remains in place under varying operating conditions, such as temperature fluctuations and mechanical stresses.
- **Enhance performance**: Improve properties like water resistance, oxidation stability, and load-carrying capacity.

The choice of thickener depends on the specific requirements of the application that include operating temperature, load conditions, and environmental factors.

606. OSHA

The **Occupational Safety and Health Administration (OSHA)** is a federal agency within the United States Department of Labor, established in 1970 under the Occupational Safety and Health Act. OSHA's primary mission is to ensure safe and healthful working conditions for employees by setting and enforcing standards, and by providing training, outreach, education, and assistance.

Key Functions of OSHA:

- **Setting standards**: OSHA develops and enforces regulations to protect workers from various workplace hazards, including chemical exposures, physical dangers, and ergonomic risks.
- **Training and education**: The agency offers training programs and resources to educate employers and employees about workplace safety and health practices.

- **Compliance assistance**: OSHA provides guidance to help employers comply with safety standards, including consultation services for small businesses.
- **Enforcement**: Through inspections and investigations, OSHA ensures compliance with safety standards and can issue citations and penalties for violations.

607. Osmolality

Osmolality refers to the **concentration of solute particles in a solution** and is a critical parameter for certain specialized lubricants, particularly those used in medical or personal care settings. For most industrial lubricants, osmolality is not a standard consideration.

608. Over Lubrication

Over-lubrication occurs when **too much lubricant** (oil or grease) is applied to a component or system—**more than what is needed for proper operation**. Over lubrication is just as detrimental as under lubrication.

Effects of Over-Lubrication

1. **Increased Heat and Pressure**
    - Excess grease in bearings causes **churning** (mixing under pressure), leading to **heat buildup**.
    - This can degrade the lubricant and reduce component life.
2. **Seal Damage or Failure**
    - Too much lubricant can **burst or displace seals**, allowing dirt or moisture in—or lubricant out.
    - Once seals are compromised, contamination and wear accelerate.
3. **Lubricant Waste**
    - Over-lubrication leads to **excessive consumption** and higher maintenance costs.
    - Also poses **environmental and housekeeping hazards** due to leaks or drips.

4. **Increased Friction**

   ○ In some cases, too much lubricant creates **hydrodynamic drag**, especially in high-speed components like electric motor bearings.

   ○ This can reduce efficiency and increase energy consumption.

5. **Contamination Risks**

   ○ Excess lubricant may **trap contaminants** instead of flushing them out.

   ○ It can also **splash or leak** into areas where lubrication is not intended (e.g., brake linings or electrical components).

Signs of Over-Lubrication

- Leaking or purged grease from seals – See Figure O11
- Excessive heat near bearing housings
- Loud or irregular bearing noise after relubrication
- Grease splatter in or around machinery

Figure O11 Over lubricated Flange Bearing. Note how the grease has acted as a dirt collection medium - Courtesy ENG**TECH** Industries Inc.

609. Overpressure Indicator

An **overpressure indicator** in lubrication systems is a device designed to monitor and signal when the pressure within the system exceeds a predetermined threshold. This is crucial for preventing damage to components, ensuring optimal lubrication, and maintaining system integrity. See Figure M2.

Applications:

- **Progressive lubrication systems**: In these systems, overpressure indicators are installed on divider blocks to detect blockages or malfunctions. When a blockage occurs, the pressure rises, causing the indicator to signal the issue.

610. Oxidation

Oxidation in lubricants refers to a chemical reaction where the lubricant's base oil reacts with oxygen mixed with heat, light, water, acid, contaminants, and metal catalysts. These work together negatively to form a sludge that thickens the oil's viscosity and depletes additives, leading to the formation of new compounds that degrade the oil's performance. This process is a primary cause of lubricant degradation over time.

Effects of Oxidation on Lubricants:

- **Viscosity increase**: Oxidation can cause the oil to thicken, impairing its ability to flow and lubricate effectively.
- **Formation of sludge and deposits**: Oxidation leads to the creation of sludge and varnish, which can clog filters and obstruct oil passages, reducing the efficiency of the lubrication system.
- **Additive depletion**: The process consumes antioxidant additives, diminishing the oil's capacity to resist further oxidation.
- **Acid formation**: Oxidation produces acidic compounds that can corrode metal surfaces, leading to increased wear and potential equipment failure.

Factors Accelerating Oxidation:

- **High temperatures**: Elevated operating temperatures can significantly accelerate the oxidation process.

- **Presence of metals**: Metals such as iron and copper can act as catalysts, speeding up oxidation reactions.
- **Contamination**: The presence of water, air, or other contaminants can promote oxidation.

611. Oxidation Inhibitor

See "Antioxidant Additive."

612. Oxidization Stability

Oxidation stability refers to a lubricant's ability to resist chemical reactions with oxygen over time. These reactions can lead to the formation of acids, sludge, and varnish-like deposits, which degrade the lubricant's performance and can cause damage to engine components.

P

613. P–F Curve

The **P–F curve**, also known as the Prevention–Failure Curve, is a graphical tool used in maintenance management to visualize the progression of equipment degradation over time, in order to prevent actual failure occurrence. (See Figure P1) It illustrates the interval between the initial signs of potential failure (Point P) and the point where the equipment can no longer perform its intended function (Point F).

Key Components:

- **Point P (potential failure)**: This is the stage where early indicators of equipment degradation are detectable. Identifying this point allows maintenance teams to implement preventive measures before the issue escalates.
- **Point F (functional failure)**: This represents the point at which the equipment fails to meet performance standards, leading to unplanned downtime or operational inefficiencies.
- **P–F interval**: The duration between Points P and F, known as the P–F interval, is crucial for scheduling maintenance activities. A longer P–F interval provides more time for preventive actions, reducing the risk of unexpected failures.

Applications:

- **Maintenance planning**: By understanding the P–F curve, maintenance teams can schedule inspections and interventions at optimal times, balancing the costs of preventive maintenance against the risks and costs associated with unexpected failures.
- **Reliability-centered maintenance (RCM)**: The P–F curve is integral to RCM strategies, helping organizations determine the most effective maintenance approaches based on the nature of equipment failures.

DOI: 10.1201/9788743806608-628

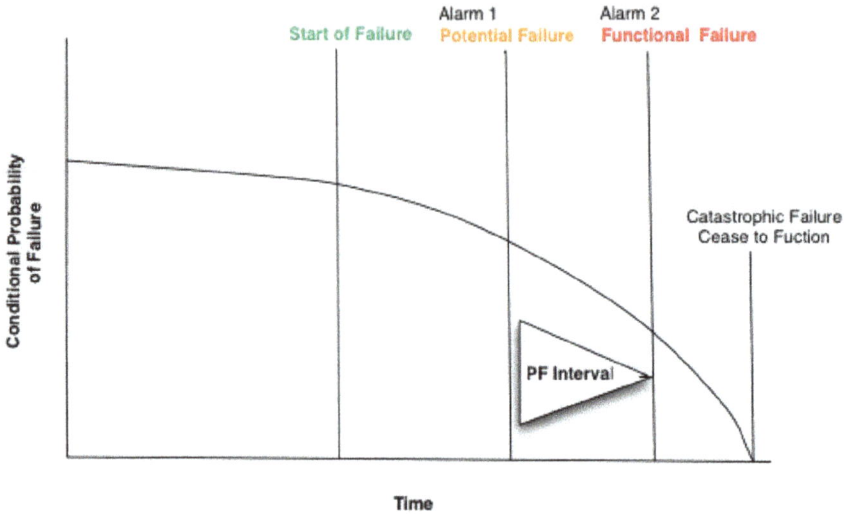

Figure P1 Typical P–F curve. Courtesy: ENG**TECH** Industries Inc.

Benefits:

- **Proactive maintenance**: Utilizing the P–F curve enables organizations to transition from reactive to proactive maintenance strategies, enhancing equipment reliability and operational efficiency.
- **Resource optimization**: By accurately predicting failure points, resources can be allocated more effectively, ensuring that maintenance efforts are focused on equipment most in need of attention.

614. Packing Seal

A **packing seal** in lubrication systems is a mechanical seal designed to prevent the leakage of fluids along a rotating or reciprocating shaft. It consists of braided or molded packing materials – such as graphite, polytetrafluoroethylene (PTFE), aramid fibers, or other synthetic fibers – inserted into a stuffing box around the shaft. When compressed by a gland, these packing materials create a seal that controls leakage and retains the lubricating fluid.

Key Components:

- **Packing materials**: The choice of material depends on the application's specific requirements, including temperature, pressure, and chemical compatibility. Common materials include:
  - **Graphite**: Offers high-temperature resistance and is chemically inert.

- ○ **PTFE**: Provides excellent chemical resistance and low friction properties.

- **Aramid fibers**: Known for their strength and resistance to abrasion.
- **Stuffing box**: The housing that contains the packing material and provides the necessary compression to form a seal.
- **Gland**: A component that applies pressure to the packing material, ensuring it maintains contact with the shaft and the stuffing box walls.

Functionality:
Packing seals function by creating a controlled leakage path that allows a small amount of fluid to escape, which lubricates the shaft and packing, preventing overheating and wear. This controlled leakage is essential for the seal's longevity and effective operation.

615. Pail

A lubricant pail refers to a container typically holding 5 gallons (approximately 18.9 liters) of lubricant. This size is commonly used for bulk applications, providing a cost-effective solution for industries requiring larger quantities of lubricant. Pails are typically made from high-density plastics or metal. See Figure P2.

Figure P2 Typical 5 gal. lubricant pails for oils and greases. Courtesy: ENGTECH Industries Inc.

616. PAG

See "Poly Alkaline Glycol Lubricants."

617. PAO

See "Polyalphaolefin Lubricants."

618. Paper Chromatography

Paper chromatography is an analytical technique used to separate and identify components within a mixture, including those in lubricants. In the context of lubrication, this method is employed to detect changes in the lubricant's composition, such as additive depletion, contamination, or degradation.

Application in Lubricant Analysis:
In lubricant analysis, paper chromatography is utilized to monitor the condition of the oil by identifying and quantifying its components. This technique can reveal the presence of contaminants, degradation products, and changes in additive concentrations, providing valuable insights into the lubricant's performance and the health of the machinery it serves.

Procedure:

- **Sample preparation**: A small drop of the lubricant is applied to a strip of chromatography paper.
- **Development**: The paper is placed in a solvent, allowing capillary action to draw the solvent up the paper, carrying the components of the lubricant with it.
- **Visualization**: After the solvent has traveled a sufficient distance, the paper is removed and dried. The separated components are visualized, often by exposing the paper to ultraviolet light or by applying chemical reagents that produce colored spots.
- **Analysis**: The distance each component has traveled is measured, and the retention factor (Rf) is calculated for each component. The Rf value is the ratio of the distance traveled by the component to the distance traveled by the solvent front.

Advantages:

- **Simplicity**: Paper chromatography is straightforward and requires minimal equipment, making it accessible for routine field analysis.

- **Cost-effective**: It is an inexpensive method, suitable for quick assessments without the need for specialized instruments.
- **Sensitivity**: The technique can detect small changes in the lubricant's composition, providing early indicators of potential issues.

619. Paraffinic Crude Oil

Paraffinic crude oil is a type of crude oil characterized by a high concentration of paraffin – saturated straight-chain hydrocarbons with the general formula C_nH_{2n+2}. These paraffins are the primary constituents of substances like paraffin wax, commonly used in products such as candles.

Key Characteristics:

- **Hydrocarbon composition**: Paraffinic crude oil predominantly contains alkanes, which are straight-chain hydrocarbons.
- **Viscosity and density**: This type of crude oil generally exhibits higher viscosity and lower density compared to naphthenic crude oils. These properties influence its flow characteristics and ease of extraction.
- **Color**: Paraffinic crude oil is often lighter in color, ranging from light yellow to golden brown.

Refining and Applications:

The high paraffin content of this crude oil makes it particularly valuable for producing high-quality lubricants, waxes, and other specialty products. The refining process involves separating the paraffins to create products with desired properties, such as low volatility and excellent lubricating capabilities.

Paraffinic crude oils are generally found in the Mid Continental United States, the Middle East, and the British North Sea, and are favored for the manufacture of crankcase oils, Gear oils, bearing oils, turbine oils, and most hydraulic fluids. Paraffinic crudes will constitute up to 60% paraffin and up to 10% wax content giving the oil an excellent Viscosity Index rating in the range of 95 to 105.

620. Particle

In lubrication, the term **particle** refers to **a solid contaminant** – such as a single piece of dirt, dust, metal fragment, soot, or rust – that could be present within a lubricant. These particles can originate from various sources, including environmental exposure, equipment wear, and improper handling during storage and transfer. With a hydrodynamic film of four microns, even minute particles, often smaller than 15 micrometers, can significantly impact machinery performance and longevity.

621. Particlc Contamination

Particle contamination in lubricants refers to the presence of solid particles – such as dirt, metal fragments, soot, and rust – within a lubricant in a quantity level that is unacceptable for use, based on the precision fit of the mechanical machinery being lubricated.

Implications of Particle Contamination:

- **Increased wear**: Particles can act as abrasives, leading to accelerated wear of critical components like bearings, gears, and seals.
- **Lubricant degradation**: Contaminants can alter the chemical composition of the lubricant, reducing its effectiveness and potentially causing chemical reactions that degrade the oil.
- **System malfunction**: Particles can clog filters, restrict fluid flow, and interfere with the operation of valves and pumps, leading to system failures.

Sources of Particle Contamination:

- **Environmental exposure**: Dust, dirt, and other airborne particles can enter the lubrication system through vents, seals, and open ports.
- **Equipment wear**: Normal operation can generate metal particles from the wear of moving parts.
- **Improper handling**: Contamination can occur during the storage, transfer, and application of lubricants if proper cleanliness protocols are not followed.

Monitoring and Control:

- **Particle counting**: Regular monitoring of particle concentration and size distribution in lubricants helps assess contamination levels. Techniques like laser-based particle counting can detect particles down to a few micrometers.
- **Filtration systems**: Implementing effective filtration can remove particles from the lubricant, maintaining its cleanliness and extending equipment life.

- **Contamination control practices**: Proper storage, handling, and application of lubricants are essential to prevent contamination. This includes using clean containers, ensuring equipment is free from debris, and sealing systems to prevent ingress of contaminants.

622. Particle Count

Particle count refers to the measurement of solid contaminants – such as dirt, metal fragments, soot, and rust – present within a lubricant. Monitoring particle count is essential for assessing lubricant cleanliness, as even minute particles can significantly impact machinery performance and longevity.

The **ISO 4406:1999 standard** is commonly used to classify particle contamination in lubricants. This system categorizes particles based on their size and concentration, providing a cleanliness code that reflects the number of particles of specific sizes present in a given volume of fluid. See **Fluid Target Cleanliness Levels** reference.

623. Particle Impingement Erosion

Particle impingement erosion is a wear mechanism where solid particles, such as dust, sand, or other abrasive materials, collide with and remove material from a surface. This process is prevalent in various industries, including aerospace, power generation, and manufacturing, where equipment is exposed to particulate-laden environments.

Mechanism:
When particles are entrained in a fluid flow, they can impact surfaces at high velocities. The energy from these impacts can cause material removal through mechanisms like cutting, indentation, or fatigue, leading to surface degradation over time. The severity of erosion depends on factors such as particle size, velocity, hardness, and the angle at which they strike the surface.

Factors Influencing Erosion:

- **Particle characteristics**: The size, shape, hardness, and concentration of particles significantly affect the erosion rate. For instance, harder particles can cause more damage upon impact.

- **Impact velocity**: Higher particle velocities increase the kinetic energy upon impact, leading to more severe erosion.

- **Impact angle**: The angle at which particles strike the surface influences the erosion pattern. For ductile materials, maximum wear rates often occur at impact angles around 30°, while for brittle materials, the maximum wear rate is typically at normal (90°) impact angles.
- **Material properties**: The hardness, ductility, and toughness of the target material determine its resistance to erosion.

Implications:
Particle impingement erosion can lead to significant damage to components such as turbine blades, piping systems, and heat exchangers. This damage can result in reduced efficiency, increased maintenance costs, and potential system failures.

Mitigation Strategies:
- **Material selection**: Using materials with higher hardness and toughness can enhance resistance to erosion.
- **Protective coatings**: Applying coatings that are resistant to erosion can prolong the lifespan of components.
- **Design modifications**: Altering component designs to minimize areas where particles can impinge at high velocities can reduce erosion.
- **Operational adjustments**: Controlling flow velocities and particle concentrations can help mitigate erosion rates.

624. Particle Ingression

Solid contaminants that can include metal swarf, dirt, silicon, silt, and **particle ingress** refer to the **entry of solid contaminants** – such as dust, dirt, metal fragments, and other particulates – into lubrication systems. This contamination can originate from various sources, including environmental exposure, equipment wear, and improper handling during storage and transfer. Even minute particles, often smaller than 15 micrometers, can significantly impact machinery performance and longevity.

Implications of Particle Ingress:
- **Increased wear**: Particles can act as abrasives, leading to accelerated wear of critical components like bearings, gears, and seals.

- **Lubricant degradation**: Contaminants can alter the chemical composition of the lubricant, reducing its effectiveness and potentially causing chemical reactions that degrade the oil.
- **System malfunction**: Particles can clog filters, restrict fluid flow, and interfere with the operation of valves and pumps, leading to system failures.

Sources of Particle Ingress:

- **Environmental exposure**: Dust, dirt, and other airborne particles can enter the lubrication system through vents, seals, or open ports.
- **Equipment wear**: Normal operation can generate metal particles from the wear of moving parts.
- **Improper handling**: Contamination can occur during the storage, transfer, and application of lubricants if proper cleanliness protocols are not followed.

Figure P3 Nonfunctioning hydraulic pump unit located in a foundry that should have been protected from falling debris. Courtesy: ENG**TECH** Industries Inc.

Mitigation Strategies:

- **Effective sealing**: Implementing high-quality seals and gaskets can prevent the ingress of contaminants. In dirty working environments such as casting and foundry operations, lubrication pumping stations should be protected from process contaminants such as sand and silica through the use of deflectors or pacing the lubricator in a dustproof cabinet.
- **Proper storage and handling**: Ensuring that lubricants are stored in clean, sealed containers and that equipment is maintained to prevent wear can reduce contamination risks.
- **Regular monitoring**: Conducting routine inspections and analyses of lubricant samples can help detect contamination early, allowing for timely corrective actions.
- **Cleanliness**: Keeping machine surfaces clean can significantly reduce the risk of particle ingression. See Figure P3, which shows a nonrunning hydraulic pump system due to contamination clogging. Once cleaned and restored, the pump unit was redesigned to fit into a protective enclosure that was vacuumed on a regular basis.

625. Particulate Matter

See "Particle."

626. Particulate Size

Particulate size refers to the dimensions of solid particles suspended in a medium, such as air, water, or lubricants. Understanding particulate size is crucial to assessing lubricant cleanliness levels and its ability to lubricate without causing surface damage.

Particle Measurement Techniques:

Several methods are employed to measure particle size, each suited to specific applications and particle size ranges:

- **Laser diffraction (LD)**: This technique analyzes the pattern of light scattered by particles when illuminated by a laser. The scattering pattern is used to determine particle size distribution. It is commonly used for particles ranging from approximately 0.1 micrometers to several millimeters.

- **Dynamic light scattering (DLS)**: DLS measures the fluctuations in scattered light caused by the Brownian motion of particles suspended in a fluid. By analyzing these fluctuations, the hydrodynamic size of particles can be determined. This method is ideal for nanoparticles and submicron particles, typically in the range of 1 nanometer to 1 micrometer.
- **Sieving**: In this method, a sample is passed through a series of mesh screens with progressively smaller openings. The amount of material retained on each sieve is weighed to determine particle size distribution. It is suitable for larger particles, typically above 40 micrometers.
- **Microscopy**: Optical or electron microscopy involves visualizing particles under magnification to measure their size directly. This method is effective for particles larger than 1 micrometer and provides detailed images, allowing for the assessment of particle shape and morphology.

627. PAS 55 Standard

PAS 55 is a **Publicly Available Specification** developed by the British Standards Institution (BSI) in 2004, to provide a framework for the optimized management of physical assets. It offers a comprehensive approach to physical asset management, covering the entire lifecycle of assets from acquisition to disposal.

The specification is divided into two parts:

- **Part 1**: Specification for the optimized management of physical infrastructure assets.
- **Part 2**: Guidelines for the application of Part 1.

PAS 55 outlines a practical **28-point checklist** of requirements for effective asset management, emphasizing the integration of business processes and practices across multiple departments.

In 2014, PAS 55 served as the foundation for the development of the ISO 55000 series of international standards on asset management that is designed to take a more holistic and value-added approach to asset management, over a family of three standards that include:

- **ISO 55000**: Overview, principles, and terminology.
- **ISO 55001**: Management systems – Requirements.
- **ISO 55002**: Guidelines for the application of ISO 55001.

These standards represent a global consensus on asset management practices, aiming to increase the value generated by organizations through effective asset management.

PAS 55 has been widely adopted across various industries, including utilities, transport, manufacturing, and natural resources, as a benchmark for best practices in asset management.

628. Pascal

In lubrication systems, **Pascal (Pa)** is the SI unit of pressure, defined as one newton per square meter (N/m^2). It quantifies the force exerted per unit area within the system. Given that 1 Pa is a relatively small unit, practical applications often use its multiples:

- **Kilopascal (kPa)**: 1000 Pa
- **Megapascal (MPa)**: 1,000,000 Pa

These units are commonly used to measure pressures in lubrication systems, such as those in hydraulic machinery, where pressures can range from several hundred to thousands of kilopascals.

Understanding and monitoring pressure in Pascals is essential for ensuring the proper function of lubrication systems, as it influences the formation of lubricant films, the prevention of metal-to-metal contact, and the overall efficiency of machinery.

Formula for converting Pa to imperial psi:

One Pascal (Pa) = 145.04 × 10(-6)psi

629. Pascal's Law

Pascal's law, also known as Pascal's Principle, is a principle established by French mathematician Blaise Pascal in the 17th century that states "Any change in pressure applied to an enclosed, incompressible fluid is transmitted undiminished throughout the fluid and to the walls of its container."

Mathematical Expression:
The relationship can be expressed as:

$$\Delta p = \rho g \cdot \Delta h.$$

where:

- Delta p is the change in pressure (in Pascals, Pa)

- ρ (rho) is the fluid density (in kilograms per cubic meter, kg/m^3)
- g is the acceleration due to gravity (in meters per second squared, m/s^2)
- Delta h is the change in height (in meters, m)

This equation illustrates how pressure changes with elevation in a fluid column.

Applications:
Pascal's Law is fundamental in various hydraulic systems that leverage Pascal's Law to efficiently transmit and amplify forces within hydraulic systems, these applications can include:

- **Hydraulic presses**: Devices that amplify force using a confined fluid, enabling the lifting of heavy objects with minimal input force.
- **Hydraulic brakes**: Systems in vehicles where a small force applied to a brake pedal is transmitted through brake fluid to apply braking force at the wheels.
- **Hydraulic jacks**: Tools that lift heavy loads by applying a small force to a piston, which is transmitted through fluid to lift the load.

630. Paste Lubricant

Paste lubricants are thick, semisolid substances designed to reduce friction and wear between moving parts. They are particularly effective in applications involving high pressures, slow-moving components, or extreme temperatures.

Key Characteristics of Paste Lubricants:
- **High solid content**: Paste lubricants contain a significant proportion of solid lubricants, such as molybdenum disulfide (MoS_2), graphite, or PTFE. These solids form a protective layer on metal surfaces, reducing direct metal-to-metal contact and minimizing wear.
- **Thick consistency**: Their semisolid nature allows for easy application to specific areas, ensuring that the lubricant stays in place during operation.
- **Temperature resistance**: Many paste lubricants are formulated to withstand extreme temperatures, making them suitable for high-temperature applications.

Common Applications:

- **Assembly processes**: Applied to threaded connections, bolts, and nuts to prevent seizing, galling, and corrosion during assembly and disassembly.
- **High-pressure equipment**: Used in situations where conventional oils or greases may not provide adequate protection under extreme pressure conditions.
- **Extreme temperature environments**: Suitable for applications exposed to high or low temperatures where other lubricants may fail.

When selecting a paste lubricant, it is essential to consider factors such as operating temperature, load conditions, and compatibility with materials to ensure optimal performance and longevity of the equipment.

631. Patch Test

An **oil patch test kit** is a portable tool designed for on-site analysis of oil cleanliness, enabling maintenance professionals to assess particulate contamination levels in lubricants and hydraulic fluids. (See Figure P4) By conducting these tests, you can identify contamination sources, monitor filtration system performance, and make informed decisions to enhance equipment reliability.

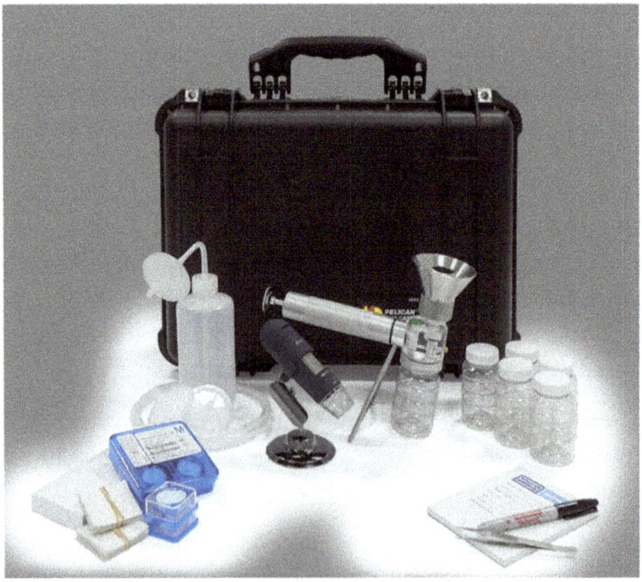

Figure P4 Typical oil patch field test kit. Source: Donaldson Hy-Pro.

Key Features of Oil Patch Test Kits:

- **Rapid analysis**: Obtain results in as little as ten minutes, facilitating prompt maintenance decisions.
- **Ease of use**: Designed for straightforward operation, allowing personnel to perform tests without extensive training.
- **Comprehensive components**: Typically include necessary tools such as a carry case, minimizer oil sample pump, sample bottles, microscope, patch filters, solvents, and reference materials for thorough analysis.

632. Pathway Flush

In lubrication systems, **pathway flushing** refers to the process of circulating a cleaning fluid through the system's pathways – such as pipes, valves, and filter housings – to remove contaminants like debris, sludge, and varnish. This procedure is essential for maintaining optimal system performance and preventing damage to sensitive components.

Key Aspects of Pathway Flushing:
Purpose: The primary goal is to eliminate contaminants that can impair fluid flow, cause wear, or lead to system failures.

Methods:

- **High-velocity oil flushing**: Involves circulating oil at elevated speeds to dislodge and remove contaminants. Achieving turbulent flow, typically with a Reynolds number of 4000 or higher, is crucial for effective flushing.
- **Chemical cleaning**: Utilizes chemical agents to dissolve and remove deposits, often used when mechanical methods are insufficient.

Applications:

- **Precommissioning**: Performed before starting new equipment to ensure cleanliness and prevent initial wear.
- **Maintenance**: Conducted during routine maintenance or after repairs to restore system cleanliness.
- **Varnish mitigation**: Used to remove varnish and sludge buildup, which can impair system performance.

Considerations:

- **Flow velocity**: Achieving the appropriate flow rate is essential for effective flushing. For instance, targeting a minimum of two to three times the normal system flow rate can enhance the flushing process.
- **Fluid compatibility**: Selecting a flushing fluid compatible with the system's materials and existing lubricants is vital to prevent damage.
- **Monitoring**: Regular inspection and analysis of the flushed fluid help assess the effectiveness of the process and ensure system cleanliness.

Implementing pathway flushing is a proactive measure to maintain the integrity and efficiency of lubrication and hydraulic systems, thereby extending equipment lifespan and reducing the risk of unexpected failures.

633. PBM

See "Perimeter Based Maintenance."

634. PDI

See "Positive Displacement Injector."

635. PdM

See "Predictive Maintenance."

636. PEMAC

Plant Engineering Maintenance Association of Canada.

637. Penetration (Grease)

In lubrication, penetration refers to the depth, measured in tenths of a millimeter, that a standard cone penetrates a sample of lubricating grease under specified conditions of cone weight, time, and temperature. This measurement is crucial for determining the grease's consistency, which influences its performance in various applications. See "Cone Penetration Test."

638. Penetrometer

A grease **penetrometer** is an instrument used to assess the consistency or hardness of lubricating greases by measuring the depth to which a standard cone penetrates the grease under specified conditions. This measurement is crucial for determining the grease's suitability for various applications, as it indicates its resistance to deformation.

How It Works:

1. **Sample preparation**: A grease sample is placed in a standardized container at a controlled temperature, typically 25 °C (77 °F).
2. **Penetration measurement**: A standard cone, with a specific weight, is released from a fixed height and allowed to drop freely into the grease for a set duration, usually 5 seconds.
3. **Recording penetration depth**: The depth to which the cone sinks into the grease is measured in tenths of a millimeter.

The depth in millimeters is then used to determine the grease rating number. For more information see "Cone Penetration Test" and Figure C13.

639. Perfluorinated Polyether (PFPEs)

Perfluorinated polyethers (PFPEs) are a class of synthetic lubricants renowned for their exceptional thermal stability, chemical inertness, and low friction properties. These lubricants are particularly valuable in applications where conventional lubricants may fail due to extreme conditions.

Key Properties of PFPEs:

- **Thermal stability**: PFPEs maintain their performance across a wide temperature range, typically from –70 °C to over 300 °C, making them suitable for both cryogenic and high-temperature environments.
- **Chemical inertness**: They are highly resistant to acids, bases, and most organic solvents, ensuring longevity and reliability in aggressive chemical environments.
- **Low friction**: PFPEs exhibit low coefficients of friction, which reduces wear and extends the lifespan of mechanical components.

Applications:

PFPE-based lubricants are utilized in various industries, including:

- **Aerospace**: Used in aircraft fuel systems and mechanical components exposed to high vacuum and cryogenic temperatures.
- **Semiconductor manufacturing**: Employed as vacuum greases and in plasma etching equipment, where chemical inertness and low outgassing are critical.

- **Electronics**: Applied in computer disk drives and scanning electron microscopes to provide lubrication in high-vacuum conditions.

640. Perimeter-based Maintenance (PBM)

Perimeter-based maintenance (PBM) is a proactive maintenance strategy that focuses on designing assets to facilitate basic preventive and diagnostic maintenance tasks without necessitating equipment shutdowns. This approach enhances operational efficiency and reduces downtime by enabling maintenance activities while equipment remains in operation.

Key Components of PBM Design:

- **Accessible maintenance points**: Incorporating easily accessible points for tasks such as fluid level checks, pressure monitoring, and temperature assessments allows operators to perform routine inspections without halting operations.
- **Automated systems**: Implementing automated lubrication delivery systems positioned outside the "asset's working envelope" that can be monitored and replenished during operation ensures continuous lubrication, thereby extending component lifespan and reducing manual intervention.
- **Integrated monitoring tools**: Utilizing integrated monitoring tools, such as Go/No-Go gages (see **Gages**: Figure G1) and automated filtration systems, enables real-time assessment of equipment condition, facilitating timely maintenance actions.

Benefits of PBM:

- **Reduced downtime**: By allowing maintenance tasks to be performed without shutting down equipment, PBM minimizes operational interruptions, leading to increased productivity.
- **Enhanced equipment reliability**: Regular, nondisruptive maintenance activities contribute to the early detection of potential issues, thereby improving overall equipment reliability.
- **Cost efficiency**: Proactive maintenance reduces the likelihood of unexpected failures and the associated repair costs, leading to more efficient resource utilization.

Implementing PBM requires thoughtful design considerations during the asset planning phase to ensure that maintenance activities can be conducted effectively without compromising operational continuity. This strategy aligns with the principles of Maintenance Prevention, aiming to create equipment that operates efficiently with minimal maintenance requirements.

641. Permeability

In lubrication filtration, **permeability** refers to the capacity of a filter medium to allow fluid to pass through it under a given pressure gradient. This property is crucial for determining the efficiency and performance of filtration systems, as it influences the flow rate and pressure drop across the filter.

Relationship Between Flow Per Unit Area and Differential Pressure: The relationship between the flow per unit area (flux) and the differential pressure across a filter medium is governed by Darcy's Law, which states:

$$J = \frac{\Delta P}{\mu \cdot R}.$$

where:

- *J* is the flux (flow per unit area)
- *Delta P* is the differential pressure across the filter medium
- *μ (mu)* is the dynamic viscosity of the fluid
- *R* is the resistance of the filter medium

642. PFPE

See "Perflourinated Polyether."

643. pH

An abbreviation for **potential hydrogen**. pH is the measure of acidity or basicity (alkalinity) and signifies the negative logarithm (base 10) of the hydrogen-ion concentration found in a solution or liquid and is expressed in grams equivalent per liter of solution.

Water at a temperature of **77 °F (22 °C)** is said to be neutral with a **pH level of 7**; solutions having a pH level less than 7 are defined as acidic, (see:

Acid) whereas solutions greater than 7 are defined as basic, or alkaline. See **Alkaline**. Figure P5 shows the pH scale from 0 to 14.

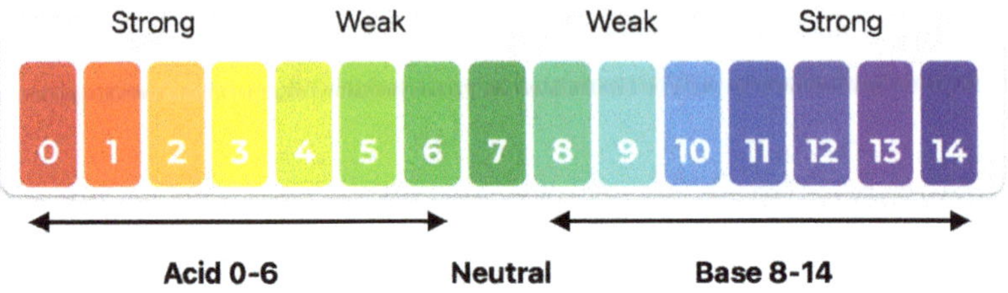

Figure P5 pH scale.

644. Phosphate Ester

Phosphate esters are a class of synthetic compounds widely utilized in lubrication due to their exceptional fire resistance, thermal stability, and antiwear properties. These compounds are synthesized by reacting phosphoric acid with alcohols, resulting in esters that serve as effective lubricants in various industrial applications.

Key Characteristics of Phosphate Esters in Lubrication:

- **Fire resistance**: Phosphate esters are among the most fire-resistant nonaqueous synthetic base stocks available. Their high ignition temperatures and low vapor pressures make them difficult to ignite, providing enhanced safety in high-temperature environments.
- **Thermal stability**: These lubricants maintain their performance across a wide temperature range, making them suitable for both cryogenic and high-temperature applications.
- **Antiwear properties**: Phosphate esters exhibit excellent antiwear characteristics, reducing friction and extending the lifespan of mechanical components.
- **Hydrolytic stability**: Phosphate esters can degrade through hydrolysis, especially in the presence of water, leading to the formation of acids that can damage equipment.
- **Chemical aggressiveness**: These lubricants can be chemically aggressive toward certain seal and coating materials, necessitating careful selection of compatible materials.

- **Cost**: Phosphate esters are generally more expensive than conventional mineral oils, which can impact operational costs.

Common Applications:

- **Hydraulic systems**: Phosphate esters are commonly used in hydraulic systems where fire resistance is critical, such as in aerospace and military applications.
- **Turbines and compressors**: Due to their thermal stability and antiwear properties, phosphate esters are utilized in turbines and compressors operating under extreme conditions.
- **Metalworking fluids**: Phosphate esters serve as effective emulsifiers and lubricants in metalworking fluids, providing excellent lubrication and corrosion inhibition.

In summary, phosphate esters are valuable lubricants in applications requiring fire resistance, thermal stability, and antiwear properties. However, their use requires careful management to address challenges related to hydrolytic stability, chemical compatibility, and cost.

645. Pillow Block

A **pillow block**, also known as a plummer block is a type of bearing mount designed to support a rotating shaft. It consists of a housing that contains a bearing, allowing the shaft to rotate smoothly while maintaining proper alignment. The housing is typically bolted to a foundation, providing stability and support. See Figure P6.

Key Components:

- **Housing**: The outer casing that encases the bearing and provides support. It is commonly made from materials such as cast iron, cast steel, or thermoplastics.
- **Bearing**: The internal component that facilitates the rotation of the shaft. Bearings can be of various types, including ball, cylindrical roller, spherical roller, tapered roller, or bushings.
- **Seals**: Protective elements that prevent contaminants from entering the bearing and retain lubrication. Seals are essential for maintaining bearing performance and longevity.

Types of Pillow Block Bearings:

- **Split housing**: Consists of two pieces that can be separated, allowing for easier installation and maintenance.
- **Solid housing**: Made from a single piece, offering increased strength and durability.

Figure P6 Typical pillow block – solid housing. Courtesy: ENG**TECH** Industries Inc.

646. Pilot Tube Level Gauge

A **pilot tube level gauge** is a specialized instrument designed to measure the liquid level within a tank or reservoir. It combines a sight glass with a pilot tube to provide both visual indication and the capability to extract representative fluid samples. This dual functionality is particularly beneficial in applications where monitoring fluid levels and ensuring sample integrity are critical – Figure P7.

Key Features:

- **Sight glass**: The transparent tube allows operators to visually inspect the liquid level, facilitating quick assessments without the need for additional tools.

- **Pilot tube**: Integrated into the sight glass, the pilot tube enables the extraction of fluid samples from specific depths within the tank. This ensures that samples are representative of the fluid's condition at various levels, which is essential for accurate analysis.
- **Protective cap**: Many pilot tube level gauges are equipped with a protective cap to prevent contamination and maintain sample integrity. Some models feature a vented cap to allow for pressure equalization, while others may include a filter cap to minimize contamination.

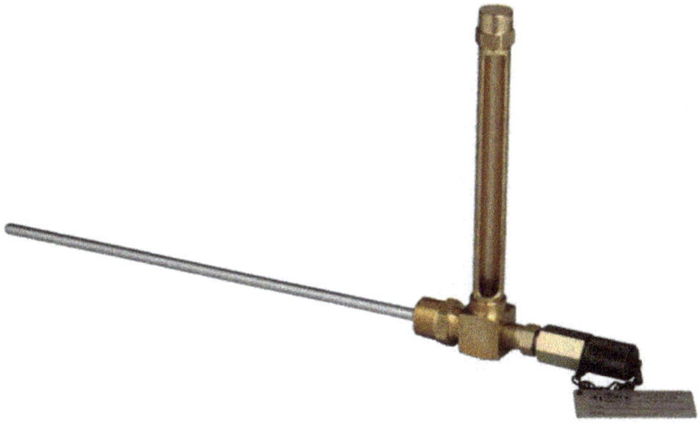

Figure P7 Trico pilot tube level gauge with sample port.

Applications:
Pilot tube level gauges are commonly used in industries such as:

- **Lubrication systems**: They allow for the monitoring of lubricant levels and the collection of oil samples for analysis, aiding in the maintenance of machinery and equipment.
- **Hydraulic systems**: These gauges help in assessing fluid levels and quality, ensuring the proper functioning of hydraulic machinery.
- **Chemical processing**: In chemical plants, pilot tube level gauges are used to monitor and sample various liquids, assisting in quality control and process optimization.

647. Pitting

Pitting in bearings refers to the formation of small, localized depressions or cavities on the rolling elements or raceways due to surface fatigue. These pits can compromise the bearing's performance, leading to increased vibration, noise, and potential failure if not addressed promptly.

Causes of Pitting:

- **Debris contamination**: Foreign particles, such as metal chips or dirt, can become trapped in the lubricant, leading to abrasive wear and pitting on bearing surfaces.
- **Moisture exposure**: Contact with moisture can cause rust formation, which, over time, leads to pitting on the raceway surfaces.
- **Inadequate lubrication**: Insufficient or improper lubrication can result in metal-to-metal contact, increasing the risk of pitting.
- **Electrical discharges**: In electrical machinery, faulty insulation or grounding can cause electric discharges through the oil film, leading to severe pitting of bearing or shaft surfaces. See Figure P8.

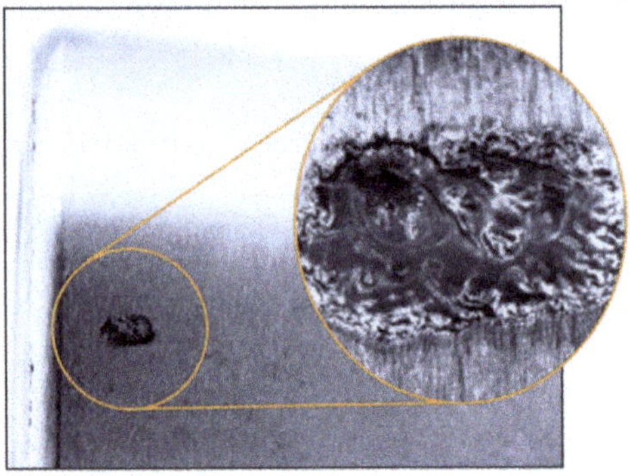

Figure P8 Pitting caused by electrical discharge. Source: Timken Bearing Company.

Preventive Measures:

- **Enhance sealing**: Improving the sealing mechanisms can prevent contaminants from entering the bearing system.
- **Thorough filtration**: Implementing effective filtration systems ensures that lubricants are free from harmful particles.
- **Proper lubricant selection**: Using the appropriate lubricant type and maintaining optimal lubrication levels are crucial for bearing longevity

648. Plain Bearing

A **plain bearing**, also known as a sliding contact bearing or friction bearing, is the simplest bearing design that consists of a bearing surface without any rolling elements. In this design, the shaft or component slides directly over the bearing surface, relying on lubrication to minimize friction and wear. See Figure P9.

Figure P9 Examples of plain bearings.

Key Features:

- **Design simplicity**: Plain bearings are straightforward in design, making them cost-effective and easy to manufacture.
- **High load capacity**: They can handle substantial loads, especially when properly lubricated, due to the large contact area between the bearing surface and the shaft.
- **Compactness**: Their simple structure allows for compact designs, which is advantageous in space-constrained applications.

Common Types:

- **Sleeve bearings**: Also known as bushings, these are cylindrical in shape and fit over the shaft.
- **Flanged bearings**: Similar to sleeve bearings but with a flange at one end to prevent axial movement.
- **Thrust bearings**: Designed to support axial loads, these bearings have a flat bearing surface.

649. Planned Maintenance

Planned maintenance is a proactive approach to equipment upkeep, involving scheduled inspections, servicing, and repairs to ensure optimal performance and prevent unexpected failures. By adhering to a structured maintenance schedule, organizations can enhance equipment reliability, extend asset lifespan, and reduce downtime.

Key Aspects of Planned Maintenance:

- **Scheduled inspections**: Regularly planned checks to monitor equipment condition and identify potential issues before they escalate.
- **Preventive actions**: Implementing measures such as lubrication, cleaning, and part replacements to maintain equipment functionality.
- **Resource allocation**: Ensuring the availability of necessary tools, spare parts, and personnel to perform maintenance tasks efficiently.

Benefits:

Planned maintenance delivers numerous benefits to both the maintenance and production departments. These include:

- **Reduced downtime**: By addressing issues proactively, planned maintenance minimizes unexpected equipment failures, leading to increased operational uptime.
- **Cost savings**: Regular maintenance can prevent major repairs and extend the lifespan of equipment, resulting in long-term financial savings.
- **Enhanced safety**: Maintaining equipment in optimal condition reduces the risk of accidents and ensures a safer working environment.

Five-step Implementation Plan:

1. **Asset assessment**: Identify and catalog all equipment assets requiring maintenance.
2. **Develop maintenance schedule**: Create a timeline for inspections and servicing based on manufacturer recommendations and operational requirements.
3. **Build a job plan**: Once work has been scheduled, a work order defining a step-by-step set of work instructions that include a list of any tools, spare parts, or permissions must be developed

4. **Assign responsibilities**: Designate qualified personnel to perform and oversee maintenance activities.

5. **Monitor and adjust**: Continuously evaluate the effectiveness of the maintenance plan and make adjustments as needed.

650. Plastic Deforming Lubricant

A plastic deforming (PD) lubricant is a highly specialized lubricant used to not only enhance the performance and longevity of mechanical components under high-pressure conditions but also to repair pitted and scarred bearing surfaces.

This microflux technology utilizes active molecules that adapt to varying operating conditions. These molecules form a protective film on metal surfaces before friction occurs. Under severe conditions, such as high pressure, vibration, and shock, the active molecules diffuse into exposed metal surfaces, creating a micro-smoothing layer. This process has the effect of resurfacing the bearing area to reduce wear and enhance the durability of the lubricated components.

Benefits of MFT-PD Technology:

- **Enhanced protection**: The micro-smoothing layer formed by MFT-PD reduces surface fatigue, pitting, spalling, and scuffing, leading to improved component reliability.

- **Extended maintenance intervals**: By reducing wear and tear, MFT-PD technology allows for longer relubrication intervals, thereby decreasing maintenance costs and downtime.

- **Adaptive performance**: The technology adjusts to different operating conditions, providing consistent protection across various environments.

651. Plasticity

In lubrication, **plasticity** refers to the ability of a lubricant to undergo permanent deformation under stress without fracturing. This property is crucial in applications where the lubricant must conform to the surfaces of contacting components, ensuring a consistent and protective film that reduces friction and wear.

Key Aspects of Plasticity in Lubrication:

- **Surface conformability**: A lubricant with adequate plasticity can adapt to the microscopic contours of mating surfaces, maintaining a continuous film even under varying loads.
- **Load-bearing capacity**: Plastic deformation allows the lubricant to support substantial loads without breaking down, which is essential in high-pressure environments.
- **Wear reduction**: By effectively distributing stresses and preventing direct metal-to-metal contact, a plastic lubricant minimizes wear and extends the lifespan of mechanical components.

Applications:

- **Metal forming processes**: In operations like forging and extrusion, lubricants with suitable plasticity facilitate material flow and prevent surface damage.
- **High-pressure machinery**: Equipment such as gears and bearings operating under heavy loads benefit from lubricants that can deform plastically to maintain a protective film.
- **Self-lubricating bearings**: Materials engineered with inherent plasticity can transfer microscopic amounts of material to mating surfaces, creating a lubricating film that reduces friction over time.

652. Pleated Filter

Pleated filters are widely used in lubrication systems to more effectively remove contaminants from oils and lubricants than nonpleated filter media, thereby enhancing equipment performance and extending component lifespan. See Figure P10.

Key Features of Pleated Filters in Lubrication:

- **Increased surface area**: Because the filter media is pleated like a concertina, more media can be fitted into the same space. As a result, the pleated design provides a much larger surface area within a compact space, allowing for higher dirt-holding capacity and improved filtration efficiency over a longer time period of service.
- **High filtration efficiency**: Pleated filters are designed to trap smaller particles, making them ideal for capturing contaminants that could cause wear and damage to engine components.

- **Durability**: Constructed from materials such as paper or glass fiber, pleated filters offer high dirt-holding capacity and long service life, reducing the frequency of replacements.

Figure P10 Typical pleated oil filter design.

653. Plunger Lubricants

In die casting, **plunger lubricants** are essential for reducing wear on the plunger tip and sleeve, thereby extending their lifespan and minimizing downtime for replacements. These lubricants also help prevent molten metal leakage around the plunger tip, which can lead to mechanical issues and additional wear.

Types of Plunger Lubricants:

1. **Dry lubricants**:
   - **Composition**: Typically consists of high-melting-point waxes with extreme pressure additives.
   - **Application**: Suitable for cold chamber aluminum and magnesium die casting.

- **Benefits**: Provide effective lubrication without the need for heating, simplifying the application process.

2. **Oil-based lubricants**:

    - **Composition**: Made from mineral, synthetic, or vegetable oils, with or without graphite.
    - **Application**: Used in various die casting applications, including aluminum, magnesium, and zinc alloys.
    - **Benefits**: Offer high lubricity and can withstand high temperatures, enhancing the performance of the die casting process.

3. **Water-soluble lubricants**:

    - **Composition**: Water-based formulations, often containing additives to improve performance.
    - **Application**: Used in die-casting processes where water-based solutions are preferred.
    - **Benefits**: Provide easy cleanup and are often more environmentally friendly.

Benefits of Using Plunger Lubricants:

- **Reduced wear**: Minimize the wear on plunger tips and sleeves, leading to longer service life and reduced maintenance costs.
- **Improved casting quality**: Help in achieving smoother filling into the die, reducing energy consumption during the shot, and improving the quality of the castings.
- **Enhanced productivity**: By reducing downtime for maintenance and replacements, plunger lubricants contribute to higher productivity in the die-casting process.

654. PM

See "Preventive Maintenance."

655. PM Checklist

A **preventive maintenance (PM) checklist for lubrication** is a structured inspection used to ensure that all lubrication points in machinery and equipment are properly maintained and lubricated. Regular adherence to this checklist helps prevent equipment failures, reduces downtime, and extends the lifespan of components.

Sample Machine Operator or Lubricator Daily/Weekly Lubrication PM Checklist:

- **Activate system**: Manually activate the lubrication system (if a centralized system is in use)and observe the completion of the cycle.
- **Check lubricant levels**: Visually inspect the oil/grease levels in reservoirs. If the lubricant has reached the low-level mark, actuate a lubricant replenishment action.
- **Check for system leaks**: Visually inspect all lines, fittings, hoses, and reservoirs for leaks.
- **Inspect oil condition**: Examine the lubricant for signs of contamination or degradation.
- **Clean filters**: Check all filters, strainers, or screens are operational and identify any requiring clean/change if required.
- **Inspect piping**: Visually check piping and hoses for physical damage or wear.
- **Verify metering devices**: Ensure that metering devices are operating correctly.
- **Test safety devices**: Check the operation of warning and safety devices.

If any issues are found as a result of the regular lubrication system inspection, these are noted on the PM checklist work order and passed back to the planning/scheduling group so the appropriate repair work/action can be planned accordingly and performed in a timely manner.

Regularly following a comprehensive lubrication PM checklist is essential for maintaining equipment reliability and performance. Tailoring the checklist to the specific requirements of your machinery and operating conditions will yield the best results.

656. PNA

In lubrication, **PNA** refers to the **Paraffin**, **Naphthenes**, **and Aromatics** analysis, a method used to determine the composition of lubricating oils.

This analysis categorizes the hydrocarbons present in the oil into three main groups:

- **Paraffins (P)**: Also known as alkanes, these are saturated hydrocarbons characterized by straight or branched chains.
- **Naphthenes (N)**: These are saturated hydrocarbons with ring structures, also referred to as cycloalkanes.
- **Aromatics (A)**: These are unsaturated hydrocarbons containing one or more aromatic rings.

Understanding the PNA composition is crucial because it influences the physical properties and performance characteristics of the lubricant, such as viscosity, oxidation stability, and low-temperature fluidity.

Importance of PNA Analysis:

- **Predicting lubricant behavior**: By analyzing the PNA composition, manufacturers can predict how a lubricant will perform under various operating conditions.
- **Quality control**: Ensuring the desired PNA composition helps maintain consistent quality and performance of lubricating oils.
- **Formulation optimization**: Understanding the PNA composition allows for the optimization of lubricant formulations to meet specific performance requirements.

657. Pneumatic Lubrication

In pneumatic systems, **pneumatic lubrication** is essential for maintaining the performance and longevity of pneumatic components that include actuators, valves, and tools. Proper lubrication reduces friction, minimizes wear, and prevents corrosion, ensuring efficient operation.

Airline Lubricators:

- These devices introduce a fine mist of oil into the compressed air stream, providing continuous lubrication to downstream components. They are typically installed in the airline and adjusted to deliver the appropriate amount of oil.

Considerations for Pneumatic Lubrication:

- **Component compatibility**: Not all pneumatic components require lubrication. Some are designed to operate without it, and introducing oil can lead to issues like contamination or malfunction. It is crucial to consult manufacturer guidelines to determine lubrication needs.

- **Environmental factors**: In environments with high temperatures, moisture, or chemical exposure, lubrication may be compromised. Regular monitoring and maintenance are necessary to ensure effective lubrication under such conditions.
- **Health and safety**: Oil mist in the atmosphere can pose health hazards. Implementing exhaust cleaners can help mitigate these risks by removing oil mist at microscopic levels.

658. Poise

In lubrication terms, **Poise** is a unit of dynamic viscosity, a fundamental property that quantifies a fluid's internal resistance to flow. Dynamic viscosity is crucial in lubrication because it determines how well a lubricant can form a protective film between moving parts, thereby reducing friction and wear.

Understanding Poise and Viscosity:

- **Poise (P)**: Named after French physicist Jean Léonard Marie Poiseuille, the poise is the SI unit of dynamic viscosity. One poise is defined as one dyne-second per square centimeter (dyne·s/cm²). In practical terms, 1 poise equals 0.1 pascal-seconds **(Pa·s)**.
- **Centipoise (cP)**: Since the poise is relatively large, the centipoise is commonly used, where 1 cP equals 0.001 Pa·s. Water at 20 °C has a viscosity of approximately 1 cP.

659. Poiseuille's law

Poiseuille's Law (1844) describes the laminar flow of an incompressible and Newtonian fluid through a long, cylindrical pipe. It states that the volumetric flow rate (*Q*) is directly proportional to the fourth power of the pipe's radius (*r*), the pressure difference (*Delta P*) between the two ends, and inversely proportional to the fluid's dynamic viscosity (*n*) and the pipe's length (*L*).

The equation is expressed as:

$$Q = \frac{\pi r^4 \Delta P}{8\eta L}.$$

where:

- Q = Volumetric flow rate

- r = Radius of the pipe
- *Delta P* = Pressure difference between the two ends of the pipe
- n = Dynamic viscosity of the fluid
- L = Length of the pipe

This law is fundamental in fluid dynamics and is used to design piping systems for optimal fluid transport. See Poise.

660. Polar Additive

In lubrication, **polar additives** are chemical compounds that possess both hydrophilic (water-attracting) and hydrophobic (water-repelling) properties. These additives are designed to interact with metal surfaces, forming protective films that reduce friction and wear under boundary lubrication conditions.

Key Characteristics of Polar Additives:

- **Surface interaction**: The polar nature of these additives allows them to adsorb onto metal surfaces, creating a protective layer that minimizes direct metal-to-metal contact. This interaction is particularly beneficial in boundary lubrication regimes where the lubricant film is thin or absent.
- **Friction reduction**: By forming a tenacious film on metal surfaces, polar additives effectively reduce friction, leading to decreased wear and improved efficiency of mechanical components.
- **Corrosion protection**: Some polar additives also offer corrosion inhibition properties, protecting metal surfaces from oxidative damage and extending the lifespan of equipment.

Common Types of Polar Additives:

- **Antiwear (AW) agents**: These additives chemically bond to metal surfaces, forming a protective layer that reduces wear during boundary lubrication. A well-known example is zinc dialkyldithiophosphate (ZDDP).
- **Extreme pressure (EP) additives**: More aggressive than AW agents, EP additives react under high pressure and temperature to form sacrificial films that prevent welding and seizure of metal surfaces. They often contain sulfur and phosphorus compounds.

- **Friction modifiers**: These additives reduce friction by adsorbing onto metal surfaces and forming smooth films, thereby enhancing fuel efficiency and performance.

661. Polar Compound

See "Polar Additive."

662. Polishing

Polishing refers to the process of refining and enhancing the surface finish of components to reduce friction, minimize wear, and improve overall performance. This is achieved by using lubricants that facilitate the polishing of surfaces during operations such as grinding, honing, or lapping.

Key Aspects of Polishing in Lubrication:

- **Surface refinement**: Polishing lubricants aid in smoothing the surfaces of components, leading to a mirror-like finish that reduces friction and wear. This is particularly beneficial in applications where low friction is essential for optimal performance.
- **Application in metallography**: In metallographic polishing, lubricants are essential as they reduce friction and wear between the sample and the polishing paper, thus avoiding damage and ensuring a high-quality finish.
- **Abrasive processes**: Polishing is often part of abrasive machining processes, where the material is removed from a workpiece using a multitude of small abrasive particles. Lubricants in these processes help in cooling, reducing friction, and flushing away debris, leading to a finer surface finish.

In summary, polishing in lubrication involves using specialized lubricants to refine the surfaces of components, thereby enhancing their performance and longevity. Selecting the appropriate lubricant is essential for achieving the desired surface finish and operational efficiency.

663. Polyalkylene Glycol (PAG)

Polyalkylene glycol (PAG) is a class of synthetic lubricants known for their exceptional performance across various applications. They are synthesized by polymerizing alkylene oxides such as ethylene oxide, propylene oxide, and butylene oxide. The specific properties of PAGs depend on the ratio and sequence of these oxides, resulting in variations like water-soluble, water-insoluble, and oil-soluble PAGs.

Key Properties of PAGs:

- **High lubricity**: PAGs offer excellent lubrication, reducing friction and wear in mechanical systems.
- **Temperature stability**: They maintain performance over a wide temperature range, making them suitable for both high and low-temperature environments.
- **Water solubility**: Certain PAGs are water-soluble, facilitating easy cleanup and reducing environmental impact.
- **Biodegradability**: Many PAGs are biodegradable, aligning with environmental sustainability goals.

Common Applications:

- **Hydraulic fluids**: PAGs are used in fire-resistant hydraulic fluids, offering high flash points and fire points.
- **Metalworking fluids**: They serve as quenchants and coolants in metal heat-treating processes.
- **Compressor lubricants**: PAGs are utilized in rotary screw air compressors, especially those employing low global warming potential refrigerants.
- **Food-grade lubricants**: Certain PAGs are approved for use in food processing equipment due to their nontoxic nature.

Considerations:

- **Seal compatibility**: PAGs may not be compatible with all seal materials, necessitating careful selection to prevent leaks.
- **Incompatibility with mineral oils**: Water-soluble PAGs are incompatible with petroleum oils, requiring thorough cleaning when transitioning between different lubricants.

664. Polyalphaolifin (PAO)

Polyalphaolefin (PAO) is a widely used synthetic base oil in the lubrication industry, known for its superior performance compared to traditional mineral oils. Manufactured through the polymerization of alpha-olefins, such as 1-decene, PAOs offer several advantages:

Key Properties of PAOs:

- **High viscosity index**: PAOs maintain consistent viscosity across a broad temperature range, ensuring reliable lubrication in both high and low-temperature environments.
- **Thermal and oxidative stability**: They exhibit excellent resistance to thermal degradation and oxidation, leading to longer oil life and reduced formation of harmful deposits.
- **Low volatility**: PAOs have lower evaporation rates at elevated temperatures, contributing to reduced oil consumption and emissions.
- **Compatibility**: They are compatible with mineral oils, allowing for easy blending and transitioning between different oil types.

Common Applications:

- **Automotive engine oils**: PAOs are utilized in high-performance engine oils, providing enhanced protection and efficiency.
- **Industrial lubricants**: They are used in gear oils, bearing oils, and compressor oils, offering reliable performance in various industrial machinery.
- **High-temperature greases**: PAOs serve as base oils in greases designed for high-temperature applications, ensuring stability and performance under heat stress.

Considerations:

While PAOs offer numerous benefits, they also have some limitations:

- **Seal compatibility**: PAOs can cause certain seals to shrink, potentially leading to leaks.
- **Additive solubility**: They may have limited solubility for some additives, necessitating careful formulation to achieve desired performance characteristics.

665. Polymer

A **polymer** is a **large molecule, or macromolecule**, composed of repeating structural units called **monomers,** which are connected by covalent bonds. The term "polymer" originates from the Greek words "poly," meaning "many," and "meros," meaning "part," reflecting the structure of these substances as chains of many parts. These polymer chains play a crucial role in enhancing the performance and functionality of lubricants across various applications. See Figure P11. Incorporating polymer additives into lubricants significantly improves such properties as viscosity, wear resistance, and stability.

Polymer Additives in Lubricants:

- **Viscosity modifiers**: Polymers are commonly used to adjust the viscosity of lubricants, ensuring optimal performance under different operating conditions. These additives help maintain a consistent lubricant film, reducing friction and wear.
- **Improved water resistance**: The addition of polymer additives can enhance the water resistance of greases, making them more effective in environments where water exposure is expected. This improvement helps prevent washout and maintains the grease's protective properties.
- **Enhanced wear resistance**: Polymers can form protective films on metal surfaces, reducing friction and minimizing wear. This is particularly beneficial in metal-forming processes, where maintaining tool integrity is essential.

Applications of Polymer-based Lubricants:

- **Metal forming**: Polymer-based lubricants are developed to improve the metal forming process, offering better lubrication and reducing friction between tools and workpieces.
- **High-performance greases**: Incorporating polymers into grease formulations can enhance properties such as water resistance and wear protection, making them suitable for demanding applications.

In summary, polymers are integral to the development of advanced lubricants, providing enhanced performance characteristics that meet the evolving demands of various industries.

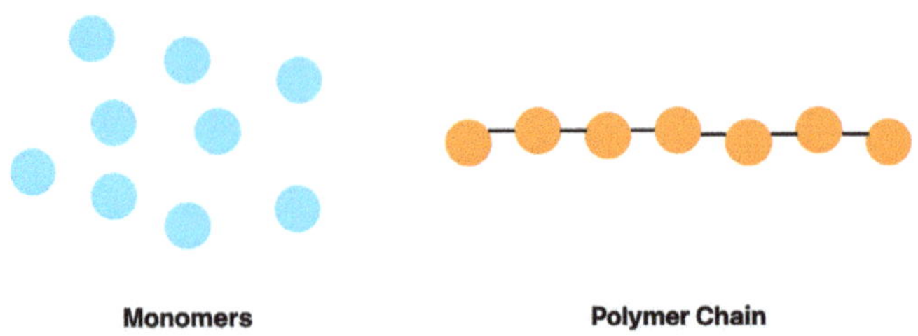

Figure P11 Monomers versus polymer chains.

666. Polymerization

Polymerization is a chemical process in which small molecules, known as monomers, chemically bond to form large, chain-like or networked molecules called polymers. This process is fundamental in creating a wide array of materials, including plastics, rubbers, fibers, and lubricant additives.

Types of Polymerizations:

1. **Addition polymerization (chain-growth polymerization)**:
   - **Mechanism**: Monomers with unsaturated bonds (like alkenes) react to form a polymer chain without the loss of any atoms.
   - **Example**: The polymerization of ethylene to form polyethylene.
2. **Condensation polymerization (step-growth polymerization)**:
   - **Mechanism**: Monomers with two or more reactive groups combine, releasing small molecules such as water or methanol.
   - **Example**: The formation of nylon from hexamethylenediamine and adipic acid

667. Polyphenyl Ester (PPEs)

Polyphenyl ethers (PPEs) are synthetic lubricants renowned for their exceptional thermal stability, low vapor pressure, and resistance to oxidation and radiation. These properties make them ideal for specialized applications where conventional lubricants may not suffice.

Key Properties of Polyphenyl Ether Lubricants:

- **High thermal stability**: PPEs maintain their integrity at elevated temperatures, making them suitable for high-heat environments.
- **Low vapor pressure**: Their low vapor pressure ensures minimal evaporation, which is crucial in vacuum systems.
- **Chemical resistance**: PPEs exhibit resistance to oxidation and radiation, enhancing their longevity and reliability in demanding conditions.

Applications in Lubrication:

1. **Ultra-high vacuum systems**: PPEs are utilized as lubricants in vacuum systems due to their low vapor pressure and thermal stability, ensuring consistent performance in high-vacuum environments.

2. **Electronic connectors**: In electronic connectors, PPE-based lubricants reduce insertion forces, prevent corrosion, and extend the lifespan of connectors by forming a protective film on metal surfaces.
3. **High-temperature applications**: PPEs are employed in applications requiring resistance to high temperatures, such as in aerospace and defense sectors, where conventional lubricants may fail.

Considerations:

- **Compatibility**: While PPEs offer numerous advantages, it is essential to ensure compatibility with materials used in specific applications to prevent degradation or adverse reactions.
- **Cost**: Due to their specialized properties, PPE-based lubricants may be more expensive than conventional lubricants.

668. Polytetrafluoroethylene (PTFE)

Polytetrafluoroethylene (PTFE), commonly known by the brand name Teflon, is a synthetic fluoropolymer renowned for its exceptional chemical resistance, high thermal stability, low friction properties, and resistance to acids, alkalis, and oxidizing agents. These characteristics make PTFE a valuable material in various applications, including lubrication.

Properties of PTFE:

- **Chemical resistance**: PTFE is inert to most chemicals, making it ideal for use in aggressive chemical environments.
- **Thermal stability**: It can withstand temperatures up to approximately 260 °C (500 °F) without significant degradation.
- **Low friction**: PTFE has one of the lowest coefficients of friction among solid materials, reducing wear and energy consumption in mechanical systems.

Applications in Lubrication:

1. **Industrial lubricants and release agents**: In industrial settings, PTFE-based lubricants are employed to reduce friction and wear in machinery, especially in high-temperature and chemically aggressive environments. They also make excellent mold-release agents.

2. **Seals and gaskets**: Due to their chemical resistance and low friction, PTFE is used in seals and gaskets to prevent leaks in various systems.

669. Polyurea

Polyurea is a nonsoap thickener used in lubricating greases, offering several advantages over traditional soap-based thickeners. These greases are synthesized by reacting isocyanates with amines, resulting in a metal-free thickener.

Advantages of Polyurea Greases:

- **High-temperature performance**: Polyurea greases maintain their stability and performance at elevated temperatures, making them suitable for applications where high thermal resistance is essential.
- **Water resistance**: These greases exhibit excellent resistance to water washout and spray-off, ensuring reliable lubrication in wet environments.
- **Mechanical stability**: Polyurea greases offer superior mechanical stability, maintaining their consistency and performance under high shear conditions.
- **Long service life**: Due to their inherent antioxidant properties, polyurea greases can extend the service life of bearings and other components.

Common Applications:

- **Electric motors**: Polyurea greases are often used in electric motor bearings, especially in sealed-for-life applications, due to their high-temperature stability and long service life.
- **Automotive bearings**: They are suitable for automotive wheel bearings and chassis components, providing reliable lubrication under varying operating conditions.
- **Industrial equipment**: Polyurea greases are utilized in steel mill continuous casters and other heavy-duty industrial machinery, offering excellent pumpability and protection against contamination.

Considerations:

- **Compatibility**: Polyurea greases are generally incompatible with other thickener types, such as lithium-based greases. Mixing different types can lead to grease hardening or softening, potentially causing lubrication failures.

- **Cost**: While offering superior performance, polyurea greases may be more expensive than traditional greases, which could influence their selection based on application requirements and budget constraints.

670. Pore Blockage Particle Counting

Pore blockage particle counting is a technique used to assess the concentration and size distribution of particles in lubricants, particularly in-service machinery oils. This method involves passing a lubricant sample through a calibrated mesh screen and monitoring the flow rate. As particles accumulate and block the mesh pores, the flow rate decreases, and this change is used to estimate particle concentration and size distribution.

Advantages of Pore Blockage Particle Counting:

- **On-site analysis**: This method allows for rapid, on-site analysis of lubricant samples, facilitating timely maintenance decisions.
- **Cost-effectiveness**: Pore blockage particle counters are generally more affordable compared to optical particle counters, making them accessible for routine monitoring.

Considerations:

- **Sensitivity**: While effective for larger particles, pore blockage particle counting may be less sensitive to very fine particles compared to optical methods.
- **Calibration**: Accurate calibration is essential to ensure reliable results, as variations in mesh size and pore structure can affect measurements.

671. Porosity

Porosity refers to the presence of tiny, interconnected voids or pores within a material. In lubrication, porosity is a critical factor influencing the storage and release of lubricants, thereby affecting the performance and longevity of mechanical components.

Porous Materials in Lubrication:

- **Self-lubricating bearings**: Materials like sintered metals and certain polymers are engineered with controlled porosity to retain lubricants. These bearings gradually release lubricant during operation, reducing friction and wear without the need for external lubrication.

- **Oil-impregnated components**: Porous structures can be impregnated with oil, allowing for continuous lubrication. For instance, Oilite bearings are made from metal alloys with pores that channel lubricants between the bearing and the shaft, ensuring consistent lubrication.

Impact of Porosity on Lubrication:

- **Lubricant storage and release**: The porosity of a material determines its capacity to store lubricant and the rate at which it is released. Materials with higher porosity can retain more lubricant, providing sustained lubrication over time.
- **Lubricant distribution**: Porous materials can facilitate the even distribution of lubricant across bearing surfaces, enhancing performance and reducing the risk of localized wear.

In summary, porosity plays a vital role in lubrication by influencing the storage, release, and distribution of lubricants within mechanical components. Understanding and controlling porosity is essential for designing efficient, self-lubricating systems that enhance performance and extend the lifespan of machinery.

672. Portable Oil Analyzer

A **portable oil analyzer**, sometimes referred to as a "Field Analyzer," is a compact, portable, or handheld device designed to assess the condition of lubricating oils directly at the point of use in the field. These analyzers provide real-time data on a variety of oil parameters, enabling fast, proactive maintenance decision-making.

Key Features of Portable Oil Analyzers:

- **Real-time analysis**: Instantly measures critical oil parameters such as viscosity, acidity (TAN), water content, and contamination levels.
- **User-friendly interface**: Equipped with intuitive displays and controls for straightforward operation.
- **Portability**: Lightweight and battery-powered, facilitating on-site testing without the need for laboratory facilities.
- **Durability**: Built to withstand harsh industrial environments, ensuring reliable performance in various conditions.

Benefits of Using Portable Oil Analyzers:

- **Proactive maintenance**: Enables early detection of potential issues, allowing for timely interventions and reducing unexpected downtime.
- **Cost savings**: By monitoring oil conditions, unnecessary oil changes can be avoided, leading to significant cost reductions.
- **Enhanced equipment longevity**: Maintaining optimal oil quality helps preserve the integrity of machinery components, extending their operational life.

Figure P12 depicts two variations of portable oil analyzers. The left-hand analyzer shows a **FluidScan** 1000 Series handheld infrared analyzer that provides rapid assessment of oil conditions and contamination. The right-hand analyzer illustrates a **FieldLab Series Fluid Analyzer**, a battery-powered, portable fluid lab offering comprehensive oil analysis in the field.

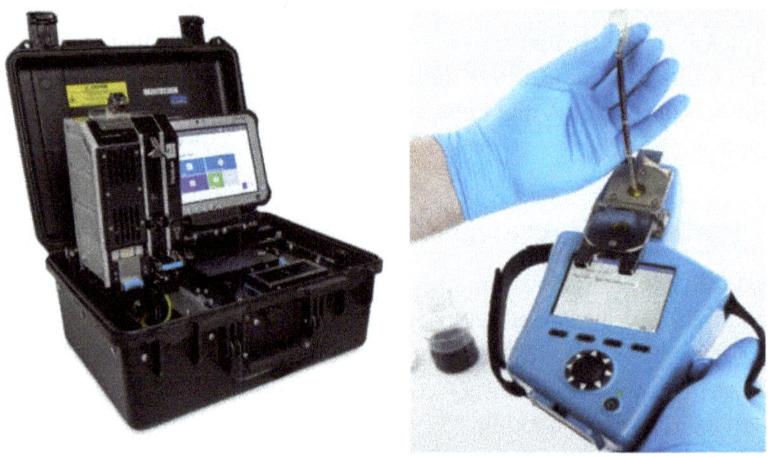

Figure P12 Two variations of portable oil analyzers fluidscan (L) and fieldlab (R). Source: Ametek Spectro Scientific.

Incorporating a portable oil analyzer into maintenance routines can empower operators to make rapid informed decisions, enhancing operational efficiency and equipment reliability.

673. Positive Displacement Injector (PDI)

The **positive displacement injector** (PDI) system, sometimes referred to as the single-line parallel system was first developed by the Lincoln Industrial Corporation in 1937 to accurately and individually displace metered quantities of oil or grease in a cyclical manner to small/medium-sized industrial equipment.

In contrast to single-line resistance and progressive divider-type systems, each metering valve, or point, can be set independently, adjusted, or easily changed, without affecting the system design. This enables additional injectors (lube points) to be added into the system at a later date, without the need to re-engineer the entire system. See Figure P13.

Key Features of Positive Displacement Injectors:

- **Precise lubricant delivery**: Each injector dispenses a fixed volume of lubricant per cycle, tailored to the requirements of individual lubrication points. This precision prevents overlubrication and underlubrication, both of which can lead to equipment issues.
- **Flexibility**: Injectors can be used to dispense oil or grease.
- **Adjustability**: Some injectors offer adjustable discharge volumes (see Figure P14) allowing for customization based on specific application needs.
- **Durability**: Designed to withstand harsh industrial environments, PDIs are built for reliability and longevity, reducing maintenance needs and downtime.

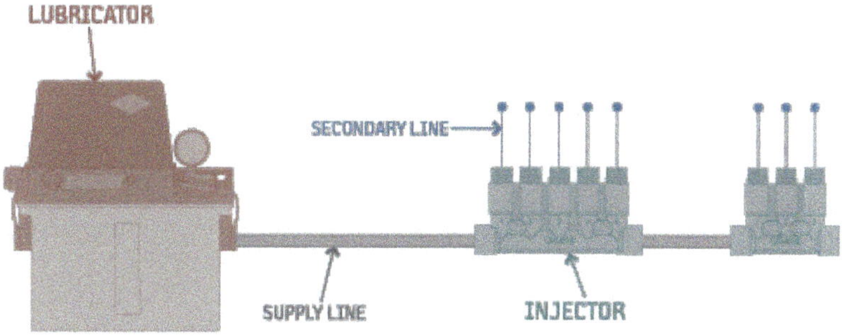

Figure P13 Typical positive displacement injector system layout. Courtesy: Bijur–Delimon Corporation.

Applications:

Positive displacement injectors are utilized across various industries, including:

- **Manufacturing**: In automated production lines, ensuring consistent lubrication of moving parts.

- **Mining**: Lubricating heavy machinery operating under extreme conditions.
- **Agriculture**: Maintaining equipment such as harvesters and tractors.

Considerations:
When selecting a positive displacement injector for an auto-lube system, consider factors such as:

- **Pressure requirements**: Choose injectors that operate effectively within the system's pressure range.
- **Maintenance needs**: Opt for injectors that are easy to service and maintain, with readily available parts.
- **Not tamper-proof**: Adjustable injectors can be easily set or reset at will. This could overfill or starve a bearing if tampered with after initial setting by a person without lubrication engineering knowledge.

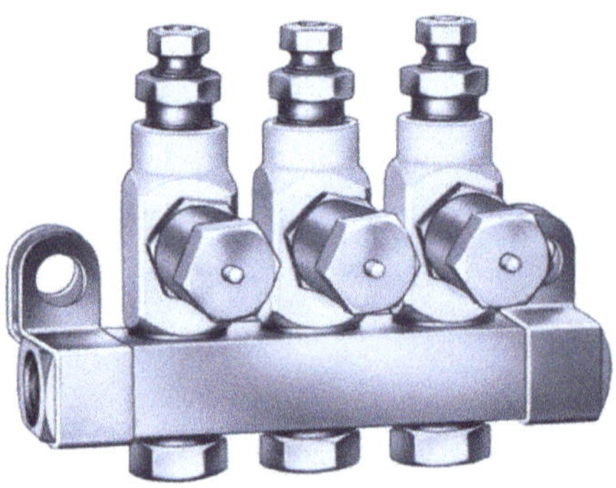

Figure P14 Typical adjustable positive displacement injectors. Courtesy: Lincoln Company.

674. Positive Seal

In lubrication systems, positive seals are essential components designed to restrict fluid bypass, ensuring that lubricants are effectively contained and directed to the intended areas. These seals maintain the integrity of the lubrication system by preventing leaks and contamination ingress, thereby enhancing equipment performance and longevity.

Types of Positive Seals in Lubrication Systems:

1. **O-rings**: Commonly used to seal static and dynamic interfaces, O-rings provide a reliable seal by compressing between two mating surfaces.

Selecting the appropriate material and size is crucial to prevent fluid bypass.

2. **Lip seals**: These seals feature a flexible lip that contacts the shaft, effectively containing lubricants and preventing leakage. They are particularly effective in high-speed applications (see Radial Lip Seal for more information).

3. **Mechanical seals**: Utilized in rotating equipment, mechanical seals consist of a stationary and a rotating face, along with a sealing element, to prevent fluid leakage. They are designed to handle high pressures and temperatures.

Considerations for Effective Sealing:

- **Material selection**: Choosing the right material is vital to ensure compatibility with the operating fluid and environmental conditions. For instance, elastomeric seals should be compatible with the specific fluid to prevent degradation.

- **Installation practices**: Proper installation is essential to maintain seal integrity. This includes ensuring correct alignment, smearing lubricant on the seal to ensure it slides into place thereby avoiding mechanical pinch damage and contamination during installation, and following manufacturer guidelines.

- **Maintenance**: Regular inspection and maintenance of seals can prevent premature failure. Monitoring for signs of wear, such as leaks or contamination, allows for timely replacement and system upkeep.

675. Potentiometric Titration Test Method

Potentiometric titration is a precise analytical method utilized to determine the concentration of acidic or basic constituents in lubricants by measuring the change in electric potential (voltage) as a titrant is added. This technique is particularly useful for assessing the **total acid number (TAN)** and **total base number (TBN)**, which are critical indicators of lubricant quality and degradation.

Total Acid Number (TAN):

TAN quantifies the amounts of acidic components in a lubricant, expressed as milligrams of potassium hydroxide **(KOH)** required to neutralize the acids present in one gram of sample. Elevated TAN values can indicate oxidation or contamination, necessitating timely analysis to prevent engine damage.

Total Base Number (TBN):
Conversely, TBN measures the concentration of basic constituents in a lubricant, reflecting its capacity to neutralize acids. A higher TBN suggests a greater reserve alkalinity, which is beneficial for combating acidic by-products in engine oils.

Potentiometric Titration Procedure for TAN and TBN:

1. **Sample preparation**:
   - Weigh a representative portion of the lubricant sample.
   - Dissolve the sample in a suitable solvent mixture to facilitate the titration process.
2. **Titration setup**:
   - Use a potentiometric titrator equipped with appropriate electrodes, such as a glass electrode for pH measurements.
   - Select a standardized titrant:

     For TAN determination, a base like potassium hydroxide (**KOH**) in isopropanol is commonly used.

     For TBN determination, an acid like perchloric acid (**$HClO_4$**) in glacial acetic acid is typically employed.
3. **Titration process**:
   - Gradually add the titrant to the sample while continuously monitoring the potential change.
   - Identify the equivalence point, where the amount of titrant added equals the amount of analyte present, by observing the inflection point in the titration curve.
4. **Calculation**:
   - Calculate the TAN or TBN based on the volume of titrant used, the concentration of the titrant, and the mass of the sample.

Advantages of Potentiometric Titration in Lubricant Analysis:

- **Precision**: Provides accurate and reproducible measurements of TAN and TBN, essential for effective lubricant management.

- **Automation**: Facilitates automated analysis, reducing manual errors and increasing throughput.
- **Versatility**: Applicable to various lubricant types, including engine oils, hydraulic fluids, and greases.

Considerations:

- **Sample preparation**: Ensure thorough homogenization of the lubricant sample to obtain representative results.
- **Electrode maintenance**: Regular cleaning and conditioning of electrodes are necessary to maintain measurement accuracy.
- **Method selection**: Choose the appropriate titration method (e.g., ASTM D664 for TAN and ASTM D2896 for TBN) based on the specific lubricant and analytical requirements.

By employing potentiometric titration, laboratories and maintenance facilities can effectively monitor lubricant quality.

676. Pour Point

The **pour point** of a lubricant is the lowest temperature at which it remains fluid and can flow under specified conditions. This property is crucial for ensuring that lubricants can circulate effectively in machinery during cold start-ups and operate efficiently in low-temperature environments.

Understanding Pour Point:

- **Definition**: The pour point is determined by cooling a lubricant sample and observing the temperature at which it no longer flows when the container is tilted. This temperature is typically recorded as 3 °C above the point where the oil shows no movement.
- **Measurement standards**: The pour point is commonly measured using **ASTM D97**, which involves cooling the sample and checking for flow at specific intervals. An alternative method, **ASTM D5949**, employs automatic apparatus for faster and more reproducible results.

Importance in Lubricant Performance:

- **Cold start-up**: A lower pour point indicates that the lubricant can flow at lower temperatures, facilitating easier machine or engine start-ups and reducing wear during initial operation.

- **Operational efficiency**: Lubricants with appropriate pour points ensure consistent performance in cold conditions, preventing issues like pump cavitation or inadequate lubrication.

Influencing Factors:

- **Base oil composition**: Naphthenic base oils typically have lower pour points compared to paraffinic oils, making them suitable for low-temperature applications.
- **Additives**: Pour point depressants are additives that can lower the pour point of lubricants, enhancing their low-temperature flow properties.

Selecting Lubricants Based on Pour Point:

When choosing a lubricant for cold environments, it is essential to select one with a pour point at least **10 °C** below the lowest expected operating temperature. This ensures reliable performance and protection against cold-related operational issues.

677. Pour Point Depressant

A **pour point depressant (PPD)** is an additive incorporated into lubricants to enhance their flowability at low temperatures. By modifying the crystallization behavior of waxes present in the oil, PPDs prevent the formation of large wax crystals that can impede fluid movement, thereby improving the lubricant's performance in cold conditions.

Mechanism of Action:

PPDs function by interacting with wax crystals in the oil, altering their size and structure. This modification inhibits the lateral growth of wax crystals, preventing them from forming a network that would otherwise restrict the oil's flow. As a result, the lubricant remains pumpable and effective at lower temperatures.

Applications:

Pour point depressants are utilized in various lubricant types, which include:

- **Crankcase engine oils**: Ensuring smooth engine operation during cold starts.
- **Automatic and power transmission fluids**: Maintaining fluidity for optimal transmission performance.

- **Gear oils**: Facilitating gear lubrication in low-temperature environments.
- **Hydraulic fluids**: Preventing pump cavitation and maintaining system efficiency.
- **Circulating oils**: Ensuring consistent lubrication in industrial machinery.

Benefits:
- **Enhanced low-temperature fluidity**: Improves the lubricant's ability to flow at reduced temperatures, facilitating easier start-ups and reducing wear.
- **Prevention of wax formation**: Inhibits the formation of wax crystals that can clog filters and restrict oil flow.
- **Improved equipment protection**: Ensures consistent lubrication, reducing the risk of component damage during cold operations.

Considerations:
While PPDs are effective in improving low-temperature performance, their concentration must be carefully controlled. Excessive amounts can lead to issues such as increased viscosity at higher temperatures, potentially affecting the lubricant's overall performance.

678. Power Transmission Fluids

Power transmission fluid, also known as **automatic transmission fluid (ATF)** is a specialized lubricant essential for the proper functioning of vehicles equipped with automatic transmissions. Similar to hydraulic fluid, it serves multiple critical roles, including power transmission, lubrication, heat dissipation, and contamination control.

Key Functions of ATF:
- **Power transmission**: Facilitates the transfer of power from the engine to the transmission, enabling smooth gear shifts and vehicle movement.
- **Lubrication**: Reduces friction between moving parts within the transmission, minimizing wear and extending component lifespan.
- **Heat dissipation**: Absorbs and dissipates heat generated during operation, maintaining optimal transmission temperatures.

- **Contamination control**: Traps and suspends contaminants, preventing them from causing damage to transmission components.

Power Transmission Fluid versus Hydraulic Oil
While both are used in fluid power systems, transmission fluid and hydraulic oil differ in composition and application:

- **Viscosity**: Transmission fluids oils are typically less viscous, suitable for lower-pressure systems.
- **Additives**: Transmission fluids contain detergents and friction modifiers, which are not present in hydraulic oils.
- **Operating conditions**: Hydraulic oils are designed for a broader temperature range and higher pressures compared to transmission fluids.

Using the correct fluid type is essential to prevent system failures and ensure optimal performance.

Types of ATF:
ATFs are formulated to meet specific auto manufacturer requirements. Common specifications include:

- **DEXRON®**: Developed by General Motors, this specification covers a range of ATFs used in various GM vehicles.
- **MERCON®**: Ford's specification for ATFs, with various iterations designed for different transmission models.
- **ATF+4®**: Chrysler's specification for ATFs, compatible with a wide range of Chrysler, Dodge, and Jeep vehicles.

679. ppb

In lubrication, **parts per billion (ppb)** is a unit of measurement used to quantify extremely low concentrations of substances within a lubricant. One ppb equates to one part of a substance per one billion parts of the lubricant, or 1 microgram per liter (µg/L).

Applications of ppb in Lubrication:

- **Contaminant detection**: Monitoring ppb levels helps identify trace contaminants, such as metals or water, that can adversely affect lubricant performance and machinery longevity. For instance, water contamination in oil can compromise the lubricant's protective film, leading to increased wear and reduced equipment lifespan.

- **Additive concentration**: Certain additives are effective at very low concentrations, often measured in ppb. For example, solid lubricants can be added at concentrations as low as 2 ppb to enhance lubricity and reduce friction in drilling fluids.

Precise ppb levels are monitored using advanced techniques like inductively coupled plasma mass spectrometry (ICP-MS) or gas chromatography-mass spectrometry (GC-MS).

680. ppm

Parts per million (ppm) is a unit of measurement used to express the concentration of a substance in a solution, typically a lubricant. One ppm represents one part of a substance for every one million parts of the mixture, which is equivalent to 1 milligram per liter (mg/L). It is often used to quantify low levels of contaminants, additives, or other substances in lubricants.

In lubrication, ppm is commonly used to measure:

- **Contaminant levels**: Monitoring contaminants like water or metal particles in lubricants to assess the health of the lubrication system and prevent damage.
- **Additive concentrations**: Tracking the concentration of additives like antiwear or friction-reducing agents, which are effective even at low ppm concentrations.
- **Water or chemical contamination**: Ensuring water content, which can degrade lubricant performance, remains below acceptable ppm levels.

Precise monitoring of ppm levels using advanced techniques like inductively coupled plasma mass spectrometry (ICP-MS) or gas chromatography-mass spectrometry (GC-MS) helps maintain lubricant effectiveness and machinery longevity.

681. Predictive Maintenance (PdM)

Predictive maintenance (PdM) for lubrication involves proactively monitoring and analyzing the condition of lubricants within machinery to forecast potential failures and optimize maintenance schedules. By assessing factors such as viscosity, contamination levels, and wear particles, this approach enables maintenance teams to address issues before they lead to equipment downtime or damage.

Key Components of Predictive Maintenance in Lubrication:

1. **Oil analysis**: Regular testing of lubricant samples to detect contami nants, degradation, and wear particles. This analysis provides insights into the health of both the lubricant and the machinery it serves.
2. **Condition monitoring**: Utilizing sensors and diagnostic tools to continuously monitor lubricant conditions, such as temperature and pressure, to identify anomalies that may indicate potential failures.
3. **Data analytics**: Employing advanced software to analyze collected data, predict equipment failures, and recommend optimal maintenance actions.

Benefits:

- **Reduced downtime**: By identifying issues early, predictive maintenance allows for timely interventions, minimizing unexpected equipment failures.
- **Cost savings**: Optimizing maintenance schedules and extending equipment lifespan lead to significant cost reductions.
- **Enhanced equipment reliability**: Regular monitoring ensures that lubricants perform effectively, maintaining machinery in optimal working condition.

682. Preparation Techniques for Lubrication

Preparing for effective lubrication involves a systematic approach to ensure machinery operates smoothly and efficiently. Key steps include:

1. **Assess equipment requirements**: Review manufacturer guidelines and operational conditions to determine the appropriate lubricant type, viscosity, and application method.
2. **Perform a lubricant consolidation exercise**: Choose lubricants that meet the specific needs of your equipment, considering factors like temperature range, load conditions, and environmental factors, minimizing the number of lubricants stored and managed on site.

3. **Establish lubrication procedures**: Develop standardized procedures detailing the correct application methods, intervals, and quantities to prevent over or underlubrication.
4. **Train and certify lubrication personnel**: Ensure that all involved staff are trained and certified in proper lubrication practices, emphasizing the importance of cleanliness and adherence to procedures.
5. **Implement contamination control measures**: Maintain a clean environment during lubrication to prevent contaminants from entering the system, which can lead to equipment damage.
6. **Monitor lubricant condition**: Regularly check the condition of lubricants through analysis to detect signs of degradation or contamination, allowing for timely interventions.
7. **Maintain lubrication equipment**: Keep lubrication tools and systems in good working order to ensure accurate and efficient application.

By following these steps, you can establish a robust lubrication program that enhances equipment reliability and extends operational lifespan.

683. Pressure

In lubrication, **pressure** refers to the force exerted by a lubricant within the gap between two surfaces in relative motion. This pressure is crucial for maintaining a lubricating film that separates the surfaces, thereby reducing friction and wear.

Key Aspects of Pressure in Lubrication:
In a lubricated bearing, pressure is required to separate the surface in three of the lubrication regimes.

- **Hydrodynamic lubrication**: In this regime, the lubricant's viscosity and the relative motion of the surfaces generate sufficient pressure to fully separate the surfaces, preventing direct contact. This separation is essential for minimizing wear and ensuring smooth operation.
- **Hydrostatic lubrication**: Here, external pressure is applied to the lubricant, forcing it into the bearing surfaces to maintain separation, even when there is no relative motion between the surfaces. This method is often used during start-up conditions to prevent metal-to-metal contact or in machining operations like plunge grinding.

- **Elastohydrodynamic lubrication**: In this high-pressure regime, the lubricant's viscosity increases significantly due to the applied pressure, allowing it to support the load despite the small contact area. This phenomenon is particularly relevant in applications like gears and rolling element bearings.

In lubrication delivery systems, **pressure** refers to the pump pressure used to move the lubricant through the delivery system to the lubrication point.

- Oil System: Lubricating oils are pumped at much lower pressures than grease due to their lower viscosities. Manual pressurized systems mobilize lubricants using lower pressures around 40–50 psi, whereas auto delivery systems develop pressures around 80–100 psi.

Figure P15 Large auto grease system using a 40:1 pneumatic drum pump. Courtesy: ENG**TECH** Industries Inc.

- Grease system: Grease requires significantly higher pressures to move it from the reservoir or gun to the lube point. A pneumatic auto-greaser pump pumps up to a 40:1 ratio that can deliver a line pressure of 4000 psi at 100 psi air pressure. See Figure P15. Manual grease guns can deliver a line pressure between 6000 and 15,000 psi!

Understanding and controlling pressure within lubrication systems is vital for optimizing performance, extending equipment lifespan, and preventing premature failure.

684. Pressure-atmospheric (Atmospheric Pressure)

In lubrication systems, **atmospheric pressure** refers to the ambient pressure exerted by the atmosphere, typically around 101.3 kPa (14.7 psi) at sea level. This baseline pressure influences various aspects of lubrication, including oil aeration and the behavior of lubricants under different operating conditions.

Impact of Atmospheric Pressure on Lubrication:

- **Oil aeration**: At atmospheric pressure, oil can dissolve a certain amount of air. However, when the oil is pressurized within the lubrication system, this dissolved air can come out of the solution, leading to entrained air or bubbles. This phenomenon can affect the oil's performance, potentially causing issues like cavitation or reduced lubrication efficiency.

- **Viscosity considerations**: The viscosity of lubricants is temperature-dependent. At atmospheric pressure, the oil's viscosity is relatively stable, but as pressure increases within the system, the oil's viscosity can change, affecting its flow characteristics and the formation of the lubricating film.

- **Cavitation risks**: In high-pressure lubrication systems, the pressure can drop below the vapor pressure of the oil, leading to the formation of vapor bubbles. When these bubbles collapse, they can cause cavitation, potentially damaging components and compromising lubrication effectiveness.

- **Rapid lubricant depletion**: Single-point lubrication devices once turned on for use will continue to discharge lubricant until exhausted. In doing so they are calibrated to operate in atmospheric pressure at sea level and rely on back pressure to control the rate of flow. Consequently, at higher elevations where atmospheric pressure is lower than 14.7 psi, the lubricant will flow faster. When setting the device delivery rate (if unit has the capability to be regulated) the flow rate setting calculation will need to take the elevation into account.

685. Pressure-back (Back Pressure)

In lubrication systems, **back pressure** refers to the resistance encountered by the lubricant as it flows through the system, including components like pipes, filters, and bearings. Excessive back pressure can impede lubricant flow, leading to a pressure build-up and a pump stall. Reduced or no lubricant flow can adversely affect the bearing and machine performance leading to full failure and downtime.

Causes of Back Pressure:

- **Clogged filters**: Accumulation of contaminants can obstruct filters, increasing resistance to lubricant flow.
- **Blocked lines**: Obstructions or kinks in lubrication lines can restrict flow, causing back pressure.
- **Overfilled bearings**: Excessive lubricant in bearings can create resistance, leading to back pressure.
- **Damaged bearings**: bearings that are poorly fitted, damaged during installation, or hit by an external force can move out of alignment or can turn in their housings restricting the lubricant opening to the bearing.

Consequences of Excessive Back Pressure:

- **Reduced lubricant flow**: Inadequate flow can result in insufficient lubrication, increasing friction and wear.
- **Seal damage**: High back pressure can cause seals to fail, leading to leaks and contamination.
- **Overheating**: Restricted flow can cause the lubricant to overheat, degrading its performance and potentially damaging components.

Managing Back Pressure:

- **Regular maintenance**: Implementing a routine maintenance schedule helps identify and address issues like clogged filters or blocked lines before they lead to significant problems.

- **Monitoring systems**: Installing pressure gauges allows for real-time monitoring of back pressure, enabling prompt detection of anomalies. Install overpressure indicators on proportioning blocks or in the delivery line system designed to "crack" open at a specific line pressure to relieve the danger of system pressure build-up.
- **Proper lubricant levels**: Ensuring that bearings and other components are filled with the correct amount of lubricant prevents overfilling, which can contribute to back pressure.

686. Pressure – Cracking (Cracking Pressure)

In lubrication systems, **cracking pressure** refers to the minimum pressure required to open a pressure-operated valve, such as a check valve, allowing fluid to pass through.

This pressure is essential for controlling fluid flow and maintaining optimal system performance.

Key Points:

- **Function**: Cracking pressure ensures that valves, such as relief or check valves, open at the correct pressure to regulate fluid flow and protect components from overpressure.
- **Adjustment**: Cracking pressure can often be adjusted to suit specific system requirements, ensuring efficient operation and preventing damage.
- **Measurement**: It is typically measured in pounds per square inch (PSI) or pascals (Pa), depending on the system's pressure units

687. Pressure Gauge

A **pressure gauge** or **pressure gage** is an instrument used to measure the pressure of a fluid – liquid or gas – within a system. In lubrication systems, pressure gauges are essential for monitoring and ensuring that the lubricant is circulating at the correct pressure, which is crucial for optimal equipment performance and longevity.

Types of Pressure Gauges:

- **Mechanical pressure gauges**: These gauges use mechanical components, such as Bourdon tubes or diaphragms, to detect pressure changes. They are widely used due to their simplicity and reliability. See Figure P16.

- **Digital pressure gauges**: These gauges employ electronic sensors to measure pressure and display the readings digitally. They often offer higher accuracy and additional features like data logging.

Applications in Lubrication Systems:

- **Monitoring lubricant pressure**: Ensuring that the lubricant is circulating at the correct pressure is vital for effective lubrication and preventing component wear.
- **Detecting blockages or leaks**: Abnormal pressure readings can indicate blockages in the lubrication lines or leaks in the system, prompting timely maintenance actions.
- **Optimizing system performance**: Regular monitoring of pressure helps in adjusting the system to maintain optimal lubrication conditions, thereby enhancing equipment efficiency and lifespan.

Figure P16 Standard mechanical, bottom mount, oil pressure gauge calibrated in PSI and kPa. Courtesy: ENG**TECH** Industries Inc.

Selecting the Appropriate Pressure Gauge: When choosing a pressure gauge for a lubrication system, consider the following factors:

- **Pressure range**: Ensure the gauge's pressure range aligns with the operating pressures of your system.

- **Accuracy**: Select a gauge with the necessary accuracy to meet your operational requirements.
- **Material compatibility**: The materials of the gauge should be compatible with the type of lubricant used to prevent corrosion or degradation.
- **Mounting options**: Choose a gauge with suitable mounting options (e.g., bottom or back mount) for your system's configuration.

688. Pressure-rated (Rated Pressure)

In lubrication systems, **pressure-rated** refers to the maximum pressure that components – such as pumps, hoses, seals, and valves –are designed to withstand during operation. Operating components within their specified pressure ratings is crucial to prevent failures, leaks, and ensure the longevity and efficiency of the lubrication system.

Key Considerations:

- **Component selection**: Choose components with pressure ratings that align with the maximum operating pressures of your system to ensure safety and reliability.
- **Regular monitoring**: Use pressure gauges to monitor system pressures, ensuring they remain within the safe operating range of all components.
- **Maintenance**: Regularly inspect and maintain components to detect signs of wear or damage that could compromise their pressure tolerance.

By adhering to pressure ratings and implementing proper maintenance practices, you can enhance the performance and lifespan of your lubrication system.

689. Pressure Switch

A **pressure switch** in lubrication systems is a device that monitors the pressure of the lubricant within the system. It activates or deactivates electrical contacts when the pressure reaches a predetermined set point, ensuring that the lubrication system operates within the designed pressure ranges.

Functions of Pressure Switches in Lubrication Systems:

- **Monitoring pressure levels**: Continuously checks the lubricant pressure to ensure it remains within the desired range.

- **Activating alarms**: Triggers alarms or notifications if the pressure deviates from the set range, indicating potential issues such as blockages or leaks.
- **Controlling system components**: Can be integrated into control circuits to start or stop pumps, valves, or other components based on pressure readings. Positive Displacement Injector systems and Dual line systems employ pressure switched to ensure enough line pressure and lubricant to set and reset the injectors.

Applications:

- **Centralized lubrication systems**: Used to monitor and control pressure in systems that distribute lubricant to multiple points.
- **Industrial machinery**: Protects equipment by ensuring that lubrication pressures are maintained, preventing damage due to inadequate lubrication
- **Protection devices**: Used to shut down systems when pressure is too low (insufficient lubricant or broken line), or when pressure is too high (restricted line or blocked bearing)

690. Pressurized Lubrication System

A **pressurized lubrication system** utilizes a pump to circulate lubricant under pressure to various components within the machinery, ensuring consistent and efficient lubrication. This method is commonly employed in mechanized industrial machinery and engines to reduce friction, cool, minimize wear, and enhance performance.

Key Features:

- **Lubricant distribution**: The pump delivers lubricant to critical areas such as bearings, pistons, and gears, maintaining a protective film between moving parts.
- **Pressure regulation**: A pressure relief valve maintains optimal pressure levels, preventing overpressurization that could lead to leaks or component damage.
- **Filtration**: Lubricant passes through filters to remove contaminants, ensuring clean lubrication and prolonging component life.

Advantages:

- **Enhanced component protection**: Consistent lubrication reduces friction and wear, extending the lifespan of machinery parts.
- **Improved heat dissipation**: The circulating lubricant aids in cooling components, preventing overheating.

Applications:

- **Industrial machinery**: Used in mechanized equipment like compressors, metal working machines, and gearboxes to maintain optimal performance.
- **Internal combustion engines**: Pressurized lubrication is standard in automotive and industrial engines to ensure all moving parts receive adequate lubrication.

691. Preventive Maintenance (PM)

Preventive maintenance (PM) is a proactive approach to equipment management that involves regularly scheduled inspections, servicing, and repairs to prevent unexpected failures and extend the lifespan of assets. By addressing potential issues before they lead to breakdowns, PM helps maintain optimal performance and reduces downtime.

Lubrication PM relates specifically to the proactive management of lubrication delivery to machines and their lubrication systems and lubricants.

Key Components of Preventive Maintenance:

- **Scheduled inspections**: Regularly examining equipment to identify signs of wear, corrosion, or other potential problems.
- **Routine servicing**: Performing tasks such as cleaning, lubricating, and calibrating machinery to ensure smooth operation.
- **Parts replacement**: Replacing components that have reached the end of their service life to prevent unexpected failures.
- **Documentation**: Keeping detailed records of maintenance activities to track equipment history and inform future maintenance decisions.

Key Aspects of Preventive Lubrication Maintenance:

- **Scheduled lubrication**: Establishing a routine for applying lubricants based on manufacturer recommendations or operational conditions to ensure consistent performance.

- **Monitoring lubricant condition**: Regularly assessing the quality and quantity of lubricants to detect contamination or degradation, ensuring they remain effective.
- **Component inspection**: Inspecting machinery parts for signs of wear or damage during lubrication intervals to identify and address issues early.
- **Record keeping**: Documenting lubrication activities, including types of lubricants used, application methods, and maintenance schedules, to track equipment health and inform future maintenance decisions.

Benefits of Preventive Maintenance:

- **Reduced downtime**: By addressing issues before they cause equipment failure, PM minimizes unexpected downtime.
- **Cost savings**: Regular maintenance can prevent costly emergency repairs and extend the lifespan of equipment.
- **Improved safety**: Well-maintained equipment is less likely to malfunction, reducing the risk of accidents.
- **Enhanced efficiency**: Properly serviced machinery operates more efficiently, leading to better overall performance.

Implementing a preventive lubrication maintenance program is essential for maintaining equipment reliability, reducing unexpected breakdowns, and optimizing maintenance resources.

692. Proactive Maintenace

Proactive maintenance is a strategic approach that focuses on identifying and addressing the root causes of equipment failures before they occur. Unlike reactive maintenance, which responds to issues after they arise, proactive maintenance aims to prevent problems by eliminating underlying conditions that lead to machine faults and degradation.

Key Components of Proactive Maintenance:

- **Root cause analysis**: Investigating and identifying the fundamental reasons behind equipment issues to implement corrective actions that prevent recurrence.

- **Condition monitoring**: Continuously assessing equipment performance through sensors and diagnostic tools to detect early signs of potential failures.
- **Scheduled inspections**: Regularly planned checks to ensure equipment is operating within optimal parameters and to identify any deviations promptly.
- **Employee training**: Educating and certifying staff on best practices and proper equipment handling to reduce human errors that can lead to equipment degradation.

693. Probe

In lubrication systems, a **probe** typically refers to a sensor or instrument designed to measure specific parameters such as temperature, pressure, or flow rate within the system. These probes provide real-time data, enabling operators to monitor and maintain optimal lubrication conditions, thereby enhancing equipment performance and longevity.

Types of Probes in Lubrication Systems:

- **Temperature probes**: Monitor the temperature of the lubricant to ensure it remains within the desired range, preventing overheating or excessive cooling. See Figure P17.
- **Pressure probes**: Measure the pressure of the lubricant, ensuring it is sufficient to maintain a proper lubricating film between moving parts.
- **Flow probes**: Assess the flow rate of the lubricant, ensuring adequate distribution to all necessary components.

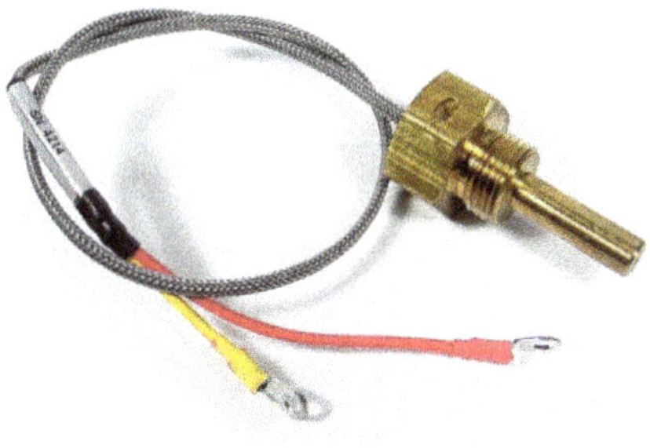

Figure P17 Typical oil reservoir temperature probe. Source: JPI EDM Probes.

By integrating these probes into lubrication systems, operators can proactively address potential issues, optimize maintenance schedules, and improve overall system efficiency.

694. Process Oil

Process oils are specialized oils utilized across various chemical and technical industries, either as raw material components or as aids in processing. They play a crucial role in enhancing the workability of materials, improving processing efficiency, and achieving desired product characteristics.

Applications of Process Oils:

- **Rubber and elastomer industry**: Process oils are integral in the production of rubber compounds, serving as plasticizers that enhance the workability and processing of rubber. They also influence the final product's properties, such as hardness, strength, and long-term stability.
- **Polymer industry**: In polymer manufacturing, process oils act as processing aids or extender oils, facilitating the blending and processing of polymer compounds. They contribute to achieving the desired consistency and performance characteristics of the final polymer products.
- **Metalworking and forming**: Process lubricants and release agents are essential in metalworking applications, such as sheet metal forming and aluminum continuous casting. They reduce friction, prevent wear, and ensure the quality of the formed metal products.

Types of Process Oils:

- **Naphthenic oils**: These oils are known for their excellent low-temperature properties and good solvency, making them suitable for applications requiring high compatibility with various polymers and additives.
- **Paraffinic oils**: Characterized by higher viscosity and better oxidative stability, paraffinic oils are often used in applications where these properties are advantageous.
- **Aromatic oils**: These oils offer high solvency and are used in applications where their ability to dissolve other substances is beneficial.

695. Properties – Lubricants

Lubricants are essential in reducing friction, minimizing wear, and enhancing the efficiency and lifespan of machinery. The effectiveness of a lubricant is determined by several **key properties**:

1. **Viscosity**:

   This property measures a lubricant's resistance to flow. A lubricant with appropriate viscosity ensures a stable film between moving parts, preventing metal-to-metal contact. Viscosity is typically evaluated using the viscosity index (VI); a higher VI indicates less change in viscosity with temperature variations.

2. **Pour point**:

   The pour point is the lowest temperature at which a lubricant remains fluid enough to flow. Lubricants with a low pour point are suitable for operations in cold environments, ensuring they flow and protect components during start-up.

3. **Flash point**:

   This is the temperature at which a lubricant emits enough vapor to ignite in the presence of a flame. A higher flash point indicates greater safety during storage and operation, reducing the risk of fire hazards.

4. **Oxidation stability**:

   Oxidation stability refers to a lubricant's ability to resist chemical breakdown when exposed to oxygen at elevated temperatures. Lubricants with high oxidation stability maintain their performance over time, preventing the formation of sludge and acids that can damage engine components.

5. **Demulsibility**:

   This property indicates a lubricant's ability to separate from water. Good demulsibility ensures that any water contamination is effectively removed, preventing rust and corrosion in machinery.

6. **Antiwear and extreme pressure properties**:

   These properties enable a lubricant to form protective films on metal surfaces, reducing wear under high-load conditions. Additives like zinc dialkyldithiophosphate (ZDDP) are commonly used to enhance these characteristics.

7. **Corrosion resistance**:

   Lubricants with good corrosion resistance protect metal surfaces from rust and degradation caused by moisture and other corrosive elements.

8. **Foam resistance**:

   Excessive foaming can lead to air entrainment, reducing lubrication efficiency. Lubricants with low foam tendency maintain their performance by preventing foam formation.

Selecting a lubricant with the appropriate combination of these properties is crucial for ensuring optimal performance and longevity of machinery. Understanding these characteristics allows for informed decisions tailored to specific operational requirements. See Table P1.

Table P1 Table comparing the considerations and their respective properties for lubricants.

Considerations	Properties
Viscosity	Resistance to flow and film strength
Pour Point	Lowest temperature at which lubricant remains fluid
Flash Point	Temperature at which lubricant emits vapor to ignite
Oxidation Stability	Ability to resist chemical breakdown under heat and oxygen exposure
Demulsibility	Ability to separate from water
Anti-Wear and Extreme Pressure	Lubricant's ability to protect under high load and extreme conditions
Corrosion Resistance	Ability to resist rust and corrosion from moisture or contaminants
Foam Resistance	Lubricant's ability to resist foam formation, ensuring efficient lubrication

696. PSI

PSI – pounds per square inch, an imperial unit of pressure commonly used to measure force applied over a specific area. It quantifies the amount of pressure exerted on a one-inch square area. PSI is widely used in various industries, including engineering, automotive, and lubrication systems.

Key Points:

- **1 PSI** equals a force of **1 pound** applied to an area of **1 square inch**.

It is commonly used to measure:

- **Air pressure** in tires
- **Fluid pressure** in hydraulic or lubrication systems
- **Water pressure** in plumbing

PSI can be converted into other pressure units such as **pascals (Pa) or bar**.

Formula:

$$\text{Pressure (PSI)} = \frac{\text{Force (pounds)}}{\text{Area (square inches)}}.$$

697. PSIA

PSIA – pounds per square inch absolute, which is a unit of pressure measurement that refers to the pressure measured relative to a perfect vacuum. It includes the atmospheric pressure in addition to the pressure exerted by the gas or fluid being measured.

Key Points:

- **1 PSIA** is the total pressure, which includes both the atmospheric pressure (14.7 PSI at sea level) and any additional pressure applied by the gas or fluid.
- **PSIA = PSI + atmospheric pressure** (typically 14.7 PSI at sea level).
- **PSIA** is commonly used in scientific, engineering, and industrial contexts where the pressure measurement needs to account for the baseline atmospheric pressure, especially in vacuum systems and other precision measurements.

Example:
If a system is showing a pressure of **5 PSIG** (pounds per square inch gauge), this would be the pressure above atmospheric pressure. To convert this to **PSIA**, you would add the atmospheric pressure:

PSIA = 5 PSIG + 14.7 PSI (atmospheric)
PSIA = 19.7 PSIA

PSIA is especially important in applications like vacuum pumps, pressure systems, and certain lubrication systems, where understanding the absolute pressure is critical.

698. PSID

PSID – pounds per square inch differential, which is a unit of measurement used to express the difference in pressure between two points in a system. It indicates how much pressure is higher at one point compared to another and is commonly used in systems like filtration, lubrication, and hydraulic systems to monitor pressure drops.

Key Points:
- **PSID** measures the **differential pressure**, or the difference in pressure, across components such as filters, pipes, or pumps.
- **PSID** is often used to monitor the condition of a filter, as an increasing pressure differential indicates that the filter is becoming clogged or is resisting flow more than when it is clean.
- **1 PSID** represents a difference of **1 pound of pressure per square inch** between two points.

Example:
In a hydraulic system, if the pressure before a filter is 50 PSI and the pressure after the filter is 40 PSI, the **differential pressure** across the filter would be:

PSID = 50 PSI – 40 PSI = 10 PSD

PSID is useful for ensuring proper functioning and for detecting blockages, clogging, or other issues in filtration systems and other pressurized systems.

699. PSIG

PSIG – pounds per square inch gauge. It is a unit of pressure that measures the pressure relative to the surrounding atmospheric pressure. Unlike **PSIA (absolute pressure)**, PSIG does not include the atmospheric pressure in its measurement. Instead, it represents the pressure above atmospheric pressure, so when the pressure is equal to atmospheric pressure, PSIG will read **0**.

Key Points:

- **1 PSIG** is the pressure difference between the pressure being measured and atmospheric pressure.
- PSIG is commonly used in systems like tires, pumps, and refrigeration systems where it is important to know how much pressure is above the ambient atmosphere.
- If the system is at **atmospheric pressure**, the **PSIG** reading will be **0** because there is no pressure above the atmospheric baseline.

Example:
If a tire has a pressure of **30 PSIG**, it means the pressure inside the tire is **30 pounds per square inch** higher than the surrounding atmosphere. If the tire pressure was **0 PSIG**, it would be at atmospheric pressure.

Conversion:

- To convert **PSIG** to **PSIA (absolute pressure)**, you would add the atmospheric pressure (typically 14.7 PSI at sea level):

So, a tire pressure of **30 PSIG** would be:

$$\text{PSID} = 30\ \text{PSIG} + 14.7\ \text{PSI} = 44.7\ \text{PSIA}$$

PSIG is widely used in everyday applications and is a key measurement for pressurized systems like gas cylinders, tires, and pumps.

700. PTFE

See "Polytetrafluoroethylene."

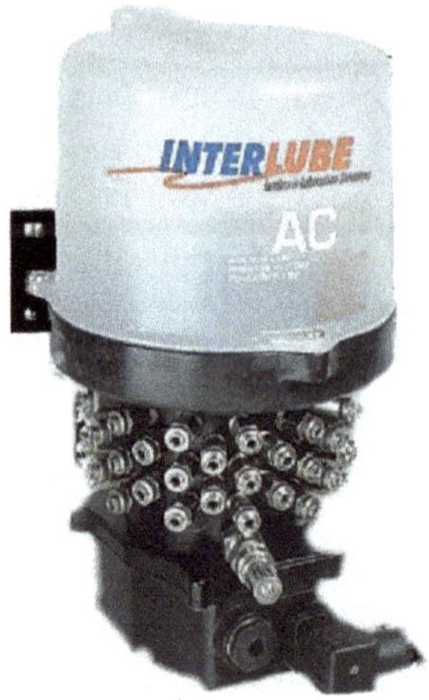

Figure P18 Modern pump-to-point lubrication unit. Courtesy: Interlube Corporation.

Key Features:

- **Direct lubricant delivery**: Metered lubricant is pumped directly to each lubrication point, ensuring that each component receives the appropriate amount of lubricant.
- **Simple design**: The system's straightforward design makes it easy to operate and maintain, reducing the complexity associated with more elaborate lubrication systems. See Figure P18.
- **Cost-effective**: By eliminating the need for extensive piping and distribution blocks, pump-to-point systems offer a more economical solution for machinery lubrication.

701. Pump-box Cam

A **box cam pump lubricator** is a mechanical device designed to deliver precise amounts of lubricant to various points within a machinery system. It operates by using a camshaft mechanism to actuate multiple pumps, each dedicated to a specific lubrication point. The camshaft can be electrically or mechanically driven (as is the case with the example in Figure P18). This design ensures that each component receives the appropriate amount of lubricant, thereby enhancing performance and extending the lifespan of the equipment.

Key Features:

- **Camshaft mechanism**: The camshaft rotates to actuate multiple pumps simultaneously, allowing for synchronized lubrication across various points (see Figure P19).
- **Modular design**: Components can be customized and expanded to meet specific lubrication requirements, providing flexibility in system configuration.
- **High-pressure capability**: Many box cam pump lubricators are engineered to handle high-pressure applications, ensuring effective lubrication even in demanding environments.

Figure P19 Victorian steam engine box cam pump to point lubricator (still in use today). Courtesy: ENG**TECH** Industries Inc.

702. Pump, Electro-mechanical

An **electro-mechanical pump lubricator** is an automatic lubrication device that utilizes an electric motor to drive a pump, delivering precise amounts of lubricant to various points within a machinery system. This mechanism ensures consistent and efficient lubrication, reducing manual intervention and enhancing equipment reliability (see Figure P20 example).

Key Features:

- **Electric motor drive**: The lubricator is powered by an electric motor, providing reliable and consistent operation.
- **Precise lubricant delivery**: It delivers accurate quantities of lubricant to each lubrication point, preventing over or underlubrication.
- **Programmable settings**: Many models offer programmable settings for lubrication intervals and quantities, allowing customization based on specific operational requirements.
- **Monitoring capabilities**: Some lubricators include monitoring features to track lubricant levels, pressure, and operational status, facilitating proactive maintenance.

Figure P20 Single point electro-mechanical pump. Courtesy: Howard Marten Company/ Memolube.

703. Pump, Gear

A **gear pump lubricator** is a mechanical pumping device that utilizes the positive displacement principle of gear pumps to deliver lubricant to various points within a machinery system. This ensures consistent lubrication.

Key Features:

- **Positive displacement mechanism**: Gear pumps operate by trapping fluid between their gears and the pump casing, forcing it through the system with each rotation.

- **Precise lubricant delivery**: The design allows for accurate control of lubricant flow, ensuring each lubrication point receives the appropriate amount.
- **Durability**: Constructed from robust materials, gear pump lubricators are designed to withstand the demands of industrial environments.

704. Pumpability

Pumpability refers to a lubricant's ability to flow through the delivery system – such as pumps, lines, and nozzles – under the operating conditions of the machinery. This property is crucial for ensuring that the lubricant reaches all necessary components, thereby reducing friction and wear.

Key Factors Influencing Pumpability:

- **Viscosity**: The internal resistance of the lubricant to flow. Higher viscosity oils are thicker and may resist flow, especially at low temperatures. Conversely, oils with lower viscosity flow more easily but may not provide adequate protection under high-load conditions.
- **Temperature**: Lubricant viscosity decreases with increasing temperature, enhancing pumpability. However, at low temperatures, increased viscosity can impair the lubricant's ability to flow, potentially leading to inadequate lubrication during start-up. This condition can be combatted through the use of a heater placed around or in the lubricant reservoir that serves to warm the oil making it more viscous under cold conditions.
- **Additives**: The presence of certain additives can affect the lubricant's flow characteristics. For example, viscosity modifiers can alter the oil's behavior under different temperature conditions, impacting its pumpability.

Importance of Pumpability:

- **Efficient lubrication**: Ensures that the lubricant reaches all critical components promptly, reducing wear and extending equipment lifespan.
- **Energy efficiency**: Improves the efficiency of the lubrication system by reducing the energy required to pump the lubricant through the system.
- **Operational reliability**: Prevents issues such as cavitation or pump starvation, which can lead to equipment damage and downtime.

705. Purpose of a Lubricant

The **primary purpose of a lubricant** is to reduce friction, prevent wear, and ensure the longevity of mechanical systems. By selecting the right lubricant for a given application, industries can improve efficiency, reduce maintenance costs, minimize energy loss, and enhance the reliability of machinery.

Key Purposes of a Lubricant

- **Reduces friction**: Forms a protective film between surfaces, decreasing metal-to-metal contact and improving efficiency.
- **Minimizes wear and extends equipment life**: Reduces surface degradation, preventing abrasion, scuffing, and fatigue failure.
- **Cools moving parts**: Absorbs and dissipates heat generated by friction, preventing overheating.
- **Prevents corrosion and rust**: Creates a barrier against moisture and contaminants, protecting surfaces.
- **Seals against contaminants**: Prevents dust, dirt, and debris from entering critical components.
- **Transmits power and motion**: In hydraulic systems, lubricants serve as the working fluid to transfer energy.
- **Dampens shock and vibrations**: Absorbs impact loads and reduces mechanical stress, improving durability.
- **Enhances efficiency and energy savings**: Reducing friction and wear leads to lower power consumption and higher system performance.

Types of Lubricants Based on Function

1. **Oils**: Provide fluid lubrication in engines, gearboxes, and hydraulic systems.
2. **Greases**: Semisolid lubricants that adhere to surfaces, used in bearings, bushings, and high-load applications.
3. **Solid lubricants**: Such as graphite or molybdenum disulfide (MoS_2), used where liquid lubricants fail (e.g., extreme temperatures or vacuum environments).
4. **Synthetic lubricants**: Engineered fluids offering superior stability, thermal resistance, and extended lifespan.

Conclusion: Lubricants are truly engineering marvels and are worthy of true respect!

706. Push Lube Truck

A **push-type lube truck** is a specialized vehicle designed for on-site lubrication and maintenance of heavy machinery and equipment. Unlike traditional lube trucks that may require a CDL (commercial driver's license) to operate, push-type lube trucks are typically built to be operated without a CDL, making them more accessible for various operators. See Figure P21.

Key Features of Push-type Lube Trucks:

- **Non-CDL operation**: Engineered to comply with weight regulations that allow operation without a CDL, facilitating easier access for operators.
- **Customizable design**: Tailored to meet specific operational needs, including tank sizes, storage compartments, and equipment configurations.
- **Enhanced mobility**: Designed for easy maneuverability, allowing operators to access and service equipment in various locations efficiently.

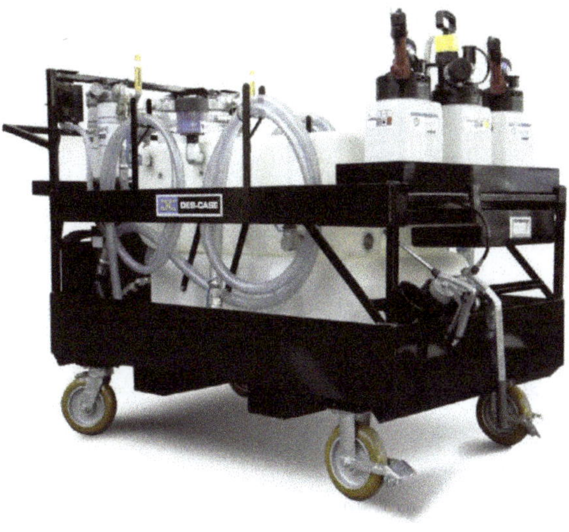

Figure P21 Purpose-built manual push lube truck. Courtesy: Des–Case Corporation.

Benefits:

- **Cost-effective**: Eliminates the need for a CDL, reducing operational costs and simplifying logistics.
- **Operational efficiency**: Streamlined design and functionality enhance the speed and effectiveness of on-site maintenance tasks.
- **Flexibility**: Suitable for a wide range of industries, including construction, mining, and agriculture, due to their adaptable configurations.

707. Pycnometer

A **pycnometer** is a precise laboratory instrument used to determine the density of liquids, including lubricants. Typically made of glass, it consists of a flask with a known volume, a stopper, and a capillary tube to allow air bubbles to escape. See Figure P22. By measuring the mass of the lubricant sample and the volume of the pycnometer, the density can be calculated using the Formula:

$$\text{Density} = \frac{\text{Mass of Sample}}{\text{Volume of Pycnometer}}.$$

Figure P22 Typical pycnometer vessels in different sizes.

Applications in Lubricant Analysis:

- **Quality control**: Determining the density of lubricants helps assess their composition and purity, ensuring they meet specified standards.
- **Formulation development**: Understanding the density aids in formulating lubricants with desired properties for specific applications.
- **Regulatory compliance**: Accurate density measurements are essential for compliance with industry regulations and standards.

Procedure:

1. **Preparation**: Clean and dry the pycnometer thoroughly to prevent contamination.
2. **Filling**: Fill the pycnometer with the lubricant sample, ensuring no air bubbles are trapped.
3. **Weighing**: Weigh the filled pycnometer using an analytical balance to determine the mass.
4. **Calculation**: Calculate the density by dividing the mass by the known volume of the pycnometer.

708. Pyrolysis

Pyrolysis in the context of lubrication refers to the thermal decomposition of lubricating oils at elevated temperatures, typically in the absence of oxygen. This process breaks down complex hydrocarbons into simpler compounds, resulting in the production of valuable products such as fuel oils and gases.

Applications of Pyrolysis in Lubrication:

- **Recycling waste lubricating oils**: Pyrolysis offers an effective method for recycling spent lubricating oils, converting them into usable fuels, and reducing environmental hazards associated with oil disposal.
- **Fuel production**: The pyrolysis of waste lubricating oils can yield diesel-like fuels, providing an alternative energy source and contributing to waste reduction.
- **Environmental impact**: By transforming waste oils into valuable products, pyrolysis helps mitigate environmental pollution and supports sustainable waste management practices.

In summary, pyrolysis serves as a crucial process in the lubrication industry for recycling waste oils and producing alternative fuels, thereby promoting environmental sustainability.

709. Pyrometer

A **pyrometer** is an inexpensive and highly accurate noncontact temperature measurement device that detects infrared radiation emitted by an object to determine its temperature. In lubrication systems, electronic hand-held pyrometers, similar to the one shown in Figure P23 are utilized to monitor the temperature of the lubricant itself and components such as gearboxes and engine parts, ensuring they operate within optimal thermal conditions.

Applications in Lubrication:

- **Lubricant temperature**: Invaluable for spot-checking reservoir lubricant temperatures.
- **Gearbox testing**: During gearbox testing, insufficient lubrication can lead to increased friction, causing elevated temperatures. Pyrometers enable quick detection of temperature variations, allowing for timely adjustments to prevent damage.
- **Engine monitoring**: In diesel engines, pyrometers help monitor exhaust temperatures, which can influence lubrication oil performance. Maintaining proper exhaust temperatures ensures better fuel economy and prolongs the life of lubrication oil.

By providing accurate, real-time temperature data, pyrometers assist in maintaining the integrity of lubrication systems, thereby enhancing equipment longevity and performance.

Figure P23 Typical handheld pyrometer.

Q

710. Quality (Lubricant)

Identifying **quality lubricants** is essential for establishing a quality lubrication program, ensuring optimal machinery performance and longevity. Several markers can help distinguish quality lubricants from inferior ones:

1. **Compliance with industry standards**:
   Quality lubricants adhere to recognized industry standards and certifications, such as those set by the American Petroleum Institute (API) or the International Lubricant Standardization and Approval Committee (ILSAC). These standards ensure the lubricant meets specific performance and quality criteria.
2. **Clear and consistent appearance**:
   High-quality lubricants typically have a clear, consistent color and texture. Discoloration, cloudiness, or the presence of particulates can indicate contamination or degradation, suggesting poor quality.
3. **Proper packaging and labeling**:
   Reputable manufacturers provide lubricants in well-sealed, properly labeled containers. The label should include essential information such as the lubricant type, viscosity grade, manufacturer details, and compliance marks. Lack of proper labeling or damaged packaging can be a red flag.
4. **Manufacturer reputation**:
   Purchasing lubricants from established and reputable manufacturers or suppliers can be a reliable indicator of quality. These companies often have a history of producing high-quality products and offer customer support for their products.
5. **Performance in application**:
   A high-quality lubricant will perform effectively under the specific operating conditions of your machinery, including temperature stability,

DOI: 10.1201/9788743806608-726

wear protection, and corrosion resistance. If a lubricant fails to meet these performance expectations, it may be of inferior quality.

Oil analysis can also provide insights into the lubricant's condition and performance, helping to identify potential issues before they lead to equipment failure.

By considering these markers, an informed decision can be made when selecting lubricants, ensuring they meet the necessary quality standards for your equipment and your lubrication program.

711. Quantitative Analysis

Quantitative analysis in lubrication involves measuring specific properties and components of lubricants to assess their performance, composition, and suitability for particular applications. This analysis is crucial for ensuring that lubricants meet the required standards and function effectively within machinery systems.

Key Aspects of Quantitative Lubricant Analysis:

1. **Additive concentration measurement**: Determining the levels of additives such as detergents, dispersants, antiwear agents, and antioxidants is essential. Techniques like X-ray fluorescence (XRF) spectrometry are commonly used for this purpose. For instance, ASTM D6443-14 outlines methods for analyzing additive elements in lubricating oils using XRF.

2. **Wear debris analysis**: Identifying and quantifying wear particles in used oil can indicate the condition of engine components. Techniques such as ferrography and spectrometric analysis are employed to detect and measure these particles.

3. **Contaminant detection**: Monitoring for contaminants like water, fuel, or coolant in the lubricant is vital. Methods such as infrared spectroscopy can detect the presence of water or fuel in oil samples.

4. **Viscosity measurement**: Assessing the viscosity of lubricants ensures they provide adequate film strength under operating conditions. Viscosity measurements are typically performed using viscometers or rheometers.

5. **Oxidation stability testing**: Evaluating a lubricant's resistance to oxidation helps predict its service life and performance degradation.

Oxidation stability can be assessed through accelerated aging tests or by measuring the increase in acid number over time.

6. **Elemental analysis**: Determining the concentration of elements like calcium, magnesium, phosphorus, sulfur, and zinc provides insights into the lubricant's formulation and potential performance characteristics. Techniques such as inductively coupled plasma (ICP) spectroscopy are commonly used for this purpose.

Benefits of Quantitative Lubricant Analysis:

- **Predictive maintenance**: Regular analysis helps predict potential equipment failures by identifying issues such as contamination or additive depletion before they lead to significant problems.
- **Quality assurance**: Ensures that lubricants meet the necessary specifications and standards, thereby maintaining equipment performance and longevity.
- **Cost efficiency**: By monitoring lubricant conditions, unnecessary oil changes can be avoided, and the optimal drain intervals can be determined, leading to cost savings.

712. Quantity Measurement – Grease

Grease quantity is typically measured by volume or weight. Accurately measuring the quantity of grease applied during lubrication is crucial for ensuring optimal equipment performance and longevity. Over-or-under lubrication can lead to increased wear, energy consumption, and potential equipment failure.

Methods for Measuring Grease Quantity:

1. **Manual measurement**: Using syringes with volume markings allows for determination of the grease gun delivery volume per "shot." To establish this volume measurement, pump ten grease gun shots into a syringe and read off the total volume, Next, divide the total by ten to establish the actual volume per shot of the particular grease gun being tested. This shot size value should then be marked on the grease gun body in large indelible print. Precise measurement.

2. **Auto grease meter**: This device accurately measures grease discharge in both volume and weight, providing readings in metric (cm^3 or g) and US units (US fl. oz or oz). It ensures consistent application of the correct grease quantity.

3. **Ultrasonic sensors**: By emitting ultrasonic waves, these sensors detect the presence and thickness of the grease film on bearing surfaces. This noninvasive method helps determine the adequacy of lubrication without disassembling components.

4. **Programmable automated lubrication systems**: These systems can be designed and set to deliver specific grease quantities at predetermined intervals, ensuring consistent lubrication and reducing human error.

Calculating Grease Quantity for Bearings:
A commonly used method involves the **SKF** Formula:

$$\text{Grease Quantity (oz)} = 0.114 \times (\text{Bearing Outside Diameter (inches)}) \times (\text{Bearing Width (inches)}).$$

This formula provides an estimate of the grease volume required for a bearing.

Considerations:

- **Operating conditions**: Factors such as temperature, speed, and load can influence the required grease quantity.
- **Environmental factors**: Exposure to contaminants or moisture may necessitate adjustments in lubrication practices.

713. Quantity Measurement – Oil

Accurate measurement of oil quantities is essential for effective lubrication, maintenance, and operational efficiency across various industries. The choice of measurement method depends on factors such as the volume of oil, the specific application, and the required precision.

Common Methods for Measuring Oil Quantities:

1. **Flow meters**:
    - **Positive displacement flow meters**: These meters measure the volume of oil by capturing fixed amounts and counting the number of captures. They are highly accurate and suitable for precise measurements.

- **Turbine flow meters**: Utilize the flow of oil to rotate a turbine; the rotational speed correlates with the flow rate. They are commonly used for continuous monitoring in pipelines.

2. **Weighing Systems**: **Load Cells**: Installed under storage tanks or containers, load cells measure the weight of the oil, which can be converted to volume using the oil's density.

3. **Level sensors**:
   - **Ultrasonic sensors**: Emit ultrasonic waves to determine the level of oil in a tank, which can then be used to calculate the volume.
   - **Capacitance sensors**: Measure the change in capacitance as the oil level changes, providing continuous level monitoring.

4. **Dipsticks and sight gauges**: Traditional methods where a dipstick is inserted into the oil to measure the level or sight gauges provide a visual indication of the oil level. These methods are simple but less precise and are typically used for smaller volumes or less critical applications.

Considerations for Accurate Oil Measurement:

- **Temperature compensation**: Oil expands and contracts with temperature changes. Accurate measurement systems often include temperature sensors to adjust readings accordingly.
- **Viscosity effects**: The viscosity of oil can affect the accuracy of flow meters. It is important to select measurement equipment compatible with the specific viscosity of the oil being used.
- **Maintenance and calibration**: Regular maintenance and calibration of measurement equipment are crucial to ensure ongoing accuracy and reliability.

Calculating Oil Quantities for Bearings:

Volume = **A**rea (bearing surface) x **R**eplenishment of lubricant film

V = A x R is a simple, but effective calculation used by many of the lubrication delivery system manufacturers to determine volume requirements of individual bearings so the total lubricant requirement can be calculated for pump and reservoir sizing and actuation control cycles per hour determined.

Bearing type surface area, A is different for each bearing type and is determined using the following calculations:

- **Plain bearing**: where = 3.14, D = Shaft Diameter, L = Bearing Length
- **Rolling bearing**: where D = Shaft Diameter, R = Number of Bearing Rows
- **Flat bearing surface**: = Area of Contact Surface
- **Gear (each single gear)**: where = 3.14, D = Pitch Diameter of Gear, W = Gear Face Width

Replenishment rate depends on lubricant type, delivery cycle and film thickness as shown in Table Q1.

Selecting the appropriate oil measurement method depends on the specific requirements of the application, including the volume of oil, required accuracy, and operational conditions. Implementing the right measurement system contributes to efficient operations, reduces waste, and ensures the longevity of equipment.

Table Q1 Lubricant replenishment rate guide.

Lubricant and delivery method	Film thickness	Time
Manual applied: Terminating grease	.002" /.050mm Film	4 hours
Automatic applied: Terminating grease	.001" /.025mm Film	4 hours
Automatic applied: Terminating oil	.001" /.025mm Film	1 hour
Automatic applied: Circulating oil	.001" /.025mm Film	1 minute

Source: Lubrication for Industry, Kenneth E. Bannister, Industrial Press.

714. Quasi-static Lubrication

Quasi-static lubrication refers to a lubrication condition where **lubricant film formation and distribution change very slowly over time**, meaning that dynamic effects such as inertia and rapid flow changes are negligible. This lubrication concept is commonly applied in slow-moving machinery, precision instruments, and computational models where lubrication conditions can be assumed to be steady or slowly evolving rather than highly transient.

Key Characteristics of Quasi-static Lubrication

- **Slowly varying lubrication film**: Lubricant film thickness changes gradually, allowing for stable performance and minimal turbulence effects.
- **Negligible inertial effects**: The forces acting on the lubricant due to acceleration or motion are small enough to be ignored, simplifying analysis.
- **Dominated by viscous and pressure forces**: In quasi-static lubrication, viscosity and pressure gradients primarily dictate fluid behavior rather than rapid motion.
- **Common in low-speed, high-precision systems**: Found in applications where lubrication conditions remain nearly constant over time.

715. Quaternary Ammonium Compounds (Quats)

Quaternary ammonium compounds (**Quats**) are **cationic surfactants** used in lubricants and metalworking fluids for their antimicrobial, anticorrosion, and emulsification properties. These compounds contain positively charged nitrogen atoms, making them effective in controlling microbial growth, stabilizing emulsions, and reducing friction in industrial lubrication applications.

Key Functions of Quats in Lubrication

- **Antimicrobial protection**: Quats act as biocides, preventing bacterial and fungal growth in water-based lubricants and metalworking fluids.
- **Corrosion inhibition**: They form a protective film on metal surfaces, reducing oxidation and rust formation.
- **Emulsification and detergency**: Quats help in stabilizing oil–water emulsions, improving lubricant dispersion and performance.
- **Surface lubrication enhancement**: Their molecular structure allows them to reduce friction and wear, particularly in boundary lubrication conditions.

Applications of Quats in Lubricants and Fluids

- **Metalworking fluids**: Used in cutting, grinding, and machining fluids to prevent microbial degradation and extend fluid life.

- **Hydraulic and water-based lubricants**: Helps maintain fluid cleanliness and stability in industrial hydraulic systems.
- **Rust preventatives and corrosion protection**: Applied in storage oils and protective coatings to reduce oxidation.
- **Textile and leather lubricants**: Enhances lubrication and antistatic properties in textile manufacturing.

716. Quenching

Quenching is a heat treatment process used to rapidly cool heated metals, typically by immersing them in a quenching medium such as oil, water, or air. This rapid cooling transforms the metal's microstructure, enhancing hardness and strength. The choice of quenching medium significantly influences the cooling rate and, consequently, the mechanical properties of the metal.

Quenching Media:

- **Oil**: Provides a moderate cooling rate, balancing hardness with reduced risk of distortion. It is suitable for a wide range of steels, including carbon steels and high-alloy steels.
- **Water**: Offers the fastest cooling rate, resulting in high hardness but also increased risk of distortion and cracking. It is commonly used for low-alloy steels.
- **Air**: Offers the slowest cooling rate, leading to lower hardness but minimal risk of distortion. It is used for steels with high hardenability.

How Quenching Oil Works

Quenching oils serve two main purposes: they harden the component by controlling the heat transfer during quenching, and they improve the wetting of the component to reduce the risk of distortion and cracking.

The quenching process occurs in the following three stages:

1. **Vapor stage**: When the component is first immersed in quenching oil, a vapor blanket forms around it. This stage involves slow cooling, as heat is transferred only through radiation via the vapor layer. To accelerate this process, additives can be used to agitate the vapor layer, dispersing it more quickly.
2. **Boiling stage**: Once the vapor blanket dissipates, nucleate boiling begins. This is the fastest heat transfer stage. The speed of cooling and

heat transfer depends on the oil's molecular composition, making the oil's formulation crucial for efficient quenching.

3. **Convection stage**: In this final stage, the component-oil interface temperature drops below the oil's boiling point. Heat transfer in this stage is influenced by the oil's viscosity, which is affected by oil decomposition. The rate of heat transfer is inversely related to the viscosity of the oil, while viscosity increases with the degree of oil decomposition. As decomposition increases, sludge formation occurs, leading to uneven cooling rates and potentially affecting the quality of the quenching process

In oil quenching operations, the quenching medium itself acts as a lubricant, facilitating the rapid cooling of the metal. The choice of quenching medium and its properties – such as viscosity, thermal stability, and cooling rate – are critical in achieving the desired hardness and minimizing distortion.

Considerations:

- **Cooling rate**: The cooling rate must be carefully controlled to achieve the desired hardness without causing excessive distortion or cracking.
- **Oil maintenance**: Regular maintenance of quenching oils, including filtration and monitoring for contamination, is essential to maintain consistent cooling performance. Agitation, filtration, and periodic product sweetening are recommended to ensure optimal quenching operations.
- **Safety**: Proper handling and storage of quenching oils are crucial due to their flammability and potential health hazards.

717. Quenching Oil

Quenching oil is a specialized fluid used in the heat treatment process to rapidly cool heated metal workpieces, thereby enhancing their hardness and strength. Quenching oils are specially formulated to minimize distortion and cracking while delivering a uniform surface finish and hardening. The choice of quenching oil significantly influences the cooling rate, which in turn affects the mechanical properties of the metal.

Types of Quenching Oils:

1. **Fast quenching oils**: Designed to provide rapid cooling rates, suitable for steels with low hardenability.

2. **Medium quenching oils**: Offer moderate cooling rates, balancing hardness with reduced risk of distortion.
3. **Slow quenching oils**: Provide slower cooling rates, suitable for steels with high hardenability.

Considerations for Selecting Quenching Oil:

- **Steel composition**: The alloy's hardenability dictates the appropriate quenching oil. For instance, high-carbon steels may require faster quenching oils, while low-carbon steels may benefit from slower quenching oils.
- **Workpiece geometry**: Complex shapes or larger sections may necessitate slower quenching oils to minimize the risk of cracking or distortion.
- **Safety and environmental impact**: The flammability and toxicity of the oil should be considered. Vegetable oils like canola oil are less hazardous and more environmentally friendly compared to mineral oils.

Safety and Maintenance:

- **Handling**: Proper storage and handling are essential due to the flammability of quenching oils.
- **Maintenance**: Regular filtration and monitoring of oil conditions are necessary to maintain consistent quenching performance.

718. Quenching Rate

Quenching rate refers to the speed at which a fluid removes heat from a heated material, typically metal, during cooling or quenching processes. It is a crucial factor in heat treatment, lubrication, and metalworking, as it directly affects material hardness, residual stress, and component performance.

Key Factors Affecting Quenching Rate

- **Type of quenching fluid**: Different fluids (water, oil, polymers, or air) have varying heat dissipation capabilities, affecting cooling speed.
- **Viscosity and thermal conductivity**: Lower viscosity and higher thermal conductivity fluids enhance heat transfer, increasing the quenching rate.
- **Fluid agitation and flow rate**: Circulating or agitating the fluid improves heat dissipation and reduces uneven cooling.

- **Workpiece material and geometry**: The size, shape, and composition of the metal influence how quickly heat is extracted.
- **Bath temperature**: Higher initial fluid temperatures slow down quenching, while cooler fluids increase heat removal efficiency.

Common Quenching Fluids and Their Cooling Rates

1. **Water-based quenchants** – Fastest quenching rate, used for high-hardness applications but with a risk of cracking due to thermal shock.
2. **Oil-based quenchants** – Moderate cooling rate, widely used in precision heat treatment where controlled cooling prevents excessive brittleness.
3. **Polymer quenchants** – Adjustable cooling rates, offering better control over metal hardness while reducing cracking risks.
4. **Air and gas quenching** – Slowest cooling rate, used for gradual stress relief in high-performance alloys.

719. Quick-change Filter

A **quick-change filter** is a filtration system designed for easy and efficient replacement of filter cartridges without the need for specialized tools or extensive downtime. For example, a cartridge element filter employing a cartridge housing, O-ring, and long bolt can easily be replaced with an all-in-one "spin-on" disposable filter similar to those now utilized on automobiles. This design simplifies maintenance procedures, making it particularly beneficial in applications where frequent filter changes are necessary.

Key Features of Quick-change Filters:

- **Tool-free replacement**: Spin-on filters can be replaced quickly without the use of specialized tools, reducing maintenance time and labor costs.
- **Automatic shut-off**: Some systems incorporate an automatic shut-off mechanism that halts the flow of fluid when the filter is removed, preventing spills and contamination.
- **Versatility**: Quick-change filters are available in various configurations to suit different filtration needs, including options for water, air, and oil filtration.

For "zero-downtime" filter changes on circulating systems, a duplex filter station can be piped into the main line from the pump, if sufficient is available.

Often, each line will have a pressure gauge before and after each filter to establish any pressure differential. Reduced pressure on the outside of the filter signifies the filter pores are becoming blocked and need to be changed. To operate, the following procedure is utilized to perform a zero-downtime filter change.

Zero-downtime Filter Change Procedure while Machine is Running:

1. Open valves on the nonoperational or dormant (clean filter) line (see Figure Q1)
2. Close valves on the dirty filter line
3. Spin off the dirty filter and discard properly
4. Install new spin-on filter
5. Clean off any oil residue

Applications:

- **Industrial processes**: In manufacturing and processing industries, quick-change filters facilitate continuous operations by minimizing maintenance interruptions.

By incorporating quick-change filters, systems can achieve higher efficiency, reduced maintenance time, and improved overall performance.

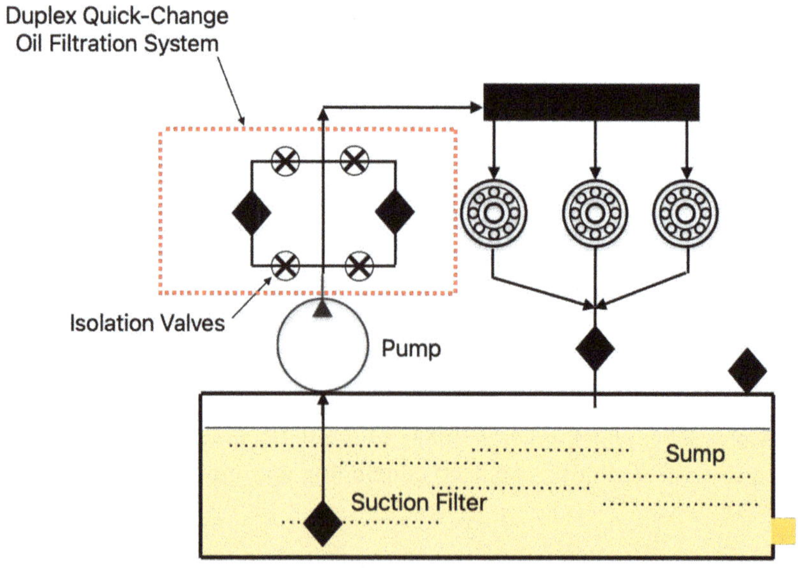

Figure Q1 Quick change oil-filter system.

720. Quick Connect/Disconnect Oil Fittings

Quick connect/disconnect oil fittings are specialized connectors designed to facilitate the rapid and secure connection and disconnection of oil lines in various systems. These fittings are essential for applications requiring frequent maintenance, fluid changes, or equipment servicing, as they allow for swift and efficient fluid transfer without the need for tools. In addition, the use of these connectors minimizes any contamination transfer while in use.

Key Features:

- **Ease of use**: Quick connect/disconnect fittings enable operators to connect and disconnect oil lines quickly, reducing downtime and enhancing operational efficiency.
- **Leak prevention**: Many quick connect/disconnect fittings are equipped with self-sealing valves that prevent oil leakage during connection and disconnection, ensuring a clean and safe working environment.
- **Durability**: Constructed from robust materials, these fittings are designed to withstand the pressures and conditions typical in oil systems, ensuring long-term reliability.

Applications:

- **Industrial equipment**: In industrial settings, these fittings are utilized to connect and disconnect oil lines in machinery, or can be used for reservoir filling and oil extraction thereby facilitating maintenance and reducing downtime.
- **Waste oil systems**: Quick connect/disconnect fittings are used in waste oil evacuation systems to quickly disconnect diaphragm pumps and suction discharge hoses, streamlining the oil disposal process.

Considerations:

- **Compatibility**: It is crucial to ensure that the quick connect/disconnect fittings are compatible with the specific oil lines and equipment in use, including matching thread sizes and pressure ratings.
- **Maintenance**: Regular inspection and maintenance of quick connect/disconnect fittings are necessary to prevent wear and tear, ensuring they function correctly and safely.

Incorporating quick connect/disconnect oil fittings into oil systems enhances efficiency, safety, and ease of maintenance, making them a valuable component in various applications.

721. Quick-change Lubrication System

A **quick-change lubrication system** is a fast and efficient method for replacing or refilling lubricants in machinery and equipment without significant downtime. These systems are designed to simplify maintenance, reduce contamination risks, and improve lubrication efficiency in industrial, automotive, and heavy equipment applications.

Key Functions of Quick-change Lubrication Systems

- **Minimizes downtime**: Allows for rapid lubricant changes without the need for full system shutdowns, maximizing equipment availability.
- **Prevents contamination**: Sealed fittings, couplings, and quick-connect systems help prevent dust, moisture, and dirt ingress.
- **Improves maintenance efficiency**: Designed for easy access and rapid fluid exchange, reducing the time and labor required for lubrication maintenance.
- **Ensures proper lubricant delivery**: Provides controlled lubricant flow to prevent overlubrication or underlubrication issues.

Types of Quick-change Lubrication Systems

1. **Quick-disconnect coupling systems**: Used in grease lines, hydraulic systems, and oil reservoirs, allowing for fast connection and disconnection of lubrication fittings.
2. **Automatic quick-change lubrication units**: Prefilled lubrication reservoirs that can be swapped out without interrupting operation, commonly used in wind turbines, conveyors, and automated machinery.
3. **Cartridge-based grease systems**: Features easy-to-replace grease cartridges that prevent contamination and allow clean, mess-free lubricant replenishment.
4. **Centralized quick-change lubrication manifolds**: A system that enables fast oil or grease refill for multiple lubrication points in industrial or fleet applications.
5. **Quick-drain and refill systems for oil sumps**: Allows for rapid draining and refilling of oil reservoirs without needing to remove or disassemble components.

R

722. 4Rs

The **4Rs of lubrication** are fundamental principles designed to optimize lubrication practices and enhance machinery performance. They are:

1. **The right amount**: Applying the correct quantity of lubricant is crucial. Overlubrication can lead to overheating and increased friction, while underlubrication may result in inadequate protection and premature wear. Striking the right balance ensures efficient operation and prolongs equipment life.
2. **The right lubricant**: Selecting the appropriate lubricant type and grade is essential. Factors such as viscosity, load-bearing capacity, and operating temperature must align with the specific requirements of the machinery to provide optimal protection and performance.
3. **The right place**: Lubricant must be applied directly to the intended lubrication points. Proper application ensures that all critical components receive adequate lubrication, preventing wear and potential failures.
4. **The right time**: Adhering to the correct lubrication intervals is vital. Lubrication should occur at times that align with the equipment's operating conditions and manufacturer's recommendations to maintain optimal performance and prevent premature wear.

By diligently applying these 4Rs, maintenance teams can enhance machinery reliability, reduce downtime, and extend the lifespan of equipment.

723. R and O Oils

RandO oils, short for **rust and oxidation-inhibited oils**, are specialized lubricants formulated to prevent rust formation and oxidation in machinery and equipment. These oils are essential in applications where metal surfaces are exposed to moisture and air, as they provide a protective barrier against corrosion and degradation.

DOI: 10.1201/9788743806608-739

Key Features of RandO Oils:

- **Rust protection**: RandO oils contain additives that form a protective film on metal surfaces, preventing moisture and air from causing rust formation.
- **Oxidation inhibition**: These oils are formulated to resist chemical reactions with oxygen, thereby extending the lifespan of the lubricant and the equipment.
- **Antiwear properties**: Many RandO oils include antiwear additives that reduce friction between moving parts, minimizing wear and tear.

Common Applications:

- **Steam turbine systems**: RandO oils were initially designed specifically for steam systems to prevent corrosion that could prove very dangerous in a steam chamber if unabated.
- **Hydraulic systems**: RandO oils are used in hydraulic pumps and motors to ensure smooth operation and prevent corrosion.
- **Circulating systems**: They are employed in circulation systems, centrifugal air compressors, and lightly loaded enclosed gear drives to maintain equipment efficiency.
- **Industrial equipment**: RandO oils are utilized in various industrial applications, including turbines and gear drives, to protect against rust and oxidation.

724. Rabiniwicz, Ernest

Ernest Rabinowicz (1927–2006) was a distinguished American mechanical engineer renowned for his pioneering contributions to tribology – the study of friction, wear, and lubrication. He spent a significant portion of his career at the Massachusetts Institute of Technology (MIT), where he advanced his understanding of mechanical surface interactions.

Professional Contributions:

Seminal work:

- Authored "Friction and Wear of Materials," first published in 1965, a foundational text in tribology that has been cited over 10,000 times. , through his research, he discovered that 70% of all bearings lost their usefulness due to surface degradation, of which 50% was due to mechanical wear and the remaining 20% to corrosion. From this, he went on to determine the impact of the bearing failure on the United States's gross domestic product, commonly referred to as the **"Rabiniwicz Law,"** it states that "Every year, 6% of the GDP (gross domestic product) [in the United States] is lost through mechanical wear."

- **Popular science articles**: Wrote extensively for **Scientific American**, including notable pieces on polishing and the stick-slip phenomenon.
- **Educational impact**: Developed and taught a widely attended extension course on Friction and Wear, reaching engineers and scientists globally.

Awards and honors:
- **Mayo D. Hersey Award**: American Society of Mechanical Engineers, 1985.
- **International Award**: Society of Tribologists and Lubrication Engineers, 1988.
- **Tribology Gold Medal**: Institution of Mechanical Engineers, 1998.

Rabinowicz's work has left an indelible mark on the field of tribology, influencing both theoretical research and practical applications in engineering.

725. Rabiniwicz Scale

Ernest Rabinowicz, a prominent figure in tribology, introduced a criterion in 1958 to determine the minimum size of wear particles during adhesive wear. This criterion, often referred to as the "**Rabinowicz Scale**" is based on the interplay between the elastic energy stored in surface asperities and the work required to separate a wear particle from the contact surface.

Key Aspects of the Rabinowicz Criterion:
- **Elastic energy versus work of separation**: The criterion suggests that wear particles will form when the elastic energy stored in the asperities exceeds the work required to separate the particle from the surface. This balance determines whether a particle will detach and contribute to wear.
- **Critical length scale**: Rabinowicz introduced a characteristic length scale, known as the Rabinowicz length, which is a function of material

properties such as hardness and elastic modulus. This length scale helps predict the formation of wear particles under specific loading conditions.

- **Influence on wear mechanisms**: The criterion emphasizes the role of adhesion in wear processes, highlighting that the detachment of wear particles is influenced by the adhesive forces between contacting surfaces.

Rabinowicz's work has significantly advanced the understanding of adhesive wear mechanisms, providing a theoretical foundation for predicting wear particle formation based on material properties and contact conditions.

726. Radial Lip Seal

A **radial lip seal**, commonly known as a **radial shaft seal** or **oil seal**, is a sealing device designed to prevent the leakage of lubricants and the ingress of contaminants in rotary applications. These seals are essential and highly effective in machinery where rotating shafts pass through stationary components, such as in gearboxes, pumps, and automotive engines. See Figure R1.

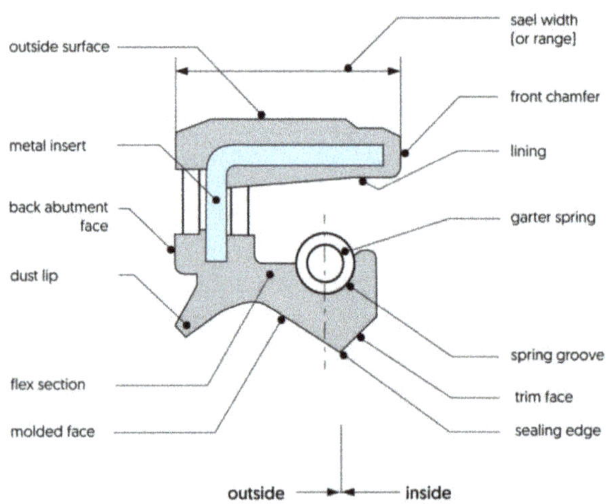

Figure R1 Cross-section of a radial lip seal. Source: Erik's Industrial.

Construction and Function:

Radial lip seals typically consist of three primary components:

- **Sealing lip**: A flexible elastomeric lip that maintains contact with the shaft, creating a barrier against fluid leakage.

- **Metal case**: Provides structural support and facilitates installation into the housing.
- **Spring (garter spring)**: An internal spring that applies pressure to the sealing lip, ensuring consistent contact with the shaft and compensating for wear over time.

Advantages:

- **Effective sealing**: Prevents lubricant loss and contamination ingress, enhancing equipment longevity.
- **Versatility**: Available in various materials and designs to suit different operating conditions.
- **Cost-effective**: Relatively inexpensive and easy to replace, reducing maintenance costs.

Considerations:

Selecting the appropriate radial lip seal involves considering factors such as operating temperature, shaft speed, and the nature of the fluid being sealed. Proper installation and regular maintenance are crucial to ensure optimal performance and prevent premature seal failure.

727. RAG Indicator

The **RAG indicator** is a color-coded system – **Red**, **Amber**, and **Green** – used in lubrication to rapidly assess and communicate lubricant levels in reservoirs. See Figure R2. This method offers a straightforward visual representation of when to add lubricant, facilitating quick assessments and informed decision-making.

Understanding RAG Indicators:

- **Red**: Indicates the lubricant is at the lowest operable level and that the lubricant pump is in immediate danger of stalling and can fail to pump at any time.
- **Amber**: Denotes to the machine operator or maintainer that a work order should be released immediately to fill the reservoir to prevent escalation to a red alert.
- **Green**: Signifies that the lubricator level is OK; no immediate issues or risks are present. Note: Green is also used to signify the highest level so as not to overfill the reservoir.

Figure R2 A simple RAG indication on a lube pump using colored tape. Courtesy: ENG**TECH** Industries Inc.

Benefits:

- **Clarity**: Offers a simple and intuitive way to convey complex information, making it accessible to all stakeholders.
- **Efficiency**: Enables quick identification of issues, allowing for prompt corrective actions.
- **Consistency**: Provides a standardized approach to status reporting, enhancing communication across teams.

By implementing RAG indicators, organizations can enhance transparency, improve decision-making, and maintain better control over their lubrication requirements.

728. Rate-of-change Limits

In lubrication management, **rate-of-change limits** are critical for monitoring the progression of lubricant degradation, contamination, and wear within machinery. By establishing thresholds for the rate at which these parameters change, maintenance teams can proactively identify and address potential issues before they lead to equipment failure.

Application of Rate-of-change Limits:

- **Particle counting**: Monitoring the rate at which particle counts increase can indicate escalating contamination levels. Setting rate-of-change limits allows for early detection of contamination trends, enabling timely interventions.
- **Elemental wear metals**: Tracking the rate of increase in wear metals helps assess the wear rate of engine components. Establishing rate-of-change limits for these metals can signal when wear is accelerating, prompting maintenance actions.
- **Acid number (AN) and RPVOT**: The rate at which the acid number and RPVOT (Rotary Pressure Vessel Oxidation Test) values change can indicate the degradation of the lubricant. Implementing rate-of-change limits for these parameters helps in monitoring lubricant health and scheduling timely oil changes.

Benefits of Implementing Rate-of-change Limits:

- **Early detection**: Identifying trends in lubricant degradation or contamination before they reach critical levels allows for proactive maintenance, reducing the risk of unexpected equipment failures.
- **Optimized maintenance scheduling**: By understanding the rate at which lubricant conditions change, maintenance activities can be scheduled more effectively, balancing the need for intervention with operational demands.
- **Enhanced equipment longevity**: Maintaining lubricant quality within optimal parameters through rate-of-change monitoring helps in extending the lifespan of machinery components.

Incorporating rate-of-change limits into lubrication management practices enhances the ability to maintain equipment reliability and performance.

729. Rate of Shear

In lubrication, the **rate of shear** refers to the rate at which adjacent layers of lubricant move relative to each other under applied stress. This concept is crucial for understanding how lubricants perform under various operating conditions, as it influences viscosity, film formation, and overall lubrication effectiveness. See Figure R3.

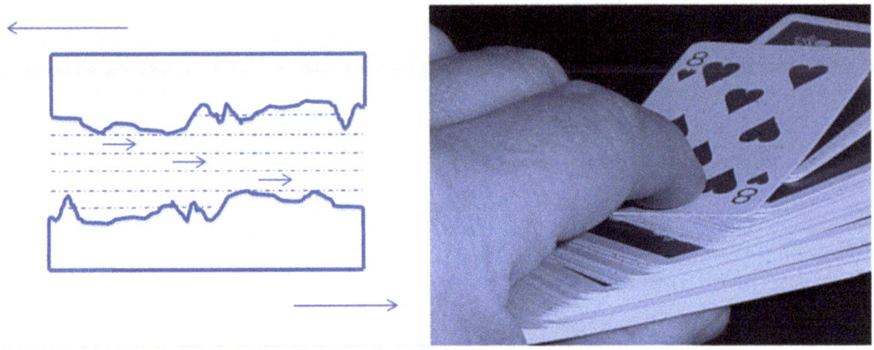

Figure R3 Full film lubrication depicting how a lubricant shears along multiple planes similar to a deck of cards. Courtesy: ENG**TECH** Industries Inc.

Shear Rates in Lubrication: At high shear rates, lubricants often exhibit shear-thinning behavior, where the viscosity decreases as the shear rate increases. This property is essential for ensuring adequate film thickness and reducing friction under dynamic conditions.

Measuring Shear Rates: Understanding shear rates is vital for designing lubricants that perform optimally under specific conditions. Advanced rheological testing methods, such as using a modified torsional Kolsky bar, can provide detailed insights into a lubricant's shear response.

730. RCAF

See "Root Cause Analysis of Failure."

731. RCL

Reliability-centered lubrication (RCL) is a strategic approach aimed at optimizing lubrication practices to enhance equipment reliability and minimize downtime. Unlike traditional methods that treat lubrication as a routine task, RCL focuses on how lubrication can be leveraged to improve asset performance.

Key Components of RCL:

- **Lubricant selection**: Choosing the appropriate lubricant type and grade based on specific operational requirements and environmental conditions.

- **Contamination control**: Implementing measures to prevent the ingress of contaminants into the lubrication system, thereby maintaining lubricant integrity.
- **Standardized procedures**: Developing and adhering to standardized lubrication and inspection procedures to ensure consistency and precision in maintenance activities.
- **Condition monitoring**: Establishing a robust oil analysis program to monitor lubricant health and detect potential issues early.
- **Training and development**: Providing comprehensive training to maintenance personnel to enhance their understanding and execution of lubrication best practices.

Benefits of Implementing RCL:

- **Reduced maintenance costs**: By optimizing lubrication practices, facilities can achieve significant cost savings through reduced equipment failures and extended asset life.
- **Enhanced equipment reliability**: Proper lubrication is crucial for the longevity and reliability of machinery, leading to improved operational efficiency.
- **Improved environmental impact**: Effective lubrication management can reduce the risk of spills and leaks, contributing to environmental protection.

732. RCM

Reliability-centered maintenance (RCM) is a strategic approach to maintenance that focuses on ensuring the reliability and performance of equipment by systematically identifying and addressing potential failure modes. Originating in the aviation industry in the 1960s, and based on the original Nowland and Heap MSG-1 report, RCM has since been adopted across various sectors, including manufacturing, to optimize maintenance practices and enhance operational efficiency.

Key Components of RCM:

1. **Function identification**: Clearly define the primary and secondary functions of each asset within the system.

2. **Failure mode analysis**: Determine how each asset can fail to fulfill its functions, considering various failure modes.
3. **Consequence assessment**: Evaluate the consequences of each failure mode, including safety, environmental, operational, and economic impacts.
4. **Criticality analysis**: Prioritize failure modes based on their severity and likelihood to focus maintenance efforts on the most critical issues.
5. **Maintenance strategy development**: Develop and implement maintenance tasks aimed at preventing or mitigating the identified failure modes.
6. **Continuous improvement**: Regularly review and adjust maintenance strategies based on performance data and evolving operational conditions.

RCM and Lubrication:
In the context of lubrication, RCM emphasizes the importance of selecting appropriate lubricants, monitoring lubricant conditions, and implementing effective lubrication practices to prevent failures related to inadequate lubrication. By integrating lubrication management into the RCM framework, organizations can enhance equipment reliability and extend asset life.

Benefits of Implementing RCM:

- **Enhanced equipment reliability**: By focusing on critical failure modes, RCM helps ensure that equipment operates as intended, reducing unexpected downtime.
- **Cost optimization**: RCM enables organizations to allocate maintenance resources effectively, performing tasks that provide the highest return on investment.
- **Improved safety and compliance**: By addressing potential failure modes proactively, RCM contributes to a safer working environment and helps maintain compliance with industry regulations.

Implementing RCM requires a thorough understanding of equipment functions, failure modes, and operational contexts. It involves collaboration across various departments, including maintenance, operations, and engineering, to develop and execute effective maintenance strategies.

733. Rerefining

Rerefining is the process of restoring used lubricating oil to a condition comparable to that of new oil by removing contaminants and impurities and replenishing additives. This process involves several key steps:

1. **Dehydration**: The used oil is heated to remove water content.
2. **Distillation**: The oil undergoes vacuum distillation to separate lighter fractions, such as fuels, from heavier base oils.
3. **Hydrotreating**: The base oil is treated with hydrogen to eliminate sulfur, nitrogen, and other impurities, enhancing its stability and performance.
4. **Additive blending**: The purified base oil is blended with additives to restore the oil's original properties, ensuring it meets industry standards.

Rerefined oil is subject to the same stringent refining, compounding, and performance standards as virgin oil. Lubricants made from rerefined base stocks must meet the same standards as virgin lubricants in order to receive the American Petroleum Institute's (API) certification. Most industrial machine manufacturers and major auto and engine manufacturers, including Mercedes-Benz, Ford, General Motors, Chrysler, and Detroit Diesel, allow the use of rerefined oils as long as they are API certified.

The rerefining process is energy-efficient, requiring less energy to produce a gallon of rerefined base stock than to produce a base stock from crude oil. Additionally, rerefined oil prices are competitive with virgin oil products. Rerefined oils that are API-certified meet warranty requirements for new cars and trucks.

By choosing rerefined oil, consumers contribute to environmental conservation, reduce dependence on foreign oil, and support sustainable practices in the automotive industry.

734. Reactive Maintenance

Reactive maintenance is a maintenance strategy where repairs are performed only after equipment has failed or malfunctioned. This approach focuses on restoring assets to their standard operating conditions following a breakdown or performance issue and is often not thought of as a positive asset management strategy.

Types of Reactive Maintenance:

1. **Emergency maintenance**: Immediate action taken to restore an asset's operational capacity, often in response to urgent safety requirements.

2. **Corrective maintenance**: Performed after a defect or failure has been detected, aiming to restore the asset to its optimal condition.

Advantages:

- **Lower initial costs**: No upfront expenses for maintenance planning, stocked spare parts, or preventive measures.

- **Simplicity**: Maintenance is straightforward, addressing issues as they arise without the need for complex schedules.

Disadvantages:

- **Higher long-term costs**: Unplanned downtime and emergency repairs can lead to increased expenses.

- **Unpredictable downtime**: Unexpected equipment failures can disrupt operations and affect production schedules. Note that downtime costs often far outweigh the cost of any planned maintenance strategy making this approach a very poor alternative.

- **Shortened equipment lifespan**: Running equipment to failure can reduce its overall lifespan and increase the frequency of replacements.

While reactive maintenance may seem cost-effective initially, it often results in higher long-term expenses due to unplanned downtime and emergency repairs. Implementing a balanced maintenance strategy that includes both reactive and proactive measures can help optimize operational efficiency and reduce overall costs.

735. Reciprocating Compressor Lubrication

Proper lubrication is essential for the efficient operation and longevity of reciprocating compressors. These compressors utilize various lubrication systems to ensure that moving components, such as pistons, cylinders, and bearings, operate smoothly and with minimal wear.

Lubrication Systems in Reciprocating Compressors:

1. **Splash lubrication**: Common in small to medium-sized compressors, this system relies on oil in a reservoir that splashes onto moving parts, providing lubrication.

2. **Mist lubrication**: Typically found in larger compressors, this system atomizes the lubricant into a mist, which is then injected into the cylinders for even distribution.

3. **Forced lubrication**: Used in large compressors operating under high loads and speeds, this system employs a pump to pressurize and deliver lubricant to critical components, ensuring consistent lubrication.

Importance of Proper Lubrication:

- **Reduction of wear and tear**: Adequate lubrication minimizes friction between moving parts, reducing wear and extending the lifespan of the compressor.
- **Enhanced efficiency**: Proper lubrication ensures that components operate smoothly, leading to improved compressor efficiency and performance.
- **Prevention of contamination**: Effective lubrication systems help prevent contaminants from entering the compressor, maintaining the integrity of the compressed air or gas.

Maintenance Considerations:

- **Regular monitoring**: Consistently check lubricant levels and quality to ensure optimal performance.
- **Scheduled oil changes**: Adhere to manufacturer-recommended oil change intervals to maintain lubrication effectiveness.
- **System inspections**: Regularly inspect lubrication systems for leaks, blockages, or other issues that could impair performance.

736. Reciprocating Engine

Proper lubrication is essential for the efficient operation and longevity of reciprocating engines, commonly found in automobiles, aircraft, and industrial machinery. Lubrication serves multiple critical functions, including reducing friction, dissipating heat, sealing combustion chambers, and carrying away contaminants.

Lubrication Systems in Reciprocating Engines:

1. **Wet-sump systems**: In these systems, oil is stored in a sump integral to the engine. An oil pump circulates the oil through the engine, and

after lubricating the components, it returns to the sump. This system is commonly used in smaller engines. See Figure R4.

2. **Dry-sump systems**: Here, oil is stored in a separate tank and circulated through the engine by pumps. This design allows for a greater volume of oil to be supplied, making it suitable for larger engines.

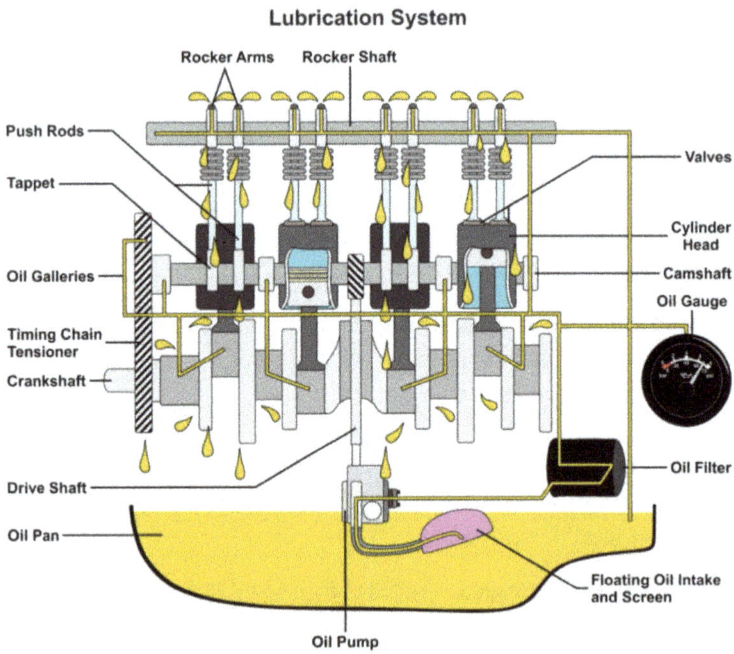

Figure R4 Reciprocating engine cross section demonstrating lubricant path. Source: Gulf Oil Company.

Types of Lubricants:

- **Straight mineral oil**: Typically used during the break-in period after a major engine overhaul or new cylinder replacement.
- **Ashless-dispersant oil**: The industry standard for normal operations, helping minimize the formation of engine sludge, varnish, and combustion chamber deposits.
- **Synthetic oil**: Represents the latest technology, offering less frictional drag that increases performance characteristics and lengthens oil change intervals.

Maintenance Considerations:

- **Regular oil changes**: Adhering to manufacturer-recommended oil change intervals is crucial to maintain engine health.
- **Monitoring oil levels and quality**: Consistently checking oil levels and quality ensures optimal performance and early detection of potential issues.
- **System inspections**: Regularly inspecting the lubrication system for leaks, blockages, or other issues is essential to prevent operational disruptions.

By implementing and maintaining appropriate lubrication systems and practices, operators can ensure the reliable and efficient operation of reciprocating engines, thereby reducing downtime and maintenance costs.

737. Recirculative Lubrication System

See "Circulating Oil System."

738. Refinery

An **oil refinery** is an industrial facility that processes crude oil into a variety of valuable products, including gasoline, diesel, kerosene, jet fuel, heating oils, and petrochemical feedstocks (see Figure R5). This transformation involves several key processes:

1. **Distillation**: Crude oil is heated and separated into different components based on their boiling points.

Figure R5 Typical oil refinery. Courtesy: Pexels – Tom Fisk.

2. **Conversion**: Heavier fractions are broken down into lighter, more valuable products through processes like cracking and reforming.

3. **Treatment**: The refined products are treated to remove impurities, ensuring they meet quality standards.

Refineries play a crucial role in the energy sector, supplying fuels for Industry, transportation, heating, and electricity generation, as well as raw materials for the petrochemical and lubricating oil industry. In 2020, the total capacity of global refineries for crude oil was about 101.2 million barrels per day.

739. Refraction

In lubrication, **refraction** refers to the bending of light as it passes through lubricating fluids. This optical property is utilized to assess the concentration of dissolved solids or additives in the lubricant. By measuring the refractive index of a lubricant, one can determine its concentration, which is crucial for maintaining optimal performance in various applications.

Refractometers in Lubrication:
A refractometer (see Figure R6) is an essential tool in lubrication management. It measures the refractive index of a liquid, which correlates with the concentration of dissolved solids or additives. In metalworking fluids, for example, refractometers help determine the concentration of coolant additives, ensuring that the fluid maintains its desired performance characteristics.

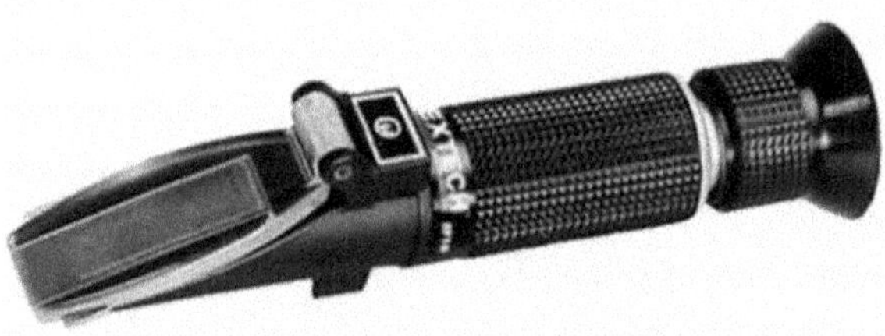

Figure R6 Typical hand-held refractometer.

Applications:

- **Monitoring lubricant concentration**: Regular use of refractometers allows for consistent monitoring of lubricant concentration, ensuring it remains within the optimal range for effective performance.

- **Quality control**: By assessing the refractive index, manufacturers can maintain the quality of lubricants, ensuring they meet the required specifications for various applications.

Understanding and utilizing the refractive properties of lubricants enables industries to optimize lubrication processes, leading to improved efficiency and extended equipment lifespan.

740. Refractive Index

The **refractive index** of a lubricant is a critical optical property that indicates how light propagates through the fluid. It is defined as the ratio of the speed of light in a vacuum to the speed of light in the lubricant. This property is particularly useful in assessing the composition and purity of lubricants, as it can reveal information about the molecular structure and concentration of additives within the fluid.

Refractive Index of Common Lubricants:

- **Mineral oils**: Typically exhibit a refractive index ranging from 1.47 to 1.48.
- **Sperm oil**: Has a refractive index of approximately 1.4649 at 15.6 °C.
- **Whale oil**: Exhibits a refractive index of about 1.4760 at 15 °C.
- **Naphthenic oils**: Demonstrate a refractive index around 1.56.
- **Biodiesel**: Typically has a refractive index of 1.56.
- **Immersion oils**: Commonly used in microscopy, these oils have a refractive index of 1.51.

Understanding the refractive index of lubricants is essential for quality control and formulation processes. It aids in identifying the presence of specific additives, ensuring consistency in lubricant production, and verifying the authenticity of the lubricant.

741. Relative Oil Thickness

Relative oil thickness is a fundamental parameter in tribology that quantifies the relationship (ratio) between the lubricant film thickness and the combined surface roughness of two contacting surfaces. It is commonly defined as the film thickness ratio or **lambda ratio** (λ),

$$\lambda = \frac{h}{\sigma}.$$

where:

- h is the central lubricant film thickness.
- σ (sigma) is the combined root mean square (RMS) surface roughness of the two contacting surfaces.

The lambda ratio is crucial for distinguishing between different lubrication regimes:

- **Boundary lubrication**: Occurs when $\lambda \leq 0.1$, indicating that the lubricant film is insufficient, and surfaces are in direct contact, leading to higher friction and wear.
- **Mixed lubrication**: Occurs when $0.1 < \lambda < 3$, where the lubricant film partially separates the surfaces, resulting in both asperity contact and hydrodynamic support.
- **Hydrodynamic lubrication**: Occurs when $\lambda \geq 3$, where a full lubricant film separates the surfaces, minimizing asperity contact and reducing friction.

Understanding and calculating the lambda ratio is essential for selecting appropriate lubricants and optimizing operating conditions to enhance equipment performance and longevity. See also "Stribeck Curve."

742. Release Agent

A **release agent** is a chemical substance applied to surfaces to prevent other materials from adhering to them. In lubrication, release agents serve as a barrier between surfaces, facilitating the separation of components during manufacturing processes and reducing friction between moving parts. They are commonly used in molding, casting, and forming operations to ensure that molded or cast parts can be easily removed from molds without sticking.

Types of Release Agents:

- **Barrier release agents**: These create a physical barrier between the mold and the material being molded, preventing adhesion. Common examples include waxes, fatty esters, silicones, and metallic soaps.

- **Reactive release agents**: These chemically interact with the material being molded to form a nonstick surface, facilitating easy release.

Applications in Lubrication:

- **Molding and casting**: Release agents are essential in processes like die-casting, plastic molding, and concrete forming, ensuring that parts can be removed from molds without damage.
- **Food processing**: In baking and cooking, release agents prevent food from sticking to pans and molds, facilitating easy removal and cleaning.
- **Rubber and plastics manufacturing**: Release agents are used to prevent rubber or plastic products from sticking to molds during production, ensuring smooth ejection and maintaining mold integrity.

By selecting appropriate release agents, manufacturers can enhance product quality, improve production efficiency, and extend the lifespan of molds and equipment.

743. Reliability

Reliability in lubrication refers to the consistent and effective performance of lubricants in reducing friction, minimizing wear, and extending the lifespan of machinery components. Proper lubrication is essential for maintaining equipment reliability, as it ensures smooth operation and prevents premature failures.

Key Aspects of Lubrication Reliability:

1. **Optimal lubricant selection**: Choosing the right lubricant type and grade based on equipment specifications and operating conditions is crucial. This selection helps in achieving the desired performance and longevity of machinery.
2. **Regular lubricant maintenance**: Implementing a routine schedule for lubricant changes and monitoring lubricant quality ensures that the lubricant remains effective throughout its service life. This practice helps in maintaining equipment reliability and preventing unexpected failures.

3. **Contamination control**: Protecting lubricants from contaminants such as dirt, water, and metal particles is vital. Contaminated lubricants can lead to increased wear and potential equipment damage.

4. **Proper lubricant application**: Applying the correct amount of lubricant at the appropriate intervals and using suitable methods ensures optimal performance. Overlubrication or underlubrication can both negatively impact equipment reliability.

5. **Training and standardization**: Educating maintenance personnel on best practices and standardizing lubrication procedures contribute to consistent and reliable lubrication practices across the organization.

By focusing on these aspects, organizations can enhance equipment reliability, reduce downtime, and lower maintenance costs. Implementing a comprehensive lubrication reliability program tailored to specific operational needs is a strategic approach to achieving these goals.

744. Reservoir

In lubrication systems, a **reservoir** serves as the storage vessel for the lubricant, ensuring a consistent supply to various machine components. Proper management of the lubricant reservoir is crucial for maintaining equipment performance and longevity.

Key Functions of a Lubricant Reservoir:

1. **Storage**: Holds the lubricant until it is needed for distribution throughout the system.

2. **Filtration**: Often equipped with a suction filter (oil) and breather filter, ensuring only clean lubricant circulates through the pump into the delivery system.

3. **Temperature**: Some reservoirs incorporate cooling or heating mechanisms to maintain the lubricant at optimal temperatures, preventing overheating and degradation and ensuring the lubricant will stay at the correct viscosity and flow correctly through the system.

4. **Monitoring**: Includes level indicators and sensors to monitor lubricant levels and quality, facilitating timely maintenance actions.

Best Practices for Reservoir Management:

- **Regular inspections**: Consistently check lubricant levels and quality to ensure optimal performance.
- **Contamination control**: Implement measures to prevent contaminants from entering the reservoir, such as using desiccant breathers and maintaining proper sealing.
- **Proper filling**: When topping up or changing the lubricant, ensure cleanliness to prevent introducing contaminants.

745. Reservoir Breather

See "Oil Breather."

746. Reservoir Headspace

In lubrication systems, **headspace** refers to the air volume above the lubricant within a reservoir. Proper management of this headspace is crucial for maintaining lubricant quality and extending equipment lifespan. See Figure R7.

Key Considerations for Headspace Management:

1. **Contamination prevention**: The headspace can be a conduit for contaminants like moisture and particulate matter entering the lubricant. Implementing desiccant breathers or membrane filtration systems can effectively remove moisture and particles, preserving lubricant integrity.
2. **Pressure regulation**: Temperature fluctuations and fluid level changes can cause pressure variations within the reservoir. Utilizing expansion chambers or pressure relief valves helps maintain optimal pressure, preventing air ingress and potential contamination.
3. **Moisture control**: Moisture in the headspace can lead to water contamination in the lubricant, promoting oxidation and degradation. Systems like the TMR®-N_2 Water Removal Nitrogen System introduce dry nitrogen into the headspace, effectively removing water and maintaining low moisture levels.

By effectively managing the headspace in lubricant reservoirs, organizations can enhance equipment reliability, reduce maintenance costs, and extend the operational lifespan of machinery.

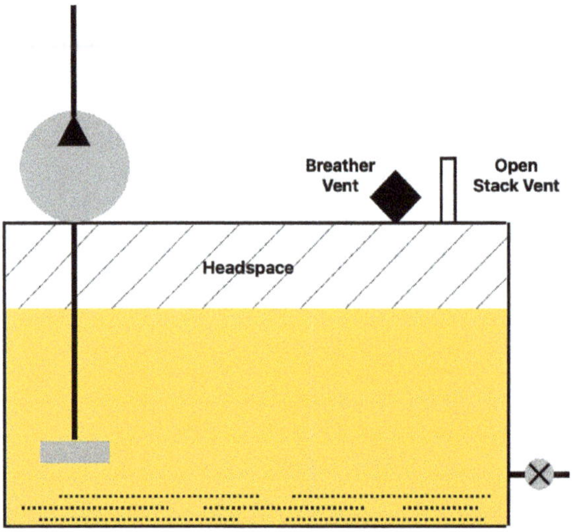

Figure R7 Headspace example in a lubricant reservoir.

747. Reservoir Vent

In lubrication systems, **reservoir vents** are essential components that facilitate the exchange of air between the reservoir and the atmosphere, accommodating pressure changes due to fluid level fluctuations and temperature variations. Proper venting is crucial for maintaining optimal fluid conditions and preventing contamination. Both vent types described below are illustrated in Figure R7.

Types of Reservoir Vents:

1. **Breather vents**:
   - **Function**: Breather vents allow air to enter and exit the reservoir, equalizing internal and external pressures. They often incorporate filters to prevent contaminants from entering the reservoir.
   - **Design**: Typically features a filter element that captures particles and moisture from the incoming air, ensuring only clean air enters the reservoir. Some models include desiccant materials to absorb moisture, maintaining the quality of the lubricant.

- **Applications**: Commonly used in hydraulic systems, gearboxes, and other machinery where maintaining clean and dry lubricant is essential.

2. **Open stack vents**:
   - **Function**: Open stack vents provide a direct path for air to enter and exit the reservoir without filtration.
   - **Design**: Consists of a simple pipe or stack extending from the reservoir, allowing air exchange. This design lacks filtration, making it susceptible to contamination.
   - **Applications**: Less common in modern systems due to the risk of contaminant ingress. They may still be found in applications where contamination is less of a concern or where simplicity is prioritized.

Considerations for Choosing a Reservoir Vent:

- **Contamination control**: Breather vents with filtration are preferable in environments where maintaining lubricant purity is critical.
- **Environmental conditions**: In humid or dusty environments, breather vents with desiccant elements can prevent moisture and particulate contamination.
- **Maintenance requirements**: Breather vents with replaceable filters or desiccant elements require regular maintenance to ensure optimal performance.

Selecting the appropriate reservoir vent is vital for the longevity and efficiency of lubrication systems. Incorporating a breather vent with filtration can significantly reduce the risk of contamination, thereby enhancing equipment reliability.

748. Resistivity/Conductivity Test

Measuring the resistivity and conductivity of lubricants is essential for assessing their electrical properties, which can influence equipment performance and longevity. These measurements help identify contaminants, monitor oil degradation, and prevent issues like electrostatic discharge.

Resistivity and Conductivity in Lubricants:

- **Resistivity** quantifies a lubricant's ability to resist electrical current flow. High resistivity indicates fewer free ions and contaminants, suggesting better-insulating properties. Conversely, low resistivity points to higher conductivity, which can lead to electrostatic charging and potential damage.

- **Conductivity** is the inverse of resistivity, representing how easily a lubricant allows electrical current to pass through. Elevated conductivity can result from contaminants like water or particulate matter, increasing the risk of electrostatic discharge.

Testing Methods:

1. **ASTM D1169 – specific resistance (resistivity) of electrical insulating liquids**:
   - The standard measures the resistivity of insulating oils, providing insights into their insulating capabilities.
2. **ASTM D2624 – electrical conductivity of aviation distillate fuels**:
   - While primarily for fuels, this method can be adapted for lubricants to assess their conductivity levels.
3. **Dissipation factor and specific resistivity testing**:
   - This approach evaluates the dielectric properties of lubricants, including their resistivity and dissipation factor, which are indicative of oil condition and contamination levels.

Importance of Monitoring:

- **Electrostatic discharge prevention**: Low-resistivity oils are more susceptible to electrostatic charging, which can lead to equipment damage. Regular monitoring helps mitigate this risk.
- **Contamination detection**: Increased conductivity often signals the presence of contaminants like water or particulate matter, necessitating timely maintenance actions.
- **Oil degradation assessment**: Changes in resistivity and conductivity can indicate oil degradation, prompting necessary interventions to maintain equipment health.

749. Return line

A **lubricant return line** is a crucial component in lubrication systems, designed to collect and transport used or excess lubricant from lubricated machine parts and bearings back to the reservoir or storage tank. This process ensures that the lubricant is properly filtered, cooled, and reconditioned before being recirculated, maintaining optimal lubrication conditions and extending the equipment lifespan. See Figure R8.

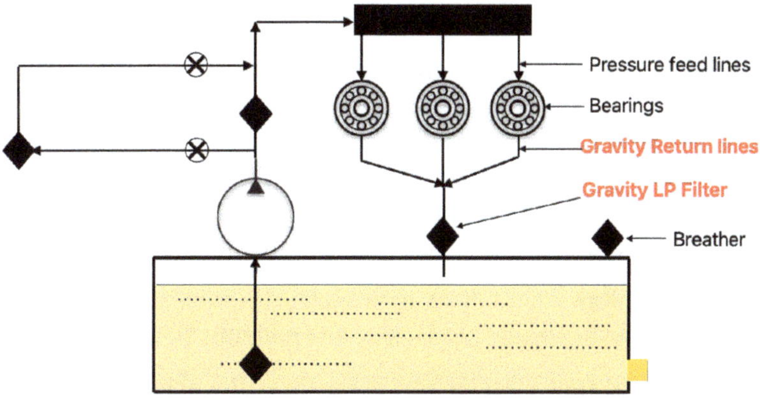

Figure R8 Lubricant return line and low-pressure filter.

Key Functions of Lubricant Return Lines:

1. **Contaminant removal**: By directing the used lubricant back to the reservoir, return lines facilitate the removal of contaminants such as metal particles, dirt, and moisture. This is typically achieved through filtration systems integrated into the return line.

2. **Temperature regulation**: As lubricant circulates through the system, it can absorb heat. Return lines often pass through cooling units to dissipate excess heat, ensuring the lubricant remains within optimal operating temperatures.

3. **Pressure management**: Return lines help maintain appropriate pressure levels within the lubrication system, preventing overpressurization and ensuring consistent lubricant flow.

750. Reynolds Number

In lubrication theory, the **Reynolds number (Re)** is a dimensionless quantity that characterizes the flow regime of a lubricant between two surfaces. It represents the ratio of inertial forces to viscous forces within the fluid and is defined as

$$\mathrm{Re} = \frac{\rho \cdot U \cdot L}{\mu}.$$

where:

- ρ (rho) is the fluid density.
- U is the characteristic velocity of the fluid.
- L is the characteristic length (e.g., the gap between surfaces).
- μ (mu) is the dynamic viscosity of the fluid.

The Reynolds number is crucial in determining the nature of the lubricant flow:

- **Laminar flow**: Occurs at low Reynolds numbers, where viscous forces dominate, resulting in smooth, orderly fluid motion.
- **Turbulent flow**: Occurs at high Reynolds numbers, where inertial forces dominate, leading to chaotic and eddying fluid motion.
- **Transitional flow**: Occurs at intermediate Reynolds numbers, where the flow can fluctuate between laminar and turbulent states.

In the context of lubrication, the Reynolds number helps predict the behavior of the lubricant film between contacting surfaces:

- **Hydrodynamic lubrication**: At high Reynolds numbers, the lubricant forms a full film separating the surfaces, minimizing direct contact and reducing friction.
- **Elastohydrodynamic lubrication (EHL)**: At moderate Reynolds numbers, the lubricant film is thin, and surface asperities may come into contact, leading to elastic deformation of the surfaces.
- **Boundary lubrication**: At low Reynolds numbers, the lubricant film is insufficient, and surfaces are in direct contact, leading to higher friction and potential wear.

Understanding the Reynolds number in lubrication is essential for designing and optimizing bearing systems, gears, and other mechanical components to ensure efficient operation and longevity.

751. Rheology

In lubrication, **rheology** refers to the study of how lubricants deform and flow under various forces. Understanding the rheological properties of lubricants is essential for predicting their behavior in different operating conditions, which directly impacts the performance and longevity of mechanical systems.

Key Rheological Properties in Lubrication:

1. **Viscosity**: This is a measure of a lubricant's resistance to flow. High viscosity indicates thicker, more resistant fluids, while low viscosity denotes thinner, more fluid-like substances. The appropriate viscosity ensures a sufficient lubricant film between moving parts, reducing friction and wear.

2. **Shear stress and shear rate**: Shear stress is the force per unit area required to deform the lubricant, while the shear rate is the rate at which the lubricant deforms. The relationship between these two factors helps in understanding how the lubricant will behave under different operational stresses.

3. **Yield stress**: This is the minimum stress required to initiate flow in a lubricant. Lubricants with higher yield stress can maintain a protective film under varying loads, preventing metal-to-metal contact.

4. **Thixotropy**: Some lubricants exhibit thixotropic behavior, where they become less viscous over time under constant shear stress and return to a more viscous state when the stress is removed. This property is particularly important in greases, affecting their pumpability and stability.

Importance of Rheology in Lubrication:

- **Performance prediction**: By understanding a lubricant's rheological properties, engineers can predict its performance under different operating conditions, ensuring optimal protection and efficiency.

- **Design optimization**: Knowledge of rheology aids in selecting or formulating lubricants that meet specific requirements, such as high load-carrying capacity or low-temperature fluidity.

- **Quality control**: Regular rheological testing ensures that lubricants maintain consistent performance throughout their service life, preventing unexpected failures.

In summary, rheology provides critical insights into the behavior of lubricants under various conditions, enabling the design of more efficient and reliable lubrication systems.

752. Rheopectic Grease

Rheopectic greases are specialized lubricants that exhibit a time-dependent increase in viscosity when subjected to shear forces. Unlike thixotropic greases, which become less viscous over time under shear, rheopectic greases thicken or solidify as they are worked or agitated.

Key Characteristics of Rheopectic Greases:

- **Time-dependent viscosity increase**: These greases become thicker with prolonged shear, providing enhanced stability and resistance to leakage under dynamic conditions.
- **Enhanced stability**: The solidifying nature of rheopectic greases helps maintain a consistent lubricant film, reducing the risk of component wear and extending equipment lifespan.

Applications:
Rheopectic greases are particularly beneficial in scenarios where:

- **High shear environments**: Components experience continuous movement or vibration, necessitating a lubricant that can maintain its consistency under dynamic conditions.
- **Leak prevention**: The increased viscosity helps prevent grease from leaking out of bearings and seals, ensuring effective lubrication over extended periods.

Considerations:
While rheopectic greases offer advantages in specific applications, it is essential to ensure compatibility with the operating conditions and materials involved.

753. Ring Lubrication

Ring lubrication refers to a method of lubricating rotating shafts using a ring oiler system. In this setup, a metal ring is placed around a horizontal shaft adjacent to a bearing. The ring dips into an oil sump beneath the shaft, and as the shaft rotates, the ring picks up oil and deposits it onto the shaft, which then migrates into the bearing to provide lubrication. See Figure R9.

Key Components of a Ring Oiler System:

- **Ring**: A metal ring that rotates with the shaft, picking up oil from the sump.

- **Oil sump**: A reservoir beneath the shaft containing the lubricating oil.
- **Bearing**: The component that receives lubrication from the oil transferred by the ring.

Figure R9 Oil ring lubrication.

Advantages of Ring Lubrication:

- **Simplicity**: The system is straightforward and easy to maintain.
- **Cost-effective**: It is a low-cost method suitable for various applications.
- **Effective lubrication**: Provides a continuous supply of oil to bearings, reducing friction and wear.

Considerations:

- **Alignment**: The shaft system must be nearly horizontal to ensure proper oil pickup by the ring.
- **Oil viscosity**: The lubricant's viscosity should be appropriate for the operating conditions to ensure effective lubrication.
- **Maintenance**: Regular inspection is necessary to ensure the system functions correctly and to prevent issues such as oil contamination.

Ring lubrication is commonly used in applications where simplicity and cost-effectiveness are priorities, such as in certain types of bearings and machinery.

754. Rolling Contact Bearings

Rolling contact bearings, commonly known as rolling-element bearings, are mechanical components that facilitate rotational or oscillatory motion by reducing friction between moving parts. They achieve this by incorporating rolling elements – such as balls or rollers – between two concentric, grooved rings called races. This design allows the rolling elements to roll with minimal resistance and little sliding, effectively supporting loads and enabling smooth movement.

Figure R10 Typical rolling element bearing (Timken double spherical roller). Source: Timken Bearing Company.

Key Components:

- **Inner race**: The ring that fits onto the shaft, providing a surface for the rolling elements to contact.
- **Outer race**: The ring that fits into the housing, also providing a contact surface for the rolling elements.
- **Rolling elements**: Balls or rollers that roll between the inner and outer races, facilitating movement and load support.
- **Cage (retainer)**: A component that separates and guides the rolling elements, ensuring even distribution and preventing contact between them.

Types of Rolling Contact Bearings:

1. **Ball bearings**:
   - **Deep groove ball bearings**: Designed to handle both radial and axial loads, suitable for high-speed applications.
   - **Angular contact ball bearings**: Capable of supporting higher axial loads in one direction, often used in pairs for bidirectional axial load handling.
2. **Roller bearings**:
   - **Cylindrical roller bearings**: Offer high radial load capacity and are suitable for high-speed applications.
   - **Spherical roller bearings**: Designed to accommodate misalignment between the shaft and housing, suitable for heavy radial and axial loads. See Figure R10.
   - **Tapered roller bearings**: Capable of handling both radial and axial loads, commonly used in automotive wheel hubs.

Advantages:

- **Reduced friction**: Minimizes energy loss and heat generation.
- **High load capacity**: Supports substantial radial and axial loads.
- **Durability**: Designed for long service life with proper maintenance.

Considerations:

- **Lubrication**: Essential for reducing wear and preventing corrosion.
- **Sealing**: Protects against contaminants and retains lubricant.
- **Alignment**: Proper installation ensures optimal performance and longevity.

755. Rolling Contact Lubrication

Rolling contact lubrication is essential for the efficient operation and longevity of rolling-element bearings, such as ball and roller bearings. Proper lubrication minimizes friction, reduces wear, and prevents corrosion by forming a protective film between the rolling elements and races.

Lubrication Methods:

1. **Oil lubrication**:
    - **Application**: Suitable for high-speed and high-temperature environments.
    - **Advantages**: Provides effective heat dissipation and can be circulated to remove contaminants.
    - **Considerations**: Requires sealing to prevent leakage and contamination.
2. **Grease lubrication**:
    - **Application**: Commonly used in applications where oil lubrication is impractical.
    - **Advantages**: Simplifies sealing requirements and reduces leakage.
    - **Considerations**: May have limitations in high-speed applications due to the potential for churning and heat generation.

Lubrication Objectives:

- **Friction reduction**: Establishing a lubricant film to separate contact surfaces, thereby minimizing friction and wear.
- **Heat dissipation**: Transferring heat away from contact areas to prevent overheating.
- **Contaminant removal**: Transporting debris and particles away from bearing surfaces.
- **Corrosion prevention**: Protecting metal surfaces from moisture and corrosive elements.

756. Rolling Friction

Rolling friction, also known as **rolling resistance**, is the **force that resists the motion of a rolling object** over a surface. It occurs due to the **deformation of the wheel, bearing, or surface** as they come into contact, causing energy loss. Rolling friction is significantly lower than **sliding friction**, making it ideal for applications involving **wheels, ball bearings, and rollers** in transportation and industrial machinery. See **"Friction"** reference.

Key Factors Affecting Rolling Friction

- **Material properties**: Softer materials (e.g., rubber) experience **higher rolling resistance**, while harder materials (e.g., steel) roll more easily.
- **Surface deformation**: Uneven or soft surfaces increase resistance by causing **more deformation**.
- **Load on the rolling object**: Higher loads **increase contact area** and resistance.
- **Lubrication**: Proper lubrication reduces **internal resistance in bearings and rollers**.
- **Rolling speed**: At very high speeds, **heat generation and material properties** can affect rolling resistance.

757. Rolling Oil

Rolling oil is a specialized lubricant formulated for metal rolling processes, where metal stock is passed through rollers to reduce metal thickness and achieve desired shapes. This lubricant serves multiple functions, including reducing friction, dissipating heat, and
preventing corrosion during the rolling process.

Key Functions of Rolling Oil:

- **Lubrication**: Minimizes friction between the metal and rollers, facilitating smoother deformation and reducing wear on equipment.
- **Cooling**: Absorbs and dissipates heat generated during the rolling process, maintaining optimal metal temperatures and preventing overheating.
- **Corrosion protection**: Forms a protective film on metal surfaces, shielding them from rust and corrosion.

Types of Rolling Oils:

1. **Cold rolling oils**: Designed for rolling metals at ambient temperatures, these oils provide excellent lubrication and cooling properties. They are often formulated with additives to enhance performance and extend tool life.
2. **Hot rolling oils**: Formulated for high-temperature applications, these oils offer superior thermal stability and oxidation resistance, ensuring effective lubrication under extreme conditions.

Applications:

- **Ferrous metals**: Used in rolling processes for steel and iron to achieve the desired thickness and surface finish.
- **Nonferrous metals**: Applied in rolling aluminum, brass, copper, and their alloys, requiring specific formulations to prevent staining and ensure surface quality.

Benefits:

- **Enhanced surface quality**: Improves the finish and appearance of rolled metal products.
- **Extended equipment life**: Reduces wear and tear on rolling mills and associated machinery.
- **Operational efficiency**: Facilitates smoother rolling operations, leading to higher productivity and reduced downtime.

Selecting the appropriate rolling oil is crucial for optimizing the rolling process, ensuring product quality, and maintaining equipment longevity.

758. RON

See "Octane Number."

759. Root Cause Analysis of Failure

Root cause analysis of failure (RCAF) is a systematic approach used to identify the fundamental causes of problems or failures within a product, process, or system. By uncovering these root causes, organizations can implement corrective actions that address the underlying issues, thereby preventing recurrence and enhancing overall performance.

Key Steps in Root Cause Analysis:

1. **Problem identification**: Clearly define the issue or failure that needs to be addressed.
2. **Data collection**: Gather relevant information, such as equipment history, maintenance records, and process data.
3. **Root cause determination**: Analyze the data to identify the underlying factors that led to the failure, rather than just the immediate cause.

4. **Corrective action development**: Design and implement solutions to address the root causes and prevent recurrence.
5. **Verification**: Monitor the effectiveness of the corrective actions and make adjustments as needed.

Implementing RCAF is crucial for organizations aiming to improve reliability, quality, and operational efficiency. It supports a culture of continuous improvement by providing a structured approach to problem-solving, empowering teams to proactively address issues before they escalate.

760. Rope Lubrication

See "Wire Rope Pressure Lubrication Systems."

761. RPVOT (Rotating Pressure Vessel Oxidization Test)

The **Rotating Pressure Vessel Oxidation Test (RPVOT)**, formerly known as the Rotating Bomb Oxidation Test (RBOT), is a laboratory procedure designed to assess the oxidation stability of lubricants, particularly turbine and circulating oils. This test evaluates how well a lubricant resists oxidation under controlled, accelerated conditions, providing insights into its potential service life and performance in real-world applications.

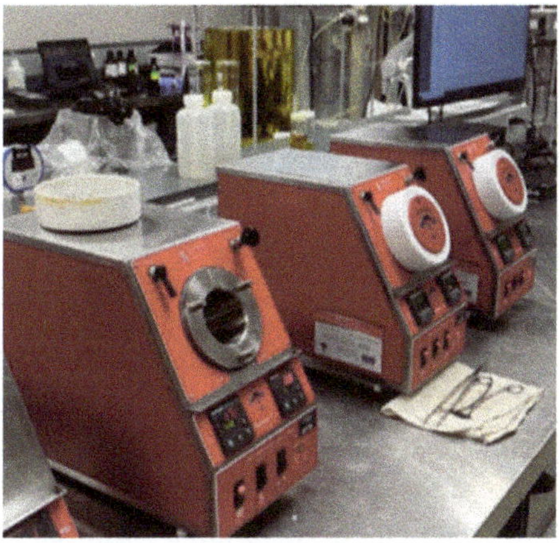

Figure R11 Typical RPVOT test apparatus in a laboratory. Source: Wear Check Corporation.

Test Procedure:
In the RPVOT test, a sample of the lubricant is combined with distilled water and a copper catalyst within a sealed pressure vessel, see Figure R11. The vessel is pressurized with pure oxygen and heated to a specific temperature, typically around 150 °C (302 °F). The vessel is then rotated at a constant speed to promote mixing. As the test progresses, the lubricant's resistance to oxidation is monitored by observing the pressure changes within the vessel. The test concludes when the pressure drops by a predetermined amount, indicating that the lubricant has reached its oxidation endpoint.

Advantages:

- **Rapid results**: RPVOT provides relatively quick assessments of a lubricant's oxidation stability compared to other methods, such as the Turbine Oil Stability Test (TOST), which can take months or years.
- **Quality control**: This test allows manufacturers to compare different batches of the same lubricant formulation, ensuring consistency and quality across production lots.

Limitations:

- **Oxidation by-products not measured**: RPVOT does not directly measure the formation of oxidation by-products like varnish, sludge, or acids, which can lead to operational issues in machinery.
- **Test conditions versus real-world conditions**: The test operates at higher temperatures and pressures than those typically found in actual equipment, which may not accurately simulate in-service conditions.
- **Reproducibility**: The test has a reproducibility of approximately ±22%, meaning results can vary between laboratories and even between tests in the same lab.

Interpretation of Results:
RPVOT results are expressed in minutes, indicating the time the lubricant withstood the accelerated oxidative conditions before reaching the endpoint. A higher RPVOT value suggests better oxidation stability. However, it is important to consider these results alongside other analyses, such as acid number (AN) and varnish potential, to obtain a comprehensive understanding of the lubricant's condition.

762. Route Lubrication

Route lubrication involves systematically planning and executing lubrication tasks across machinery and equipment within a facility. This approach ensures that all necessary lubrication points are addressed efficiently, maintaining optimal performance and extending the lifespan of equipment.

Key Components of Route Lubrication:

1. **Identification of lubrication points by machine asset and lubricant**: Determine all components requiring lubrication, such as bearings, gears, and motors by machine asset number. Next, separate the list by lubricant specification for both oils and grease types in use.
2. **Lubricant consolidation**: If a lubricant consolidation program is not already in place to reduce the amounts of different lubricants in use, now is a perfect time to put one in place.
3. **Scheduling**: Establish a series of lubrication routes based on lubricant types, lubrication interval requirements, machine locations, asset availability, and resource availability.
4. **Documentation**: Maintain records of completed lubrication work tasks, including details like date, time, lubricant type, and any observations.
5. **Training**: Ensure personnel are trained and certified in proper lubrication techniques and understand the importance of adhering to established routes.

Benefits of Route Lubrication:

- **Consistency**: Regularly scheduled lubrication prevents over or under-lubrication, reducing the risk of equipment failure.
- **Efficiency**: A well-planned route minimizes downtime and optimizes maintenance efforts.
- **Cost savings**: Proper lubrication reduces wear and tear, leading to fewer repairs and extended equipment life.

763. Rust Inhibitor

See "Corrosion Inhibitor."

S

764. Sacrificial Solid Film Barrier

A **sacrificial solid film barrier** is a protective lubrication layer designed to wear away over time, preventing direct contact between moving surfaces and minimizing friction, wear, and surface damage. These films are commonly used in extreme operating conditions where conventional liquid lubricants may fail.

Key Functions of a Sacrificial Solid Film Barrier

Friction reduction: Forms a low-friction interface between surfaces, minimizing wear and improving efficiency.

Wear protection: Absorbs mechanical stress and gradually wears away, preserving the integrity of underlying components.

Extreme condition performance: Provides reliable lubrication in high-temperature, high-load, or vacuum environments where liquid lubricants are ineffective.

Types of Sacrificial Solid Film Barriers

1. **Dry film lubricants**: Coatings such as molybdenum disulfide (MoS_2), graphite, and PTFE (polytetrafluoroethylene) that reduce friction and wear in extreme conditions.
2. **Antiwear coatings**: Engineered layers like diamond-like carbon (DLC) or ceramic-based coatings that enhance surface durability and longevity.
3. **Corrosion-resistant films**: Protective layers that prevent oxidation and degradation in harsh environments, extending component lifespan.

Applications

- **Aerospace and defense**: Used in aircraft components, space equipment, and military applications where lubrication failure is not an option.

DOI: 10.1201/9788743806608-782

- **Automotive and industrial machinery**: Applied to gears, bearings, and sliding surfaces to improve durability under high loads.
- **Medical and precision equipment**: Utilized in surgical instruments and micro-mechanical devices where contamination from liquid lubricants is undesirable.

Benefits

- **Extended component life**: Protects surfaces from excessive wear, reducing the need for frequent replacements.
- **Reliable performance in harsh environments**: Functions effectively in extreme temperatures, high pressures, and vacuum conditions.
- **Reduced maintenance requirements**: Minimizes lubrication intervals and improves overall system efficiency.

765. Sacrificial Wear

Sacrificial wear is a controlled wear mechanism where a designated bearing surface material is intentionally designed to wear out, protecting critical machine components from excessive damage. This approach extends the lifespan of essential parts by allowing replaceable elements to absorb wear and prevent costly failures.

Key Functions of Sacrificial Wear

- **Component protection**: Shields vital machine parts by allowing a softer or designated material to wear instead.
- **Wear management**: Controls the rate and location of wear to maintain system performance and reliability.
- **Cost reduction**: Reduces maintenance costs by enabling the replacement of low-cost sacrificial components instead of expensive machinery parts.

Types of Sacrificial Wear

1. **Abrasive sacrificial wear**: Occurs when a softer material gradually wears down due to contact with harder surfaces or abrasive particles, preventing damage to critical components.
2. **Adhesive sacrificial wear**: Involves the gradual loss of a coating or surface treatment to protect underlying structures from excessive friction and wear.

3. **Corrosive sacrificial wear**: Utilizes materials like sacrificial anodes in cathodic protection systems, which deteriorate over time to prevent corrosion of essential metal surfaces.

Applications

- **Bearings and bushings**: Soft metal or polymer bearings wear down to protect shafts and housings from excessive damage.
- **Brake pads and clutch linings**: Designed to wear out gradually, ensuring consistent braking and transmission performance.
- **Sacrificial anodes**: Used in marine, pipeline, and infrastructure applications to prevent corrosion of metal structures.

Benefits

- **Extended equipment life**: Protects expensive machinery components from premature failure and costly repairs.
- **Predictable maintenance**: Enables scheduled part replacements, reducing unplanned downtime.
- **Cost efficiency**: Minimizes repair costs by focusing wear on replaceable elements rather than critical machine parts.

Implementing sacrificial wear strategies enhances reliability, optimizes maintenance planning, and extends equipment lifespan in various industrial applications.

766. SAE International

SAE International (formerly the Society of Automotive Engineers) is a global professional association and standards-developing organization dedicated to advancing mobility technology across various industries, including automotive, aerospace, and commercial vehicles. It establishes engineering best practices, publishes technical standards, and fosters innovation in transportation and manufacturing.

Key Functions of SAE International

- **Standards development**: Creates and maintains industry standards for automotive, aerospace, and commercial vehicle systems, ensuring safety, reliability, and performance.
- **Technical research and publications**: Provides peer-reviewed technical papers, journals, and industry reports to support engineering advancements.

- **Professional certification and training**: Offers credentialing programs and educational resources to enhance technical expertise and workforce development.

Major SAE Lubrication Standard

- **SAE J300 – engine oil viscosity classification**: Defines viscosity grades for engine oils, widely used in automotive lubrication.

Benefits

- **Enhanced safety and reliability**: Ensures consistent engineering and lubrication practices that improve product performance and durability.
- **Global industry standardization**: Facilitates interoperability and compliance across international markets.
- **Technological advancement**: Supports research and innovation in mobility, sustainability, and emerging engineering fields.

SAE International plays a critical role in shaping the future of mobility, setting industry benchmarks, and promoting engineering excellence worldwide.

767. Safety Data Sheets – SDS

A **safety data sheet (SDS)** for lubricants is a comprehensive standardized document that provides essential information about the composition, hazards, safe handling, and regulatory compliance of a lubricant product. Required by global regulations such as OSHA's Hazard Communication Standard (HCS) and the Globally Harmonized System (GHS), SDS documents ensure workplace safety and environmental protection.

Key Functions of Safety Data Sheets for Lubricants

- **Hazard identification**: Details potential risks, such as flammability, toxicity, or environmental impact.
- **Safe handling and storage**: Provides guidelines on proper use, storage conditions, and personal protective equipment (PPE) requirements.
- **Emergency and first aid measures**: Outlines procedures for spills, accidental exposure, and fire hazards.

Location of SDS in a Plant

Safety data sheets are typically found in a collective binder and placed in numerous conspicuous places within a plant (see Figure S1). Obviously, one of these locations would include the storage area for lubricants and chemicals, as well as any first aid station.

SDS sheets can also typically be found electronically on a company's internal website for any person to view.

Figure S1 Typical safety data sheet (SDS) binder displayed in a conspicuous manner.

Sections of a Lubricant Safety Data Sheet

1. **Identification**: Product name, manufacturer details, and intended uses.
2. **Hazard identification**: Warnings on health, environmental, and physical hazards.
3. **Composition/ingredients**: Chemical formulation, including hazardous components.
4. **First-aid measures**: Instructions for treating exposure through inhalation, skin contact, or ingestion.
5. **Fire-fighting measures**: Recommended extinguishing agents and fire hazards.

6. **Accidental release measures**: Spill containment, cleanup procedures, and environmental precautions.
7. **Handling and storage**: Safe storage conditions and compatibility information.
8. **Exposure controls/PPE**: Recommended protective equipment and workplace exposure limits.
9. **Physical and chemical properties**: Appearance, odor, viscosity, and flash point.
10. **Stability and reactivity**: Conditions to avoid, potential hazardous reactions, and decomposition products.
11. **Toxicological information**: Effects of exposure, including acute and chronic toxicity.
12. **Ecological information**: Environmental impact, biodegradability, and aquatic toxicity.
13. **Disposal considerations**: Proper waste disposal and recycling guidelines.
14. **Transport information**: Classification for shipping and regulatory compliance.
15. **Regulatory information**: Compliance with local, national, and international safety regulations.
16. **Other information**: Revision date and additional safety notes.

Applications

- **Workplace safety**: Ensures employees understand the risks and proper handling of lubricants.
- **Regulatory compliance**: Helps organizations comply with OSHA, GHS, and other safety regulations.
- **Emergency response**: Provides critical information for first responders and environmental agencies.

Benefits

- **Enhanced worker protection**: Reduces health risks associated with lubricant exposure.

- **Regulatory adherence**: Ensures compliance with safety and environmental laws.
- **Improved risk management**: Helps prevent workplace accidents, spills, and hazardous incidents.

Safety data sheets are essential for maintaining workplace safety, regulatory compliance, and responsible lubricant handling in industrial and commercial settings.

768. Sample Bottle

A lube sample bottle is a specialized container used to collect, store, and transport oil samples for analysis. These bottles are designed to maintain sample integrity, preventing contamination and degradation while ensuring accurate test results for condition monitoring and predictive maintenance programs. Typical bottle sizes used are between 4 and 6 oz fluid or 100 and 150 mL (see Figure S2).

Key Functions of a Lube Sample Bottle

- **Contamination control**: Ensures the collected oil remains free from external contaminants, preserving the accuracy of analysis.
- **Secure storage and transport**: Provides a leak-proof, durable solution for safe handling and shipping of lubricant samples.
- **Standardized sampling**: Allows for consistent and repeatable oil sampling, critical for trend analysis and reliability assessments.

Types of Lube Sample Bottles

1. **Plastic bottles**: Typically made of high-density polyethylene (HDPE) or polypropylene, offering chemical resistance and durability.
2. **Glass bottles**: Used for applications requiring minimal permeability and high chemical stability, often preferred for fuel and volatile fluid samples.
3. **Precleaned bottles**: Designed to meet laboratory standards, ensuring no pre-existing contamination before sample collection.

Applications

- **Lubrication condition monitoring**: Used in industrial, automotive, and heavy equipment applications to assess lubricant health.

Figure S2 Sample bottles are usually between 2 and 6 fluid Oz (60 to 180 mL) in size.

- **Wear debris analysis**: Helps detect early signs of mechanical wear, contamination, and additive depletion.
- **Predictive maintenance programs**: Supports data-driven decision-making for optimizing oil change intervals and equipment life.

Benefits

- **Accurate oil analysis**: Preserves sample integrity for precise laboratory results.
- **Extended equipment life**: Enables early detection of issues, reducing unplanned downtime and costly failures.
- **Regulatory compliance**: Meets industry standards for lubricant sampling and environmental safety.

Selecting the right lube sample bottle is crucial for obtaining reliable oil analysis results, enhancing predictive maintenance strategies, and ensuring optimal equipment performance.

769. Sample Bottle Cleanliness

See "Signal to Noise Ratio."

770. Sample 10× Flush Rule – Lubricant

The lubricant sample flush rule is a best practice guideline ensuring that oil samples accurately reflect the condition of the lubricant in service. It specifies that a designated amount of oil must be flushed from the sampling point before collecting a sample to eliminate stagnant, contaminated, or residual oil that could compromise analysis results.

Key Functions of the Lubricant Sample 10× Flush Rule

- **Contamination removal**: Eliminates stagnant oil and debris from the sampling valve or tubing, preventing false contamination readings.
- **Sample accuracy**: Ensures the collected oil represents the actual condition of the lubricant in circulation.
- **Consistency in sampling**: Standardizes the process for trend analysis and reliability monitoring.

Flush Volume Guidelines

1. Flush the sample point oil pathway
2. Flush the sample tube with **10× volume space** with system oil being sampled into a nonsample bottle
3. Now take the real sample in a virgin sample bottle for laboratory testing

Applications

- **Oil condition monitoring**: Used in predictive maintenance for industrial machinery, automotive, and heavy equipment.
- **Wear debris and contamination analysis**: Ensures accurate detection of particles, moisture, and lubricant degradation.
- **Compliance with industry standards**: Supports oil sampling best practices outlined in ISO 18436 and ASTM D7917.

Benefits

- **Accurate lubricant assessment**: Prevents misleading results caused by residual or stagnant oil.
- **Enhanced equipment reliability**: Provides precise data for proactive maintenance decisions.
- **Optimized lubrication management**: Improves oil change intervals and reduces unnecessary lubricant waste.

Following the lubricant sample flush rule is critical for obtaining reliable oil analysis results, supporting effective maintenance strategies, and ensuring long-term equipment performance.

771. Sampling – Lubricant

Lubricant sampling is the process of systematically collecting an oil sample from a machine or lubrication system to assess the condition of the lubricant and detect potential equipment issues. Proper sampling techniques are essential for accurate oil analysis, supporting predictive maintenance and reliability programs.

Key Functions of Lubricant Sampling

- **Condition monitoring**: Provides insight into lubricant health, contamination levels, and wear debris.
- **Early fault detection**: Identifies abnormal wear, contamination, and oil degradation before major failures occur.
- **Data-driven maintenance decisions**: Supports optimized lubrication schedules, oil change intervals, and proactive maintenance actions.

Types of Lubricant Sampling Methods

1. **Live zone sampling**: Collecting oil from a dynamic, circulating system to ensure the sample represents actual operating conditions.
2. **Drain port sampling**: Taking a sample from the oil drain plug, typically less accurate due to potential contamination from settled debris.
3. **Vacuum pump sampling**: Using a vacuum pump and tubing to extract oil from nonpressurized reservoirs or sumps. See Figure S3.
4. **Inline sampling**: Utilizing a permanently installed sampling port to collect representative oil samples consistently.

Applications

- **Industrial machinery**: Used in rotating equipment, compressors, and hydraulic systems to monitor lubricant performance.
- **Automotive and heavy equipment**: Helps track engine oil, transmission fluid, and gear oil condition.
- **Turbines and high-performance systems**: Ensures precision monitoring for critical assets in power generation and aerospace applications.

Benefits

- **Improved equipment reliability**: Detects potential failures before they cause costly downtime.

Figure S3 Taking a vacuum pump oil sample. Source: Lubeflex.

- **Accurate lubricant analysis**: Provides a true representation of lubricant health for informed decision-making.
- **Optimized maintenance efficiency**: Reduces unnecessary oil changes and minimizes operational disruptions.

772. Sampling – Drop Tube Method

Drop tube oil sampling is a method used to extract lubricant samples from nonpressurized reservoirs, sumps, and tanks using a flexible tube and a vacuum pump. This technique is ideal when a dedicated sampling port is unavailable or when accessing oil from a specific depth is necessary to obtain a representative sample.

Key Functions of Drop Tube Oil Sampling

- **Versatile sampling access**: Allows oil extraction from deep reservoirs, tanks, and sumps where other methods may not be feasible.
- **Representative sampling**: Enables collection from targeted depths to avoid contaminants that settle at the bottom or floating debris on the surface.
- **Contamination-free extraction**: Ensures clean and accurate samples when using a dedicated, contamination-free drop tube.

Drop Tube Oil Sampling Process

1. **Select the correct drop tube length**: The tube should reach the desired oil level without touching the bottom to avoid sludge contamination.
2. **Connect up to vacuum pump**: Connect the drop tube to a clean vacuum pump.
3. **Insert the drop tube into the reservoir**: Carefully place the tube into the oil approximately 1/3 from the bottom of the reservoir, ensuring it does not disturb settled debris. See Figure S4.
4. **Extract the sample**: Use the vacuum pump to draw oil into the sampling bottle, maintaining a steady flow to prevent aeration.
5. **Seal and label the sample**: Secure the bottle and document relevant details such as equipment ID, date, and sampling location.

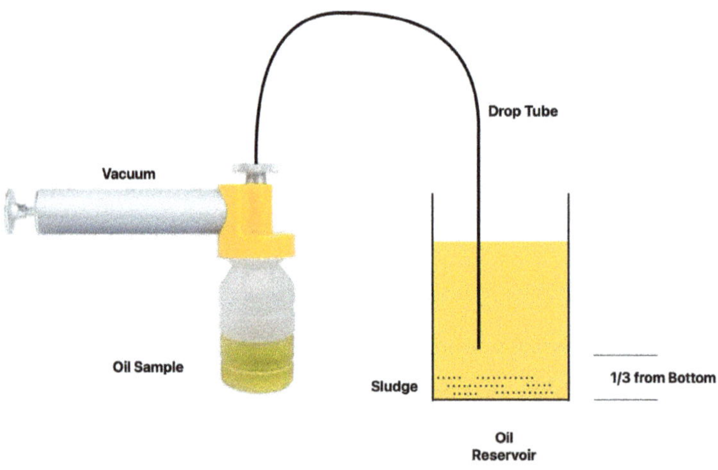

Figure S4 Typical drop tube sampling.

Applications

- **Hydraulic and lubrication systems**: Commonly used in reservoirs for hydraulic fluids, gear oils, and turbine lubricants.
- **Large oil sumps and tanks**: Ideal for applications where direct sampling ports are unavailable.
- **Predictive maintenance programs**: Supports oil analysis for early detection of wear, contamination, and lubricant degradation.

Benefits

- **Nonintrusive sampling**: Eliminates the need for additional fittings or modifications to the system.
- **Improved sample accuracy**: Allows targeted sampling from the most representative zone of the lubricant.
- **Cost-effective**: Reduces the need for permanent sampling ports and specialized equipment.

Drop tube oil sampling is a valuable technique for obtaining reliable oil samples in systems without pressurized flow, helping optimize lubrication management and predictive maintenance strategies.

773. Sampling – Live Zone Method

See "Live Zone Sampling."

774. Sampling Location Point

Selecting the right lubricant: Sampling location point is critical to obtaining a representative oil sample for accurate condition monitoring and predictive maintenance. The prime sampling points are strategically chosen to ensure oil is collected from active flow areas, minimizing contamination from stagnant oil or external sources.

Key Functions of Prime Sampling Location Selection

- **Accurate lubricant condition monitoring**: Ensures the sample represents real-time lubricant performance and contamination levels.
- **Early fault detection**: Captures wear debris and contaminants before they pass through filtration or settle in reservoirs.
- **Consistency and repeatability**: Enables reliable trend analysis by standardizing sampling locations.

Prime Sampling Locations

1. **Pressurized return line (before filtration)**

   Best for: Hydraulic systems, gearboxes, and circulating lubrication systems.

   Why? Captures any bearing wear debris, contaminants, and oil degradation before they are removed by filters, providing a true reflection of system conditions.

2. **Main supply line (after pump, before components)**

   Best for: Engines, compressors, and turbines.

   Why? Ensures a sample from the actively circulating lubricant that has gone through initial filtration but before entering critical components.

3. **Drain port or sump (last resort option)**

   Best for: Gearboxes and static oil reservoirs when other options are unavailable.

   Why? May contain settled contaminants, providing some insights but potentially leading to misleading results if not properly managed.

4. **Oil reservoir (using a drop tube sampling method)**

   Best for: large oil tanks, sumps, and nonpressurized systems.

   Why? Allows targeted sampling at specific depths to avoid contamination from the surface or settled debris at the bottom.

5. **Filter inlet or outlet**

   Best for: Systems with high contamination risk.

   Why? Helps determine filter efficiency by comparing contamination levels before and after filtration.

Choosing the right sampling prime location is essential for reliable oil analysis, enabling proactive maintenance and maximizing equipment uptime.

775. Sampling Methods

See "Sampling – Lubricant."

776. Sampling Trap

A **sampling trap** is a controlled collection area within a cyclic lubrication/hydraulic system designed to capture wear debris, contaminants, or lubricant samples for condition monitoring and oil analysis. The typical location is in a vertical nonpressurized, nonflooded return pipe to the reservoir tank. A deflector plate slows down the return and traps information-rich lubricant in a dead space that can be accessed with a vacuum tube to take a sample. See

Figure S5. It ensures that samples accurately reflect the operating conditions of the equipment while minimizing external contamination.

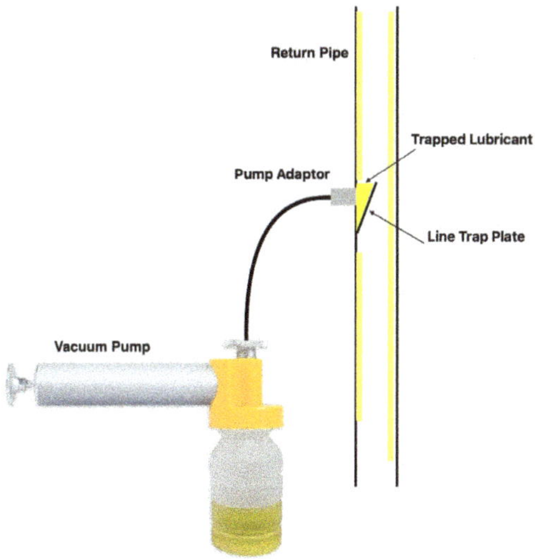

Figure S5 Return line sampling trap.

Key Functions of a Sampling Trap

- **Wear debris collection**: Captures metal particles and contaminants to assess equipment wear trends.
- **Accurate oil sampling**: Provides a stable and controlled point for collecting representative lubricant samples.
- **Contamination control**: Prevents cross-contamination by isolating the sample from environmental exposure.

Types of Sampling Traps

1. **Inline sampling traps**: Installed in pressurized systems, such as hydraulic and circulating oil systems, to capture real-time wear debris and contaminants.
2. **Magnetic sampling traps**: Use magnets to collect ferrous particles from the lubricant, commonly used in gearboxes and rotating equipment.
3. **Filter sampling traps**: Positioned before or after a filtration system to assess filter efficiency and contamination levels.

777. Sampling Vacuum Pump

A **sampling vacuum pump** can be a manual or powered device used to extract oil samples from nonpressurized reservoirs, sumps, or hard-to-reach lubrication systems. It creates a vacuum to draw lubricant through a sampling tube into a clean sample bottle, ensuring a contamination-free and representative sample for oil analysis. See Figures S3-S5 for examples in use.

Key Functions of a Sampling Vacuum Pump

- **Nonpressurized sampling**: Allows oil extraction from reservoirs and sumps where no sampling port is available.
- **Contamination-free collection**: Ensures a clean and controlled sample by preventing external contaminants from entering the bottle.
- **Consistent and repeatable sampling**: Provides a reliable method for trend analysis and condition monitoring.

Applications

- **Hydraulic and lubrication systems**: Used when no inline sampling ports are available.
- **Gearboxes, sumps, and tanks**: Allows targeted oil extraction from specific depths.
- **Predictive maintenance programs**: Supports oil analysis for wear debris, contamination, and additive depletion monitoring.

778. Saponification

Saponification is a chemical reaction in which a fat or oil (triglyceride) reacts with an alkali, such as sodium hydroxide (NaOH) or potassium hydroxide (KOH), to produce soap and glycerol. In lubrication, saponification is an important process in the formulation of grease, where metallic soaps act as thickening agents.

Key Functions of Saponification in Lubrication

Grease thickening: Metallic soaps formed through saponification (e.g., lithium, calcium, or aluminum soaps) help thicken lubricating grease, providing stability and consistency.

Water resistance: Certain metallic soaps enhance the grease's ability to repel water, improving performance in wet conditions.

Lubricant stability: Properly saponified greases maintain their structure under varying temperatures and loads, ensuring consistent lubrication.

Saponification Process

1. **Reaction with an alkali**: A triglyceride (fat or oil) reacts with a strong base (e.g., NaOH or KOH).
2. **Soap and glycerol formation**: The reaction breaks down the ester bonds in the triglyceride, producing soap (fatty acid salt) and glycerol.
3. **Grease thickening**: Metallic soaps are incorporated into lubricants to create structured grease formulations.

Applications

- **Lubricating grease manufacturing**: Used to produce lithium, calcium, and aluminum-based greases.
- **Industrial and automotive lubrication**: Enhances grease stability and performance under extreme conditions.
- **Soap-based metalworking fluids**: Helps emulsify oils for improved cooling and lubrication in machining operations.

Saponification plays a crucial role in lubricant and grease formulation, contributing to improved lubrication properties, water resistance, and overall performance in demanding applications.

779. Saponification Number

The **saponification number** (SAP number) is a measure of the amount of alkali (typically potassium hydroxide, KOH) required to saponify a specific quantity of oil or fat. It is expressed in milligrams of KOH per gram of oil and is used to determine the molecular weight and presence of ester-based compounds in lubricants and greases.

Key Functions of the Saponification Number

- **Lubricant identification**: Helps distinguish between mineral oils and ester-based lubricants, as only esters react with alkali to undergo saponification.

- **Quality control in grease manufacturing**: Ensures the correct formulation of soap-thickened greases.
- **Oxidation and degradation monitoring**: Detects oxidation by-products in used oil, which may form acidic compounds that increase the saponification number.

Interpretation of Saponification Number

1. **High saponification number**: Indicates the presence of esters or fatty acids, common in biodegradable lubricants and synthetic esters.
2. **Low or zero saponification number**: Found in most mineral-based lubricants, as they do not contain saponifiable esters.
3. **Changes in used oil**: An increase in saponification number over time may signal contamination with esters, oxidation by-products, or degradation of additives.

Applications

- **Synthetic and biodegradable lubricants**: Helps in characterizing ester-based oils.
- **Lubricating grease analysis**: Assists in verifying soap content in thickened greases.
- **Condition monitoring in oil analysis**: Used to track lubricant degradation and contamination.

Benefits

- **Accurate lubricant classification**: Differentiates ester-based oils from mineral-based lubricants.
- **Predictive maintenance tool**: Helps detect oil degradation and contamination.
- **Ensures grease quality**: Confirms proper formulation and performance of saponified grease thickeners.

780. Saturation Point (Lubricant)

In lubrication systems, the saturation point refers to the maximum amount of dissolved water a lubricant can hold at a specific temperature. Exceeding this saturation level leads to the formation of emulsified water droplets within the oil, which can cause significant damage to both the lubricant and the machinery.

The saturation point varies depending on the lubricant's composition and operating conditions. For instance, mineral oils typically have a saturation level of around 100 parts per million (ppm) at 60 °C, while some hydraulic fluids can tolerate up to 5000 ppm under similar conditions.

Monitoring and maintaining water content below the saturation point is crucial to prevent emulsification, which can lead to increased wear, corrosion, and reduced lubricant effectiveness. Regular testing and implementing water removal strategies are essential practices to ensure optimal lubrication performance.

781. Saybolt Second Universal – SSU

Saybolt second universal (SSU) is a unit of measurement used to determine the kinematic viscosity of petroleum-based fluids, including lubricants and fuels, at specific temperatures. It represents the time (in seconds) required for 60 mL of oil to flow through a calibrated Saybolt Universal Viscometer under controlled conditions.

Key Functions of Saybolt Second Universal (SSU) Viscosity Measurement

- **Viscosity classification**: Helps in selecting lubricants with appropriate flow characteristics for machinery and engine applications.
- **Oil performance evaluation**: Determines how a lubricant behaves at operating temperatures, impacting film strength and fluidity.
- **Fuel and oil specification compliance**: Used in fuel grading, ensuring proper combustion characteristics in engines.

Standard Testing Conditions

1. **Temperature measurement points**:
    - **100 °F (37.8 °C)**: Common for fuel oils and lighter lubricants.
    - **210 °F (98.9 °C)**: Standard for engine and industrial lubricants.
2. **Test procedure**: Oil is heated to the required temperature and allowed to flow through the Saybolt Universal orifice, with the flow time recorded in seconds.

Applications

- **Automotive and industrial lubricants**: Ensures proper oil viscosity for engine and gearbox efficiency.
- **Fuel quality assessment**: Determines viscosity characteristics for diesel, fuel oils, and kerosene.
- **Hydraulic and turbine oils**: Helps select lubricants for pumps, turbines, and hydraulic systems based on flow properties.

782. Saybolt Universal Viscosity

Saybolt universal viscosity (**SUV**) is a measure of a liquid's resistance to flow, expressed in Saybolt Universal Seconds (SUS or SSU). It quantifies the time (in seconds) required for 60 mL of fluid to pass through a calibrated Saybolt Universal Viscometer under controlled conditions at specific temperatures. This test is widely used for petroleum-based fluids, including lubricating oils and fuels.

783. Scar Diameter

Scar diameter is a key measurement used in tribology and lubricant testing to evaluate the antiwear properties of a lubricant. It refers to the width of the wear scar formed on a metal surface due to friction in standardized tests, such as the four-ball wear test (ASTM D4172). (See "Four-ball Method") A smaller scar diameter indicates better wear protection, while a larger scar suggests poor lubrication performance.

Key Functions of Scar Diameter Measurement

- **Assessing lubricant performance**: Determines the effectiveness of a lubricant in minimizing metal-to-metal contact and wear.
- **Comparing antiwear additives**: Helps evaluate and optimize lubricant formulations for improved wear resistance.
- **Predictive maintenance indicator**: Provides insights into lubricant degradation and potential failure risks in machinery.

Measurement Process

1. **Four-ball wear test (ASTM D4172)**:
    - Three fixed steel balls are immersed in the test lubricant.

- A rotating steel ball is pressed against it under a specified load, speed, and duration.
- The wear scar left on the stationary balls is measured using a microscope.

2. **Scar diameter evaluation**:
   - Reported in **millimeters (mm)**.
   - A typical range is **0.3 to 1 mm**, with smaller values indicating superior antiwear properties.

784. Schematic – Lubrication

A **lubrication schematic** is a visual representation of a lubrication system, illustrating the flow of lubricant, key components, and lubrication points within machinery or equipment. It serves as a blueprint for understanding how lubricants are delivered, distributed, and circulated to ensure optimal equipment performance and longevity.

Key Functions of a Lubrication Schematic

- **Lubrication flow visualization**: Displays how lubricant moves through the system, ensuring all critical components receive proper lubrication.
- **Maintenance and troubleshooting aid**: Helps technicians identify lubrication points, filters, reservoirs, and potential failure areas.
- **System design and optimization**: Assists engineers in designing efficient lubrication systems for minimal wear and maximum efficiency.

Common Components in a Lubrication Schematic

1. **Lubricant reservoir**: Stores oil or grease before distribution.
2. **Pumps**: Deliver lubricant to various points at controlled pressures.
3. **Filters**: Remove contaminants to ensure clean lubricant circulation.
4. **Lubrication lines and manifolds**: Direct oil or grease to bearings, gears, and friction points.
5. **Metering devices and valves**: Regulate lubricant flow to prevent over or underlubrication.

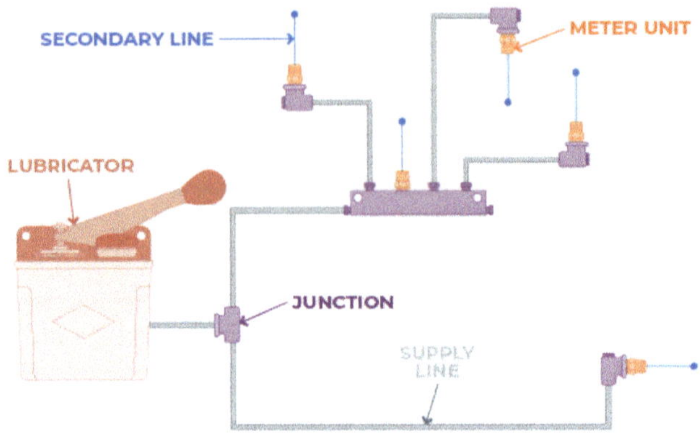

Figure S6 Manual pump five-point centralized SLR lubrication system schematic. Courtesy: Bijur–Delimon International.

6. **Return lines (in circulating systems)**: Allow oil to flow back to the reservoir for reuse.

Types of Lubrication Schematics

- **Manual lubrication schematic**: Shows grease or oiling points for hand lubrication.
- **Centralized lubrication system schematic**: Illustrates automatic lubrication via single-line, dual-line, or circulating systems. See Figure S6 showing a five-point manual pump centralized SLR system.
- **Hydraulic lubrication schematic**: Includes hydraulic-powered lubrication for high-load applications.

785. Scoring (Wear)

Scoring wear is a severe form of surface damage that occurs when metal-to-metal contact takes place due to ineffective lubrication. It results from excessive friction and heat buildup, leading to material transfer, surface tearing, and deep grooves or scratches on the contacting surfaces. Scoring is often seen in high-load, high-speed applications where inadequate lubrication fails to provide sufficient film strength. See Figure S7.

Key Causes of Scoring Wear

- **Insufficient lubricant film**: When the lubricant film is too thin or absent, direct metal contact causes surface damage.

Figure S7 Typical scoring wear on a bearing surface. Source: NSK Bearing Company.

- **High operating temperatures**: Excessive heat leads to lubricant breakdown, reducing its ability to prevent wear.
- **Contaminated lubricant**: The presence of dirt, debris, or metal particles accelerates abrasive wear, worsening scoring.
- **Inadequate lubricant viscosity**: A lubricant that is too thin for the load conditions fails to provide proper separation between moving parts.
- **Overloaded or misaligned components**: High pressure or misalignment increases localized stress, leading to surface failure.

Common Affected Components

- **Gears**: Heavily loaded gears, especially in transmissions and industrial gearboxes, are prone to scoring if lubrication fails.
- **Bearings**: Insufficient lubrication in rolling or sliding bearings can cause scoring on raceways and rollers.
- **Pistons and cylinders**: Poor lubrication in engines can lead to scoring on cylinder walls and piston skirts.
- **Hydraulic and pump systems**: Scoring on pump surfaces and valve spools can result from inadequate lubrication and contamination.

Effects of Scoring Wear

- **Increased friction and heat**: Accelerates further wear and leads to premature failure.
- **Reduced equipment efficiency**: Scored surfaces create higher resistance, decreasing performance.
- **Higher maintenance and repair costs**: Damaged components require frequent replacement or resurfacing.
- **Potential system failure**: Severe scoring can lead to catastrophic failure if left unaddressed.

Prevention Strategies

- **Use the correct lubricant**: Ensure the lubricant has the appropriate viscosity and additives for the load and operating conditions.
- **Maintain proper lubricant film thickness**: Select lubricants with extreme pressure (EP) additives to prevent metal-to-metal contact.
- **Monitor lubricant condition**: Regular oil analysis helps detect wear particles and contamination early.
- **Optimize lubrication intervals**: Ensure timely relubrication to prevent lubricant depletion.
- **Control operating conditions**: Manage temperature and load to minimize stress on the lubricated component.

786. Scuffing (Wear)

See "Adhesive Wear."

787. Seal

A **seal** is a mechanical rotary device used to encapsulate or surround a shaft where it passes through a closed, lubricated environment into the atmosphere. Seals are designed to protect any lubricated environment from harmful outside contaminants and to maintain a positive pressure. See "Packing Seal," "Radial Lip Seal," and "Labyrinth Seal."

788. Seal Compatibility

Seal compatibility refers to the ability of a lubricant to function effectively with sealing materials without causing degradation, swelling, hardening, or leakage. The interaction between lubricants and seals is crucial for maintaining proper sealing performance, preventing fluid leaks, and ensuring the longevity of both the seal and the lubrication system.

Key Factors Affecting Seal Compatibility

- **Chemical compatibility**: Some lubricants contain additives that may react with seal materials, leading to deterioration.
- **Temperature effects**: High temperatures can accelerate seal degradation or cause excessive swelling or shrinkage.
- **Lubricant base oil type**: Mineral oils, synthetics, and bio-based lubricants have different interactions with elastomers.
- **Seal material composition**: The type of elastomer (e.g., Nitrile, Viton, and EPDM) determines its resistance to different lubricants.

Common Seal Materials and Their Compatibility

1. **Nitrile rubber (NBR)**:
   - **Compatible with**: Mineral oils and petroleum-based lubricants.
   - **Not recommended for**: Synthetic esters, glycol-based fluids.
2. **Fluorocarbon (Viton®)**:
   - **Compatible with**: Synthetic oils, esters, and phosphate esters.
   - **Not recommended for**: Some amine-based additives, hot water, steam.
3. **Ethylene propylene diene monomer (EPDM)**:
   - **Compatible with**: Water-based fluids and some synthetic lubricants.
   - **Not recommended for**: Petroleum-based oils, greases.
4. **Silicone rubber**:
   - **Compatible with**: Some synthetic oils and water-based lubricants.
   - **Not recommended for**: Hydrocarbon-based lubricants and high-friction applications.
5. **Polytetrafluoroethylene (PTFE)**:
   - **Compatible with**: Most oils and greases, extreme chemical environments.
   - **Not recommended for**: Highly abrasive applications, some thickened greases.

Effects of Incompatible Lubricants on Seals

- **Swelling**: Excessive absorption of lubricant can cause the seal to expand, leading to leakage or extrusion.
- **Hardening and cracking**: Some lubricants extract plasticizers from elastomers, making them brittle and prone to failure.
- **Softening and deformation**: Weakens the seal, causing poor sealing performance and leaks.
- **Chemical breakdown**: Additives in lubricants may react with the seal material, reducing its lifespan.

Preventing Seal Compatibility Issues

- **Check manufacturer recommendations**: Always verify lubricant compatibility with the seal material before application.
- **Conduct compatibility testing**: Perform lab tests or field trials to assess the effects of lubricants on seals.
- **Use lubricants with compatible additives**: Select lubricants formulated to protect and extend seal life.
- **Monitor seal condition in service**: Inspect seals periodically for signs of swelling, hardening, or degradation.

Applications of Seal Compatibility Considerations

- **Hydraulic systems**: Ensuring seals withstand hydraulic fluids without degradation.
- **Gearboxes and bearings**: Preventing oil leaks due to seal softening or swelling.
- **Automotive and industrial equipment**: Maintaining proper sealing in engines, transmissions, and lubrication systems.

Ensuring seal compatibility with lubricants is essential for system reliability, leak prevention, and extended seal life. Proper selection of lubricants and seal materials reduces maintenance costs and improves overall equipment performance.

789. Seal Swell

See "Seal Compatibility."

790. Seal Swell Additive

Seal swell additives are chemical compounds included in lubricants to moderately expand and condition elastomer seals, improving their sealing effectiveness and preventing leaks. These additives work by causing a controlled amount of swelling in the seal material, ensuring a tight fit against metal surfaces, especially in aged or worn seals.

Key Functions of Seal Swell Additives

Leak prevention: Expands seals slightly to improve contact and prevent lubricant leakage.

- **Seal conditioning**: Helps maintain the flexibility and elasticity of rubber seals, reducing brittleness.
- **Extended seal life**: Prevents shrinkage and hardening caused by aging, heat, and chemical exposure.

How Seal Swell Additives Work

1. **Mild chemical interaction**: The additive interacts with elastomer molecules, causing a controlled degree of swelling.
2. **Seal expansion and tightening**: This slight expansion restores the seal's ability to prevent leaks in worn or aged components.
3. **Lubricant compatibility**: The additive is carefully formulated to avoid excessive swelling that could weaken the seal.

Common Applications of Seal Swell Additives

- **Engine and transmission oils**: Used in automatic transmission fluids (ATFs) and motor oils to prevent leaks in aging seals.
- **Hydraulic fluids**: Helps maintain the integrity of rubber seals in hydraulic systems under high pressure.
- **Gear and differential oils**: Enhances sealing in gearboxes and differentials, reducing oil loss.

791. Sediment

Sediment in lubricants refers to solid or semisolid contaminants that accumulate over time due to wear debris, oxidation by-products, contamination, or additive breakdown. Excessive sediment can impair lubrication performance, accelerate wear, and lead to system failures if not properly monitored and controlled.

Key Sources of Sediment in Lubricants

- **Wear debris**: Metal particles generated from gear, bearing, or engine component wear.
- **Oxidation by-products**: Formation of sludge and varnish due to oil degradation at high temperatures.
- **Contamination**: Dirt, dust, water, and external particles introduced from the operating environment.
- **Additive precipitation**: Insoluble residues formed when lubricant additives break down or react with contaminants.

Effects of Sediment in Lubricants

- **Increased abrasive wear**: Solid particles act as abrasives, accelerating wear on metal surfaces.
- **Reduced lubricant effectiveness**: Clogged filters and oil passages lead to poor lubrication and overheating.
- **Seal and component damage**: Sediment buildup can cause seal degradation and block lubrication flow.
- **Shortened oil life**: Accelerates lubricant degradation, increasing the frequency of oil changes.

Detection and Measurement of Sediment

1. **ASTM D2273 – sediment in lubricating oils**: A standard test method to quantify sediment levels in new and used oils.
2. **Centrifuge testing**: Separates sediment from the oil to measure its concentration.
3. **Particle count analysis**: Determines the number and size distribution of solid contaminants in the lubricant.
4. **Fourier transform infrared spectroscopy (FTIR)**: Identifies chemical composition of sediment and degradation by-products.
5. **Microscopic analysis**: Examines sediment morphology and elemental composition using SEM-EDX.

Prevention and Control Strategies

- **Regular oil analysis**: Monitors sediment levels and identifies sources of contamination.
- **Proper filtration and contaminant control**: Using high-quality filters and breathers prevents debris accumulation.
- **Temperature and oxidation management**: Avoid excessive heat exposure to slow down oil breakdown and sludge formation.
- **Scheduled oil changes**: Replacing oil at optimal intervals prevents sediment buildup.
- **Proper storage and handling**: Ensures clean lubricant storage to prevent contamination before application.

792. SEM EDX

Scanning electron microscopy with energy-dispersive X-ray spectroscopy (SEM-EDX) is an advanced analytical technique used in lubrication studies to examine wear particles, surface degradation, and lubricant contamination at a microscopic level. It provides both high-resolution imaging and elemental composition analysis, making it a powerful tool for diagnosing lubrication-related failures.

Key Functions of SEM-EDX in Lubrication Analysis

- **Wear particle identification**: Analyzes shape, size, and elemental composition of debris to determine wear mechanisms.
- **Surface damage examination**: Detects scuffing, scoring, pitting, and material transfer caused by ineffective lubrication.
- **Contamination analysis**: Identifies contaminants such as dirt, oxidation by-products, and additive depletion residues.

How SEM-EDX Works in Lubrication Analysis

1. **Scanning electron microscopy (SEM)**: Provides high-magnification imaging (up to nanometer resolution) of wear particles and lubricated surfaces.
2. **Energy dispersive X-ray spectroscopy (EDX or EDS)**: Determines the elemental composition of wear debris, helping identify material sources (e.g., iron from bearings and copper from bushings).

3. **Failure mode diagnosis**: By combining imaging and elemental data, SEM-EDX helps pinpoint lubrication failures such as **adhesive wear, abrasive wear, oxidative wear, and surface fatigue**.

Applications in Lubrication

- **Oil and grease analysis**: Evaluates the composition of wear debris found in used lubricant samples.
- **Component failure investigation**: Identifies lubrication-related failure mechanisms in bearings, gears, and sliding surfaces.
- **Tribology research**: Studies lubricant performance and additive interactions at the microscopic level.
- **Contaminant detection**: Helps identify sources of abrasive particles, water ingress, and oxidation residues in lubrication systems.

Benefits of SEM-EDX in Lubrication Studies

- **High-resolution wear analysis**: Detects early-stage wear before macroscopic failure occurs.
- **Precise material identification**: Determines the elemental composition of debris to trace wear sources.
- **Improved lubrication optimization**: Helps refine lubricant formulations and maintenance strategies.

Example Findings from SEM-EDX in Lubrication Analysis

✓ Iron (Fe) and Chromium (Cr) wear particles = bearing or gear wear

✓ Copper (Cu) and Tin (Sn) particles = bushing or bearing alloy degradation

✓ Silicon (Si) and Aluminum (Al) = dirt or sand contamination

✓ Sulfur (S) and Phosphorus (P) depletion = additive breakdown in lubricants

793. Semisynthetic Base Oil Stock

Semisynthetic base oil stocks are a blend of Synthetic base oil and mineral base oil products. These blends generally contain up to 20% pure synthetic product and are less expensive than pure synthetic oils. Because of the lack of standardization and disclosure regarding the blended ratio with this product their engineered value continues to be debated.

794. Semisolid

In lubrication, a **semisolid** refers to a substance that has properties between a solid and a liquid, meaning it retains its shape under static conditions but flows under applied pressure. Greases, gels, and certain thickened lubricants fall into this category, providing both lubrication and sealing functions in various applications.

Key Characteristics of Semisolid Lubricants

- **Shear-thinning behavior**: Remains in place when at rest but flows when mechanical force is applied.
- **Thickened structure**: Contains a base oil suspended in a thickening agent, giving it a gel-like consistency.
- **Good adhesion and retention**: Stays on surfaces for long-lasting lubrication, reducing the need for frequent reapplication.
- **Water and contaminant resistance**: Acts as a protective barrier in harsh environments.

Common Types of Semisolid Lubricants

1. **Grease**: A mixture of **base oil, thickener (e.g., lithium, calcium, or polyurea), and additives** designed for long-term lubrication in bearings, gears, and sliding components.
2. **Lubricating pastes**: High-viscosity lubricants with solid additives (e.g., molybdenum disulfide and graphite) for extreme pressure applications.
3. **Gels and synthetic thickeners**: Used in specialty lubrication for high-speed bearings, electrical contacts, and food-grade applications.
4. **Sealant lubricants**: Semisolid compounds that provide both lubrication and sealing, preventing leaks and contamination.

Applications of Semisolid Lubricants

- **Bearings and gears**: Ensures prolonged lubrication, reducing wear and friction.
- **Automotive and industrial machinery**: **Used** in high-load and high-temperature applications where liquid lubricants may not be effective.
- **Sealing applications**: Prevents moisture, dust, and contaminants from entering critical components.

- **High-temperature and extreme environments**: Provides stability where liquid lubricants would evaporate or break down.

795. Separation (Grease)

Grease separation refers to the process where the base oil separates from the thickener over time, leading to changes in consistency, reduced lubrication performance, and potential equipment failure. This phenomenon, also called oil bleed or oil separation (see Figure A9 and Bleeding reference), occurs due to storage conditions, mechanical stress, or environmental factors.

Key Causes of Grease Separation

- **Gravity separation**: Over time, base oil can migrate out of the thickener, especially in poorly stored grease containers.
- **Temperature fluctuations**: High temperatures can accelerate oil separation, while cold temperatures can alter grease consistency.
- **Mechanical work and shear stress**: Repeated mechanical forces (e.g., in bearings or gears) can break down the grease structure, causing excess oil to bleed.
- **Incompatible thickener-oil blend**: Some grease formulations have poor thickener stability, leading to rapid oil separation.
- **Prolonged storage**: Older grease may exhibit increased separation due to aging and thickener degradation.

Types of Grease Separation

1. **Static oil bleed**: Occurs during storage when oil naturally separates from the thickener and pools on the surface.
2. **Dynamic oil bleed**: Happens under mechanical stress when the grease releases too much or too little oil during operation.
3. **Thickener breakdown**: Extreme mechanical forces or incompatible formulations cause the thickener structure to collapse, leading to excessive separation.

Effects of Grease Separation

- **Reduced lubrication effectiveness**: Insufficient oil content can lead to dry-running conditions and increased friction.

- **Grease consistency changes**: Loss of base oil alters the grease texture, making it too stiff or too soft for proper application.
- **Contamination risk**: Excess-separated oil can leak from grease fittings, attracting dirt and debris.
- **Shortened grease service life**: High separation rates can lead to frequent regreasing and increased maintenance costs.

Prevention and Control Strategies

- **Use high-quality, well-formulated grease**: Greases with stable thickener structures (e.g., lithium complex, calcium sulfonate, and polyurea) resist separation better.
- **Proper storage conditions**: Store grease in a cool, dry place with tightly sealed containers to minimize separation.
- **Choose the right consistency (NLGI grade)**: Using grease with the correct NLGI grade ensures it remains stable under operating conditions.
- **Monitor temperature exposure**: Avoid extreme heat or freezing conditions that accelerate oil separation.
- **Periodic inspection and regreasing**: Regularly check stored grease for oil separation before application and mix gently if minor separation occurs.

796. Separation Additive – Grease

A **separation additive** in grease formulations is a chemical compound designed to reduce oil separation and maintain grease consistency over time. These additives help stabilize the thickener structure, ensuring that the base oil remains uniformly distributed within the grease matrix under storage, mechanical stress, and temperature variations.

Key Functions of Separation Additives

- **Prevents oil bleed**: Enhances the ability of the thickener to retain base oil, minimizing excessive leakage.
- **Improves mechanical stability**: Strengthens grease structure to withstand shear forces in bearings and gears.
- **Enhances grease longevity**: Reduces the rate of oil separation, extending relubrication intervals.

- **Optimizes consistency retention**: Helps maintain the intended NLGI grade and flow characteristics.

Types of Separation Additives

1. **Polymeric stabilizers**: High-molecular-weight polymers that enhance the thickener's oil-holding capacity.
2. **Surface-modified clays**: Used in **bentonite or silica-thickened greases** to reinforce the structure and prevent oil separation.
3. **Metallic complex agents**: Calcium sulfonate and lithium complex thickeners often include stabilizers to minimize oil bleed.
4. **Gel structure enhancers**: Synthetic additives that help improve the interaction between the thickener and base oil for better retention.

Applications of Separation Additives

- **High-temperature greases**: Prevents excessive oil separation in extreme heat conditions.
- **Heavy-duty industrial and automotive greases**: Ensures stability under continuous load and vibration.
- **Storage-sensitive greases**: Maintains uniformity in long-term stored lubricants to prevent preuse separation.
- **Food-grade and specialty lubricants**: Ensures compliance with strict regulations by minimizing oil migration.

Benefits of Separation Additives

- **Maintains lubrication performance**: Ensures consistent oil release for effective lubrication.
- **Reduces leakage and contamination**: Prevents excess oil from seeping out and attracting dirt.
- **Extends grease service life**: Minimizes degradation, reducing the need for frequent reapplication.
- **Improves reliability in critical applications**: Ensures grease stability in bearings, gears, and high-load machinery.

797. Series Progressive Lubrication System

A **series progressive divider lubrication system**, also referred to as a Trabon System after the originating company, is an automatic lubrication system that distributes lubricant (oil or grease) in a controlled manner to multiple lubrication points through a series of progressive divider valves. These systems ensure that each lubrication point receives the precise amount of lubricant required for optimal equipment performance and longevity. See Figures S8 and S9.

Key Functions of a Series Progressive Divider System

- **Controlled lubricant distribution**: Dispenses lubricant in a precise sequence to each lubrication point.
- **Ensures metered lubrication**: Guarantees that all lubrication points receive the designated amount of grease or oil.
- **Monitored lubrication flow**: Can be equipped with monitoring systems to detect blockages or flow failures.

How a Series Progressive Divider System Works

1. **Lubricant supply**: A centralized **pump** delivers lubricant under pressure to the **progressive divider block**.
2. **Divider valve operation**: The valve consists of **pistons** that move sequentially, ensuring lubricant is delivered to each outlet in a set order.

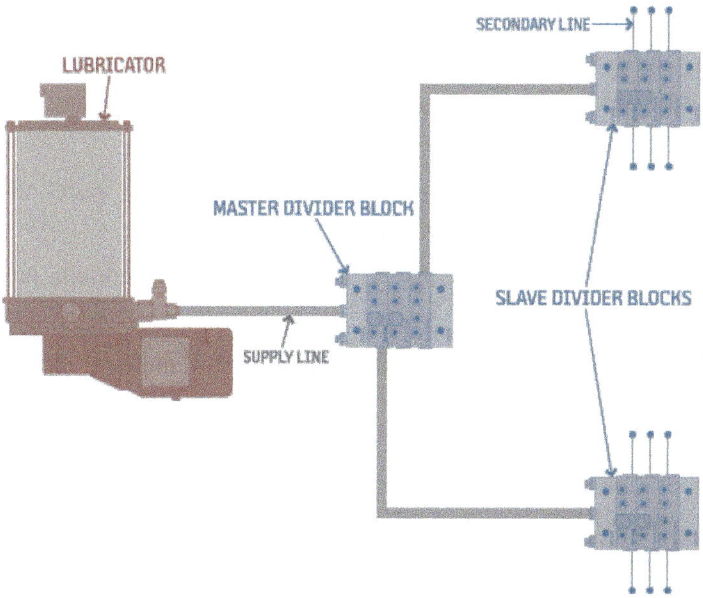

Figure S8 Progressive divider system schematic. Courtesy: Bijur–Delimon International.

3. **Cycle completion**: Once all pistons have completed their strokes, the cycle resets and lubrication begins again with the next pump activation.
4. **Flow monitoring (optional)**: Some systems include **cycle indicators, proximity sensors, or pressure switches** to confirm proper lubrication flow.

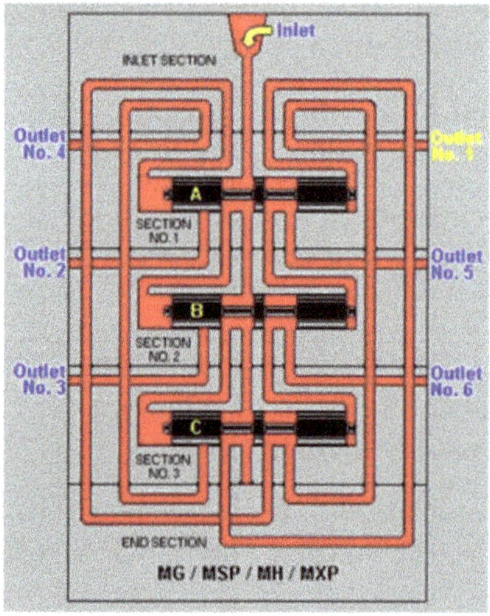

Figure S9 Divider block cutaway. Courtesy: Howard Marten Company.

Applications of Series Progressive Divider Systems

- **Industrial machinery**: Used in conveyors, presses, and automated production lines.
- **Heavy equipment**: Ensures lubrication for bearings, gears, and joints in construction, mining, and agricultural machinery.
- **Wind turbines and energy equipment**: Provides reliable lubrication to rotating components.
- **Steel mills and paper processing plants**: Maintains lubrication in extreme temperature and load conditions.

Benefits of a Series Progressive Divider Lubrication System

✓ **Precise lubrication control**: Ensures consistent lubricant delivery to all critical components.

- ✓ **Reduced downtime**: Automated lubrication prevents premature wear and failures.
- ✓ **Enhanced equipment life**: Minimizes friction, extending the lifespan of machinery components.
- ✓ **Monitoring and verification**: Allows real-time monitoring of lubrication flow, preventing dry-running conditions.
- ✓ **Efficient lubricant usage**: Reduces lubricant waste by delivering only the required amount.

A series progressive divider lubrication system is renowned as an efficient, reliable, and cost-effective method for maintaining proper lubrication in industrial, automotive, and heavy-duty machinery, improving performance and reducing maintenance costs.

798. Settling

Settling in lubrication refers to the process where solid particles, contaminants, or additive precipitates in a lubricant separate and accumulate at the bottom of a container, reservoir, or system due to gravity, temperature variations, or prolonged inactivity. Settling can lead to uneven lubricant distribution, clogged filters, and potential machinery damage.

Key Causes of Settling in Lubricants

- **Contaminant accumulation**: Dirt, wear debris, and rust particles settle over time in static lubrication systems.
- **Water and emulsion separation**: Water contamination in oil can lead to phase separation, causing sediment to accumulate at the bottom.
- **Additive precipitation**: Some lubricant additives (such as antiwear or detergent additives) may settle if the lubricant is not properly formulated or stored.
- **Extended inactivity**: Lubricants in systems that remain idle for long periods can experience phase separation or sedimentation.
- **Improper storage conditions**: Exposure to extreme temperatures or prolonged storage can accelerate the settling of additives or contaminants.

Effects of Settling in Lubrication Systems

- **Lubrication starvation**: Settled contaminants can block oil passages, reducing lubrication effectiveness.
- **Filter and component blockage**: Thick sediment may clog filters, pumps, and valves, leading to lubrication failure.
- **Increased wear and corrosion**: Settled debris can form sludge, promoting abrasive wear and corrosion.
- **Inconsistent lubricant performance**: Uneven additive distribution affects lubricant effectiveness, leading to premature degradation.

Prevention and Control of Settling in Lubrication

- **Regular oil circulation and agitation**: Periodically circulate oil in reservoirs to keep additives and contaminants suspended.
- **Proper storage practices**: Store lubricants in temperature-controlled environments and avoid excessive storage durations.
- **Filtration and sediment removal**: Use offline filtration or sediment removal techniques to keep lubricants clean.
- **Use of dispersant additives**: Some lubricants contain dispersants that help keep particles suspended and prevent settling.
- **Routine oil analysis and maintenance**: Periodically check for sedimentation in reservoirs and sumps to assess lubricant health.

Applications Where Settling is a Concern

- **Hydraulic and circulating oil systems**: Ensuring consistent oil quality in reservoirs.
- **Large oil sumps and tanks**: Preventing sludge buildup in static lubrication storage.
- **Industrial and marine lubrication systems**: Avoiding contamination-related failures in high-load machinery.

Proper lubricant selection, storage, and maintenance can help minimize settling, ensuring consistent lubrication performance and extending equipment life.

799. Shear Rate

See “Rate of Shear.”

800. Shear Stability

Shear stability refers to a lubricant's ability to resist viscosity loss under mechanical stress. It is a critical property for lubricants, especially those used in high-load, high-speed, and high-shear applications, where excessive thinning of the lubricant film can lead to increased wear and failure.

Key Functions of Shear Stability

- **Maintains lubricant viscosity**: Prevents excessive thinning in gear oils, hydraulic fluids, and engine oils.
- **Ensures consistent lubrication performance**: Keeps oil film thickness stable under varying operating conditions.
- **Prevents premature additive breakdown**: Protects viscosity modifiers from mechanical degradation.

How Shear Forces Affect Lubricants

Shear forces occur when layers of lubricant move at different velocities, such as in bearings, gears, and hydraulic pumps. Lubricants containing polymer-based viscosity index (VI) improvers can experience polymer chain breakdown, leading to reduced viscosity and poor lubrication performance.

Shear Stability Testing Methods

1. **KRL Tapered Roller Bearing Test (CEC L-45-A-99)**: Measures viscosity loss in gear oils under high shear conditions.
2. **ASTM D5275 (Sonic Shear Test)**: Simulates polymer degradation in engine oils using ultrasonic vibrations.
3. **ASTM D7109 (Diesel Injector Shear Test)**: Evaluates shear stability of heavy-duty engine oils in high-pressure fuel injectors.
4. **ASTM D5621 (Hydraulic Shear Stability Test)**: Assesses viscosity loss in hydraulic fluids under extreme pressure.

Factors Affecting Shear Stability

- **Viscosity modifier type**: Some polymer-based viscosity improvers degrade faster than others under shear.
- **Operating conditions**: Higher speeds, loads, and pressures increase shear stress on the lubricant.

- **Lubricant formulation**: Synthetic oils and shear-stable additives improve resistance to viscosity loss.
- **Temperature fluctuations**: High operating temperatures can accelerate shear-induced viscosity loss.

Applications Where Shear Stability is Critical

- **Engine oils**: Multigrade oils (e.g., 5W-30 and 15W-40) must maintain viscosity under extreme shear conditions in engines.
- **Gear oils**: Lubricants in transmissions, differentials, and industrial gears must resist thinning under high loads.
- **Hydraulic fluids**: Hydraulic systems rely on shear-stable fluids to maintain pressure and efficiency.
- **Turbine and compressor oils**: Must withstand long-term mechanical stress without significant viscosity loss.

Benefits of High Shear Stability

✓ **Extends equipment life**: Prevents metal-to-metal contact by maintaining adequate lubrication film.

✓ **Reduces maintenance costs**: Lowers frequency of lubricant changes due to viscosity loss.

✓ **Enhances fuel and energy efficiency**: Maintains optimal lubricant flow, reducing power losses in hydraulic and transmission systems.

✓ **Improves performance in extreme conditions**: Ensures reliable lubrication in high-load, high-speed, and high-temperature environments.

Shear stability is a key factor in lubricant performance and longevity. Selecting lubricants with high shear stability ensures consistent protection, reduced wear, and extended maintenance intervals, making it essential for engines, gears, hydraulics, and industrial applications.

801. Shear Strength

Shear strength in lubrication refers to a lubricant's ability to resist shear forces and maintain its structural integrity under mechanical stress. It determines how well a lubricant can sustain its protective film without breaking down, ensuring effective lubrication in high-load and high-speed applications.

Key Functions of Shear Strength in Lubrication

- **Prevents film breakdown**: Ensures the lubricant maintains separation between moving parts under extreme stress.
- **Supports load-bearing capacity**: Helps lubricants withstand heavy loads in gears, bearings, and hydraulic systems.
- **Reduces friction and wear**: Maintains consistent viscosity and prevents metal-to-metal contact.

How Shear Strength Affects Lubrication

1. **High shear strength**: Indicates the lubricant can maintain a stable film under heavy loads, reducing wear and energy losses.
2. **Low shear strength**: Causes lubricant film breakdown, leading to increased friction, heat generation, and component wear.

Factors Affecting Shear Strength

- **Viscosity and additives**: **Higher** viscosity oils and extreme pressure (EP) additives enhance shear strength.
- **Base Oil type**: Synthetic lubricants generally have superior shear strength compared to mineral oils.
- **Temperature and load conditions**: High temperatures can reduce shear strength, requiring viscosity index improvers or thickening agents.
- **Lubricant structure**: Greases with strong thickeners (e.g., lithium complex and calcium sulfonate) resist shear-induced breakdown better.

Shear Strength Testing Methods

- **Four-ball extreme pressure Test (ASTM D2783)**: Measures lubricant film resistance under high stress.
- **KRL shear stability test (CEC L-45-A-99)**: Evaluates viscosity loss due to mechanical shear in gear oils.
- **Ryder gear test**: Assesses load-carrying capacity in industrial gear lubricants.

Applications Where Shear Strength is Critical

- **Heavy-duty gearboxes**: Prevents film collapse in industrial gears, differentials, and transmissions.
- **Hydraulic systems**: Ensures hydraulic fluid maintains pressure and lubricity under constant movement.

- **Bearings and turbines**: Supports rolling element bearings and high-speed machinery.
- **Grease applications**: Helps maintain structure in high-shear environments like CV joints and couplings.

802. Shear Stress

Shear stress in lubrication refers to the **force per unit area** exerted on a lubricant when adjacent fluid layers move at different velocities. It plays a crucial role in determining **lubricant film thickness, viscosity behavior, and overall performance in reducing friction and wear**.

Key Functions of Shear Stress in Lubrication

- **Determines lubricant flow behavior**: Affects how oil or grease spreads and forms a protective film.
- **Influences viscosity response**: High shear stress can cause viscosity changes in **non-Newtonian** lubricants.
- **Affects load distribution**: Helps maintain even pressure and prevents excessive wear on lubricated surfaces.

Formula for Shear Stress (τ)

$$\tau = \eta \cdot \dot{\gamma}.$$

where:

- τ **(Shear Stress, Pa or N/m^2)** = Force per unit area within the lubricant
- η **(Viscosity, Pa·s)** = Lubricant viscosity (dynamic)
- $\dot{\gamma}$ **(Shear Rate, s^{-1})** = Velocity gradient between fluid layers

Types of Lubricants Based on Shear Stress Response

1. **Newtonian fluids**: Maintain a constant viscosity regardless of shear stress (e.g., mineral oils).
2. **Non-Newtonian fluids**:
    - **Shear-thinning (pseudoplastic)**: Viscosity decreases as shear stress increases (e.g., multigrade engine oils, polymer-thickened lubricants).

- **Shear-thickening (dilatant)**: Viscosity increases with shear stress (e.g., specialty greases and impact-resistant lubricants).

Effects of Shear Stress on Lubrication Performance:
Excessive Shear Stress:

- Causes viscosity breakdown in **low shear-stable lubricants**.
- Leads to **thin film failure**, increasing friction and wear.
- Accelerates **additive depletion** in high-load applications.

Controlled Shear Stress:

- Maintains proper **film thickness** in bearings and gears.
- Ensures **optimal viscosity stability** for effective lubrication.
- Reduces **power losses** in hydraulic and mechanical systems.

Applications Where Shear Stress is Critical

- **Hydrodynamic bearings**: Ensures proper oil film formation under dynamic loads.
- **Engine oils**: Multigrade oils must maintain viscosity under **high shear conditions** in combustion engines.
- **Gear and transmission lubricants**: Prevents lubricant breakdown under **extreme pressure (EP) loads**.
- **Hydraulic fluids**: Maintains consistent performance in **high-pressure hydraulic systems**.

803. Shearing (Wear)

Shearing wear refers to material loss or deformation caused by excessive shear stress between contacting surfaces in lubricated machinery. It occurs when lubricant film thickness is insufficient to prevent direct metal-to-metal contact, leading to material displacement, surface tearing, and wear particle formation.

Key Causes of Shearing Wear
Inadequate lubrication film: If the lubricant is too thin, it fails to separate moving surfaces, causing high shear forces.

Excessive shear stress: High loads, speeds, or temperature fluctuations can weaken the lubricant film.

Viscosity breakdown: Shear-induced degradation of viscosity modifiers results in reduced film strength.

Misalignment or overloading: Increases localized stress, making surfaces more prone to shear-induced wear.

Contamination: Hard particles (dirt, wear debris) embedded in the lubricant act as abrasives, accelerating shear damage.

Effects of Shearing Wear

Surface fatigue: Repeated shearing weakens material surfaces, leading to microcracks and failure.

Loss of dimensional accuracy: Excessive wear alters tolerances, reducing mechanical efficiency.

Increased friction and heat generation: Worn surfaces produce more friction, accelerating lubricant degradation.

Higher maintenance costs: Requires frequent repairs or part replacements due to excessive wear.

Types of Shearing Wear in Lubricated Systems

1. **Adhesive shearing wear**: Occurs when surfaces momentarily weld together due to extreme shear stress, leading to material transfer and surface damage.
2. **Abrasive shearing wear**: Happens when contaminants or hard particles cause cutting or plowing of the surface under high shear forces.
3. **Fatigue shearing wear**: Repeated shear stress weakens the material, causing microscopic cracks and eventual spalling.

804. Sheen Rule

The **Sheen Rule** is a regulatory standard established by the US Environmental Protection Agency (EPA) under the Clean Water Act (CWA). It is used to determine whether a lubricant, oil, or petroleum product spill into navigable waters requires reporting and regulatory action. The rule is based on the visual appearance of a sheen on the water surface, which indicates potential environmental contamination. See Figure S10.

Figure S10 Typical rainbow oil sheen on water.

Key Criteria of the Sheen Rule

According to 40 CFR 110.3, a discharge of oil or lubricants is reportable if it:

- Causes a sheen on the water's surface (rainbow-colored or silvery reflection).
- Violates water quality standards (affecting aquatic life or drinking water).
- Leaves sludge or emulsion deposits on shorelines or water bodies.

How the Sheen Rule Applies to Lubricants

- **Detecting leaks and spills**: Lubricant spills in industrial sites, marine vessels, or hydraulic systems can form a visible sheen.
- **Environmental compliance**: Companies using lubricants in offshore drilling, marine transport, and heavy industry must monitor discharges.
- **Biodegradable and low-sheen lubricants**: Some environmentally acceptable lubricants (EALs) are designed to minimize sheen formation in case of accidental spills.

Monitoring and Reporting Requirements

- **Visual inspection**: Regular monitoring of stormwater runoff, marine discharges, and industrial outflows for oil sheen.
- **Immediate reporting**: If a sheen is observed, the spill must be reported to the National Response Center (NRC) at 1-800-424-8802.

- **Remediation and cleanup**: Companies must take corrective actions to contain, clean up, and prevent further contamination.

The Sheen Rule is a critical environmental regulation ensuring that lubricant spills and discharges are properly managed to protect water quality. Industries using lubricants must adopt preventive measures, monitor for sheen formation, and follow proper spill reporting procedures to maintain compliance and minimize environmental impact.

805. Shell Bearing

See "Babbitt Bearing."

806. Shock Absorption

Shock absorption in lubrication refers to a lubricant's ability to dampen impact forces and cushion mechanical shocks between moving surfaces. This property is critical in applications where sudden loads, vibrations, or extreme pressure can cause metal-to-metal contact, excessive wear, or mechanical failure.

Key Functions of Shock Absorption in Lubrication

Dampens sudden load impacts: Reduces stress on gears, bearings, and hydraulic components during shock loading.

Prevents metal-to-metal contact: Forms a protective film that prevents surface damage during impact forces.

Enhances load distribution: Spreads out force over a larger area, reducing localized stress points.

Minimizes noise and vibration: Reduces mechanical noise caused by high-impact operations.

How Shock Absorption Works in Lubricants

1. **Film strength and viscosity**: Lubricants with high viscosity and strong film-forming properties resist compression under shock loads.
2. **Elasticity of base oil and additives**: Some lubricants contain polymers or **viscoelastic modifiers** that help absorb and dissipate energy.
3. **Extreme pressure (EP) and antiwear additives**: Special additives **react under load**, forming protective layers that prevent surface damage during impact.

4. **Grease cushioning effect**: Greases with **high tackiness and cohesive strength** provide an extra layer of shock-dampening protection in high-impact environments.

Lubricants Designed for Shock Absorption

- **High-viscosity gear oils**: Used in heavy-duty gearboxes and transmissions to prevent gear tooth damage.
- **Extreme pressure (EP) greases**: Contain solid lubricants (e.g., molybdenum disulfide and graphite) for extra cushioning.
- **Hydraulic fluids with antifoaming agents**: Ensure smooth operation in shock-loaded hydraulic systems.
- **Damping fluids and shock-absorber oils**: Specialized lubricants used in suspension systems and vibration control applications.

Applications Where Shock Absorption is Critical

- **Heavy-duty gearboxes**: Protects gears from shock loads during startup, braking, or sudden torque changes.
- **Mining and construction equipment**: Prevents damage to bearings and joints exposed to impact loading.
- **Automotive and industrial shock absorbers**: Ensures consistent damping performance in suspension systems.
- **Presses and stamping machines**: Lubricates high-impact metal forming processes.

807. SI

See "System International."

808. Sight Gauge

A **sight gauge** is an external **transparent tube** or indicator mounted on a lubricant reservoir or tank to visually monitor fluid levels. It allows operators to quickly assess lubricant quantity without opening the system. See Figure S11.

Key Features of Sight Gauges:

- **Mounted externally**: Typically attached vertically to tanks or reservoirs.

- **Fluid level indication**: Shows real-time lubricant levels for quick inspection.
- **Used for large volume monitoring**: Common in **industrial reservoirs, hydraulic systems, and storage tanks**.
- **Can include float indicators**: Some designs include floats or markers for better level visibility.

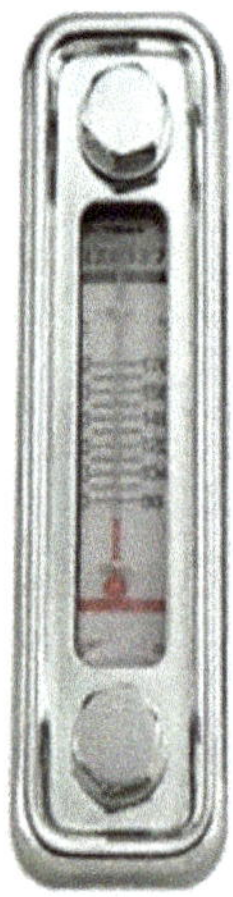

Figure S11 External sight gauge.

Common Applications:

- Hydraulic and lubrication reservoirs.
- Large gearboxes and sumps.
- Storage tanks for industrial fluids.

Advantages of Sight Gauges:

- Easy-to-read, real-time monitoring of oil levels.
- Simple installation and maintenance.
- Reduces the need to open reservoirs, preventing contamination.

809. Sight Glass

A **sight glass** is a **transparent window or dome** installed directly in the lubrication system, allowing operators to observe **both oil level and**

oil condition (e.g., clarity, contamination, bubbles, or foaming). See Figure S12.

Key Features of Sight Glasses:

- **Mounted directly on equipment**: Typically installed on gearboxes, bearing housings, and sumps.
- **Oil condition monitoring**: Allows for **visual inspection of lubricant clarity, contamination, bubbles, and oxidation**.
- **Pressure or vacuum rated options**: Some sight glasses are designed for **pressurized or vacuum systems**.

Figure S12 Dome sight glass. Courtesy: Fluid Defense Systems.

Common Applications:

- Gearboxes, pumps, and compressors.
- Bearing housings and circulating oil systems.
- Steam turbines and hydraulic systems.

Advantages of Sight Glasses:

- Provides **oil level and condition** visibility.

- Helps detect **contaminants**, **bubbles**, **and foam**, indicating potential lubrication issues.
- Allows for quick **troubleshooting** of lubrication system problems.

810. Signal-to-noise Ratio (SNR)

Signal-to-noise ratio (**SNR**) in oil sample bottles refers to the measurement quality and clarity of data obtained from oil analysis, specifically in detecting wear debris, contaminants, and additive elements. A higher SNR indicates a strong and accurate detection of meaningful particles or chemical signatures, while a low SNR suggests that background noise (such as contamination, air bubbles, or bottle material interference) is affecting the accuracy of the results. Therefore, the cleaner the bottle, the higher the SNR will be.

Key Factors Affecting Signal-to-noise Ratio in Oil Sampling

- **Contamination from sampling process**: Dirt, improper bottle cleaning, or residue can introduce unwanted noise.
- **Sample homogeneity**: Inadequate mixing can lead to uneven particle distribution, affecting data accuracy.
- **Background interference**: Some plastics used in oil sample bottles may introduce spectral interference during X-ray Fluorescence (XRF) or ICP (Inductively Coupled Plasma) analysis.
- **Particle overload or dilution**: Too many wear particles or excessive dilution can distort results.
- **Instrument sensitivity and calibration**: Proper calibration of spectrometers, particle counters, or ferrography instruments ensures a high-quality signal.

How to Improve Signal-to-noise Ratio in Oil Sampling

✓ **Use high-quality, contamination-free sample bottles:** Ensure bottles are precleaned and made from inert materials to minimize background interference.

✓ **Follow proper sampling procedures:** Prevent environmental contamination by using clean sampling tools and following best practices.

✓ **Ensure proper sample homogenization:** Shake or agitate samples before testing to ensure uniformity, especially for wear debris analysis.

✓ **Optimize testing instrument settings**: Adjust spectrometer or particle counter sensitivity and calibration to enhance signal clarity.

✓ **Select appropriate analytical methods**: Choose ferrography, ICP, or XRF methods that provide high sensitivity for detecting wear elements with minimal background noise.

Applications Where a High SNR is Critical in Oil Analysis

- **Wear debris analysis (Ferrography, SEM-EDX)**: Detecting low-concentration metal particles from bearings, gears, or pistons.
- **Contaminant identification (ISO 4406, Particle Counting)**: Differentiating between external contaminants (dirt, silica) and internal wear.
- **Additive element analysis (ICP, XRF)**: Ensuring the correct detection of zinc, phosphorus, calcium, and other key lubricant additives.
- **Oxidation and degradation monitoring (FTIR, TAN Analysis)**: Measuring changes in oil chemistry without interference.

Benefits of a High Signal-to-noise Ratio in Oil Sampling

- **More accurate oil analysis results**: Ensures reliable detection of wear metals, contaminants, and additives.
- **Better predictive maintenance decisions**: Improves the ability to detect early signs of machine failure.
- **Reduced false positives or negatives**: Minimizes the risk of misinterpreting data due to noise interference.
- **Enhanced equipment protection**: Leads to better lubrication monitoring and optimized maintenance intervals.

811. Silt

Silt in lubricants refers to microscopic solid particles (typically 1–10 microns in size) that accumulate in oil due to contamination, wear debris, or external dirt ingress. These fine particles can lead to abrasive wear, filter clogging, and lubricant degradation, affecting equipment performance and reliability.

Sources of Silt in Lubricants

- **External contaminants**: Dust, dirt, and atmospheric particles entering through seals, breathers, or improper lubricant handling.

- **Wear debris**: Metal fines from bearings, gears, or hydraulic components.
- **Oxidation and additive breakdown**: Varnish, sludge, and by-products of lubricant degradation forming microscopic solids.
- **Filter inefficiency**: If filtration systems fail to remove submicron particles, silt accumulates in the lubricant.
- **Water contamination**: Water in oil can lead to soft silt formation (emulsified contaminants), accelerating wear.

812. Single-line Parallel Lubrication System

See "Positive Displacement Injector" (PDI) Lubrication System.

813. Single-line Resistance (SLR) Lubrication System

A **single-line resistance (SLR) lubrication system** is a **low-pressure lubrication system** designed as the first-ever centralized lubrication system for specific use on automobile chassis by Joseph Bijur In 1924. The system was designed to deliver oil to multiple lubrication points through a **single main supply line** using **metering restrictors**. It is now commonly used in **light-duty industrial equipment**, **machine tools**, **textile machinery**, **and food processing systems**, ensuring consistent oil distribution to critical components. See Figure S6 for an example of an SLR system.

Key Functions of Single-line Resistance Lubrication

- **Continuous or intermittent lubrication**: Provides a steady flow of lubricant to moving parts, reducing wear.
- **Metered lubricant distribution**: Uses **restrictor orifices** to control oil flow to each lubrication point.
- **Simple, gravity or low-pressure operation**: Functions with **manual**, **electric**, **or gravity-fed pumps** for oil distribution.

Standard Operating Conditions

1. Pressure and flow rate

- **Low-pressure system**: Operates at **10–50 psi (0.7–3.5 bar)**.
- **Oil flow control**: Regulated through fixed **metering restrictors**.

2. Lubricant compatibility

- **Designed for oil lubrication only**: Works with **light- to medium-viscosity oils (ISO VG 10–100)**.

- **Not suitable for grease**: Cannot handle **high-viscosity lubricants**.

Applications of Single-line Resistance Lubrication Systems

- **Machine tools**: Lubricates **spindles, slides, and bearings** in CNC machines.
- **Textile machinery**: Ensures **continuous lubrication** for high-speed components.
- **Food and beverage industry**: Provides **precise oil dosing** for conveyor systems and food-grade equipment.
- **Printing presses**: Keeps **rollers and guides lubricated**, reducing downtime.
- **Small industrial equipment**: Used in **packaging machines and light-duty automation systems**.

Conclusion

A **single-line resistance** lubrication system is an efficient, low-maintenance solution for automated oil lubrication in machine tools, textile machines, and food processing equipment. Its simple design, metered oil distribution, and continuous operation make it ideal for light-duty applications requiring consistent and reliable lubrication.

814. Single-pass Test

A **single-pass test** is a laboratory or field test used to evaluate the **performance of a lubricant** – specifically its ability to **remove** contaminants, suspend particles, and maintain film strength in a single flow cycle through a filtration or lubrication system. This test is commonly used to assess lubricant cleanliness, filtration efficiency, and additive retention in industrial and automotive applications.

Key Functions of a Single-pass Test in Lubrication

- **Evaluates contaminant removal**: Measures how well a lubricant carries away **wear debris, dirt, and particles** in a single pass through a system.
- **Determines filtration efficiency**: Tests how effectively a **filter removes solid contaminants** in one flow cycle.
- **Assesses lubricant performance in extreme conditions**: Helps analyze lubricant behavior under **high load, temperature, or shear stress** in a single pass.

Standard Testing Conditions

1. **Test setup**
   - **Lubricant sample**: A controlled volume of oil or grease is passed through a test setup.
   - **Filtration stage (if applicable)**: The sample flows through a **test filter** to measure contaminant retention.
   - **Wear or stress application**: Some tests include **contact with metal surfaces under pressure** to analyze wear performance.
2. **Measurement parameters**
   - **Particle removal efficiency**: Determines the **percentage of particles removed per pass** (ISO 4406).
   - **Wear particle analysis**: Measures **ferrous/nonferrous debris levels** using ferrography.
   - **Viscosity stability**: Evaluates if the lubricant maintains **film thickness and flow properties**.
   - **Additive retention**: Assesses if essential additives **degrade or remain stable** after a single pass.

Applications of Single-pass Lubrication Testing

- **Hydraulic systems**: Evaluates **oil cleanliness and filter efficiency** in high-pressure systems.
- **Gear and bearing lubrication**: Assesses **lubricant film formation** under load.
- **Engine oils and fuel testing**: Determines **contaminant control and additive stability** in combustion systems.
- **Turbine and compressor oils**: Verifies **lubricant flow properties and varnish resistance** in critical machinery.

Benefits of the Single-pass Test

✓ **Fast and reliable evaluation**: Provides **quick insights into lubricant and filtration performance**.

✓ **Enhances predictive maintenance**: Helps detect **early signs of wear, contamination, or oil degradation**.

✓ **Optimizes filtration and lubricant selection**: Ensures the **right oil and filter combination** for a specific application.

✓ **Supports industry compliance**: Meets standards such as **ISO 4406, ASTM D7647, and SAE J1488** for lubricant cleanliness and filtration.

815. Single-point Lubricator

A **single-point lubricator (SPL)** is an automatic lubrication device designed to deliver a controlled amount of lubricant (oil or grease) to a single lubrication point at regular intervals. It ensures consistent lubrication, reducing the risk of wear, friction, and equipment failure in bearings, gears, chains, and other moving components.

Key Functions of a Single-point Lubricator

- **Automated lubrication**: Provides **continuous or interval-based lubrication** without manual intervention.
- **Precise lubricant dispensing**: Delivers a **preset quantity of lubricant**, preventing over or underlubrication.
- **Enhanced equipment reliability**: Reduces friction, wear, and contamination, ensuring optimal performance.

Standard Operating Conditions

1. **Lubrication frequency and volume**
   - **Adjustable dispensing rate**: Typically set for **1–12 months per cycle**. See Figure S13.
   - **Back pressure-controlled flow**: Ensures a **steady supply of lubricant** to the lubrication point based on viscosity. See Figure S14.

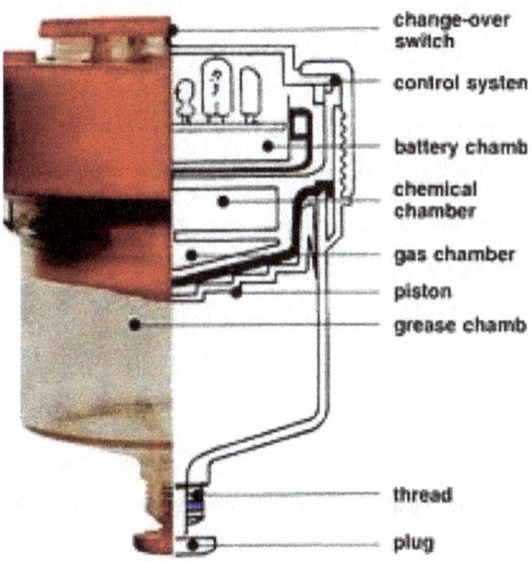

Figure S13 Programmable electrochemical SPL. Courtesy: ATS Electrolube.

2. Compatible lubricants

- Works with **oil or grease** (up to **NLGI 2 viscosity**).
- Select lubricant based on **machine specifications** and environmental conditions.

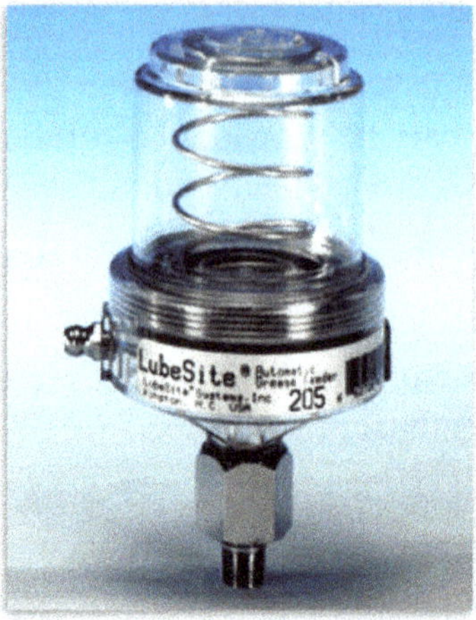

Figure S14 Spring actuated SPL.

Benefits of Using a Single-point Lubricator

✓ **Reduces manual lubrication needs**: Eliminates the risk of **missed lubrication intervals**.

✓ **Improves workplace safety**: Reduces the need for workers to access **hard-to-reach lubrication points**.

✓ **Enhances equipment longevity**: Prevents **wear, overheating, and contamination-related failures**.

✓ **Minimizes lubricant waste**: Precise dispensing **reduces excess lubricant usage and mess**.

816. Sintered Bearing

A **sintered bearing is** a type of self-lubricating bearing made from powdered metal that is compacted and sintered (heated without melting) to create a porous structure. These porous bearings retain lubricants within their structure, allowing for continuous lubrication without the need for external grease or oil application. Sintered bearings are widely used in small motors, automotive applications, and industrial machinery due to their low maintenance and high wear resistance.

Key Functions of Sintered Bearings

- **Self-lubrication**: The porous structure absorbs and releases lubricant, reducing friction and wear.
- **Smooth and quiet operation**: Minimizes noise and vibration in high-speed applications.
- **High load capacity**: Designed to handle **moderate to high loads** with excellent durability.
- **Corrosion and contaminant resistance**: Suitable for **harsh environments and low-maintenance applications**.

817. Sleeve Bearing

See "Plain Bearing."

818. Sliding Bearing

See "Plain Bearing."

819. Sliding Contact Lubrication

See "Sliding Friction."

820. Sliding Friction

Sliding friction is the resistance force that occurs when two solid surfaces slide against each other. Unlike rolling friction, which involves minimal contact, sliding friction results in direct surface interaction, leading to higher resistance, increased heat generation, and material wear. Proper lubrication and surface treatments are essential in reducing sliding friction to improve equipment efficiency and longevity. See also "Friction."

Key Characteristics of Sliding Friction

- **Higher than rolling friction**: More energy is required to overcome resistance.

- **Surface contact dependent**: Rougher surfaces experience greater resistance.
- **Affected by lubrication**: Proper lubrication reduces wear and heat generation.
- **Influenced by load and material type**: Heavier loads and softer materials increase friction.

Formula for Sliding Friction

$$F_f = \mu_k F_n.$$

where:

- **Ff** = Sliding friction force
- **μk** = Coefficient of kinetic friction (material-dependent)
- **Fn** = Normal force (force pressing the surfaces together)

Factors Affecting Sliding Friction

- **Material composition**: Harder materials generally produce **less friction** than softer ones.
- **Surface roughness**: Smoother surfaces result in **lower friction**, while rough ones increase resistance.
- **Lubrication and coatings**: Applying lubricants or antifriction coatings **reduces direct metal-to-metal contact**.
- **Load and pressure**: Higher loads increase **surface interaction**, raising friction.

Feature	Sliding Friction	Rolling Friction
Surface Contact	Direct surface interaction	Minimal surface deformation
Friction Level	Higher	Lower
Wear Rate	High	Low
Lubrication Need	Essential for reducing heat	Moderate
Applications	Brakes, bearings, machining	Wheels, ball bearings, rollers

- **Temperature effects**: Heat buildup from sliding can lead to **thermal expansion and surface damage**.

821. SLR

See "Single-line Resistance" lubrication system.

822. Sludge

Sludge in lubrication systems refers to the thick, semisolid deposit formed by the accumulation of degraded oil, contaminants, oxidation by-products, and dirt in engines, hydraulic systems, and industrial machinery. It can reduce lubrication efficiency, block oil passages, and lead to increased wear, overheating, and equipment failure.

Causes of Sludge Formation

- **Oil oxidation**: Exposure to high temperatures and oxygen causes the oil to break down into sludge-forming deposits.
- **Contaminant accumulation**: **Dirt**, **dust**, **fuel residues**, **and coolant leaks** mix with oil, forming sludge.
- **Water and moisture contamination**: Leads to **emulsification**, accelerating sludge formation.
- **Additive breakdown**: The depletion of **dispersants and detergents** allows sludge particles to clump together.
- **Infrequent oil changes**: Overused or degraded oil thickens and forms sludge deposits.

Prevention and Removal of Sludge

- **Use high-quality lubricants**: Select **oxidation-resistant**, **detergent-rich oils** to minimize sludge formation.
- **Maintain proper oil change intervals**: Regular oil changes prevent oil degradation and deposit buildup.
- **Monitor oil condition**: Conduct **oil analysis (FTIR**, **viscosity tests**, **and particle count analysis)** to detect sludge precursors.
- **Improve filtration and contaminant control**: Use **high-efficiency oil filters** to remove particulates and water.
- **Perform system flushing**: Use **chemical flushes or fresh oil circulation** to clean sludge deposits from systems.

823. Slumpability (Grease)

Slumpability is a property used to describe the **ability of grease or semi-solid lubricants to flow or deform under gravitational force** without external pressure. It indicates how well a lubricant maintains its shape or spreads when applied to bearings, gears, or other mechanical components. This characteristic is crucial in ensuring proper lubrication distribution, adhesion, and stability in various industrial and automotive applications.

Key Functions of Slumpability Measurement

- **Lubricant flow behavior**: Determines how grease **spreads or remains in place** when subjected to gravity.
- **Application suitability**: Helps in selecting **greases for vertical surfaces, enclosed gear systems, or open applications**.
- **Storage and handling assessment**: Evaluates whether a lubricant **remains stable in containers or migrates over time**.

Standard Testing Conditions

1. **Temperature and environment factors**:
   - **Room temperature (25 °C)**: Evaluates slumpability under **normal operating conditions**.
   - **Elevated temperatures (e.g., 100 °C or higher)**: Tests the grease's tendency to **soften and spread under heat exposure**.
2. **Test procedure**:
   - A sample of grease is placed on a vertical or inclined surface.
   - The **degree of movement or deformation** is observed over a specific time period.
   - The **extent of slumping** is recorded, indicating its suitability for different lubrication applications.

824. SNR

See "Signal-to-noise Ratio."

825. Soap

Soap in lubricant manufacturing refers to the metallic soap thickener used in grease production. It is formed by reacting a fatty acid with a metal hydroxide, creating a soap-based structure that provides stability, adhesion, and lubrication properties in grease formulations. Soaps are essential for thickening lubricating oils and enhancing grease performance in industrial, automotive, and machinery applications.

Key Functions of Soap in Lubricant Manufacturing

- **Grease thickening agent**: Converts base oil into a **semisolid** grease for improved adhesion and stability.
- **Lubrication enhancement**: Provides **film-forming properties** for wear protection.
- **Temperature and water resistance**: Some soaps offer **high-temperature stability and water washout resistance**.
- **Load-bearing capability**: Certain metallic soaps improve grease **extreme pressure (EP) performance**.

Types of Soap Thickeners in Lubricants

1. **Simple soaps (single-metal soaps)**
    - Formed using **one metal hydroxide** and a fatty acid.
    - Example: **Lithium, calcium, sodium, and aluminum-based greases**.
2. **Complex soaps (multimetal soaps)**
    - Made using **a metal hydroxide with additional complexing agents**, improving **thermal stability**.
    - Example: **Lithium complex, calcium sulfonate, aluminum complex greases**.
3. **Nonsoap thickeners (alternative to soaps)**
    - Used in specialized applications requiring **higher thermal or chemical resistance**.
    - Example: **Polyurea, clay, or silica-based greases**.

Manufacturing Process of Soap-based Greases

1. **Saponification reaction**:
    - **Fatty acids (e.g., stearic acid) react with metal hydroxides (e.g., lithium hydroxide)** used to form metallic soap.

2. **Soap dispersion in base oil**:
   - The soap is dissolved in **hot lubricating oil**, forming a thickened grease matrix.
3. **Cooling and additive blending**:
   - The grease is cooled and blended with **antiwear**, **extreme pressure (EP), and antioxidant additives**.
4. **Final homogenization**:
   - The grease is milled and processed to ensure **uniform consistency** and performance.

Soap-based lubricants are the backbone of grease manufacturing, providing thickening, stability, and wear protection across various industries. Lithium, calcium, aluminum, and complex soaps offer unique properties, ensuring efficient lubrication in automotive, industrial, and extreme environment applications.

826. Society of Automotive Engineers

See **SAE International**. The Society of Automotive Engineers is one of the oldest engineering societies still going strong. Formed originally in North America in 1911 to set standards for the emerging automobile they recently changed their name to SAE International

827. Society of Tribologists and Lubrication Engineers (STLE)

The **Society of Tribologists and Lubrication Engineers (STLE)** is a professional organization dedicated to advancing the fields of tribology, lubrication engineering, and fluid technology. Originally incorporated in 1944 under the American Society for Lubrication Engineers (ASLE), it later changed its name in 1987 to the STLE to better adapt to and reflect evolving market and audience needs. The STLE's mission is to advance the science of tribology and the practice of lubrication engineering in order to foster innovation, improve the performance of equipment and products, conserve resources, and protect the environment by providing education, certification, research, and networking opportunities for professionals in industries related to friction, wear, and lubrication.

Key Functions of STLE

- **Technical knowledge advancement**: Promotes research and innovation in lubrication science and tribology.
- **Professional certification programs**: Offers a full body of knowledge and full certification programs for individuals
- **Industry standards and best practices**: Develops guidelines for lubricant performance, tribological testing, and condition monitoring.

Core Programs and Services

- **Certification and education**: Includes three specific certification levels: Certified Lubrication Specialist (CLS), Oil Monitoring Analyst (OMA), and Metalworking Fluid Specialist (CMFS) programs designed to validate expertise in lubrication, oil analysis, and metalworking fluid management.
- **Research and technical resources**: Publishes the **Tribology and Lubrication Technology (TLT) magazine**, supports research in wear analysis, fluid dynamics, and lubrication performance, and hosts the Annual STLE Conference for global lubrication experts.

Applications of STLE in Industry

- **Automotive and Aerospace**: Advances engine lubrication, transmission fluids, and aerospace coatings.
- **Manufacturing and Industrial Equipment**: Improves lubrication for bearings, hydraulics, and metalworking operations.
- **Energy and Power Generation**: Supports wind turbines, nuclear plants, and gas turbines with optimized lubricants.
- **Research and Academia**: Provides a platform for tribology and lubrication engineering studies.

Benefits of STLE Membership and Standards

- **Industry recognition**: Enhances credibility through globally recognized certifications.
- **Access to cutting-edge research**: Connects professionals with technical papers, case studies, and expert knowledge.
- **Networking opportunities**: Facilitates collaboration through conferences, technical webinars, and local chapters.

- **Improved equipment efficiency**: Promotes best practices in lubrication reliability and maintenance.

Conclusion: The **Society of Tribologists and Lubrication Engineers (STLE)** plays a crucial role in advancing tribology, lubrication technology, and condition monitoring. Through certifications, research, and industry collaboration, STLE helps professionals enhance equipment reliability, efficiency, and sustainability across multiple industries.

828. Sodium Grease

Sodium grease is a **soap-based grease** that uses **sodium soap** as a thickener to provide lubrication, water resistance, and high-temperature stability. It is commonly used in industrial machinery, automotive applications, and heavy-duty equipment where high mechanical stability and adhesion are required.

Key Functions of Sodium Grease

- **High mechanical stability**: Maintains consistency under heavy loads and high shear conditions.
- **Good adhesion to metal surfaces**: Clings well to moving components, reducing relubrication frequency.
- **Water resistance**: Offers moderate water resistance but is less effective in humid or wet conditions compared to lithium or calcium greases.
- **Thermal stability**: Performs well at **moderate to high temperatures**, typically up to **120 °C (250 °F)**.

Composition of Sodium Grease

- **Base oil**: Typically, mineral oil, though synthetic variants exist for specialized applications.
- **Sodium soap thickener**: Formed by reacting **fatty acids (such as tallow-based acids) with sodium hydroxide**, giving the grease its thickening properties.
- **Additives (optional)**: May include **antiwear agents, corrosion inhibitors, and oxidation stabilizers** to enhance performance.

Applications of Sodium Grease

- **Automotive bearings and chassis lubrication**: Provides **long-lasting lubrication** in moderate-speed wheel bearings.

- **Industrial machinery**: Used in **open gears, bearings, and mechanical linkages** that require strong adhesion.
- **Heavy-duty equipment**: Applied in **farm machinery**, **mining equipment, and construction machinery** for **shock-load resistance**.
- **Railroad and steel industries**: Suitable for **high-temperature**, **high-load environments** in rail applications.

Benefits of Sodium Grease

- **Excellent load-carrying capacity**: Protects components under **moderate to high loads**.
- **Good temperature stability**: Performs well in applications requiring **moderate heat resistance**.
- **Strong adhesion and film strength**: Ensures **prolonged protection** against metal-to-metal contact.
- **Resists oil separation**: Maintains consistency over time, reducing lubricant loss.

Table S1 Grease comparison chart.

Feature	Sodium Grease	Lithium Grease	Calcium Grease
Temperature Stability	Moderate (up to 120°C)	High (up to 180°C)	Low to moderate (up to 80°C)
Water Resistance	Moderate	Excellent	Very High
Load-Carrying Capacity	High	Very High	Moderate
Adhesion to Metal	Strong	Strong	Moderate
Mixing Compatibility	Limited	Compatible with most	Limited

Limitations of Sodium Grease

- **Limited water resistance**: Not ideal for **high-moisture or submerged applications**.
- **Not compatible with other grease types**: Cannot be mixed with **lithium, calcium, or aluminum greases** without performance degradation.
- **Oxidation sensitivity**: Requires **oxidation inhibitors** to prevent hardening over time.

829. Solid Friction

Solid friction refers **to the resistance to motion that occurs when two solid surfaces slide or attempt to slide against each other**. Unlike fluid friction, which involves a lubricant or liquid medium, solid friction is caused by direct contact between surfaces and is influenced by surface roughness, material properties, and normal force. It plays a critical role in mechanical systems, lubrication science, and wear prevention.

Types of solid friction

- **Static friction**: The force that resists the initiation of motion between two solid surfaces in contact. It must be overcome for movement to begin.
- **Kinetic (sliding) friction**: The resistance encountered once the surfaces are already moving relative to each other.
- **Rolling friction**: The resistance to motion when an object rolls over a surface, typically lower than sliding friction.

See Also "**Friction**."

830. Solid Lubricant

Solid lubricant refers to a material that reduces friction and wear between two sliding surfaces without the need for liquid or grease-based lubrication. These lubricants form a thin, protective film that prevents direct metal-to-metal contact, ensuring smooth operation in extreme conditions such as high temperatures, vacuum environments, and heavy loads.

Key Functions of Solid Lubricants

- **Reduces friction**: Forms a low-shear film that minimizes resistance between surfaces.
- **Prevents wear and surface damage**: Protects components from excessive wear, scuffing, and galling.
- **Performs in extreme environments**: Functions under **high temperatures, vacuum conditions, and radiation exposure** where traditional lubricants fail.
- **Resistant to contamination**: Does not attract dust, dirt, or moisture, making it ideal for **cleanroom and dry environments**.

Types of Solid Lubricants

- **Graphite**: Provides excellent lubrication under high temperatures, often used in **bearings, furnaces, and electrical contacts**.
- **Molybdenum disulfide (MoS_2)**: A high-pressure lubricant with superior wear resistance, commonly used in **aerospace, automotive, and heavy-duty applications**.
- **Polytetrafluoroethylene (PTFE or Teflon®)**: Offers ultra-low friction and chemical resistance, widely used in **medical devices, food processing, and nonstick coatings**.
- **Hexagonal boron nitride (h-BN)**: A thermal and chemical-resistant lubricant used in **high-temperature applications and coatings**.
- **Soft metals (e.g., silver, lead, and gold)**: Used in **space applications** where extreme conditions require high durability and conductivity.

Applications of Solid Lubricants

- **Aerospace and defense**: Lubricates **satellites, spacecraft, and high-temperature jet engine components**.
- **Automotive industry**: Applied in **CV joints, engine coatings, and braking systems**.
- **Industrial machinery**: Used in **gears, bearings, and high-load sliding mechanisms** where conventional lubricants degrade.
- **High-temperature environments**: Essential for **furnaces, metal forming, and heat-treated surfaces**.
- **Vacuum and cleanroom equipment**: Ideal for **semiconductor manufacturing, medical devices, and laboratory equipment**.

Benefits of Solid Lubricants

- **Performs where liquid lubricants fail**: Works in extreme **temperatures, pressures, and vacuum conditions**.
- **Long-lasting and low-maintenance**: Does not evaporate or degrade like conventional lubricants.
- **Prevents contamination**: Suitable for **dust-sensitive, radiation-exposed, and cleanroom applications**.
- **Reduces energy consumption**: Lowers friction, improving efficiency in mechanical systems.

831. Solids Contamination

Solids contamination in lubricants refers to the presence of **unwanted solid particles** such as dirt, metal shavings, soot, wear debris, and process contaminants in lubricating oils and greases. These particles can cause abrasive wear, increased friction, and reduced lubrication efficiency, leading to premature equipment failure and costly maintenance. Proper contamination control through filtration, monitoring, and preventive maintenance is essential for ensuring lubricant cleanliness and system reliability.

Key Functions of Monitoring Solids Contamination in Lubricants

- **Prevents abrasive wear**: Reduces **metal-to-metal contact** that can damage bearings, gears, and hydraulic components.
- **Maintains lubricant performance**: Ensures **oil viscosity, film strength, and antiwear properties** remain effective.
- **Extends equipment life**: Protects **critical components from accelerated wear and breakdown**.
- **Improves filtration and maintenance planning**: Helps determine **when filters need replacement and oil needs cleaning**.

Sources of Solids Contamination in Lubricants

- **Wear debris**: Generated from **bearing, gear, and metal surface wear** over time.
- **Dust and dirt ingress**: Enters through **breathers, seals, or poor storage conditions**.
- **Soot and combustion residues**: Common in **engine lubricants due to incomplete fuel combustion**.
- **Manufacturing and process contaminants**: Leftover particles from **machining, casting, or assembly processes**.
- **Rust and corrosion by-products**: Forms due to **moisture contamination and oxidation reactions**.

Effects of Solids Contamination on Lubrication Performance

- **Increased component wear**: Hard particles act as **abrasives**, causing surface damage.

- **Clogging of filters and passages**: Leads to **restricted oil flow and lubrication starvation**.
- **Accelerated oil degradation**: Contaminants trigger **oxidation and sludge formation**, reducing oil lifespan.
- **Reduced heat dissipation**: Sludge and deposits impair the lubricant's ability to **dissipate heat effectively**.

Table S2 Contamination levels based on ISO 4406 code.

ISO Range	Description	Application	Life Extension	Dirt p.a.*
22/20/17	**Very Dirty**	• Unsuitable for any lubrication system (H,L,G)	.5x (H) .75x (G)	600 kg
19/17/14	**New Oil**	• Medium/low pressúre systems (H,L) • Non-critical gear systems (G)	1	140kg
17/15/12	**Light Contamination**	• Standard hydraulic and lube oil systems (H,L) • Standard Gear system (G)	1.6x (H) 1.3x (G)	34kg
16/14/11	**Clean**	• Servos and high pressure hydraulics (H,L) • Critical gear systems (G)	3x (H) 1.7x (G)	17kg
<14/12/10	**Very Clean**	• All oil systems (H,L,G)	4(H)-2x(G)	8.5kg

H-Hydraulic Oil, L - Lube Oil, G - Gear Oil
* - Dirt thru pump at 200ltr/min, 230 work day, single shift in 1 year

Courtesy: ENG**TECH** Industries Inc.

Methods to Detect and Control Solids Contamination

- **Oil analysis and particle count testing**: Identifies **contaminant levels using ISO 4406 cleanliness codes**.
- **High-efficiency filtration systems**: Removes **fine particles before they cause damage**.
- **Magnetic separators and centrifuges**: Capture **metallic wear debris in lubricating systems**.
- **Breathers and seals maintenance**: Prevents **dust and airborne particles from entering the system**.
- **Proper lubricant handling and storage**: Avoids **contamination during transfer and application**.

832. Soluble Oil

Soluble oil is a type of water-miscible lubricant that forms an oil-in-water emulsion when mixed with water. It is commonly used in metalworking, machining, and cooling applications, providing lubrication, cooling, and corrosion protection during cutting, grinding, and forming operations. See Figure S15.

Key Functions of Soluble Oil

- **Lubrication**: Reduces friction and wear between cutting tools and workpieces.
- **Cooling**: Dissipates heat generated during machining, preventing overheating and tool damage.
- **Corrosion protection**: Forms a protective film on metal surfaces to prevent rust and oxidation.
- **Chip removal**: Helps flush away metal chips and debris during machining operations.

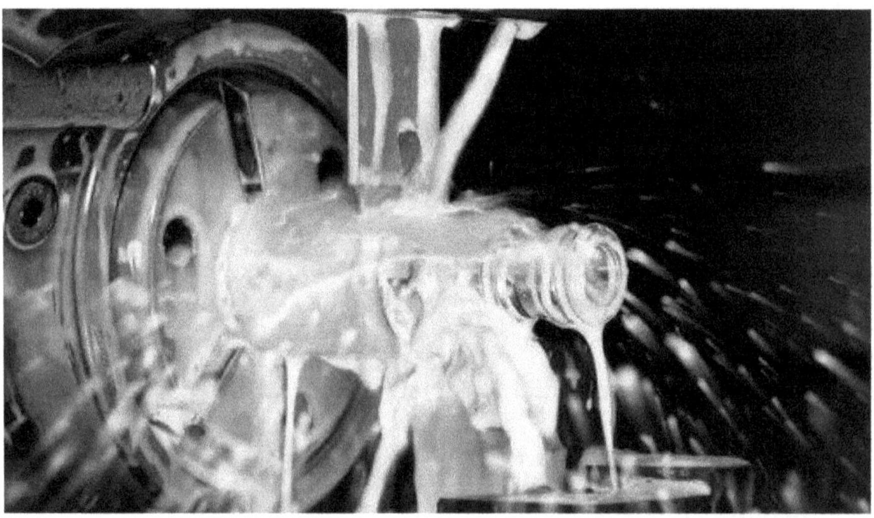

Figure S15 Soluble oils are a must in any machining operation. Source: Indiamart.

Composition of Soluble Oil

- **Base oil**: Typically, **mineral oil** (50%–80%) that provides lubrication.

- **Emulsifiers**: Enable the oil to **disperse in water**, forming a stable emulsion.
- **Additives**: Include **rust inhibitors, antiwear agents, biocides, and stabilizers** to enhance performance and longevity.

Types of Soluble Oils

1. **Emulsifiable cutting oils**: Form a **milky white emulsion** when mixed with water, commonly used in machining.
2. **Semisynthetic fluids**: Contain a lower percentage of oil and **offer better cooling and cleanliness.**
3. **Synthetic fluids**: Oil-free, water-based formulations providing **superior cooling with minimal residue**.

Applications of Soluble Oil

- **Metal cutting and machining**: Used in **turning, milling, drilling, and grinding** to improve tool life and surface finish.
- **Forming and stamping operations**: Reduces friction and wear in **sheet metal forming and pressing**.
- **Heat dissipation in industrial processes**: Used in **quenching, die casting, and rolling mills**.
- **Automotive and aerospace manufacturing**: Essential in **precision machining and high-speed metal processing**.

Benefits of Soluble Oil

- **Effective cooling and lubrication**: Enhances **machining precision and tool longevity**.
- **Cost-effective**: reduces **oil consumption** while maintaining high performance.
- **Corrosion protection**: Prevents **rust on machine components and workpieces**.
- **Improves workplace safety**: Nonflammable and **reduces smoke and mist formation** compared to straight oils.

Limitations of Soluble Oil

- **Bacterial growth and degradation**: Requires **biocides and proper maintenance** to prevent microbial contamination.

- **Foaming issues**: Can generate **foam in high-speed applications**, affecting lubrication efficiency.
- **Water quality sensitivity**: Hard water can **destabilize the emulsion**, leading to performance issue

833. Solvent

A **solvent** is a **substance that dissolves other materials** (solutes) to form a **homogeneous solution** without chemically altering either component. Solvents are commonly used in **cleaning, chemical processing, lubrication, and industrial applications** to dissolve, dilute, or disperse substances.

Key Functions of Solvents

- **Dissolution**: Breaks down solids, liquids, or gases into a uniform solution.
- **Cleaning and degreasing**: Removes oils, greases, and contaminants from surfaces.
- **Lubrication carrier**: Acts as a temporary carrier in **oil-based sprays, coatings, and metalworking fluids**.
- **Reaction medium**: Provides a medium for **chemical reactions** in industrial and pharmaceutical processes.

Types of Solvents

1. **Polar solvents**: Have strong molecular dipoles and dissolve **water-soluble (hydrophilic) substances**.
   - **Examples**: Water, alcohols, acetone.
2. **Nonpolar solvents**: Lack strong dipoles and dissolve **oil-based (hydrophobic) substances**.
   - **Examples**: Hexane, benzene, mineral spirits.
3. **Semipolar solvents**: Exhibit intermediate polarity, dissolving both **polar and nonpolar** compounds.
   - **Examples**: Ethyl acetate, diethyl ether.

Solvents play a crucial role in cleaning, lubrication, chemical processing, and industrial applications. Their ability to dissolve, dilute, and disperse substances makes them indispensable in manufacturing, maintenance, and scientific research. Proper selection, handling, and disposal of solvents are essential to maximize efficiency and ensure safety.

834. Solvent Extract

Solvent extract refers to a product or substance obtained through solvent extraction, a process that uses a liquid solvent to separate specific components from a mixture. In lubrication, petroleum refining, and industrial applications, solvent extracts are commonly derived from crude oil, and used for improving lubricant base stocks, isolating valuable compounds, and removing impurities.

Key Functions of Solvent Extracts

- **Lubricant base stock production**: Enhances **viscosity, oxidation stability, and performance** in lubricants.
- **Selective component separation**: Extracts **aromatic compounds, resins, or impurities** from crude oil or industrial mixtures.
- **Improved lubricant quality**: Removes **undesirable fractions**, enhancing thermal and oxidation stability.

Solvent Extraction Process

1. **Solvent selection**: A specific **polar or nonpolar solvent** is chosen based on the desired separation.
2. **Mixing and dissolution**: The solvent is introduced to dissolve **targeted compounds** while leaving others behind.
3. **Phase separation**: The mixture is separated, allowing the **solvent-rich phase** (extract) to be isolated.
4. **Solvent recovery**: The solvent is removed and recycled, leaving behind a purified extract.

Applications of Solvent Extracts

- **Petroleum refining**: Used in the **manufacturing of high-quality lubricant base oils**.
- **Metalworking and industrial lubricants**: Produces **cleaner, high-viscosity index base stocks** for cutting oils and greases.
- **Pharmaceutical and food processing**: Extracts **essential oils, vitamins, and bioactive compounds**.

- **Chemical processing and paints**: Isolates **desired components** for coatings, adhesives, and polymers.

Conclusion

Solvent extracts are **critical** in lubrication, petroleum refining, and industrial applications, enabling the production of high-quality base oils and purified chemical components. By using solvent extraction technology, manufacturers can improve lubricant performance, remove impurities, and enhance industrial processes.

835. Solvent Refining Process

The **solvent refining process** is a critical method used in petroleum refining to improve the quality of lubricant base oils, fuels, and other petroleum products. This process involves the use of selective solvents to remove undesirable components such as aromatics, sulfur compounds, and unstable hydrocarbons from crude oil distillates. By extracting these impurities, solvent refining enhances the thermal stability, viscosity index, and oxidation resistance of the final product, making it suitable for high-performance lubrication applications.

The process typically consists of three main stages:

- **Extraction**: First, the feedstock is mixed with a polar or nonpolar solvent, depending on the type of impurities targeted for removal. The solvent selectively dissolves unwanted components while leaving behind purified base oil.

- **Separation**: The solvent-rich phase (extract) is separated from the treated oil (raffinate).

- **Solvent recov**ery: The solvent is removed from both streams and recycled for reuse, ensuring cost efficiency and environmental sustainability.

Solvent refining is widely used in the production of high-quality lubricating oils, transformer oils, and waxes. It plays a crucial role in extending the service life of lubricants, reducing deposits in engines, and improving fuel efficiency. Compared to other refining methods, such as hydrotreating, solvent refining is energy-efficient and cost-effective, making it a preferred choice for many lubricant manufacturers. However, due to evolving environmental regulations and increasing demand for ultra-low sulfur products, many refiners are adopting hydro-processing techniques as an alternative or complementary method to solvent refining.

836. Soot Contamination

Soot contamination refers to the accumulation of fine carbon particles in lubricating oil due to incomplete fuel combustion in diesel engines, gasoline engines, and industrial machinery. These microscopic soot particles, typically less than 1 micron in size, can agglomerate, leading to increased viscosity, wear, and deposit formation in the lubrication system.

Causes of Soot Contamination

- **Incomplete combustion**: Poor fuel combustion produces excess carbon particles.
- **Blow-by gases**: Soot enters the lubricant when exhaust gases bypass the piston rings.
- **Extended oil drain intervals**: Prolonged use of oil increases soot accumulation.
- **Cold engine operation**: Low temperatures prevent complete fuel burning, generating more soot.

Effects of Soot Contamination

- **Increased oil viscosity**: Thickened oil reduces flow efficiency and lubrication.
- **Wear and abrasion**: Agglomerated soot particles cause increased friction and metal wear.
- **Sludge and deposits**: Soot buildup leads to sludge formation and filter clogging (see Figure S16).
- **Reduced lubrication film strength**: Thickened oil compromises proper lubrication.
- **Potential engine failure**: Excessive soot contamination can cause severe damage to engine components.

Monitoring and Detection

- **Thermogravimetric analysis (TGA)**: Measures soot concentration in oil samples.
- **Fourier transform infrared (FTIR) spectroscopy**: Detects soot particles based on infrared absorption.
- **Particle count analysis**: Quantifies soot particle size and distribution in the oil.

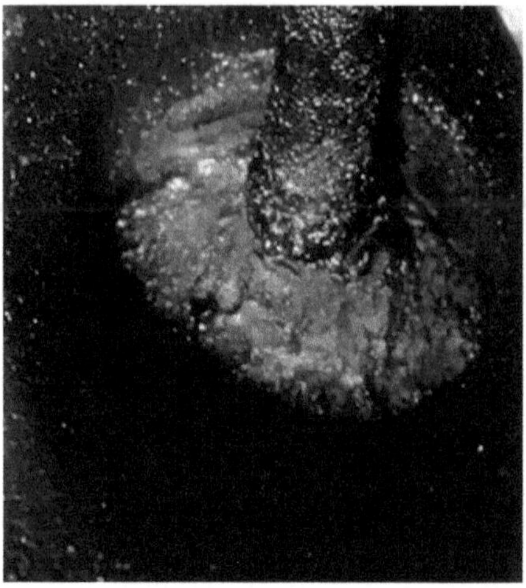

Figure S16 Typical soot contamination on a valve stem.

- **Particle count analysis**: Quantifies soot particle size and distribution in the oil.

Prevention and Control

- **Use high-quality dispersant additives**: Prevents soot particles from agglomerating.
- **Regular oil changes**: Removes accumulated soot and maintains lubrication quality.
- **Proper engine tuning**: Ensures optimal fuel combustion to minimize soot formation.
- **Efficient air and fuel filtration**: Reduces contaminants that contribute to incomplete combustion.

Conclusion: Soot contamination is a critical factor in engine wear and lubricant degradation. Proper oil analysis, high-quality dispersant additives, and regular maintenance help mitigate its impact, ensuring optimal lubrication, reduced wear, and extended engine life.

837. Spalling (Wear)

Spalling is a type of surface failure that occurs when small fragments or layers of material break away from a solid surface due to stress, fatigue, or rolling contact pressure. It is commonly observed in bearings, gears, and other load-bearing components in industrial and automotive applications. Spalling leads to roughened surfaces, increased friction, and premature component failure if not addressed.

Causes of Spalling

- **Rolling contact fatigue (RCF)**: Repeated cyclic stress causes subsurface cracks that eventually lead to surface material detachment.
- **Overloading**: Excessive loads generate high stress on metal surfaces, accelerating wear and material failure.
- **Poor lubrication**: Inadequate lubrication results in increased friction and localized overheating, leading to material breakdown.
- **Contaminant-induced wear**: Dirt, debris, or hard particles in the lubricant cause abrasive wear, weakening the material.
- **Manufacturing defects**: Improper heat treatment, material inconsistencies, or surface roughness increase susceptibility to spalling.

Effects of Spalling

- **Increased friction and heat generation**: Rough surfaces cause higher resistance, leading to overheating. See Figure S17.
- **Loss of dimensional accuracy**: Surface deterioration affects precision in moving components.
- **Reduced load-carrying capacity**: Spalled surfaces cannot distribute loads efficiently, leading to further wear.
- **Vibration and noise increase**: Uneven surfaces generate excessive noise and operational instability.
- **Premature component failure**: If left unchecked, spalling can cause catastrophic equipment breakdown.

Applications Affected by Spalling

- **Bearings and gears**: Common in **high-load rotating components** where fatigue and wear occur over time.
- **Railroad tracks and wheels**: Metal fatigue and impact stress cause rail surface degradation.
- **Turbines and compressors**: Rolling elements in **high-speed machinery** are vulnerable to fatigue spalling.

Figure S17 Spalling on an inner roller bearing ring. Source: NTN Bearings.

Prevention and Control

- **Optimize lubrication**: Use the right **viscosity and extreme pressure (EP) additives** to reduce friction and wear.
- **Control operating loads**: Avoid overloading by ensuring **proper design and material selection**.
- **Use high-quality materials**: Heat-treated, wear-resistant alloys reduce susceptibility to fatigue failure.
- **Improve contaminant control**: Use **efficient filtration systems** to keep lubricants clean.
- **Regular condition monitoring**: Conduct **vibration analysis, oil analysis, and surface inspections** to detect early signs of spalling.

Conclusion: Spalling is a **serious wear mechanism** that affects **bearings, gears, and rolling components** due to fatigue, overloading, or inadequate lubrication. Implementing **preventive maintenance**, **proper lubrication**, **and material selection** helps minimize spalling, extending the service life of critical machinery and reducing downtime.

838. Specific Gravity

Specific gravity (SG) in lubricants refers to the **ratio of the density of a lubricant to the density of water** at a specified temperature, usually **60 °F**

(15.6 °C). It is a **dimensionless value** used to assess the heaviness of a lubricant compared to water. Specific gravity is crucial in **lubricant formulation, storage, and application**, as it affects **lubricant separation, compatibility, and flow properties**.

Key Functions of Specific Gravity in Lubricants

- **Determines lubricant density**: Helps compare lubricants of different formulations and viscosities.
- **Affects lubricant separation**: Lubricants with **SG < 1.0** float on water, while those with **SG > 1.0** sink, impacting contamination control and water separation efficiency.
- **Influences storage and handling**: Helps in **pipeline flow calculations, tank volume determination, and blending processes**.
- **Correlates with base oil type**: Different base oils have varying SG values, influencing **lubricant performance and application suitability**.

Typical Specific Gravity Values of Lubricants

- **Mineral oil-based lubricants**: **0.85–0.90** (lighter than water, floats on water).
- **Synthetic lubricants**: **0.90–1.05** (varies depending on formulation; some synthetics may sink in water).
- **Heavy gear oils and greases**: **0.90–1.00** (higher density due to thickening agents and additives).
- **Chlorinated and phosphate ester-based lubricants**: **>1.0** (denser than water, may sink).

Applications of Specific Gravity in Lubrication

- **Oil–water separation**: Determines how a lubricant behaves when mixed with water, affecting **filtration and contamination control**.
- **Compatibility in fluid blending**: Ensures proper **mixing of base oils and additives** during formulation.
- **Leak detection in cooling systems**: Identifies whether an oil or coolant leak is present based on SG differences.
- **Pipeline transport and storage calculations**: Helps in **determining volume expansion and flow rate adjustments**.

839. Spectral Subtraction

Spectral subtraction is an analytical technique used in lubrication and oil analysis to isolate and identify contaminants, additives, and degradation by-products in used lubricants. It involves comparing the infrared (IR) spectrum of a used oil sample to a baseline spectrum of new (unused) oil and mathematically subtracting the two to reveal chemical changes. This method is commonly applied in Fourier transform infrared (FTIR) spectroscopy to detect oxidation, nitration, sulfation, water contamination, and additive depletion in lubricants. See Figure S18.

Key Functions of Spectral Subtraction in Lubrication

- **Detects lubricant degradation**: Identifies oxidation, nitration, and sulfation by analyzing chemical changes in the lubricant.
- **Monitors additive depletion**: Tracks the reduction of antiwear, detergent, and dispersant additives in oil.
- **Identifies contaminants**: Highlights foreign substances such as water, glycol, fuel, or soot in the lubricant.
- **Improves oil condition monitoring**: Provides a clear picture of lubricant health and helps optimize maintenance schedules.

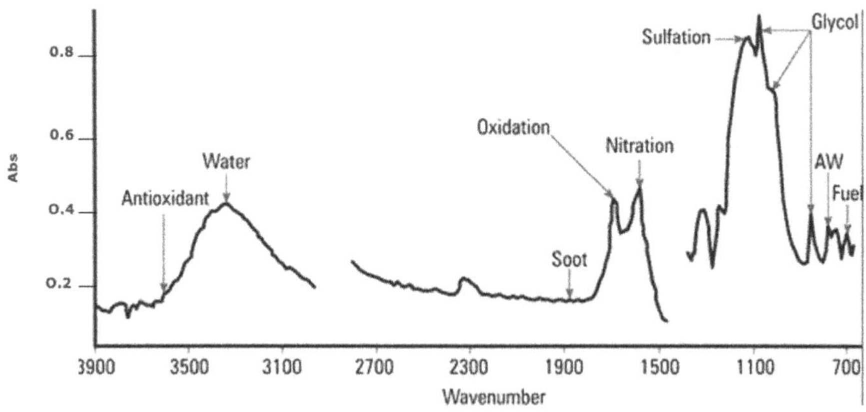

Figure S18 FTIR spectroscope difference spectrum. Source: ENG**TECH** Industries Inc.

How Spectral Subtraction Works

1. **Baseline spectrum collection**: A reference spectrum is taken from a **new, unused lubricant** to establish a benchmark.

2. **Used oil spectrum analysis**: The infrared spectrum of the **in-service lubricant** is measured.
3. **Mathematical subtraction**: The **new oil spectrum is subtracted from the used oil spectrum**, isolating changes caused by contamination, degradation, or additive loss.
4. **Interpretation of results**: The resulting spectrum highlights the chemical components that have either **increased (contaminants and degradation by-products)** or **decreased (additives and base oil components)**.

Applications of Spectral Subtraction in Lubrication

- **Industrial lubricants**: Monitors oil degradation in **hydraulic systems, turbines, and compressors**.
- **Automotive and engine oils**: Detects **oxidation**, **soot buildup**, **fuel dilution, and coolant leaks** in engine lubricants.
- **Gear and transmission oils**: Identifies **extreme pressure (EP) additive depletion and contamination**.
- **Metalworking fluids**: Ensures the stability of **coolants, cutting fluids, and anticorrosion additives**.

Conclusion: **Spectral subtraction** is a powerful diagnostic tool in lubrication analysis, allowing precise identification of oil degradation, contamination, and additive depletion. By comparing used oil spectra against new oil references, this technique enables proactive maintenance, extends equipment life, and optimizes lubricant performance in industrial and automotive applications.

840. Spectrometry

Spectrometry is an analytical technique used to measure and identify the composition of substances based on their interaction with electromagnetic radiation or charged particles. In lubrication analysis, spectrometry is widely used to detect wear metals, contaminants, and additive elements in oils, greases, and other lubricants. By analyzing how materials absorb, emit, or scatter light or other energy sources, spectrometry provides valuable insights into lubricant condition, machinery health, and potential failure risks.

Key Functions of Spectrometry in Lubrication

- **Identifies wear metals**: Detects **iron, copper**, **aluminum**, **lead**, **and other metals** from engine, gearbox, and bearing wear.

- **Measures additive levels**: Determines the presence of **zinc, phosphorus, calcium, and sulfur**, essential for lubricant performance.
- **Detects contaminants**: Identifies **water, fuel, dirt, and coolant contamination** in used oils.
- **Monitors lubricant degradation**: Tracks chemical changes such as **oxidation, nitration, and sulfation** that affect lubricant performance.

Types of Spectrometry Used in Lubrication Analysis

1. **Inductively coupled plasma (ICP) spectrometry**:
   - Detects **trace amounts of wear metals and additives** in lubricants.
   - Used in **ASTM D5185** standard oil analysis.
2. **Fourier transform infrared (FTIR) spectroscopy**:
   - Measures **oxidation, nitration, sulfation, and contamination** by analyzing molecular vibrations.
   - Used in **ASTM D7414** for monitoring lubricant degradation.
3. **X-ray fluorescence (XRF) spectrometry**:
   - Identifies **elemental composition** in lubricants, including heavy metals.
   - Common in **quality control of new and in-service oils**.
4. **Atomic absorption spectroscopy (AAS)**:
   - Detects **specific metal concentrations** in oil samples.
   - Used in wear debris and trace element analysis.

Spectrometry is a vital tool in lubricant analysis, allowing for precise detection of wear metals, additives, and contaminants in oils and lubricants. By providing early warning signs of mechanical wear and lubricant degradation, spectrometry enhances equipment reliability, reduces maintenance costs, and optimizes lubricant performance in industrial and automotive applications.

841. Speed Factor (DN)

DN speed factor is a commonly used parameter to evaluate the operational suitability of bearings and lubrication systems in high-speed applications. It is calculated as the product of the bearing bore diameter (D) in millimeters and the rotational speed (N) in revolutions per minute (RPM). The DN speed factor helps determine the appropriate lubricant, bearing design, and cooling requirements to ensure optimal performance and prevent premature wear or failure.

Formula for DN Speed Factor

$$DN = D \times N.$$

where:

- **D** = Bearing bore diameter (mm)
- **N** = Rotational speed (RPM)

Key Functions of DN Speed Factor

- **Determines lubricant selection**: Helps choose between **oil or grease lubrication** based on operating speed.
- **Predicts bearing suitability**: Identifies whether a bearing can operate efficiently at a given speed.
- **Affects heat generation**: Higher DN values indicate **higher frictional heat**, requiring enhanced cooling or lubricant management.
- **Optimizes lubrication intervals**: Ensures proper relubrication schedules to maintain bearing efficiency.

Typical DN Speed Factor Ranges for Different Bearings

- **Grease-lubricated bearings**: Up to **500,000** DN
- **Oil-lubricated ball bearings**: **500,000 to 1,500,000** DN
- **Oil mist or air–oil lubricated bearings**: **Above 1,500,000** DN
- **High-speed precision bearings (aerospace/spindle bearings)**: **3,000,000+** DN

Impact of DN Speed Factor on Lubrication

- **Low DN values**: Grease lubrication is generally sufficient for **low-to-moderate-speed applications**.
- **High DN values**: Requires **oil mist, air–oil, or circulating oil lubrication** to manage heat and minimize friction.
- **Very high DN applications**: Specialized **low-viscosity oils and advanced cooling techniques** are necessary.

Conclusion: **DN speed factor is an essential metric for selecting bearings and lubrication systems in high-speed applications**. By understanding DN limits, engineers can optimize lubricant selection, cooling strategies, and maintenance schedules, ensuring efficient, reliable, and long-lasting bearing performance in industrial and automotive systems.

842. Spill Control

Spill control in lubrication refers to the preventive measures and response strategies used to contain, manage, and clean up lubricant spills to minimize environmental impact, workplace hazards, and equipment contamination. Proper spill control is essential in industrial facilities, maintenance operations, and oil-handling processes to comply with safety regulations and environmental standards while ensuring efficient lubricant use.

Key Functions of Spill Control in Lubrication

- **Prevents environmental contamination**: Reduces the risk of soil, water, and air pollution from oil and grease spills.
- **Enhances workplace safety**: Prevents slips, falls, and fire hazards caused by spilled lubricants.
- **Ensures regulatory compliance**: Meets OSHA, EPA, and ISO 14001 environmental safety standards.
- **Minimizes equipment damage**: Prevents spilled lubricants from contaminating machines, bearings, and moving components.

Components of an Effective Spill Control Plan

- **Spill prevention measures**: Proper lubricant storage, handling procedures, and regular inspections to prevent leaks and accidental spills.
- **Containment systems**: Includes drip trays, bunds, absorbent pads, and spill containment pallets to control leaks.
- **Spill response equipment**: Quick-access spill kits, absorbents, barriers, and drain covers for immediate containment.
- **Emergency response protocols**: Training staff in spill containment, cleanup procedures, and disposal practices.
- **Proper disposal and recycling**: Ensures safe disposal of contaminated absorbents and used oil following environmental guidelines.

Spill Control Methods for Lubricants

- **Absorbent materials**: Use of spill pads, booms, and granules to soak up leaked oil or grease.
- **Secondary containment systems**: Includes drip pans, spill pallets, and bund walls for storage tanks and lubricant drums.
- **Drain protection measures**: Using spill socks, barriers, or drain covers to prevent lubricant entry into water systems.
- **Vacuum or pump recovery systems**: For large-scale spills, using vacuum pumps or oil skimmers to recover and recycle lubricants.

Benefits of Spill Control in Lubrication

- **Reduces environmental liabilities**: Minimizes fines and penalties for oil spill violations.
- **Enhances safety and compliance**: Ensures workplace safety and regulatory adherence.
- **Extends equipment life**: Prevents lubricant contamination and reduces unexpected downtime.
- **Promotes sustainable practices**: Encourages oil recycling and waste reduction.

Conclusion: **Spill control** in lubrication is essential for preventing environmental contamination, ensuring workplace safety, and maintaining regulatory compliance. By implementing preventive measures, effective containment strategies, and proper spill response procedures, industries can minimize risks, protect equipment, and promote sustainable lubricant management practices.

843. Spill Pallets

Spill pallets are secondary containment devices designed to hold and contain leaks, spills, or drips from oil drums, lubricant containers, and hazardous liquids in industrial, automotive, and environmental applications. These pallets are made from chemical-resistant materials such as polyethylene, steel, or fiberglass and feature a grated deck that allows spilled fluids to drain into a built-in sump, preventing spills from spreading and contaminating work areas. See Figure S19.

Key Functions of Spill Pallets

- **Secondary containment**: Captures accidental spills, preventing leaks from reaching the floor or drainage systems.
- **Environmental protection**: Helps comply with EPA (Environmental Protection Agency), OSHA, and ISO 14001 spill containment regulations.
- **Workplace** Safety Enhancement: Reduces the risk of slips, falls, and fire hazards caused by spilled lubricants.
- **Efficient storage and handling**: Provides a stable platform for storing drums, containers, and intermediate bulk containers (IBCs).

Types of Spill Pallets

1. **Polyethylene spill pallets**:
   - Resistant to corrosive chemicals, oils, and lubricants.
   - Lightweight and easy to move.
2. **Steel spill pallets**:
   - Designed for flammable liquids and heavy-duty industrial use.
   - High load-bearing capacity and fire resistance.
3. **IBC spill pallets**:
   - Designed for intermediate bulk containers (IBCs) used for large volumes of lubricants and chemicals.
   - Higher containment capacity compared to standard drum spill pallets.
4. **Low-profile spill pallets**:
   - Lower deck height for easy drum loading and unloading.
   - Ideal for areas with limited space.

Applications of Spill Pallets

- **Lubricant and oil storage**: Prevents contamination from leaking drums, totes, and bulk containers.
- **Chemical and hazardous waste handling**: Safely stores corrosive fluids, solvents, and acids.

- **Manufacturing and maintenance facilities**: Used in automotive workshops, power plants, and factories to maintain clean work areas.

Figure S19 Typical polyethylene spill pallet for drums and pails. Courtesy: ENG**TECH** Industries Inc.

- **Environmental compliance and spill prevention**: Ensures regulatory compliance in industrial and commercial facilities.

Benefits of Spill Pallets

- **Prevents environmental contamination**: Helps contain hazardous spills, protecting soil and water sources.
- **Enhances workplace safety**: Reduces slip-and-fall risks and maintains a cleaner work environment.
- **Regulatory compliance**: Meets EPA, OSHA, and SPCC (Spill Prevention, Control, and Countermeasure) regulations.
- **Durable and long-lasting**: Resistant to chemicals, weather conditions, and industrial wear.

Conclusion: **Spill pallets** are essential for secondary containment, protecting workplaces from lubricant and chemical spills while ensuring compliance with environmental and safety regulations. Available in polyethylene, steel, and IBC-compatible designs, they help industries safely store and manage oil, lubricants, and hazardous fluids, preventing contamination and reducing workplace hazards.

844. Spin-on Filter

A **spin-on oil filter** is a self-contained, replaceable filtration unit used to remove contaminants, debris, and wear particles from engine oil, hydraulic fluids, and lubricants in automotive, industrial, and heavy equipment applications. It consists of a metal canister housing a filtration media, a bypass valve, an antidrainback valve (in some designs), and a threaded baseplate for quick installation and removal.

Components of a Spin-on Oil Filter

1. **Outer metal canister**: Protects the internal filter media from external damage.
2. **Filter media**: A paper, synthetic, or cellulose blend that traps contaminants while allowing oil to pass through.
3. **Baseplate with threaded center hole**: Allows the filter to be screwed onto the engine or lubrication system.
4. **Antidrainback valve (in some filters)**: Prevents oil from draining out when the engine is off, ensuring immediate lubrication on startup.
5. **Bypass valve**: Allows oil to flow even if the filter becomes clogged, preventing oil starvation.
6. **Gasket seal**: Ensures a tight, leak-proof connection between the filter and the engine block.

Applications of Spin-on Oil Filters

- **Automotive engines**: Used in cars, trucks, motorcycles, and commercial vehicles to maintain engine oil cleanliness.
- **Industrial machinery**: Protects hydraulic systems, compressors, and heavy-duty engines from oil contamination.
- **Construction and mining equipment**: Ensures lubrication integrity in excavators, loaders, and drilling rigs.

- **Marine and power generation systems**: Filters oil in ship engines, turbines, and backup generators.

Limitations of Spin-on Oil Filters

- **Disposable design**: Unlike cartridge filters, spin-on filters contribute **to waste and require proper disposal**.
- **Bypass valve activation**: If clogged, the bypass valve allows unfiltered oil to circulate, potentially increasing wear.
- **Size and compatibility restrictions**: Must **match engine specifications** for optimal performance.

Conclusion: **Spin-on oil filters** provide essential lubrication system protection by effectively removing contaminants from engine and industrial oils. Their simple design, easy installation, and efficient filtration make them a preferred choice for automotive, heavy equipment, and industrial applications, ensuring clean oil, reduced wear, and extended equipment life.

845. Spindle Lubricant

Spindle lubricant is a high-performance oil or grease specifically formulated for high-speed, high-precision spindle bearings used in CNC machines, grinding equipment, lathes, milling machines, and textile machinery. It provides low friction, excellent thermal stability, and superior wear protection, ensuring smooth operation, minimal heat generation, and extended spindle life.

Key Functions of Spindle Lubricant

- **Reduces friction and wear**: Forms a thin, durable lubricating film to protect spindle bearings.
- **Enhances thermal stability**: Prevents overheating and maintains consistent lubrication under high speeds.
- **Minimizes vibration and noise**: Ensures smooth motion and precise machining performance.
- **Resists oxidation and degradation**: Offers long-term stability, preventing oil breakdown and deposits.

Types of Spindle Lubricants

1. **Spindle oils (ISO VG 2–32)**:
   - Low-viscosity oils designed for **high-speed**, **precision spindle bearings**.

- Excellent **cooling properties** and minimal drag resistance.

2. **Mist and air–oil lubricants**:
   - Used in **high-speed applications where minimal lubrication is required**.
   - Delivered in fine mist or air–oil mixtures to **reduce oil consumption**.
3. **Grease lubricants**:
   - Used in **sealed spindle bearings requiring long-term lubrication**.
   - Ensures **high load-carrying capacity** and minimal grease migration.

Applications of Spindle Lubricants

- **CNC machine spindles**: Ensures accurate positioning and reduced tool wear.
- **Grinding and milling machines**: Maintains **low friction and heat control** in precision machining.
- **Textile machinery bearings**: Lubricates **high-speed spinning and weaving machine spindles**.
- **Aerospace and medical equipment**: Provides lubrication in **high-precision machining applications**.

Benefits of Spindle Lubricant

- **Extends spindle and bearing life**: Prevents premature wear and failure.
- **Optimizes high-speed operation**: Reduces frictional losses and improves energy efficiency.
- **Improves surface finish and machining accuracy**: Minimizes thermal expansion and vibration.
- **Prevents oil contamination and deposit formation**: Ensures **clean operation** for precision equipment.

Conclusion: **Spindle lubricants** play a crucial role in maintaining the performance and longevity of high-speed, precision machinery. By providing low-viscosity, thermally stable lubrication, they ensure smooth spindle operation, reduced wear, and improved machining accuracy, making them essential in CNC machining, grinding, and textile manufacturing applications.

846. SPL

See "Single Point Lubricator."

847. Splash Lubrication

Splash lubrication is a **passive lubrication method** where rotating components, such as gears or crankshafts, dip into an oil reservoir and "splash" lubricant onto surrounding surfaces. This system is commonly used in gearboxes, small engines, compressors, and industrial machinery where a simple, low-maintenance lubrication method is required.

Key Functions of Splash Lubrication

- **Distributes lubricant automatically**: Oil is carried by rotating parts and splashed onto moving components.
- **Reduces friction and wear**: Ensures a protective oil film on bearings, gears, and shafts.
- **Cools and removes heat**: Circulating oil helps dissipate heat generated by friction.
- **Requires minimal maintenance**: Operates without pumps or external lubrication systems.

How Splash Lubrication Works

1. **Oil reservoir**: A sump or oil bath holds the lubricant at the bottom of the system.
2. **Rotating components**: Gears, crankshafts, or oil paddles dip into the oil and lift it as it rotate.
3. **Lubricant splashing**: Oil is flung onto bearings, gears, and internal surfaces.
4. **Drainage and recirculation**: The oil drips back into the sump, ready for reuse in the next cycle.

Applications of Splash Lubrication

- **Small gasoline and diesel engines**: Used in lawnmowers, motorcycles, and small industrial engines.
- **Gearboxes and transmissions**: Lubricates open and semienclosed gear systems.

- **Compressors and pumps**: Ensures proper lubrication of rotating and reciprocating parts.
- **Agricultural and construction equipment**: Used in older or low-speed machinery.

Benefits of Splash Lubrication

- **Simple and cost-effective**: Requires no pumps, filters, or external lubrication systems.
- **Reliable and self-sustaining**: Works as long as rotating components are moving.
- **Reduces maintenance needs**: No complex components to fail or maintain.
- **Provides adequate cooling and lubrication**: Ensures consistent oil distribution in low- to moderate-speed applications.

Limitations of Splash Lubrication

- **Limited to low- and medium-speed systems**: Not suitable for **high-speed or high-load applications**.
- **Inefficient at extreme temperatures**: Oil viscosity can be affected by **cold starts or high operating temperatures**.
- **Inconsistent lubrication in certain positions**: Oil distribution may be uneven in **tilted or inclined equipment**.

Conclusion: Splash lubrication is a simple, effective, and low-maintenance method for delivering oil to rotating components in small engines, gearboxes, and industrial machinery. While ideal for low- to moderate-speed applications, it may require additional cooling or circulation in high-load, high-speed environments. Proper oil level monitoring and viscosity selection are essential for maintaining efficient lubrication and long equipment life.

848. SSU

See "Saybolt Second Universal."

849. Starburst Symbol – API

The **Starburst symbol** in lubrication refers to the **API (American Petroleum Institute) Certification Mark**, commonly seen on engine oil containers.

See Figure S20. This symbol indicates that the oil meets API performance standards and has been tested for quality, performance, and compatibility with modern engines. The starburst mark is specifically associated with API "Energy Conserving" and "SP" rated motor oils, ensuring compliance with industry requirements for fuel efficiency, wear protection, and deposit control.

Image courtesy of API

Figure S20 API starburst symbol of quality.

Key Functions of the Starburst Symbol

- **Certifies oil performance**: Confirms that the lubricant meets API engine oil specifications.
- **Ensures compatibility with modern engines**: Indicates that the oil is designed for gasoline engines with emission control systems.
- **Promotes energy efficiency**: Found on low-friction, fuel-saving engine oils.
- **Validates testing and approval**: Signifies compliance with API testing procedures and industry standards.

How to Identify the Starburst Symbol

- The symbol appears as a circular starburst design with the words "**API Certified for Gasoline Engines**" inside.
- It is usually displayed on the front of oil containers from API-licensed manufacturers.
- It is different from the API Service Donut, which provides specific details on viscosity, API classification, and energy conservation properties.

Stated Benefits of Using Starburst-certified Lubricants

- **Improves fuel economy**: Helps engines run efficiently with **low-friction formulations**.
- **Protects against engine wear**: Ensures **proper lubrication, sludge control, and oxidation resistance**.
- **Reduces emissions**: Supports cleaner engine operation by meeting **low-phosphorus and low-sulfur requirements**.
- **Guarantees industry compliance**: Meets or exceeds **API, ILSAC, and OEM specifications** for modern engines.

850. Static Friction

Static friction is the **resistance to motion between two stationary surfaces in contact**. It prevents an object from moving until a force **exceeds a certain threshold**, overcoming the frictional resistance. Static friction is typically **greater than kinetic (sliding) friction** because surface irregularities have more time to interlock when at rest. See also **"Friction"** reference.

Key Characteristics of Static Friction

- **Acts only on stationary objects**: Exists between two surfaces **until motion begins**.
- **Self-adjusting force**: Increases or decreases to match the applied force, up to a maximum limit.
- **Greater than kinetic friction**: Requires more force to initiate motion than to maintain it.

Formula for Static Friction:

$$F_s \leq \mu_s F_n.$$

where:

- F_s = Static friction force
- μ_s = Coefficient of static friction (material-dependent)
- F_n = Normal force (force perpendicular to the surfaces)

Factors Affecting Static Friction

- **Surface roughness**: Rougher surfaces have higher static friction.

- **Material composition**: Different materials exhibit different friction coefficients (e.g., rubber on asphalt vs. steel on ice).
- **Normal force**: Greater pressure between surfaces increases static friction.
- **Lubrication**: Reduces static friction by forming a low-resistance boundary layer.

851. Steam Cylinder Oil

See "Cylinder Oil."

852. Stick-slip Motion

Stick-slip motion is a cyclic, jerky movement that occurs when two surfaces alternate between static friction (sticking) and kinetic friction (slipping). This phenomenon is common in machinery, bearings, hydraulic systems, and precision motion control where frictional resistance fluctuates, causing irregular or oscillatory motion. It can lead to vibration, noise, and mechanical wear, affecting equipment performance and longevity.

Causes of Stick-slip Motion

- **High static-to-kinetic friction ratio**: When static friction (\mu_s) is significantly greater than kinetic friction (\mu_k), movement starts abruptly.
- **Inconsistent lubrication**: Poor lubrication or improper viscosity causes **frictional fluctuations**, leading to stick-slip behavior.
- **Elastic deformation**: Stored energy in materials (such as rubber seals or sliding contacts) causes a sudden release when motion overcomes friction.
- **Low-speed or intermittent motion**: Common in **slow-moving actuators, hydraulic cylinders, and machine tool slides**.
- **Surface roughness and contamination**: Uneven or dirty surfaces increase resistance, causing erratic motion.

Effects of Stick-slip Motion

- **Unstable and jerky movement**: Causes **positioning inaccuracies** in CNC machines, robotics, and precision systems.

- **Increased wear and fatigue**: Leads to **premature failure of bearings, seals, and sliding components**.
- **Vibration and noise**: Generates unwanted **squeaks, chatter, or oscillations** in mechanical systems.
- **Reduced efficiency**: Interrupts **smooth energy transfer**, affecting system performance.

Methods to Reduce Stick-slip Motion

- **Optimize lubrication**: Use **low-friction, high-viscosity-index lubricants** to maintain stable friction.
- **Surface treatments**: Apply **coatings, polishing, or texturing** to smooth out irregularities.
- **Modify Load and speed conditions**: Adjust **operating parameters** to minimize sudden force variations.
- **Use damping systems**: Install **shock absorbers or vibration dampers** to reduce oscillations.
- **Replace or upgrade materials**: Use **self-lubricating materials, polymer coatings, or ceramic bearings** to lower friction differences.

853. Stiction

Stiction (static friction + adhesion) is the **resistance to initial motion** that occurs when two surfaces remain in contact and require additional force to break free. It is the force that must be overcome to initiate movement before transitioning to **kinetic friction**. Stiction is commonly observed in **bearings, seals, hydraulic actuators, machine slides, and precision motion control systems**, where it can cause **jerky motion, performance loss, and excessive wear**.

Causes of Stiction

- **High static friction** (\mu_s) **relative to kinetic friction** (\mu_k): The greater the difference, the more difficult it is to initiate movement.
- **Surface adhesion and bonding**: Molecular attraction between surfaces, particularly in **polished metal contacts, gaskets, and seals**.
- **Inadequate lubrication**: Poor lubrication or excessive viscosity prevents smooth movement.

- **Contamination and deposits**: Dust, debris, or dried lubricant residue increases sticking effects.
- **Long periods of rest**: Surfaces in prolonged contact tend to form stronger **adhesive bonds**, making initial movement harder.

Stiction is a key challenge in lubrication, precision engineering, and mechanical systems where initial motion resistance impacts efficiency and performance. By using low-friction lubricants, optimized surface finishes, and controlled operating conditions, industries can reduce stiction, improve motion control, and extend component lifespan. See also "Stick-slip Motion."

854. STLE

See "Society of Tribologists and Lubrication Engineers."

855. Stoke (St)

Stoke (St) is the **CGS (centimeter-gram-second) unit of kinematic viscosity**, representing the ratio of a fluid's dynamic viscosity to its density. It is used in lubrication and fluid dynamics to measure how easily liquid flows under the influence of gravity.

Formula for Stoke (St)

$$\text{Kinematic Viscosity } (\upsilon) = \frac{\text{Dynamic Viscosity}(\mu)}{\text{Density } (\rho)}.$$

where:

- υ (nu) = Kinematic viscosity (measured in Stokes, cm^2/s)
- μ (mu) = Dynamic viscosity (measured in Poise, g/cm·s)
- ρ (rho) = Density of the fluid (measured in g/cm^3)

Conversion of Stoke to SI Units

Since the **Stoke (St)** is part of the CGS system, the SI unit equivalent is **square meters per second (m^2/s)**.

$$1 \text{ St} = 10^{-4} \text{ m}^2/\text{s} = 100 \text{ cSt (centistokes)}.$$

Applications of Stoke in Lubrication

- **Lubricant selection**: Helps determine the correct oil viscosity for engines, **bearings, and hydraulic systems**.
- **Fluid flow analysis**: Used in pumps, pipelines, and industrial fluid systems to assess flow behavior.
- **Turbine and compressor oils**: Ensures proper lubrication and heat dissipation in high-speed machinery.
- **Oil classification**: Used in ISO VG (Viscosity Grade) and SAE viscosity standards for lubricants.

Typical Viscosity Ranges in Stokes

- **Water**: ~0.01 St at 20 °C
- **Light machine oil**: 0.1–1.0 St
- **Engine oil (SAE 30)**: ~1.5–2.5 St
- **Heavy gear oil (SAE 140)**: ~10–30 St

Benefits of Measuring Viscosity in Stokes

- **Ensures proper lubrication**: Prevents excessive wear, overheating, and fluid breakdown.
- **Optimizes equipment efficiency**: **Reduces energy losses due to fluid resistance**.
- **Improves flow control**: Helps design efficient lubrication and cooling systems.

Conclusion: Stoke (St) is a fundamental unit for measuring kinematic viscosity, crucial in lubrication, fluid mechanics, and industrial applications. By understanding viscosity in Stokes or centistokes (cSt), engineers can select the right lubricants and optimize fluid flow for enhanced equipment performance and longevity.

856. Stoke's Procedure

Stokes' procedure refers to a method used to measure the **kinematic viscosity of a fluid**, based on **Stokes' Law**, which describes the motion of a sphere falling through a viscous liquid under gravity. This procedure is commonly used in lubrication analysis, industrial fluid testing, and quality control of oils and greases.

Principle of Stokes' Procedure

Stokes' Law states that the **viscous drag force** acting on a sphere moving through a fluid is proportional to its velocity, radius, and the fluid's viscosity. By measuring the **terminal velocity** of a sphere falling through a lubricant, the **kinematic viscosity** of the fluid can be determined.

Formula Used in Stokes' Procedure

$$\eta = \frac{2r^2\left(\rho_s - \rho_f\right)g}{9\upsilon}.$$

where:

- η (eta) = Dynamic viscosity (Poise)
- **r** = Radius of the falling sphere (cm)
- ρ**s** = Density of the sphere (g/cm^3)
- ρ**f** = Density of the fluid (g/cm^3)
- **g** = Acceleration due to gravity (980 cm/s^2)
- **v** = Terminal velocity of the sphere (cm/s)

Steps in Stokes' Procedure

1. **Prepare the fluid sample**: Fill a **viscometer tube or a tall graduated cylinder** with the lubricant.
2. **Select a calibrated sphere**: Use a **small steel or glass sphere** of known diameter and density.
3. **Release the sphere**: Drop the sphere into the fluid and allow it to fall freely under gravity.
4. **Measure terminal velocity**: Record the time it takes for the sphere to travel a fixed distance.
5. **Calculate the viscosity**: Apply Stokes' Law to determine the dynamic viscosity, then divide by the fluid's density to obtain **kinematic viscosity**.

Applications of Stokes' Procedure in Lubrication

- **Determining oil viscosity**: Used in quality control of industrial and automotive lubricants.
- **Testing contaminated lubricants**: Identifies viscosity changes due to dirt, oxidation, or additive breakdown.

- **Grease and fluid formulation**: Ensures lubricants **meet ISO VG (Viscosity Grade) and ASTM standards**.
- **Comparing synthetic versus mineral oils**: Evaluates fluid flow characteristics and efficiency in different lubricants.

Advantages of Stokes' Procedure

- **Simple and cost-effective**: Requires minimal equipment compared to high-end viscometers.
- **Accurate for Newtonian fluids**: Works well for oils and low-viscosity lubricants.
- **Used in field and laboratory testing**: Helps in quick lubricant assessments without specialized instruments.

Limitations of Stokes' Procedure

- **Not suitable for non-Newtonian fluids**: Does not work accurately for thick greases or fluids with variable viscosity.
- **Temperature sensitivity**: Lubricant viscosity changes with temperature, requiring controlled conditions.
- **Requires precise measurements**: Minor errors in sphere size, density, or velocity timing affect accuracy.

Conclusion: Stokes' procedure is a fundamental method for measuring the viscosity of lubricants using a falling sphere experiment. It provides an easy and effective way to evaluate lubricant flow properties, quality, and contamination levels, making it valuable in industrial, automotive, and research applications.

857. Straight oil

Straight oil is a **nonemulsifiable, petroleum-based lubricant** used in metalworking, machining, and industrial applications where extreme pressure (EP) protection and lubrication are required. Unlike soluble oils or synthetic cutting fluids, straight oils are used undiluted and provide superior lubrication, cooling, and wear protection in high-speed, high-load machining operations.

Composition of Straight Oils

- **Base oil**: Mineral oil or highly refined petroleum oil.

- **Extreme pressure (EP) additives**: Sulfur, chlorine, or phosphorus compounds for heavy-duty applications.
- **Antiwear agents**: Improve surface protection in precision machining.
- **Antioxidants and corrosion inhibitors**: Extend fluid life and prevent oxidation.

Applications of Straight Oil

- **Metal cutting and grinding**: Used in turning, milling, drilling, and threading operations.
- **Heavy-duty machining**: Suitable for broaching, gear hobbing, and deep-hole drilling.
- **Threading and tapping operations**: Provides lubrication in high-friction metalworking.
- **Gear and bearing lubrication**: Used in some low-speed, high-load industrial gear systems.

Benefits of Straight Oil

- **Excellent lubrication and wear protection**: Ideal for heavy-duty machining and high-pressure applications.
- **High thermal stability**: Resists breakdown under high temperatures and extreme conditions.
- **prevents workpiece damage**: Minimizes metal deformation and tool burn.
- **Long fluid life**: Does not evaporate or degrade as quickly as water-based coolants.

Limitations of Straight Oil

- **Lower cooling efficiency**: Does not dissipate heat as effectively as water-based cutting fluids.
- **Potential smoke and mist formation**: May generate vapors under high-speed machining, requiring ventilation.
- **Slower chip removal**: Does not wash away metal chips as efficiently as emulsified fluids.
- **Fire hazard in high-temperature** Applications: Requires fire safety precautions in machining environments.

858. Strainer (Grease)

A **grease strainer** is a **filtration device** designed to remove solid contaminants, debris, and foreign particles from grease before it enters a lubrication system. It ensures that only clean grease reaches bearings, gears, and other lubricated components, preventing clogging, wear, and system failure in industrial, automotive, and heavy machinery applications.

Key Functions of a Grease Strainer

- **Removes contaminants**: Filters out metal particles, dirt, and hardened grease deposits.
- **Protects lubrication systems**: Prevents clogs in grease lines, injectors, and lubrication points.
- **Extends equipment life**: Reduces wear on bearings, gears, and other moving components.
- **Ensures consistent lubrication flow**: Maintains smooth, uninterrupted grease distribution.

Types of Grease Strainers

1. **Inline grease strainer (See Figure S21)**
    - Installed within grease distribution **lines** to filter contaminants greater than 145 microns before the lubricant reaches components.
    - Common in automatic lubrication systems and centralized grease distribution networks.

Figure S21 Inline Grease Strainer

2. **Grease pump strainer**
   - Positioned at the inlet of grease pumps to prevent debris from entering the system.
   - Protects manual, pneumatic, and electric **grease pumps from** contamination.
3. **Reservoir/drum grease strainer**
   - Used in bulk grease storage systems to filter grease before dispensing.
   - Prevents hardened grease or contaminants from reaching lubrication points.

859. Stribeck Curve

The **Stribeck curve** is a graphical representation of the relationship between friction and lubrication regimes in a tribological system. It shows how friction coefficient changes transition between different lubrication regimes – boundary, mixed, and hydrodynamic lubrication – which are critical for optimizing lubrication performance and reducing wear in bearings, gears, and machine components. See Figure S22.

Benefits of Understanding the Stribeck Curve

- **Optimizes lubricant performance**: Helps in choosing the right oil viscosity and film thickness.

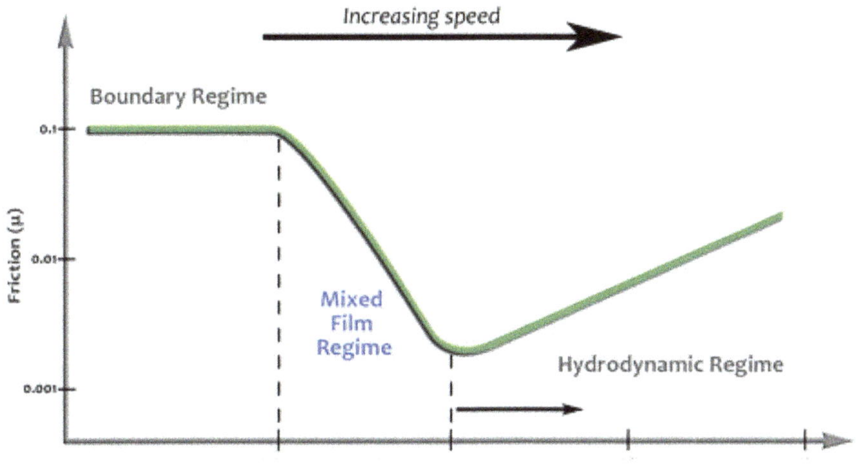

Figure S22 Stribeck curve Courtesy: ENG**TECH** Industries Inc.

- **Reduces energy loss and heat generation**: Minimizes frictional resistance, increasing machine efficiency.
- **Prevents wear and extends equipment life**: Ensures lubrication operates in the mixed or hydrodynamic region.
- **Improves predictive maintenance**: Helps diagnose lubrication failures based on friction trends.

860. Stribeck, Richard

Richard Stribeck (1861–1950) was a German mechanical engineer and tribologist best known for his groundbreaking research on friction, lubrication, and bearing performance. In charge of the Dresden Technical University's physical metallurgy department where in 1902, his work led to the development of the Stribeck Curve, which remains one of the most important concepts in tribology (the study of friction, wear, and lubrication). Stribeck's contributions have had a lasting impact on bearing design, lubrication science, and industrial machinery efficiency.

Key Contributions of Richard Stribeck

- **Developed the Stribeck curve (1902)**:
  - His research on **journal bearings** demonstrated how friction changes with lubrication conditions.
  - The Stribeck Curve illustrates the transition between boundary, mixed, and hydrodynamic lubrication.
- **Advancements in bearing lubrication**:
  - Studied the role of viscosity, speed, and load in reducing bearing wear.
  - His findings helped improve lubricant formulation and bearing materials.
- **Tribology research and machine efficiency**:
  - Investigated the relationship **between lubrication and energy losses in rotating machinery**.
  - **His work laid the foundation for modern bearing and lubrication system design**.

861. Suction Screen Filter

A reservoir **suction screen filter** is a protective filtration device installed at the inlet of a hydraulic pump or lubrication system reservoir to prevent large contaminants, debris, and foreign particles from entering the system. It ensures that only clean oil or lubricant is drawn into the pump, reducing wear, clogging, and potential pump failure in hydraulic, lubrication, and fluid transfer systems. See Figure S23.

Key Functions of a Reservoir Suction Screen Filter

- **Prevents large particles from entering the pump**: Captures metal shavings, dirt, and solid contaminants before they reach the pump or lubrication system.
- **Protects pumps and bearings**: Reduces the risk of premature wear, cavitation, and damage to critical components.
- **Ensures proper oil flow**: Maintains a steady lubricant supply while preventing clogging or pressure loss.
- **Extends system lifespan**: Minimizes unplanned maintenance and equipment downtime.

Figure S23 Typical suction screen filter.

Design and Components of a Suction Screen Filter

1. **Screen mesh or perforated metal filter**: Traps contaminants while allowing fluid to pass through.
2. **Metal or plastic support frame**: Provides structural integrity to the filter.
3. **Threaded or slip-on connection**: Attaches securely to the suction line inside the reservoir.

4. **Flow-optimized design**: Allows **smooth fluid movement** while minimizing restriction.

Types of Reservoir Suction Screen Filters

- **Inline suction strainers**: Installed directly inside the reservoir at the suction port.
- **Basket-type suction screens**: Used in larger lubrication or hydraulic reservoirs for high-flow applications.
- **Magnetic suction filters**: Equipped with magnetic elements to capture ferrous (metal) contaminants.

862. Sulfonate

Sulfonate refers to a class of chemical compounds derived from **sulfonic acid** where the hydrogen atom is replaced by a metal or organic group. Sulfonates are commonly used as detergents, corrosion inhibitors, and extreme pressure (EP) additives in lubricants, greases, and industrial fluids. Their ability to provide high detergency, metal surface protection, and thermal stability makes them essential in engine oils, metalworking fluids, and high-performance greases.

Key Functions of Sulfonates in Lubrication

- **Detergency and deposit control**: Helps keep engine and machine components clean by preventing sludge and varnish formation.
- **Corrosion protection**: Forms a protective layer on metal surfaces, preventing oxidation and rust.
- **Extreme Pressure (EP) and antiwear properties**: Used in calcium sulfonate greases to withstand high loads and shock conditions.
- **Water resistance and emulsification**: Enhances grease stability **in wet environments** and improves lubricant performance.

Types of Sulfonates in Lubrication

1. **Calcium sulfonate**:
   - Used in high-performance greases for corrosion resistance, extreme pressure protection, and water washout resistance.
2. **Magnesium and sodium sulfonates**:
   - Found in engine oils and hydraulic fluids, acting as detergents and rust inhibitors.

3. **Synthetic sulfonates**:
   - Engineered for specialized applications requiring enhanced thermal and oxidative stability.

863. Sulphation

Sulphation refers to the formation of **sulfate compounds** in lubricants, particularly in engine oils, due to the reaction of sulfur-based fuel combustion by-products with oxygen and moisture. This process leads to the accumulation of acidic deposits, which can cause oil thickening, corrosion, and reduced lubrication efficiency. Sulphation is a key indicator of oil degradation and is commonly monitored in used oil analysis.

Key Functions of Sulphation Monitoring in Lubricants

- **Detects oil degradation**: Helps determine oil condition and replacement intervals.
- **Prevents corrosion damage**: Identifies potential acidic buildup that can harm engine components.
- **Ensures engine efficiency**: Helps maintain proper lubrication flow and viscosity control.
- **Optimizes maintenance schedules**: Used in predictive maintenance programs to prevent failures.

Causes of Sulphation in Lubricants

- **Combustion of sulfur-containing fuels**: High sulfur fuel increases SO_2 (sulfur dioxide) and SO_3 (sulfur trioxide) formation.
- **Incomplete combustion**: Results in higher sulfuric acid formation, leading to lubricant contamination.
- **High operating temperatures**: Promotes oxidation reactions, accelerating sulfate deposit formation.
- **Moisture contamination**: Reacts with sulfur compounds to create acidic sludge and corrosion.

864. Surface Fatigue Wear

Surface fatigue wear is a type of material degradation that occurs when repeated cyclic loading and stress cause microcracks, pitting, or spalling on a surface. See Figure S24. This type of wear is commonly observed in bearings, gears, camshafts, and rolling-contact applications, where surfaces experience continuous contact stress under heavy loads. Over time, the accumulated damage leads to surface material loss, increased friction, and potential component failure.

Key Functions of Surface Fatigue Wear Analysis

- **Predicts component lifespan**: Helps estimate the service life of gears, bearings, and rollers.
- **Prevents catastrophic failure**: Identifies early signs of fatigue damage before severe system failure.
- **Optimizes lubrication strategies**: Ensures proper lubricant selection to reduce stress and wear.
- **Improves material selection**: Helps choose wear-resistant alloys and surface coatings.

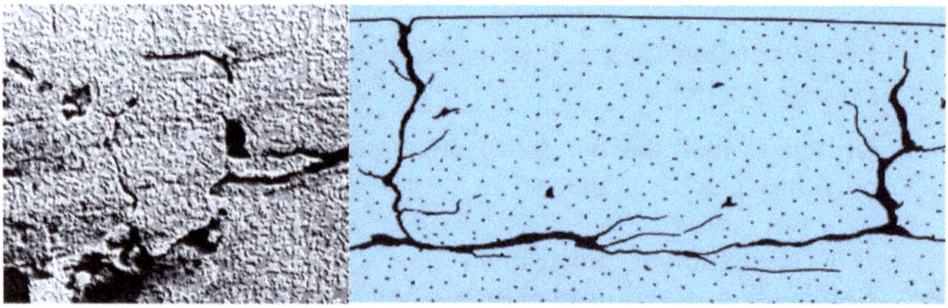

Figure S24 Surface and subsurface cracking and deformation due to fatigue.

Effects of Surface Fatigue Wear

- **Microcrack formation**: Small cracks form beneath the surface due to repeated stress cycles.
- **Pitting and spalling**: Cracks propagate to the surface, leading to material loss and pitting damage.
- **Increased friction and heat generation**: Wear causes higher resistance and temperature buildup, leading to efficiency loss.
- **Reduced load-carrying capacity**: Fatigue wear weakens bearings, gears, and load-bearing surfaces.

Common Applications Affected by Surface Fatigue Wear

- **Rolling-element bearings**: Experience pitting and fatigue under high-speed rotation and load cycles.
- **Gears and gearboxes**: Gear teeth fatigue due to continuous meshing and torque transmission.
- **Camshafts and valve train components**: Repeated contact with lifters or rollers causes fatigue wear.
- **Railway tracks and wheels**: Metal surfaces degrade from high-impact rolling loads.

Conclusion: **Surface fatigue wear** is a major failure mode in rolling and sliding contact applications, leading to microcracks, pitting, and component degradation. By implementing proper lubrication, material enhancements, and maintenance strategies, industries can significantly extend equipment life, improve efficiency, and reduce costly downtime in critical mechanical systems.

865. Surface Filter

Surface filter is a filtration device designed to **trap contaminants on the surface of the filter media**, preventing particles from passing through while allowing fluid to flow. Unlike depth filters, which capture contaminants throughout the media's thickness, surface filters rely on a thin, porous membrane or mesh structure to separate unwanted particles. These filters are widely used in lubrication systems, hydraulic systems, fuel filtration, and industrial fluid processing, ensuring cleaner fluids, reduced wear, and extended equipment life.

Key Functions of Surface Filters

- **Particle retention**: Captures **solid contaminants** on the filter's outer surface, preventing them from entering the system.
- **Maintains fluid purity**: Ensures clean lubricants, fuels, or process fluids by removing dirt, debris, and particulates.
- **Protects machinery components**: Reduces wear on bearings, pumps, injectors, and hydraulic systems.
- **Improves system efficiency**: Prevents clogging and maintains consistent flow by filtering fine and coarse contaminants.

Types of Surface Filters

- **Mesh and screen filters**: Fine metal or synthetic mesh used **in hydraulic and lubrication systems**. See Figure S25.
- **Membrane filters**: Thin polymer membranes designed for precise filtration of fine particles.
- **Pleated surface filters**: Increased surface area for high-efficiency particle removal.
- **Magnetic surface filters**: Capture **ferrous contaminants** using a magnetized surface in gearboxes and rotating equipment.

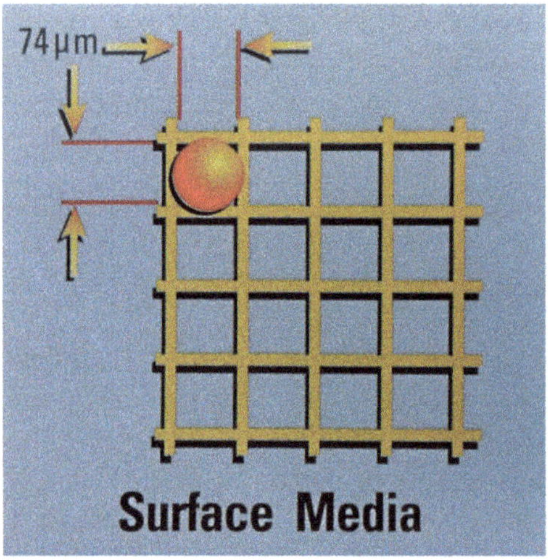

Figure S25 Typical representation of a screen/mesh filter media and how it works. Courtesy: Pall Filters.

866. Surface Tension

Surface tension in lubricants refers to the **force that holds the surface molecules of a liquid together**, creating a resistance to spreading or breaking. It is a critical property in lubrication, affecting oil film formation, wetting ability, emulsification, and contamination separation. Measured in dynes/cm or mN/m, surface tension plays a key role in how lubricants behave in bearings, gears, hydraulic systems, and industrial machinery.

Key Functions of Surface Tension in Lubrication

- **Affects oil film formation**: Determines how well a lubricant can spread across metal surfaces, ensuring **consistent coverage**.

- **Controls wettability and adhesion**: Impacts the lubricant's ability to stick to or repel surfaces, influencing lubrication efficiency.
- **Influences water separation and emulsification**: Lower surface tension promotes water shedding, while higher surface tension increases oil–water emulsion stability.
- **Affects air release and foam control**: Lubricants with optimized surface tension help prevent air entrapment and excessive foaming.

Factors Affecting Surface Tension in Lubricants

- **Base oil composition**: Synthetic oils generally have lower surface tension than mineral oils, improving lubrication efficiency.
- **Additives and surfactants**: Detergents, dispersants, and emulsifiers modify surface tension to improve oil performance.
- **Contaminants and degradation**: Oxidation by-products, dirt, and fuel dilution can alter surface tension, affecting lubrication properties.
- **Operating conditions**: High temperatures can **reduce surface tension**, impacting lubricant stability.

Applications of Surface Tension in Lubrication

- **Hydraulic and engine oils**: Controls oil–air separation, reducing foam formation and cavitation risks.
- **Metalworking fluids**: Influences coolant wetting and chip flushing efficiency.
- **Gear and bearing lubrication**: Ensures proper oil film stability under high loads and shear stress.
- **Oil–water separation systems**: Helps in **removing water contamination from lubricants**.

Benefits of Optimizing Surface Tension in Lubricants

- **Improves lubrication film integrity**: Enhances load-carrying capacity and wear protection.
- **Enhances water separation**: Reduces corrosion risks by preventing emulsification.
- **Reduces air entrapment and foam formation**: Prevents oil aeration that can impact pump efficiency.
- **Optimizes wetting and spreading**: Ensures better surface coverage for high-performance lubrication

867. Surfactant

Surfactant (surface-active agent) is a chemical compound that **reduces surface tension** (see "Surface Tension") between two liquids, or between a liquid and a solid, improving lubrication, wetting, emulsification, and contaminant dispersion. In lubrication, surfactants are used to enhance oil performance, stabilize emulsions, control foaming, and improve detergent properties in engine oils, metalworking fluids, and industrial lubricants.

Key Functions of Surfactants in Lubrication

- **Reduces surface tension**: Enhances lubricant spreading on surfaces, ensuring even film formation.
- **Improves Wetting and adhesion**: Helps lubricants adhere to metal surfaces, reducing friction and wear.
- **Enhances emulsification**: Allows oil and water to mix effectively in coolants and metalworking fluids.
- **Aids in contaminant dispersion**: Prevents sludge and deposits by keeping contaminants suspended in oil.
- **Controls foam formation**: Either reduces or stabilizes foam to maintain lubricant efficiency.

868. Synthetic Base Oil Stock

Synthetic base oil stock refers to **engineered lubricant base fluids** that are chemically synthesized rather than refined from crude oil. These base oils provide superior performance, stability, and efficiency compared to conventional mineral oils, making them ideal for high-temperature, extreme-pressure, and long-service-life applications in automotive, industrial, and aerospace lubricants.

Key Functions of Synthetic Base Oil Stock

- **Enhanced lubrication and film strength**: **Provides stronger molecular stability**, reducing friction and wear.
- **Superior oxidation and thermal stability**: Resists breakdown at high temperatures, extending oil life.
- **Improved low-temperature fluidity**: Flows efficiently in cold conditions, ensuring smooth start-ups.

- **Better contaminant resistance**: Offers reduced sludge and deposit formation, keeping systems clean.

Types of Synthetic Base Oils

1. **Polyalphaolefins (PAOs)**:
    - Highly stable, low volatility, and excellent low-temperature properties.
    - Common **in engine oils, industrial lubricants, and hydraulic fluids**.
2. **Esters**:
    - Excellent lubricity and high-temperature performance.
    - Used in aviation lubricants, racing oils, and biodegradable fluids.
3. **Polyalkylene glycols (PAGs)**:
    - Water-soluble and nonoxidizing, suitable for high-load applications.
    - Found in gear oils, compressor lubricants, and fire-resistant fluids.
4. **Silicones and fluorinated fluids**:
    - Exceptional temperature and chemical resistance.
    - Used in extreme aerospace, electrical, and high-vacuum applications.

Benefits of Synthetic Base Oil Stock

- **Longer lubricant life**: Reduces the need for frequent oil changes.
- **Improved fuel efficiency**: Lowers internal friction, enhancing energy efficiency.
- **Superior protection in extreme conditions**: Performs well in high- and low-temperature environments.
- **Better compatibility with additives**: Allows for customized formulations to meet specific performance requirements.

Conclusion: **Synthetic base oil stock** is a high-performance alternative to mineral oils, offering enhanced lubrication, thermal stability, and longevity.

869. System International (SI)

System International (SI), officially known as the **International System of Units (Système International d'Unités)**, is the globally recognized **metric measurement system** used in science, engineering, and industry, including lubrication and tribology applications. SI units provide a standardized framework for measuring physical properties such as viscosity, density, force, and temperature, ensuring consistency in lubricant formulation, performance testing, and equipment specifications.

Common SI Units in Lubrication and Tribology

- **Viscosity**: Measured in Pascal-seconds (Pa·s) or centistokes (cSt) for kinematic viscosity.
- **Density**: Expressed in kilograms per cubic meter (kg/m^3).
- **Force and pressure**: Newtons (N) and Pascals (Pa) for load-bearing and lubrication film pressure.
- **Temperature**: Degrees Kelvin (K) or Celsius (°C) for lubricant stability testing.
- **Energy and power**: Joules (J) and Watts (W) for thermal efficiency and energy loss calculations.

Applications of SI in Lubrication Engineering

- **Oil and grease specifications**: Used in ISO viscosity grades, ASTM testing, and lubricant formulation.
- **Machine and bearing design**: Ensures precise calculations of load distribution and friction reduction.
- **Tribology research and wear analysis**: Standardizes wear rate and friction coefficient measurements.
- **Industrial and automotive lubrication systems**: Ensures compliance with OEM and industry standards.

T

870. Tackifier

A **tackifier** is an additive used in lubricants, particularly in **greases, chain oils, and adhesive lubricants**, to **increase stickiness and adhesion** to surfaces. It enhances a lubricant's ability to resist sling-off, dripping, or washout, ensuring long-lasting protection in high-speed, high-impact, or outdoor applications. Tackifiers are commonly used in open-gear lubricants, wire rope greases, chain oils, and heavy-duty machinery greases where maintaining lubrication under extreme conditions is critical.

Key Functions of Tackifiers in Lubrication

- **Enhances adhesion**: Improves a lubricant's ability **to cling to metal surfaces**, reducing the need for frequent reapplication.
- **Prevents sling-off and drip loss**: Reduces lubricant loss in high-speed chains, gears, and rotating components.
- **Improves water and washout resistance**: Ensures lubricant retention in wet or outdoor environments.
- **Extends lubrication intervals**: Maintains film stability, reducing lubricant consumption and maintenance costs.

Types of Tackifiers in Lubricants

1. **Polymer-based tackifiers**:
    - Used in greases and chain oils to provide high-temperature stability.
    - Common in synthetic and high-viscosity **lubricants**.
2. **Resin-based tackifiers**:
    - Improves adhesion in wire rope and open-gear lubricants.
    - Resistant to extreme pressure (EP) and outdoor conditions.

DOI: 10.1201/9788743806608-889

3. **Natural and bio-based tackifiers**:
    - Found **in biodegradable lubricants for environmentally sensitive applications**.
    - **Used in forestry, agriculture, and marine applications**.

Applications of Tackified Lubricants

- **Chain and conveyor lubrication**: Prevents oil from flinging off high-speed chains and conveyors.
- **Open-gear lubrication**: Ensures grease remains on exposed gear teeth under heavy loads.
- **Wire rope and cable lubricants**: Prevents lubricant drip loss and enhances corrosion protection.
- **Forestry and agricultural equipment**: Keeps lubrication effective **in harvesters, saws, and farm machinery**.

871. Tackiness Additive

See “Tackifier.”

872. Tagging

Tagging in lubrication management refers to the process of **labeling, identifying, and tracking lubricants, equipment, and lubrication points** to ensure proper maintenance, prevent cross-contamination, and improve lubrication efficiency. This system is widely used in industrial plants, manufacturing facilities, and asset management programs to streamline lubricant selection, application, and storage.

Key Functions of Tagging in Lubrication

- **Prevents cross-contamination**: Ensures that the correct lubricant is applied to the right equipment.
- **Improves maintenance efficiency**: Provides clear identification of lubrication points for technicians and operators.
- **Enhances lubricant tracking**: Helps monitor lubricant consumption, storage, and replacement schedules.
- **Supports compliance and safety**: Aligns with ISO 55001 asset management and industry best practices.

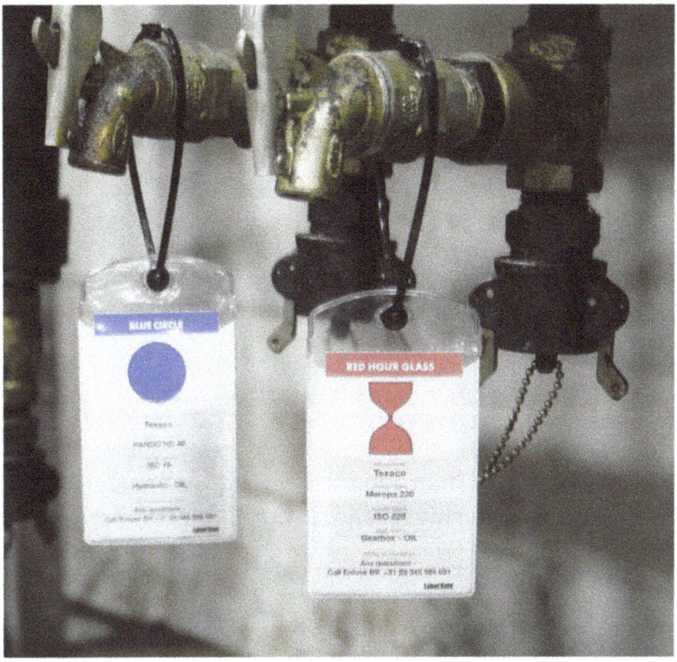

Figure T1 Tagging with shapes and color ID labels. Courtesy: Fluid Defense Systems.

Types of Lubrication Tagging Systems

1. **Color-coded tags See Figure T1**:
    - Assigns specific colors to lubricant types (e.g., hydraulic oil, gear oil, and grease).
    - Reduces the risk of misapplication and contamination.
2. **Barcode and RFID tags**:
    - Enables digital tracking of lubricant use and maintenance history.
    - Integrates with CMMS (Computerized Maintenance Management Systems).
3. **Equipment and lubrication point tags**:
    - Labels bearings, gears, pumps, and reservoirs with specific lubricant requirements.
    - Ensures standardized lubrication practices.

4. **Lubricant storage and transfer tags**:
    - Identifies oil and grease storage containers to avoid mix-ups.
    - Helps maintain lubricant integrity from storage to application.

Benefits of Implementing a Lubrication Tagging System

- **Reduces human error**: Prevents misapplication by clearly identifying lubricant requirements.
- **Increases equipment reliability**: Ensures correct lubrication, reducing premature wear and failures.
- **Enhances maintenance workflow**: Saves time by providing clear, quick lubricant identification.
- **Supports predictive maintenance and CMMS integration**: Helps in tracking lubrication intervals and oil condition monitoring.

873. Tallow

Tallow is a **natural fat-based lubricant** derived from **rendered animal fats**, primarily from cattle or sheep and was the primary lubricant used in the Industrial Revolution historically used as a base oil, thickener, and extreme pressure (EP) additive in greases, cutting fluids, and industrial lubricants. Due to its high lubricity, biodegradability, and water-resistant properties, tallow remains relevant in metalworking, leather treatment, and environmentally friendly lubrication applications.

Key Functions of Tallow in Lubrication

- **Provides natural lubricity**: Offers low friction and smooth metal-to-metal contact, reducing wear.
- **Acts as a thickener in grease formulation**: Improves consistency and stability in heavy-duty greases.
- **Enhances water resistance**: Repels moisture, making it suitable for marine and outdoor applications.
- **Serves as a base for biodegradable lubricants**: Used in eco-friendly greases and industrial lubricants.

Types of Tallow-based Lubricants

1. **Tallow-based cutting fluids**:
    - **Provides lubrication and cooling in metalworking processes**.

- **Used in threading, drilling, and machining operations**.

2. **Tallow greases**:
   - **Mixed with lime or calcium-based soaps to form water-resistant lubricants**.
   - **Common in agriculture, mining, and vintage machinery applications**.
3. **Tallow-based leather and textile lubricants**:
   - **Used to soften and preserve leather in industrial applications**.
   - **Prevents fiber wear in textile manufacturing equipment**.

Applications of Tallow in Lubrication

- **Heavy equipment and agriculture**: Provides water-resistant lubrication for bearings, gears, and exposed parts.
- **Metalworking and machining**: Acts as a coolant and antiwear agent in cutting fluids.
- **Leather processing and textile industry**: Used for conditioning and preserving industrial fabrics.
- **Eco-friendly and biodegradable lubrication**: Used in renewable, plant-based lubricants for environmental applications.

Benefits of Tallow as a Lubricant

- **Highly biodegradable**: Suitable for environmentally sensitive applications.
- **Provides excellent lubricity**: Reduces friction and metal wear in industrial applications.
- **Water and corrosion resistant**: Protects surfaces **in wet and outdoor conditions**.
- **Sustainable and renewable**: Offers an alternative to synthetic or petroleum-based lubrican**ts**.

874. Tallow Box

A **tallow box** is a lubrication system historically used for **lubricating early Industrial Textile machinery and locomotive and railcar bearings**, particularly in **plain bearings and journal boxes**. It consists of a container filled with tallow-based grease or solid lubricants, which gradually melts and wicks onto the bearing surface as the system operates. This method ensures continuous lubrication, reducing friction, wear, and overheating in rail transport, industrial machinery, and early steam-powered equipment. See Figure T2.

Key Functions of a Tallow Box

- **Provides continuous lubrication**: Supplies a steady film of lubricant to metal surfaces under load.
- **Prevents overheating and seizure**: Maintains proper lubrication, reducing bearing friction and heat buildup.
- **Extends bearing life**: Reduces wear and metal fatigue in journal bearings.
- **Self-regulating lubrication system**: The lubricant is released as needed **based on temperature and motion**.

Figure T2 Tallow box lubricator – main shaft lubrication of an 1890 Cotton Carding Machine UK.

Components of a Tallow Box

1. **Reservoir**: Holds solid tallow grease or oil-soaked waste material.

2. **Wicking mechanism**: Uses cotton, wool, or felt packing to draw lubricant onto the journal surface.
3. **Journal bearing contact area**: Ensures proper lubricant distribution to the rotating shaft.

Benefits of Tallow Box Lubrication

- **Simple and reliable lubrication system**: Requires minimal maintenance.
- **Prevents metal-to-metal contact**: Reduces friction and bearing damage.
- **Long-lasting lubrication supply**: Uses a slow-release system, extending lubrication intervals.
- **Water and dust resistant**: Suitable for outdoor and heavy-duty rail applications.

Limitations and Modern Replacements

- **Limited performance at high speeds and temperatures**: Modern synthetic lubricants and sealed roller bearings offer superior protection.
- **More frequent maintenance required**: Compared to modern lubrication systems like oil sumps and automatic greasers.
- **Environmental concerns**: Traditional tallow greases are being replaced **by biodegradable or synthetic alternatives**.

Conclusion: Though largely replaced by **modern roller bearings and synthetic lubricants**, tallow boxes remain an essential part of **historic railway preservation and vintage machinery maintenance**.

875. TAN

See "Total Acid Number."

876. Tank Farm

A **tank farm** is a **large-scale storage facility** designed to hold **liquid fuels, chemicals, lubricants, or industrial fluids** in multiple above-ground or underground tanks. These facilities are commonly used in oil refineries, petrochemical plants, lubricant blending operations, and fuel distribution centers to manage bulk storage, transfer, and distribution of various liquid products. See Figure T3.

Key Functions of a Tank Farm

- **Bulk storage and inventory management**: Provides safe and efficient storage for large volumes of oil, fuel, chemicals, and lubricants.
- **Fluid transfer and distribution**: Facilitates pumping, mixing, and distribution of stored liquids to pipelines, trucks, or processing units.
- **Contamination and safety control**: Implements filtration, separation, and spill prevention systems to maintain fluid purity and safety.
- **Regulatory compliance and environmental protection**: Ensures adherence to fire safety, emissions control, and spill containment regulations.

Components of a Tank Farm

1. **Storage tanks**:
    - **Fixed-roof tanks**: Used for nonvolatile fluids like lubricants and water.
    - **Floating-roof tanks**: Reduce evaporation losses for volatile fuels and crude oil.
    - **Pressurized tanks**: Store liquefied gases and high-pressure fluids.
2. **Pipelines and transfer pumps**:
    - Used for loading, unloading, and transferring fluids between storage tanks and distribution systems.
3. **Filtration and separation systems**:
    - Includes oil–water separators, particulate filters, and chemical treatment units to maintain fluid quality.
4. **Fire protection and safety systems**:
    - Equipped with foam suppression systems, leak detection, and bund walls for spill containment.
5. **Instrumentation and monitoring**:
    - **Uses flow meters, level sensors, and SCADA (Supervisory Control and Data Acquisition) systems** for remote monitoring and automation.

Benefits of a Well-managed Tank Farm

- **Enhances storage efficiency**: Maximizes space utilization while ensuring safe inventory control.
- **Reduces product losses and contamination**: Prevents leaks, evaporation, and cross-contamination.
- **Improves safety and environmental compliance**: Mitigates risks of fire, spills, and emissions violations.
- **Supports reliable supply chain operations**: Ensures consistent availability and distribution of critical fluids.

Figure T3 Typical tank farm oil storage facility – Amsterdam.

Conclusion: A tank farm is a critical infrastructure for the storage, handling, and distribution of bulk liquids, including fuels, lubricants, and chemicals. By incorporating advanced safety, automation, and environmental protection measures, modern tank farms ensure efficient, safe, and regulatory-compliant operations across various industries.

877. Tapered Roller Bearing

A **tapered roller bearing** is a type of rolling-element bearing designed to **handle both radial and axial (thrust) loads** due to its unique **cone-shaped**

rolling elements. The bearing consists of an inner ring (cone), an outer ring (cup), tapered rollers, and a cage that holds the rollers in place. This design allows for efficient load distribution, making tapered roller bearings ideal for high-speed, high-load applications such as automotive wheel hubs, industrial machinery, and heavy equipment transmissions. See Figure T4.

Figure T4 Taper roller bearing cross-section. Courtesy: SKF.

Key Functions of a Tapered Roller Bearing

- **Supports combined loads**: Handles **both radial and axial forces** simultaneously.
- **Reduces friction and wear**: Provides low rolling resistance for smooth operation.
- **Enhances load distribution**: The tapered shape distributes loads evenly across the contact surface.
- **Allows adjustable preload**: Can be preloaded to optimize stiffness and reduce play in mechanical systems.

Components of a Tapered Roller Bearing

1. **Inner ring (cone)**: Houses the tapered rollers and rotates with the shaft.
2. **Outer ring (cup)**: Encases the rollers and provides an outer raceway.

3. **Tapered rollers**: Positioned at an angle to support both radial and axial loads.
4. **Cage**: Maintains roller spacing and alignment within the bearing.

Benefits of Tapered Roller Bearings

- **High load-carrying capacity**: Handles heavy loads with minimal deformation.
- **Reduced friction and heat generation**: Ensures longer bearing life and efficiency.
- **Axial and radial load handling**: Supports combined loading conditions in rotating equipment.
- **Long service life and durability**: Designed for continuous operation in demanding applications.

878. TBN

See "Total Base Number."

879. Temperature Control

Temperature control in lubrication is essential for **maintaining lubricant performance, preventing thermal degradation, and ensuring reliable equipment operation**. Excessive heat can cause oil oxidation, viscosity breakdown, and increased wear, while low temperatures can lead to poor fluidity, reduced lubrication film thickness, and start-up failures. Proper temperature management enhances lubricant longevity, efficiency, and mechanical system reliability.

Key Functions of Temperature Control in Lubrication

- **Prevents lubricant degradation**: Reduces oxidation, viscosity loss, and sludge formation.
- **Ensures optimal lubrication film thickness**: Maintains sufficient oil film to prevent metal-to-metal contact.
- **Reduces friction and wear**: Keeps lubricant at an ideal viscosity for smooth operation.
- **Enhances equipment performance and longevity**: Prevents thermal stress and component failure.

Effects of Temperature on Lubrication

- **High temperatures (overheating)**:
  - Accelerates **oil oxidation**, leading to sludge and varnish formation.
  - Reduces **viscosity**, weakening lubricant film strength.
  - Causes **additive depletion**, reducing antiwear and antifoaming properties.
  - Increases **thermal expansion**, affecting component clearances.
- **Low temperatures (cold start conditions)**:
  - Increases **oil viscosity**, reducing flow and pumpability.
  - Leads to **inadequate film formation**, causing higher friction and wear.
  - Risks **wax crystallization or grease hardening**, impairing lubrication efficiency.

Methods for Temperature Control in Lubrication Systems

1. **Cooling systems**:
   - **Heat exchangers and oil coolers**: Dissipate excess heat in circulating oil systems.
   - **Air-cooled and water-cooled systems**: Regulate lubricant temperatures in industrial and automotive applications.
2. **Lubricant selection**:
   - **High-temperature synthetic oils**: Formulated for thermal stability and oxidation resistance.
   - **Cold-weather lubricants**: Include low-viscosity oils and pour point depressants.
3. **Controlled Heating in Cold Conditions**:
   - **Oil heaters and heat tracing**: Used in low-temperature environments to prevent oil thickening.
   - **Prelubrication systems**: Ensure proper flow before start-up in engines and hydraulic systems.
4. **Temperature sensors and monitoring**:
   - **Thermal probes and infrared sensors**: Provide real-time temperature tracking.

- **Automated control systems**: Adjust cooling/heating based on operating conditions.

Benefits of Effective Temperature Control in Lubrication

- **Extends lubricant and equipment life**: Reduces the rate **of oil degradation and component wear**.
- **Optimizes lubricant viscosity**: Ensures the correct flow and film thickness for protection.
- **Improves system efficiency**: Reduces energy loss due to excessive friction or inadequate lubrication.
- **Prevents downtime and maintenance costs**: Minimizes failures caused by overheating or poor lubrication.

Conclusion: Temperature control in lubrication is critical for maintaining oil stability, reducing wear, and ensuring optimal equipment performance. By implementing cooling systems, proper lubricant selection, heating solutions, and real-time monitoring, industries can maximize lubrication efficiency, minimize failures, and enhance the longevity of mechanical systems in various applications.

880. Terminating Oil System

A **terminating oil system** refers to a **lubrication system that does not recirculate oil**, meaning the lubricant is applied to a component, performs its function, and then either evaporates, drains away, or is discarded. Also referred to as a total-loss system, these systems are commonly used in applications where lubricant contamination, degradation, or loss prevention is not a primary concern. Instead, the focus is on ensuring sufficient lubrication for a specific period or process.

Key Functions of a Terminating Oil System

- **Provides one-time lubrication**: Oil is applied and not recovered or recirculated.
- **Ensures adequate lubricant film for operation**: Prevents metal-to-metal contact and reduces friction.
- **Prevents contaminant buildup in the lubricant**: Since oil is not reused, contamination does not accumulate over time.

Types of Terminating Oil Systems

1. **Drip feed lubrication**:
   - Delivers oil in controlled drops to bearings, chains, or gears, usually through a single-point lubricating device.
   - Used in low-speed or intermittently operating machinery.
2. **Oil mist and spray lubrication**:
   - Sprays a fine mist of oil onto the components for minimal but effective lubrication.
   - Common in high-speed spindles and food processing machinery.
3. **Manually applied lubricants**:
   - Applied via brushes, wick feeders, or aerosol sprays.
   - Common in industrial maintenance and temporary lubrication needs.
4. **Total loss auto-lube systems**:
   - Applied via a pump to bearings through metering devices on a controlled pump activation cycle. See: Single-line Resistance System.

881. Test Slate Selection

Test slate selection refers to the **process of choosing specific laboratory tests** to analyze the condition, performance, and contamination levels of lubricants in industrial, automotive, and machinery applications. Selecting the right test slate ensures accurate monitoring of oil health, early detection of equipment issues, and optimized maintenance planning.

Key Functions of Test Slate Selection

- **Monitors lubricant degradation**: Detects oxidation, viscosity breakdown, and additive depletion.
- **Identifies contamination sources**: Measures water, dirt, fuel, or metal particulates in the lubricant.
- **Assesses wear and component health**: Evaluates metal wear particles and frictional surface degradation.
- **Optimizes maintenance intervals**: Helps determine when to change oil, filters, or perform corrective actions.

Factors Affecting Test Slate Selection

- **Lubricant type and application**: Engine oils, hydraulic fluids, gear oils, and greases require different tests.
- **Operating environment**: High-temperature, high-load, or water-prone environments influence test choices.
- **Failure mode risk assessment**: Identifies potential risks like oil oxidation, contamination, or wear debris formation.
- **Regulatory and OEM requirements**: Ensures compliance with ISO, ASTM, and manufacturer specifications.

Common Test Slates for Lubrication Analysis

1. **Basic lubricant health tests**:
    - **Viscosity measurement (ASTM D445)**: Determines lubricant thickness and flow properties.
    - **Total acid number (TAN)/total base number (TBN)**: Assesses oil degradation and acid neutralization capacity.
    - **Oxidation and nitration testing**: Detects lubricant breakdown from heat and combustion by-products.
2. **Contamination detection tests**:
    - **Water content (Karl Fischer method)**: Measures water contamination, which can lead to rust and oxidation.
    - **Particle count (ISO 4406)**: Evaluates solid contamination affecting hydraulic and gear systems.
    - **Fuel and soot dilution (FTIR analysis)**: Identifies fuel contamination in engine oils.
3. **Wear analysis and condition monitoring tests**:
    - **Elemental spectroscopy (ICP analysis)**: Detects metal wear particles from bearings, gears, and pumps.
    - **Ferrous debris analysis (PQ index)**: Measures large ferrous wear particles.
    - **Fourier transform infrared (FTIR) spectroscopy**: Identifies chemical changes in lubricants.

882. Texture (Grease)

Grease texture refers to the **physical consistency and feel** of grease, influenced by its **thickener type**, **base oil viscosity**, **and additive formulation**. The texture of grease affects its application method, pumpability, adhesion, and performance in different operating conditions. Understanding grease texture is essential for proper lubricant selection in bearings, gears, and mechanical components where specific flow and consistency characteristics are required.

Key Functions of Grease Texture

- **Determines application suitability**: Affects how grease spreads, adheres, and moves under load.
- **Influences pumpability and flow**: Softer greases flow more easily, while firmer textures resist movement.
- **Affects lubrication retention**: Greases with fibrous or tacky textures provide better adhesion to surfaces.
- **Indicates consistency and structural integrity**: Helps predict shear stability, mechanical degradation, and resistance to thinning.

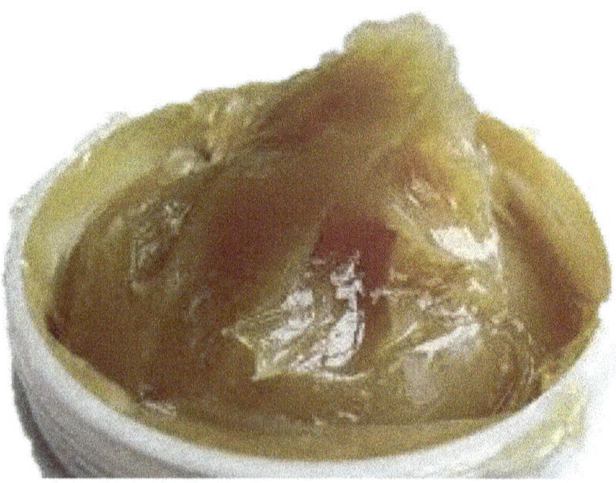

Figure T5 Grease comes in many textures such as smooth. Source: ENG**TECH** Industries Inc.

Types of Grease Textures

1. **Smooth grease**: **(See Figure T5)**

    - Fine, buttery texture with no visible fibers or particles.

- Common in automotive, industrial, and general-purpose applications.

2. **Fibrous grease**:
   - Contains visible fiber structures that enhance adhesion.
   - Used in high-load, low-speed applications such as open gears and chassis lubrication.
3. **Buttery grease**:
   - Soft, buttery consistency with good spreadability.
   - Found in light-duty bearings, automotive greases, and food-grade lubricants.
4. **Tacky grease**:
   - Sticky, high-adhesion texture that resists washout.
   - Ideal for marine, off-road, and heavy-duty applications.
5. **Granular or gritty grease**:
   - Contains solid additives or fillers like molybdenum disulfide or graphite.
   - Used in extreme pressure (EP) and wear-resistant applications.

Factors Affecting Grease Texture

- **Thickener type**: Lithium, calcium, aluminum, and polyurea thickeners impact grease firmness and fiber structure.
- **Base oil viscosity**: Higher viscosity oils lead to thicker, more adhesive greases.
- **Shear stability**: Mechanical stress can alter grease consistency over time.
- **Additives and fillers**: EP additives, solid lubricants, and tackifiers modify texture to enhance load capacity and retention.

883. Thermal Degradation

Thermal degradation in lubricants refers to the **breakdown of base oil and additives due to prolonged exposure to high temperatures**. This process leads to oxidation, viscosity increase, sludge formation, and loss of lubrication properties, which can cause accelerated wear, deposits, and mechanical failure in industrial, automotive, and hydraulic systems. Thermal degradation is also referred to as "Thermal Stability."

Key Functions of Thermal Stability in Lubricants

- **Maintains lubricant integrity**: Prevents breakdown of oil molecules under heat stress.
- **Ensures consistent viscosity**: Reduces the risk **of oil thickening or thinning**.
- **Prevents sludge and varnish formation**: Minimizes deposit buildup on components.
- **Extends lubricant and equipment life**: Enhances operational efficiency and reduces maintenance.

Causes of Thermal Degradation in Lubricants

- **Prolonged high-temperature exposure**: Lubricants exposed to excessive heat (>150 °C/300 °F) degrade faster.
- **Inadequate heat dissipation**: Poor cooling leads to localized oil overheating and molecular breakdown.
- **Shear-induced heat generation**: High loads and speeds increase frictional heat, accelerating degradation.
- **Oxidation acceleration**: Heat exposure increases chemical reactions with oxygen, leading to sludge.
- **Contaminants and catalyst reactions**: Metal particles, fuel, and water can accelerate thermal instability.

Effects of Thermal Degradation on Lubricants

- **Increase in viscosity**: Degraded oil **becomes thicker**, reducing flow and lubrication efficiency.
- **Sludge and varnish formation**: Deposits accumulate, leading to clogging of oil passages and filters.
- **Loss of additive performance**: Antiwear, antioxidation, and detergency additives break down, reducing protection.
- **Oil Volatility and evaporation**: High temperatures cause oil evaporation, leading to increased consumption.
- **Accelerated wear and component failure**: Degraded lubrication **film allows metal-to-metal contact**.

Methods to Prevent Thermal Degradation

1. **Use high-temperature-resistant lubricants**:
   - Synthetic oils (PAOs, esters, silicones) have superior thermal stability.
   - High VI (viscosity index) lubricants maintain stability across temperature ranges.
2. **Improve cooling and heat dissipation**:
   - **Use heat exchangers, oil coolers, and proper ventilation**.
   - **Prevent localized hotspots in machinery**.
3. **Monitor oil condition regularly**:
   - FTIR (Fourier transform infrared) spectroscopy: Detects thermal oxidation products.
   - Viscosity testing: Identifies oil thickening due to degradation.
   - Sludge and deposit analysis: Measures varnish formation potential.
4. **Implement proper lubrication practices**:
   - Avoid overheating components due to excessive loads or speeds.
   - Ensure adequate oil circulation and flow.
   - Replace lubricants at recommended drain intervals.

Conclusion: **Thermal degradation** is a leading cause of lubricant failure, resulting in viscosity changes, sludge formation, and loss of protective properties. By using high-temperature-resistant oils, improving heat dissipation, and conducting regular oil analysis, industries can prevent premature lubricant breakdown, extend equipment life, and optimize operational efficiency in high-heat applications.

884. Thermal Stability

See “Thermal “Degradation.”

885. Thermography

See “Infra-red Thermography.”

886. Thick Film Lubrication

See “Hydrodynamic Lubrication.”

887. Thickener (Grease)

A grease **thickener** is the component of grease that **gives it a semisolid consistency** and helps retain the base oil and additives within the grease structure. The thickener acts as a **sponge**, holding the lubricant in place and releasing it under mechanical stress. Different types of thickeners provide varying mechanical stability, temperature resistance, water resistance, and compatibility with other lubricants.

Key Functions of a Grease Thickener:

- **Holds base oil and additives in suspension**: Prevents grease from separating and ensures proper lubrication.
- **Controls grease consistency**: Determines NLGI (National Lubricating Grease Institute) grade, affecting pumpability and application.
- **Enhances load-carrying ability**: Improves shear stability under mechanical stress.
- **Influences water and temperature resistance**: Some thickeners provide better resistance to heat, moisture, and chemical exposure.

Types of Grease Thickeners:

1. **Soap-based thickeners**:
    - **Lithium and lithium complex**: Most common, with good heat, water, and mechanical stability.
    - **Calcium and calcium complex**: Excellent water resistance, used in marine applications.
    - **Sodium-based**: High-temperature resistance but poor water resistance.
    - **Aluminum complex**: High-temperature resistance with good adhesion and oxidation stability.
2. **Nonsoap thickeners**:
    - **Polyurea**: Used in electric motor bearings, with long life and high-speed performance.
    - **Clay (Bentonite)**: Heat-resistant, nonmelting grease for high-temperature applications.

- **Silica-based**: Chemically inert, used in specialized industrial lubricants.

3. **Specialty thickeners**:
   - **PTFE (polytetrafluoroethylene)**: Used in food-grade and chemically resistant greases.
   - **Organic and inorganic gels**: Found in synthetic and specialty high-performance greases.

Applications of Different Grease Thickeners

- **Lithium and lithium complex greases**: Used in automotive, industrial, and general-purpose applications.
- **Calcium-based greases**: Ideal for marine, agricultural, and wet environments.
- **Polyurea greases**: Found in electric motors and high-speed applications.
- **Clay-based greases**: Used in high-temperature bearings and kilns where soap-based greases would melt.

888. Thickener Compatibility

Thickener compatibility refers to the ability of different grease thickeners to mix without causing **performance issues such as oil separation, consistency breakdown, or loss of lubrication properties**. Incompatible grease thickeners can lead to grease softening, hardening, or loss of load-carrying ability, potentially causing equipment failure. Proper compatibility assessment is essential when switching grease types or blending lubricants in machinery. See Table T1.

Key Functions of Thickener Compatibility

- **Ensures lubrication stability**: Prevents grease structure breakdown when mixed.
- **Maintains mechanical performance**: Ensures shear stability, viscosity retention, and wear protection.
- **Reduces risk of equipment failure**: Avoids grease hardening or excessive softening that could cause lubrication failure.
- **Supports effective grease selection and changeover**: Helps in transitioning to new greases without compromising system performance.

Common Compatibility Issues and Effects

- **Softening or oil separation**: Some thickener combinations cause grease to become too fluid, leading to leakage and lubrication failure.
- **Hardening and consistency increase**: Incompatible greases may react, causing thickening and restricted flow in lubrication points.
- **Loss of extreme pressure (EP) and wear protection**: Certain thickeners react negatively, reducing load-carrying capacity and film strength.
- **Additive interactions**: Different grease chemistries may cause chemical instability, oxidation, or corrosion **issues**.

Best Practices for Managing Thickener Compatibility

1. **Refer to manufacturer recommendations**: Always consult **OEM specifications or grease supplier data** before mixing greases.
2. **Flush and purge before switching greases**: Completely remove the **previous grease** to prevent contamination.
3. **Conduct compatibility testing**: Perform **drop point tests, penetration tests, and mechanical stability assessments** before blending greases.
4. **Use single-thickener greases where possible**: Minimize the need for **cross-compatibility concerns** by standardizing lubricant selection.
5. **Monitor lubrication system performance**: After changing grease, inspect for **leakage, excessive softening, or abnormal wear**.

Table T1 Thickener compatibility.

Thickener Type	Lithium	Lithium Complex	Calcium	Calcium Complex	Sodium	Aluminum Complex
Lithium	✅ Compatible	✅ Compatible	⚠️ Limited	⚠️ Limited	❌ Incompatible	⚠️ Limited
Lithium Complex	✅ Compatible	✅ Compatible	⚠️ Limited	✅ Compatible	❌ Incompatible	⚠️ Limited
Calcium	⚠️ Limited	⚠️ Limited	✅ Compatible	✅ Compatible	❌ Incompatible	❌ Incompatible
Calcium Complex	⚠️ Limited	✅ Compatible	✅ Compatible	✅ Compatible	❌ Incompatible	❌ Incompatible
Sodium	❌ Incompatible	❌ Incompatible	❌ Incompatible	❌ Incompatible	✅ Compatible	❌ Incompatible
Aluminum Complex	⚠️ Limited	⚠️ Limited	❌ Incompatible	❌ Incompatible	❌ Incompatible	✅ Compatible
Bentonite (Clay)	❌ Incompatible	❌ Incompatible	❌ Incompatible	❌ Incompatible	❌ Incompatible	❌ Incompatible
Polyurea	❌ Incompatible	❌ Incompatible	❌ Incompatible	❌ Incompatible	❌ Incompatible	❌ Incompatible

✅ **Compatible** – Greases can be mixed with minimal performance concerns.

⚠️ **Limited Compatibility** – Mixing may cause slight degradation in performance; test before use.

❌ **Incompatible** – Mixing leads to **structural breakdown, oil separation, or loss of lubrication properties.**

889. Thin Film Lubrication

See "Boundary Layer Lubrication."

890. Thixotropy

Thixotropy is the property of certain lubricants, especially greases, where their **viscosity decreases when subjected to shear stress (such as stirring, shaking, or mechanical movement) and gradually recovers when the stress is removed**. This reversible thickening and thinning behavior ensures effective lubrication under dynamic conditions while maintaining stability at rest.

Key Functions of Thixotropic Lubricants

- **Enhances pumpability and flow**: Grease flows easily under mechanical stress and thickens when at rest, preventing leaks.
- **Provides controlled lubrication release**: Ensures consistent oil distribution without excessive runoff.
- **Improves adhesion and retention**: Stays in place after application, resisting dripping or migration.
- **Enhances load-bearing performance**: Adjusts viscosity dynamically to handle **varying loads and shear forces**.

How Thixotropic Lubricants Work

- **At rest**: The lubricant **remains thick and gel-like**, preventing separation or leakage.
- **Under shear stress (movement or agitation)**: The structure temporarily breaks down, allowing smooth flow and application.
- **After stress removal**: The lubricant rebuilds its consistency, ensuring stability and prolonged lubrication.

Benefits of Thixotropic Lubricants

- **Prevents leakage and dripping**: Ideal for vertical surfaces and high-speed rotating parts.
- **Enhances lubrication efficiency**: Ensures consistent film formation without overapplication.

- **Reduces maintenance and reapplication needs**: Long-lasting, stable lubrication under dynamic conditions.
- **Improves resistance to contamination and washout**: Thickens when idle, protecting components **from moisture and dirt ingress**.

891. Three Body Abrasion

See "Abrasive Wear."

892. Thrust Bearing

A **thrust bearing** is a type of rotary bearing designed to **support axial loads (thrust loads) and prevent shaft movement in the axial direction**. Unlike radial bearings, which handle radial forces, thrust bearings are specifically engineered to accommodate high axial loads in rotating equipment such as gearboxes, turbines, pumps, and automotive transmissions. See Figure T6.

Figure T6 Thrust ball bearing. Courtesy: SKF.

Key Functions of a Thrust Bearing

- **Supports axial (thrust) loads**: Prevents excessive axial movement in rotating machinery.
- **Reduces friction between moving surfaces**: Ensures smooth rotation with minimal energy loss.
- **Maintains shaft alignment**: Helps keep the shaft in position, reducing wear and vibration.

- **Enhances load distribution**: Spreads axial forces evenly to protect components from premature failure.

Types of Thrust Bearings

1. **Ball thrust bearings**:
    - Uses balls between bearing rings to support axial loads.
    - Found in low-speed applications such as turntables and light-duty machinery.
2. **Roller thrust bearings**:
    - Uses cylindrical or tapered rollers for higher load capacity than ball thrust bearings.
    - Used in heavy-duty applications like gearboxes and turbine shafts.
3. **Tapered roller thrust bearings**:
    - Designed for high axial loads and shock loads.
    - Common in automotive differentials and high-load industrial applications.
4. **Fluid film thrust bearings**:
    - Uses a thin film of lubricant to separate bearing surfaces and reduce friction.
    - Found in large turbines, compressors, and marine propulsion systems.
5. **Magnetic thrust bearings**:
    - Uses magnetic fields to levitate and support axial loads without contact.
    - Used in high-speed, precision applications such as aerospace and advanced turbines.

Benefits of Thrust Bearings

- **Handles high axial loads**: Designed for both light and heavy-duty applications.
- **Minimizes friction and wear**: Extends component lifespan by **reducing contact stress.**

- **Enhances equipment efficiency**: Reduces energy loss due to axial friction forces.
- **Supports reliable operation in rotating machinery**: Prevents axial misalignment and component damage.

893. Timer

A **timer in a lubrication system** is a **control device** used to regulate the timing and frequency of lubricant application in automated lubrication systems, such as single-line, dual-line, and progressive lubrication systems. Timers ensure that lubricants are delivered at the right intervals, and in the correct amounts, preventing over or underlubrication, reducing wear, and improving equipment lifespan.

Key Functions of a Timer in Lubrication Systems

- **Controls lubrication frequency**: Ensures lubricant is applied at defined intervals to avoid excessive wear or overlubrication.
- **Regulates lubricant flow duration**: Determines **how long lubricant is dispensed** per cycle.
- **Optimizes maintenance and energy efficiency**: Reduces **manual intervention and lubricant waste**.
- **Improves equipment protection and performance**: Ensures **consistent lubrication** under varying operating conditions.

Types of Timers in Lubrication Systems

1. **Fixed-cycle timers**:
    - Operate on **a** preset schedule with consistent lubrication intervals.
    - Used in basic automated lubrication systems for general-purpose applications.
2. **Adjustable interval timers**:
    - Allow flexible setting of lubrication frequency and duration based on machine operating conditions.
    - Common in progressive and dual-line lubrication systems.
3. **Pulse timers (on-demand lubrication)**:
    - Activate lubrication only when needed, based on machine signals or conditions.

- Used in condition-based lubrication systems.

4. **Programmable timers (PLC-controlled)**:
   - Integrated with PLC (programmable logic controller) for precise customized lubrication scheduling.
   - Used in high-end industrial and automated manufacturing setups.

5. **Electromechanical and digital timers**:
   - Electromechanical timers use relays or mechanical switches to control lubrication cycles.
   - Digital timers provide greater precision, memory storage, and remote control options.

894. Timken Bearing

Although Timken is a world-class bearing manufacturer the name "Timken Bearing" is synonymous with the tapered roller bearing. See "Tapered Roller Bearing."

895. Timken OK Load Test

The **Timken OK load test** is a standardized laboratory test used to evaluate the **extreme pressure (EP) performance of lubricants, particularly greases and gear oils**. It measures a lubricant's ability to prevent wear and scuffing under heavy loads in rolling and sliding contact conditions. The test is widely referenced in industrial, automotive, and heavy machinery lubrication to ensure proper protection in high-load applications.

Key Functions of the Timken OK Load Test

- **Determines extreme pressure (EP) performance**: Assesses the lubricant's ability to withstand heavy loads without failure.
- **Evaluates wear and scuffing protection**: Identifies the point where metal-to-metal contact leads to damage.
- **Ensures lubricant suitability for high-load applications**: Helps industries select the right grease or oil for bearings, gears, and other load-bearing components.

- **Standardized by ASTM D2509 (Greases) and ASTM D2782 (oils)**: Provides a consistent benchmark for lubricant comparison.

How the Timken OK Load Test Works

1. **Test setup**:
   - A rotating steel test cup is partially submerged in the test lubricant.
   - A stationary steel block (test block) is pressed against the rotating cup using an increasing load.
2. **Load application**:
   - The applied load is progressively increased in measured increments (in pounds or kilograms).
   - The lubricant's ability to maintain a protective film between the surfaces is observed.
3. **Failure detection**:
   - The test stops when visible wear, scuffing, or scoring appears on the stationary test block.
   - The maximum load at which no damage occurs is recorded as the Timken OK Load rating.

Understanding Timken OK Load Ratings

- Expressed in **pounds (lbs) or kilograms (kg)** as the maximum load the lubricant can withstand before failure.
- **Higher Timken OK load values** indicate better **extreme pressure (EP) and antiwear performance**.
- **Common OK load ranges**:
  - **Low EP performance**: Below **20 lbs (9 kg)** – suitable for **light-duty applications**.
  - **Moderate EP performance**: **20–40 lbs (9–18 kg)** – used in **general-purpose lubricants**.
  - **High EP performance**: **40+ lbs (18+ kg)** – designed for **heavy-duty applications like industrial gears and mining equipment**.

Factors Affecting Timken OK Load Performance

- **Additives in lubricants**:

  - **Extreme pressure (EP) additives** improve performance by forming **protective films**.
  - **Antiwear (AW) additives** help reduce metal surface degradation.
- **Base oil type and viscosity**: Higher viscosity lubricants provide **better film strength and load-carrying ability**.
- **Operating conditions**: High-speed, high-temperature, or contamination can affect **actual lubricant performance**.

896. Titration

Titration is a **quantitative chemical analysis method** used in lubrication testing to determine the **concentration of acids, bases, water, or additives** in a lubricant sample. It involves adding a reagent (titrant) to a lubricant sample until a chemical reaction reaches completion, typically indicated by a color change or electrical measurement. See Figure T7. Titration is widely used to monitor oil degradation, contamination, and additive depletion in industrial, automotive, and hydraulic lubricants.

Key Functions of Titration in Lubrication Analysis

- **Determines lubricant degradation**: Measures acid formation due to oxidation (total acid number, TAN).
- **Monitors additive depletion**: **Tracks alkaline reserve in engine oils** (total base number, TBN).
- **Detects water contamination**: Identifies moisture levels in hydraulic fluids and gear oils (Karl Fischer titration).
- **Ensures lubricant formulation accuracy**: Confirms correct additive concentrations in new and in-service oils.

Common Types of Titrations in Lubrication Testing

1. **Total acid number (TAN) titration**:
   - Measures acid buildup in lubricants due to oxidation.
   - Higher TAN values indicate oil degradation and potential corrosion risk.

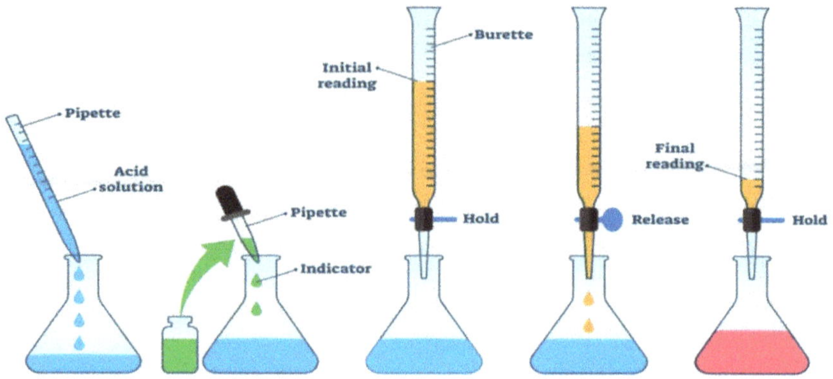

Figure T7 Titration process. Source: Chemistry Made Simple.

2. **Total base number (TBN) titration**:
    - Determines alkaline reserve in engine oils, which neutralizes acidic combustion by-products.
    - Low TBN indicates depleted additives and reduced oil life.
3. **Karl Fischer titration (water content analysis)**:
    - Measures water contamination in lubricants, which can cause wear, oxidation, and viscosity loss.
    - Detects water levels as low as ppm (parts per million).
4. **Additive concentration titration**:
    - Verifies levels of detergents, dispersants, and antiwear additives in formulated lubricants.
    - Ensures consistent oil performance and formulation integrity.

Applications of Titration in Lubrication Analysis

- **Engine and gear oil monitoring**: Tracks acidic oxidation products and additive depletion.
- **Hydraulic and turbine oil condition testing**: Identifies water contamination and oil breakdown.
- **Compressor and refrigeration lubricants**: Ensures moisture levels are within operational limits.

- **Industrial and power generation lubricants**: Maintains lubrication efficiency in turbines, compressors, and transformers.

Conclusion: **Titration** is an essential analytical technique in lubrication testing, helping industries monitor oil degradation, additive health, and contamination levels. By using TAN, TBN, Karl Fischer, and other titration methods, maintenance teams can extend lubricant life, prevent failures, and ensure optimal equipment performance in industrial and automotive applications.

897. Tooling and Mold Release Lubricant

Tooling and mold release lubricants are **specialized lubricants** designed to **prevent sticking, reduce friction, and protect molds and tooling surfaces** during manufacturing processes. These lubricants create a thin, protective film between the mold and the workpiece, ensuring smooth part release, reduced wear, and increased production efficiency in industries such as plastic injection molding, die casting, metal stamping, and rubber molding.

Key Functions of Tooling and Mold Release Lubricants

- **Prevents material sticking**: Forms a **nonstick barrier**, ensuring easy removal of molded parts.
- **Reduces friction and wear**: Protects **molds, dies, and tooling surfaces** from excessive wear and damage.
- **Improves product surface Finish**: Minimizes **imperfections, drag marks, and residues** on finished parts.
- **Enhances production efficiency**: Reduces **downtime and cleaning cycles**, allowing for **continuous operation**.

Types of Tooling and Mold Release Lubricants

1. **Silicone-based mold release lubricants**:
    - Offers excellent heat resistance and nonstick properties.
    - Used in plastic molding, rubber molding, and die-casting applications.
2. **Nonsilicone mold release agents**:
    - Leaves no residue, making them ideal for paintable or adhesive bonding applications.

  - Used in automotive, aerospace, and medical-grade molding.

3. **Water-based mold release lubricants**:
   - Environmentally friendly and low in VOC emissions.
   - Used in food-grade applications and eco-conscious manufacturing.
4. **PTFE (Teflon) and dry film mold release lubricants**:
   - Provides a dry, long-lasting lubrication film.
   - Used in high-precision molding where liquid residues must be avoided.
5. **Graphite and metal-based release agents**:
   - Suitable for high-temperature applications such as forging and die casting.
   - Used in metal stamping, aluminum casting, and extrusion processes.

Applications of Tooling and Mold Release Lubricants

- **Plastic injection molding**: Prevents plastic from **sticking to molds**, reducing cycle times.
- **Die Casting and metal forging**: Ensures **smooth metal flow and easy part release** in high-heat environments.
- **Rubber and composite molding**: Enhances **part ejection and prevents tearing** in rubber-based manufacturing.
- **Metal stamping and extrusion**: Reduces **friction and die wear** in precision metal forming.

Factors Affecting Mold Release Lubricant Performance

- **Operating temperature**: Some lubricants withstand **extreme heat**, while others degrade under high temperatures.
- **Residue and surface compatibility**: Nonsilicone and water-based options are better for **paintable and adhesive bonding applications**.
- **Lubricant longevity**: High-performance coatings provide **longer-lasting release properties**, reducing reapplication frequency.
- **Environmental and regulatory compliance**: Some industries require **low-VOC, food-grade, or biodegradable options**.

Benefits of Tooling and Mold Release Lubricants

- **Extends mold and tool Life**: Protects against **abrasion, corrosion, and wear**.
- **Improves manufacturing efficiency**: Reduces **downtime, cleaning, and rejected parts**.
- **Enhances product quality**: Produces **consistent, defect-free molded parts**.
- **Reduces energy and maintenance costs**: Minimizes **sticking, drag, and manual part removal efforts**.

898. Total Acid Number (TAN)

Total acid number (TAN) is a **measurement of the acidity level in a lubricant**, expressed in **milligrams of potassium hydroxide (mg KOH) required to neutralize the acidic components in one gram of oil**. TAN is a key indicator of oil degradation, oxidation, and contamination, helping industries assess lubricant conditions and determine oil change intervals.

Key Functions of Total Acid Number (TAN) Testing

- **Monitors oil degradation**: Detects **oxidation by-products** that form over time.
- **Identifies contamination Issues**: Helps identify **ingress of acidic contaminants**, such as **fuel, water, or process chemicals**.
- **Prevents corrosion and equipment damage**: Acidic buildup can **attack metal surfaces**, leading to premature wear.
- **Optimizes oil drain intervals**: Allows **condition-based oil changes**, reducing waste and maintenance costs.

How TAN is Measured

- **ASTM D664 Method**:
    - Uses **potentiometric titration** to determine the exact acidity level.
    - More accurate for **industrial and high-performance lubricants**.
- **ASTM D974 Method**:
    - A **color-indicator titration** that provides a simpler but less precise result.
    - Commonly used for **routine oil monitoring in general applications**.

Interpretation of TAN Values

- **New oil TAN**: Fresh lubricants have a **low initial TAN**, typically **< 1.0 mg KOH/g**.
- **Increasing TAN**: Rising TAN indicates **oxidation, additive depletion, or contamination**.
- **Critical TAN levels**: When TAN reaches a predetermined limit (varies by application), the oil should be **replaced or further analyzed**. See Table T2.

Table T2 Oil change threshold based on TAN levels.

Application	Typical Fresh Oil TAN	Oil Change Threshold
Hydraulic & Turbine Oils	0.05 – 0.5 mg KOH/g	Increase by 0.3 – 0.5 mg KOH/g
Engine Oils	1.0 – 3.0 mg KOH/g	Increase by 2.0 – 4.0 mg KOH/g
Gear & Industrial Oils	0.1 – 1.5 mg KOH/g	Increase by 1.0 – 2.0 mg KOH/g

Applications of TAN Testing in Lubrication

- **Turbine and compressor oils**: Prevents **acid-induced varnish formation and deposit buildup**.
- **Industrial hydraulic systems**: Ensures **oil longevity and corrosion protection**.
- **Automotive and diesel engine oils**: Tracks **oxidation and combustion byproducts**.
- **Transformer and insulating oils**: Detects **degradation affecting electrical performance**.

Factors Affecting TAN in Lubricants

- **Oxidation and thermal degradation**: High temperatures accelerate **acidic byproduct formation**.
- **Contaminants (water, fuel, process chemicals)**: Can introduce **acidic compounds into the oil**.
- **Additive depletion**: Some **antioxidant and antiwear additives** neutralize acids but degrade over time.

Benefits of Monitoring TAN in Lubricants

- **Prevents equipment corrosion and failures**: Avoids **acid attack on metal components**.
- **Extends oil and component life**: Enables **proactive maintenance and optimized oil changes**.
- **Reduces maintenance costs**: Prevents **unexpected breakdowns and excessive lubricant waste**.
- **Improves operational efficiency**: Supports **predictive maintenance strategies** by monitoring oil health.

899. Total Base Number (TBN)

Total base number (TBN) is a measure of a lubricant's alkaline reserve, indicating its ability to neutralize acidic contaminants formed during combustion or oxidation. It is expressed in milligrams of potassium hydroxide (mg KOH) per gram of oil. TBN is especially critical in engine oils, where it helps counteract acidic by-products from fuel combustion, sulfur contamination, and oxidation.

Applications of TAN Testing in Lubrication

- **Turbine and compressor oils**: Prevents **acid-induced varnish formation and deposit buildup**.
- **Industrial hydraulic systems**: Ensures **oil longevity and corrosion protection**.
- **Automotive and diesel engine oils**: Tracks **oxidation and combustion byproducts**.
- **Transformer and insulating oils**: Detects **degradation affecting electrical performance**.

Factors Affecting TAN in Lubricants

- **Oxidation and thermal degradation**: High temperatures accelerate **acidic by-product formation**.
- **Contaminants (water, fuel, process chemicals)**: Can introduce **acidic compounds into the oil**.
- **Additive depletion**: Some **antioxidant and antiwear additives** neutralize acids but degrade over time.

Benefits of Monitoring TAN in Lubricants

- **Prevents equipment corrosion and failures**: Avoids **acid attack on metal components.**
- **Extends oil and component life**: Enables **proactive maintenance and optimized oil changes.**
- **Reduces maintenance costs**: Prevents **unexpected breakdowns and excessive lubricant waste.**
- **Improves operational efficiency**: Supports **predictive maintenance strategies** by monitoring oil health.

Table T3 Oil change threshold based on TBN levels.

Application	Typical Fresh Oil TBN	Change Oil When TBN Drops Below
Gasoline Engine Oils	6 – 10 mg KOH/g	50% of original TBN
Diesel Engine Oils	8 – 15 mg KOH/g	2 – 3 mg KOH/g
Marine & High-Sulfur Fuels	15 – 70 mg KOH/g	Around 50% depletion

Applications of TBN Testing in Lubrication

- **Automotive and heavy-duty diesel engines**: Ensures **protection against sulfuric acid corrosion.**
- **Marine and power generation engines**: Monitors TBN in **high-TBN oils used with sulfur-rich fuels.**
- **Industrial and turbine lubricants**: Helps **assess base oil stability in long-life lubricants.**

Factors Affecting TBN Depletion

- **Fuel sulfur content**: Higher sulfur fuels generate **more acidic byproducts**, reducing TBN faster.
- **Oil oxidation and aging**: Over time, lubricant **loses its alkaline reserves.**
- **Contaminants (water, soot, fuel dilution)**: Increase **acid formation,** accelerating TBN depletion.

- **Operating conditions**: High temperatures and extended drain intervals **increase TBN consumption**.

Benefits of Monitoring TBN in Lubricants

- **Prevents corrosion and engine wear**: Maintains **sufficient alkalinity to neutralize acids**.
- **Optimizes oil change intervals**: Reduces **unnecessary oil drains**, saving costs.
- **Extends engine and lubricant life**: Protects components from **acidic damage and deposits**.
- **Supports predictive maintenance**: Enables **condition-based oil replacement** for efficient operation.

900. Tote

A **tote**, also known as an **intermediate bulk container (IBC)**, is a large, reusable, and transportable container used for storing and dispensing lubricants, oils, greases, and other industrial fluids. Totes are widely used in manufacturing, maintenance, and industrial lubrication programs due to their high capacity, durability, and ease of handling. See Figure T8.

Key Functions of a Tote in Lubrication Management

- **Bulk storage and transport**: Holds **275 to 330 gallons (1040 to 1250 liters)** of lubricant, reducing handling frequency.
- **Efficient dispensing and refilling**: Equipped with **valves, pumps, and hose connections** for controlled lubricant flow.
- **Reduces waste and spillage**: Designed to **minimize leaks and contamination**, improving fluid management.
- **Supports inventory management**: Enables **easy tracking of lubricant usage and stock levels**.

Types of Lubricant Totes

1. **Plastic (HDPE) totes**:
   - Made of high-density polyethylene (HDPE).
   - Suitable for nonreactive lubricants, oils, and coolants.
   - Common in food-grade and biodegradable lubricant storage.

Figure T8 Composite 330 gal. tote containing hydraulic oil (left hand) and metal 330 gal. tote containing water glycol. Courtesy: ENG**TECH** Industries Inc.

2. **Metal (steel) totes**:
   - Provides high durability and chemical resistance.
   - Used for heavy-duty lubricants, greases, and industrial oils.
   - Ideal for outdoor storage and harsh environments.
3. **Composite totes**:
   - Features plastic inner liners with a metal cage for protection.
   - Used for various industrial lubricants and specialty fluids.
4. **Pressurized totes**:
   - Designed for lubricants requiring sealed, pressurized dispensing.
   - Prevents air and moisture ingress, extending lubricant life.

Factors to Consider When Using Totes for Lubrication Storage and Dispensing

- **Material compatibility**: Ensure tote material is **compatible with the lubricant to prevent contamination or degradation**.
- **Proper sealing and venting**: Prevents **moisture, dust, and air ingress**, which can degrade lubricants.
- **Storage conditions**: Keep totes in **temperature-controlled environments** to avoid viscosity changes.
- **Labeling and identification**: Use **color-coded or labeled totes** to avoid lubricant cross-contamination.
- **Dispensing system integration**: Compatible with **pumps, meters, and filtration systems** for controlled lubricant delivery.

Benefits of Using Totes for Lubricant Storage

- **Reduces packaging waste**: Reusable and refillable, minimizing single-use drum disposal.
- **Enhances lubrication program efficiency**: Bulk storage reduces handling time and refilling needs.
- **Improves workplace safety**: Spill-proof design prevents accidental leaks and contamination.
- **Lowers overall lubricant costs**: Buying in bulk reduces per-gallon lubricant costs and procurement frequency.

901. Toxicity

Toxicity in lubricants refers to the **potential harmful effects of lubricants on human health, the environment, and equipment materials**. Lubricants may contain chemical additives, heavy metals, or petroleum-based compounds that can be hazardous if inhaled, ingested, or absorbed through the skin. Proper handling, disposal, and selection of low-toxicity alternatives are essential in industrial, automotive, and food-grade lubrication applications. See Table T3.

Key Factors Affecting Lubricant Toxicity

- **Base oil type**:
    - **Mineral and synthetic oils**: Can contain **polycyclic aromatic hydrocarbons (PAHs)**, which may be carcinogenic.

- **Biodegradable lubricants (esters, PAGs)**: Lower toxicity, safer for **human and environmental exposure**.

- **Additive package**:
  - **Extreme pressure (EP) additives**: Some contain **zinc dialkyldithiophosphate (ZDDP) or sulfur compounds**, which may pose health risks.
  - **Corrosion inhibitors and detergents**: Can contain **toxic metal compounds like barium or lead**.
- **Environmental impact**:
  - **Persistence in water and soil**: Petroleum-based lubricants can **contaminate groundwater and ecosystems**.
  - **Bioaccumulation risks**: Certain **metal-based additives can accumulate in organisms**, affecting wildlife.
- **Application and exposure**:
  - **Industrial and automotive Lubricants**: Risks include **skin irritation, respiratory issues, and long-term health effects**.
  - **Food-grade lubricants (NSF H1, H2, H3)**: Must meet **strict nontoxicity and FDA safety standards**.

Table T3 Potential lubricant toxins and alternatives.

Component	Potential Toxic Effects	Safer Alternative
Heavy Metals (Lead, Barium, Zinc)	Neurotoxicity, bioaccumulation	Low-metal or metal-free additives
PAHs (Polycyclic Aromatic Hydrocarbons)	Carcinogenic potential	Refined base oils, synthetic esters
Chlorinated Additives	Environmental toxicity, ozone depletion	Halogen-free EP additives
Phosphates & Sulfur Compounds	Skin/eye irritation, toxicity at high doses	Organic or ashless friction modifiers

Applications Where Lubricant Toxicity is a Concern

- **Food processing equipment**: Requires **NSF H1-approved lubricants** to avoid contamination.
- **Medical and pharmaceutical equipment**: Uses **nontoxic, biocompatible lubricants**.
- **Marine and environmental lubricants**: Must be **biodegradable and nontoxic to aquatic life**.

- **Worker safety in industrial plants**: Requires **low-VOC, nonhazardous lubricants to reduce inhalation risks**.

Reducing Toxicity in Lubrication Programs

1. **Choose low-toxicity and biodegradable lubricants**:
   - Use ester-based, PAG, or synthetic oils with nontoxic additive packages.
   - Select food-grade or environmentally acceptable lubricants (EALs) where applicable.
2. **Use proper handling and PPE (personal protective equipment)**:
   - Avoid direct skin contact, inhalation, or ingestion.
   - Use gloves, safety glasses, and ventilation systems.
3. **Monitor and control lubricant disposal**:
   - Prevent groundwater contamination by following disposal regulations.
   - Implement oil recycling and filtration systems to reduce waste.
4. **Check regulatory compliance**:
   - Follow OSHA, EPA, and REACH (Registration, Evaluation, Authorization of Chemicals) guidelines.
   - Use lubricants with NSF, USDA, and ISO 21469 certifications for food safety compliance.

Benefits of Using Low-toxicity Lubricants

- **Improves workplace safety**: Reduces **health risks for workers handling lubricants**.
- **Protects the environment**: Minimizes **soil, water, and air contamination risks**.
- **Ensures compliance with industry standards**: Meets **regulatory safety and environmental guidelines**.
- **Enhances equipment reliability**: Safer additives reduce **corrosion, wear, and long-term material degradation**.

902. Transfer Pump

A **transfer pump** is a **mechanical device used to move lubricants, oils, greases, and other fluids from one container or system to another**. These pumps are essential for bulk fluid handling, lubricant distribution, and oil circulation in industrial, automotive, and manufacturing applications.

Key Functions of a Transfer Pump in Lubrication Systems

- **Efficient fluid movement**: Transfers **oils, greases, and lubricants** between **tanks, drums, totes, and machinery**.
- **Reduces manual handling risks**: Minimizes **spillage, contamination, and worker strain**.
- **Supports automated lubrication systems**: Delivers lubricants to **gearboxes, hydraulic systems, and centralized lubrication systems**.
- **Ensures consistent flow and pressure**: Maintains **optimal lubricant viscosity and flow rates**.

Types of Transfer Pumps for Lubricants

1. **Positive displacement pumps (gear, vane, piston)**:
   - Ideal for viscous fluids like gear oils and greases.
   - Ensures precise, consistent flow at varying pressures.
2. **Centrifugal pumps**:
   - Suitable for low-viscosity fluids like hydraulic oils and light lubricants.
   - Works best in continuous flow applications.
3. **Diaphragm pumps**:
   - Handles thicker lubricants and contaminated fluids.
   - Used in industrial and hazardous fluid transfer.
4. **Drum and tote pumps**:
   - Specifically designed for pumping lubricants from barrels, IBC totes, and reservoirs.
   - Reduces waste and contamination during dispensing.

5. **Electric, pneumatic, and manual pumps**:
    - **Electric pumps**: High-efficiency operation for large-scale lubricant transfer.
    - **Pneumatic (air-operated) pumps**: Used in hazardous or explosive environments where electric motors are unsuitable.
    - **Manual pumps**: Ideal for small-scale applications like grease guns and oil dispensers.

Factors to Consider When Selecting a Lubricant Transfer Pump

- **Viscosity of the lubricant**: High-viscosity lubricants require **positive displacement pumps**.
- **Flow rate and pressure requirements**: Choose a pump that meets the **desired transfer speed and system pressure**.
- **Pump material compatibility**: Ensure pump materials are resistant to **corrosion, wear, and chemical reactions**.
- **Power source and operation type**: Consider **electric, pneumatic, or manual operation based on application needs**.
- **System integration and automation**: Some pumps integrate with **CMMS (Computerized Maintenance Management Systems) and automated lubrication setups**.

Benefits of Using a Transfer Pump for Lubrication

- **Enhances lubricant handling efficiency**: Reduces manual labor and **ensures smooth fluid transfer**.
- **Minimizes contamination risks**: Prevents **dirt, moisture, and cross-contamination** during lubricant transfer.
- **Extends equipment life**: Ensures **precise lubricant delivery**, reducing wear and failure risks.
- **Improves workplace safety**: Prevents spills, leaks, and ergonomic hazards related to manual oil handling.

Pump Use Code of Practice:

It is very easy to adopt the use of a single pump for transferring many different lubricants and chemicals. This is a dangerous practice as many lubricants are not compatible with other lubricants or chemicals and could cause great harm to the lubricated equipment. The code of safe practice is to:

- Employ dedicated transfer pumps for every lubricant and chemical type
- Clearly label the pump with the product name.
- Clean and flush the pump of all old product before being dedicated for a different lubricant or chemical product.
- Ensure seal compatibility for each pump with lubricant or chemical being transferred

903. Trending Analysis

Trending analysis is a **data-driven method used to monitor changes in lubricant condition over time**, helping predict equipment failures, optimize maintenance schedules, and extend lubricant life. By analyzing key lubricant parameters (such as viscosity, wear metals, and contamination levels) through regular oil sampling, maintenance teams can identify patterns, detect abnormalities, and take corrective action before failures occur. See Table T4.

Key Functions of Trending Analysis in Lubrication

- **Predicts lubricant and equipment health**: Identifies **gradual degradation, contamination, or wear**.
- **Optimizes maintenance intervals**: Allows **condition-based oil changes instead of fixed schedules**.
- **Detects early signs of failure**: Monitors **abnormal trends in viscosity, oxidation, and metal wear**.
- **Improves reliability and reduces downtime**: Prevents **unexpected breakdowns by tracking key indicators**.

How Trending Analysis Works

1. **Regular oil sampling and testing**:
    - Samples are collected at **consistent intervals** from critical lubrication points.
    - Tests are conducted for **viscosity, total acid number (TAN), total base number (TBN), wear metals, water contamination, oxidation, and particle count**.

2. **Data logging and comparison**:
   - Test results are recorded over time in a **lubricant analysis database or CMMS (Computerized Maintenance Management System)**.
   - Trends are plotted to **observe deviations from normal baselines**.
3. **Thresholds and alert levels**:
   - Acceptable operating ranges are established based on **OEM guidelines and historical data**.
   - Alerts are triggered when parameters **exceed safe limits**, allowing pre-emptive action.
4. **Actionable maintenance decisions**:
   - Based on trend analysis, teams **adjust lubrication schedules, change oil, filter contaminants, or inspect machinery**.

Benefits of Trending Analysis in Lubrication

- **Extends lubricant and equipment life**: Prevents unnecessary oil changes and **reduces lubricant waste**.
- **Improves predictive maintenance efficiency**: Reduces unscheduled downtime by **identifying failures before they occur**.

Table T4 Key parameters for trending analysis.

Parameter	What It Indicates	Why It's Important
Viscosity	Oil thickening/thinning	Affects lubrication film strength
Total Acid Number (TAN)	Oil oxidation & degradation	Indicates acid buildup & corrosion risk
Total Base Number (TBN)	Alkalinity reserve depletion	Predicts oil life in combustion engines
Wear Metal Analysis (ICP Spectroscopy)	Component wear	Early warning of bearing, gear, or piston wear
Water Contamination (% or ppm)	Moisture ingress	Leads to rust, oxidation, and additive breakdown
Particle Count (ISO 4406)	Solid contamination levels	Detects dirt, soot, and wear debris
Oxidation & Nitration	Oil aging & combustion byproducts	Indicates lubricant breakdown & varnish potential

- **Enhances cost savings**: Minimizes **repair costs, lost productivity, and lubrication waste**.
- **Supports data-driven decision-making**: Provides **historical insights into lubricant and machine performance**.

904. Tribology

Tribology is the **scientific study of friction, wear, and lubrication** in interacting surfaces in **relative motion**. It plays a crucial role in mechanical engineering, manufacturing, and maintenance by optimizing lubricant selection, reducing energy losses, and extending equipment life. Tribology integrates physics, chemistry, and materials science to improve the efficiency and durability of mechanical systems.

Derived from the Greek word "Tribos," meaning "to rub," the word was coined by Sir Peter H. Jost in the 1960s to describe the study of **Friction, Lubrication, and Wear**. Tribology was described by Jost as "**the science and technology of interacting surfaces in relevant motion and of related subjects and practices**."

Key Functions of Tribology

- **Reduces friction**: Minimizes energy loss between moving surfaces.
- **Prevents wear and surface damage**: Helps maintain component integrity under stress.
- **Optimizes lubrication**: Ensures proper lubricant selection and application for smooth operation.
- **Enhances equipment efficiency and lifespan**: Reduces mechanical failures, overheating, and maintenance costs.

Fundamental Aspects of Tribology

1. **Friction**:
   - The resistance to motion when two surfaces slide or roll against each other.
   - Influenced by **surface roughness, material properties, lubrication, and load conditions**.
   - Types of friction: **Static, kinetic (sliding), and rolling friction**.
2. **Wear**:
   - The gradual removal or deformation of material due to mechanical interaction.
   - Common wear types: **Adhesive, abrasive, fatigue, erosive, and corrosive wear**.

3. **Lubrication**:
   - The application of a **fluid, grease, or solid lubricant** to reduce friction and wear.
   - Types of lubrication: **Boundary, mixed, and hydrodynamic lubrication**.

905. Tribological System

A **tribological system** is a physical system consisting of **at least two contacting bodies and all environmental factors influencing their interaction**. It involves the study of friction, wear, and lubrication between surfaces in relative motion. Understanding tribological systems is essential for designing efficient, durable, and reliable mechanical components by optimizing materials, lubrication, and operating conditions.

Key Components of a Tribological System

- **Tribo-elements**: The interacting surfaces in contact, such as **gears, bearings, or seals**.
- **Interfacial elements**: The medium between surfaces, which can be **lubricant, gas, or vacuum**.
- **Environmental factors**: Conditions like **temperature, humidity, and contaminants** that influence tribological behavior.

Understanding Tribological Systems

Analyzing a tribological system requires assessing:

- **Inputs**: Load, velocity, and temperature that affect surface interactions.
- **Outputs**: Motion, torque, and force generated by the system.
- **Losses**: Friction, wear, and vibrations that reduce efficiency.
- **System structure**: Tribo-elements, kinematics, and interaction mechanisms that define performance.

A **comprehensive understanding of a tribological system** allows for the **optimization of materials, lubrication, and operational conditions** to minimize **friction, wear, and energy losses** while maximizing efficiency.

Tribological System and Lubricant Design
A tribological system defines the **operating conditions** under which a lubricant must perform, shaping the **technical specifications** required for lubricant formulation. Lubricant design involves selecting a **base oil and additives** to enhance performance characteristics, ensuring the lubricant meets **specific application requirements**.

Key Lubricant Design Parameters

1. **Lubricant type**: Designed for **gear systems, hydraulics, transmissions, compressors, or other machinery**.
2. **Application requirements**: Load capacity, speed, and bearing surface motion (**sliding, rolling, or combination**).
3. **Lubricant quality**: Includes viscosity index (**VI**), service life, and market positioning.
4. **Operating temperature range**: Ensures stability in **high-heat or extreme cold environments**.
5. **Environmental conditions**: Resistance to **moisture, chemicals, and extreme weather**.
6. **Biodegradability**: Compliance with **environmental standards and regulatory requirements**.

Conclusion: A tribological system is fundamental in determining the optimal lubrication strategy for mechanical components. By understanding the interaction between tribo-elements, environmental conditions, and lubricant properties, engineers and tribologists can develop high-performance lubricants and materials that enhance efficiency, durability, and reliability in various industrial applications.

906. TriboSystem

See "Tribological System."

907. Turbidity

Turbidity refers to the cloudiness or haziness of a lubricant due to suspended particles, contaminants, or emulsified water. It is an important parameter in lubricant condition monitoring, as increased turbidity can indicate contamination, oxidation, or additive depletion that may affect the lubricant's performance and the health of the equipment.

Key Functions of Turbidity Analysis in Lubrication

- **Identifies oil contamination**: Detects **water, dirt, metal particles, or microbial growth** in lubricants.
- **Monitors oil degradation**: Measures the presence of **oxidation by-products, varnish precursors, or sludge formation**.
- **Ensures lubricant performance**: Helps determine if the oil is **suitable for continued use or requires filtration or replacement**.
- **Supports predictive maintenance**: Alerts maintenance teams to **early signs of system contamination or wear**.

Causes of Increased Turbidity in Lubricants

- **Water contamination**: Emulsified or free water causes a **milky or cloudy appearance** in oil.
- **Solid particles and debris**: Wear metals, dirt, or rust increases **light scattering**, making the oil appear turbid.
- **Oil oxidation and degradation**: The formation of **sludge, varnish, and insoluble oxidation by-products** contributes to turbidity.
- **Additive precipitation**: Certain lubricant additives may separate under **temperature fluctuations or chemical reactions**.
- **Microbial growth**: In biodegradable and food-grade lubricants, **bacterial contamination** can cause cloudiness.

Methods for Measuring Turbidity in Lubricants

1. **Visual inspection**:
   - Simple field test by holding the oil sample against a light source to check for cloudiness.
   - A clear oil indicates low turbidity, while a hazy or milky appearance suggests contamination.
2. **Turbidimeter analysis**:
   - Uses a light beam and sensor to measure the intensity of scattered light in the lubricant.

- Provides a quantitative turbidity value in **NTU (nephelometric turbidity units)**.

3. **Spectrophotometric analysis**:

   - Measures light absorption and scattering properties to detect fine particles in oil.
   - Used for advanced oil condition monitoring in laboratories.

4. **Microscopic and particle count analysis (ISO 4406)**:

   - Identifies solid contaminants, oxidation by-products, and microbial presence in oil samples.

Ways to Reduce Turbidity in Lubricants

- **Use proper filtration and water separation systems**: Removes particulate and water contamination.
- **Monitor and maintain proper storage conditions**: Prevents moisture ingress and oxidation.
- **Select high-quality lubricants with stable additives**: Reduces the risk of additive separation or breakdown.
- **Regular oil sampling and analysis**: Detects turbidity changes early to prevent equipment damage.

908. Turbine Lubricant

A **turbine lubricant** is a **specialized oil designed to provide long-term protection, cooling, and lubrication** in high-speed and high-temperature turbine systems. It ensures efficient operation, minimizes wear, and prevents deposits or varnish formation in turbines used in power plants, aviation, marine propulsion, and industrial applications.

Key Functions of Turbine Lubricants

- **Reduces friction and wear**: Maintains a stable lubrication film to **protect bearings, gears, and moving parts**.
- **Prevents oxidation and varnish formation**: Formulated with **antioxidation additives** to resist thermal breakdown.

- **Provides heat dissipation and cooling**: Transfers heat away from **critical turbine components**.
- **Enhances contaminant control**: Includes **water-separating and anti-foaming additives** to maintain lubricant purity.
- **Ensures long service life**: Designed for **extended oil drain intervals and minimal degradation**.

Types of Turbine Lubricants

1. **Steam turbine oils**:
   - Designed to handle high moisture levels and thermal stress.
   - Used in power plants, refineries, and industrial turbines.
2. **Gas turbine oils**:
   - Withstands extreme temperatures and oxidation stress.
   - Used in aviation, jet engines, and gas-fired power turbines.
3. **Hydroelectric turbine oils**:
   - Features high oxidation stability and water resistance.
   - Used in hydropower stations and dam turbines.
4. **Aeroderivative turbine oils**:
   - Specially designed for aviation-derived gas turbines used in power generation and marine propulsion.
5. **Synthetic turbine oils**:
   - Offers superior thermal stability, lower volatility, and extended drain intervals.
   - Ideal for critical applications with extreme temperature fluctuations.

Turbine Lubricant Contamination and Maintenance

- **Water ingress**: Can cause **emulsification, corrosion, and reduced film strength**.
- **Oxidation and varnish formation**: Leads to **deposit buildup and restricted oil flow**.
- **Particulate contaminants**: Dirt and wear debris accelerate **bearing and gear damage**.

- **Regular oil sampling and analysis**: Includes tests for **viscosity, oxidation, acid number (TAN), water content, and particle contamination**.

909. Turbulent Flow

Turbulent flow refers to a type of fluid flow characterized by chaotic and irregular motion, where fluid particles move in random directions and velocities. Unlike **laminar flow**, where fluid moves smoothly in parallel layers, turbulent flow is marked by swirling eddies and a significant amount of energy dissipation.

Key Characteristics of Turbulent Flow:

- **High Reynolds number**: Turbulent flow typically occurs when the Reynolds number (a dimensionless value used to predict flow behavior) exceeds 4000. The Reynolds number depends on the fluid's velocity, density, viscosity, and the diameter of the pipe or channel.
- **Irregular and swirling motion**: Fluid particles move in a chaotic manner, creating vortices and eddies that cause mixing and energy loss.
- **Increased energy dissipation**: Due to the irregular motion, energy is lost more rapidly as heat, making turbulent flow less efficient in terms of energy use compared to laminar flow.

Impact on Lubrication:

- **Heat generation**: Turbulent flow in lubrication systems can lead to excessive heat buildup, which can degrade the lubricant and reduce its effectiveness.
- **Wear and tear**: The chaotic motion can lead to increased mechanical stress on components, accelerating wear.
- **Enhanced mixing**: While turbulent flow can have negative effects, it can also be beneficial in ensuring that additives in a lubricant are well-distributed.

Applications and Considerations:

- **Pipelines and fluid transport**: Understanding and controlling turbulent flow is crucial in designing pipelines to minimize energy loss and wear.
- **Cooling and mixing**: In systems where efficient mixing or heat transfer is needed, turbulent flow can be advantageous.

Conclusion: Turbulent flow is an important concept in fluid dynamics, with implications for system efficiency, lubrication performance, and the design of fluid-handling equipment.

910. Two-body Abrasion

See "Abrasive Wear."

911. Tyndall Effect

The **Tyndall effect** is the **scattering of light by particles in a colloidal or fine suspension**, making the path of light visible. In lubrication, this effect is observed in oil emulsions, contaminated lubricants, and synthetic fluids where microscopic particles or droplets cause light scattering. Detecting the Tyndall effect in lubricants can help identify oil degradation, contamination, or phase separation in industrial and automotive applications.

Key Functions of the Tyndall Effect in Lubrication Analysis

- **Indicates oil contamination**: Helps detect the presence of **water**, **air bubbles, or fine particulates**.
- **Identifies emulsification issues**: Determines if **oil and water have formed a stable emulsion**, affecting performance.
- **Assesses lubricant stability**: Helps monitor **colloidal additives or dispersed solid lubricants** in suspension.
- **Aids in condition monitoring**: Provides a **quick visual check** for contamination in **oil samples and hydraulic fluids**.

Examples of the Tyndall Effect in Lubricants

- **Oil–water emulsions**: Water contamination in oil causes light scattering, making the lubricant appear **hazy or milky**.
- **Air entrapment and foam formation**: Tiny air bubbles in oil can scatter light, affecting **hydraulic and engine oil performance**.
- **Colloidal additives and nanoparticles**: Some lubricants contain **solid dispersions** (e.g., molybdenum disulfide or PTFE) that exhibit the Tyndall effect.
- **Oxidation and degradation by-products**: Degraded oil may form **fine suspended particles**, leading to visible scattering.

Applications of the Tyndall Effect in Lubrication Analysis

- **Used oil condition monitoring**: A quick **visual inspection technique** to check for contamination.
- **Fuel and coolant contamination detection**: Helps determine **cross-contamination in lubricating systems**.
- **Colloidal lubricant stability testing**: Ensures **solid lubricant dispersions remain evenly distributed**.
- **Hydraulic and industrial fluid analysis**: Detects **emulsified water or trapped air**, which can affect system efficiency.

Methods to Control the Tyndall Effect in Lubricants

- **Use high-quality oil filtration systems**: Removes **fine particles that cause light scattering**.
- **Avoid water contamination**: Implement **desiccant breathers, oil purifiers, and water-separating additives**.
- **Monitor and maintain proper lubricant additives**: Ensures **colloidal stability** of solid lubricants and dispersions.
- **Perform regular oil sampling and analysis**: Helps detect **early signs of contamination or oxidation** before failure occurs.

Benefits of Understanding the Tyndall Effect in Lubricants

- **Quick and nondestructive testing**: Allows for **on-site visual assessments** of lubricant condition.
- **Early contamination detection**: Helps prevent **machine wear and system failures**.
- **Improves lubricant performance and longevity**: Identifies **issues before they cause significant degradation**.
- **Enhances predictive maintenance programs**: Supports **condition-based maintenance strategies**.

U

912. Ultrasonic Greasing

Ultrasonic greasing is a **precision lubrication technique** that uses **ultrasonic acoustic sensors** to determine the optimal amount of grease required for a bearing or rotating component. By measuring friction and vibration levels in real-time, ultrasonic greasing ensures accurate grease application, preventing overlubrication and underlubrication, which can lead to premature failure. See Figure U1.

Key Functions of Ultrasonic Greasing

- **Optimizes grease application**: Prevents **excessive or insufficient lubrication** in critical bearings.
- **Reduces friction and wear**: Ensures **adequate grease film formation** for smooth operation.
- **Extends bearing and equipment life**: Minimizes stress on **bearings, motors, and rotating machinery**.

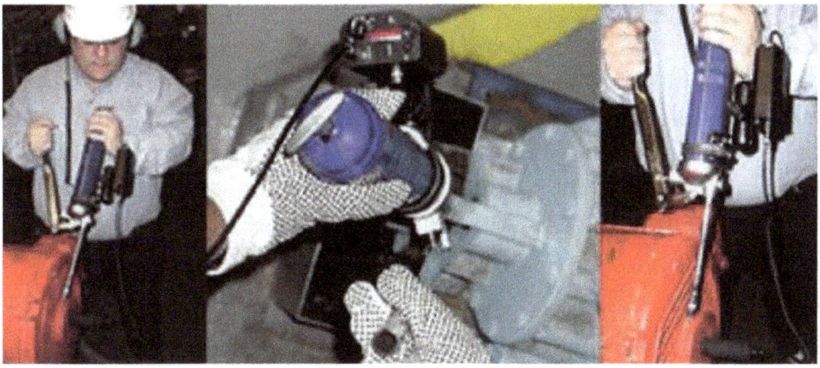

Figure U1 Operator performing manual greasing using UE systems ultraprobe 201 UT grease system. Courtesy: UE Systems Inc. Elmsford, NY, United States.

DOI: 10.1201/9788743806608-932

- **Enhances predictive maintenance**: Provides **real-time feedback** to improve lubrication scheduling.

How Ultrasonic Greasing Works

1. **Baseline measurement**:
    - A handheld **ultrasonic sensor (or acoustic grease gun)** is placed on the bearing housing.
    - The sensor **measures ultrasonic noise levels** (friction-related sound waves).
2. **Grease application**:
    - Grease is added slowly **while continuously monitoring ultrasonic signals**.
    - A drop in noise levels indicates **optimal lubrication** has been achieved.
3. **Stop greasing at optimal level**:
    - Once the sound levels **stabilize or reach the lowest point**, greasing is stopped.
    - This prevents **excess grease buildup, which can increase bearing temperature and cause failure**.

913. Unloading Filter

An **unloading filter** is a **filtration component used in lubrication and hydraulic systems** to manage pressure surges, flow fluctuations, and contaminant removal during lubricant circulation. It is designed to relieve excess pressure and prevent filter clogging, ensuring smooth system operation and prolonging the life of lubricants and equipment. It is also commonly referred to as a bypass filter (See "Bypass Filter") as depicted in Figure B12.

914. Uptime

Uptime refers to the **amount of time equipment or a system remains operational and available for use** without unplanned downtime. In lubrication and maintenance strategies, **maximizing uptime** is critical for improving productivity, reducing failures, and optimizing asset performance in industrial, automotive, and manufacturing environments.

Key Functions of Uptime in Lubrication and Maintenance

- **Ensures continuous operation**: Reduces **unplanned stoppages due to lubrication failures**.
- **Optimizes equipment reliability**: Minimizes **wear, overheating, and component failures**.
- **Reduces maintenance costs**: Extends **oil, grease, and component life**, lowering service expenses.
- **Improves production efficiency**: Enhances overall **output by maintaining machine performance**.

Factors Affecting Uptime in Lubrication Systems

1. **Lubrication best practices**:
    - Regular **oil analysis, regreasing schedules, and contamination control** prevent failures.
    - Correct **lubricant selection** based on **viscosity, load, and environmental conditions**.
2. **Predictive and condition-based maintenance**:
    - Using **real-time oil condition monitoring** helps detect early signs of degradation.
    - Technologies like **ultrasonic greasing and vibration analysis** prevent under or overlubrication.
3. **Filtration and contaminant control**:
    - **Proper oil filtration** reduces **dirt, moisture, and varnish buildup**, extending lubricant life.
    - **Water separation and particle removal systems** help maintain lubricant purity.
4. **Automated lubrication systems**:
    - **Centralized lubrication systems** ensure **consistent oil or grease supply** to moving parts.
    - Reduces human error and **prevents manual lubrication-related downtime**.
5. **Temperature and load management**:
    - **Overheating accelerates oil degradation**, causing breakdowns.

- Proper lubrication **reduces friction and heat buildup**, improving uptime.

Benefits of Maximizing Uptime

- **Increases asset lifespan**: Prevents **premature wear and unexpected breakdowns**.
- **Improves cost savings**: Reduces **repair expenses, lubricant consumption, and labor costs**.
- **Enhances productivity and output**: Keeps **machinery running efficiently without disruptions**.
- **Supports data-driven decision-making**: Using **oil analysis trends and predictive maintenance improves scheduling**.

915. Universal Lubricants

See "Multipurpose Grease."
See also "Multipurpose Oil."

V

916. V = A × R

Delivering the right amount of lubricant to a bearing requires the engineer designing the system to calculate each bearing point lubricant requirement. This allows the metering valves and pump to be sized correctly. Different bearing types require different amounts of lubricant. For example, a plain bearing has more surface contact area than a rolling element bearing and will require more lubricant. Also, a total loss system will require less lubricant than a continuous lubricated bearing. To quickly determine a metered amount of lubricant the lubrication system manufacturing companies devised a simple calculation method that has proven extremely effective over many decades of use and is still the primary calculation used today. The calculation is simply:

$$V = A \times R.$$

where:

V = Volume (Oil or Grease)

A = Bearing Surface

R = Replenishment rate of lubricant film

Area of Bearing Surface
The **bearing surface area** will change depending on the type of bearing being lubricated. These include **plain bearings, rolling element bearings, flat surface bearing area, and gears**. Bearing surface areas are calculated as follows:

1 **Plain bearings**: These include bushings, sleeve bearings.

$A = \pi DL$
$\pi = 3.142$
D = Shaft Diameter
L = Bearing Length

DOI: 10.1201/9788743806608-937

2 **Rolling element bearings**: These include ball. roller, taper roller, needle, double row, etc.

$$A = D^2R$$

D = Shaft Diameter
R = Replenishment Rate

3 **Flat bearing surfaces**: These include gibs, ways, slides, etc

A = **Contact Surface Areas**

4 **Gears;** Each single gear – Spur, Bevel, etc.

$$A = \pi DW$$

$\pi = 3.142$
D = Gear Pitch Diameter
W = Gear Face Width

Lubricant Replenishment Rate

The lubricant replenishment rate represents the lubricant film replenishment rate over a period of time for both cyclic and continuous lubricating systems delivering either oil or grease. See Table V1.

Table V1 Lubricant replenishment rate table.

Lubricant and delivery method	Film thickness requirement	Time
Manual applied terminating grease	.002"–.050 mm film	Every 4 hours
Automatic applied terminating oil	.001"–.025 mm film	Every 1 hour
Automatic applied terminating grease	.001"–.025 mm film	Every 4 hours
Automatic applied circulating oil	.001"–.025 mm film	Every minute

917. Vacuum Dehydrator

A **vacuum dehydrator** is a specialized **oil purification system designed to remove water, gases, and contaminants** from lubricating oils, hydraulic fluids, and transformer oils. It uses **vacuum distillation and low-temperature evaporation** to extract **free, emulsified, and dissolved water**, ensuring optimal lubricant performance and extending equipment life. See Figure V1.

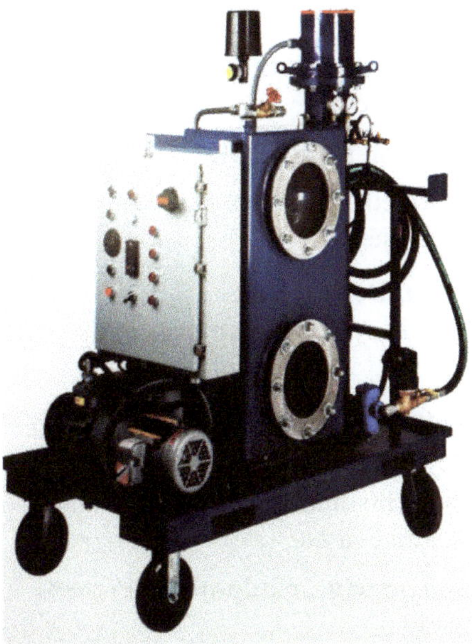

Figure V1 Vacuum dehydrator distillation unit. Courtesy: Des-Case Corporation.

Key Functions of a Vacuum Dehydrator

- **Removes water contamination**: Eliminates **free, emulsified, and dissolved water** from lubricants.
- **Prevents oil oxidation and degradation**: Reduces moisture-induced **varnish, sludge, and acid formation.**
- **Enhances lubricant performance**: Maintains **viscosity, film strength, and antiwear properties.**
- **Protects bearings and hydraulic components**: Prevents **rust, corrosion, and cavitation.**

How a Vacuum Dehydrator Works

1. **Oil intake and prefiltration**:
    - Contaminated oil passes through a **prefilter to remove large particles** before dehydration.
2. **Vacuum chamber evaporation**:
    - The oil enters a **vacuum chamber where pressure is reduced,** allowing water to **evaporate at lower temperatures.**

3. **Water and gas removal**:
   - A vacuum pump **extracts the evaporated moisture and air**, separating them from the oil.
4. **Fine filtration and oil return**:
   - The dried, purified oil passes through a final filter before being returned to the system.

Applications of Vacuum Dehydrators

- **Turbine lubrication systems**: Removes moisture buildup in steam and gas turbine oils.
- **Hydraulic and gear oils**: Prevents water contamination that causes foaming, sludge, and wear.
- **Transformer and insulating oils**: Maintains dielectric strength by eliminating water and gases.
- **Paper mills and marine equipment**: Protects lubricants in water-prone environments.

Benefits of Using a Vacuum Dehydrator

- **Extends lubricant life**: Reduces oil oxidation and contamination-related failures.
- **Prevents component wear and corrosion**: Minimizes bearing and pump failures due to moisture.
- **Improves system efficiency**: Ensures optimal lubrication and hydraulic performance.
- **Reduces maintenance costs and downtime**: Lowers oil replacement frequency and equipment failures.

918. Vacuum Distillation

Vacuum distillation is a key refining process used to produce **high-quality base oils**, which are the primary building blocks of **industrial lubricants** such as gear oils, hydraulic fluids, and turbine oils.

Purpose in Lubricant Production

After **atmospheric distillation** removes lighter fractions (e.g., gasoline, diesel) from crude oil, the remaining **residue** still contains valuable heavy hydrocarbons. However, these components:

- **Boil** at very high temperatures
- **Decompose** or degrade if heated at normal pressure

To extract these safely, vacuum distillation is used.

How Distillation Works in Lubricant Refining

1. **Low Pressure, Lower Boiling Points**: By lowering the pressure inside the distillation column, heavy fractions can vaporize at **temperatures below 400°C**, preventing thermal cracking.
2. **Separation of Heavy Fractions:**
   - **Vacuum gas oil**
   - **De-asphalted oil**
   - **Lubricating oil distillates**

   Once separate the heavy fractions are sent for further refining steps (like solvent extraction and hydrocracking).
3. **Base Oil Production:**
   - The distillates become **Group I or Group II** base oils after further treatment.
   - Heavier residues may be used for asphalt or further processed into **Group III base oils** using advanced techniques.

Vacuum Distillation Process
In Figure V2 below we see the following:

1. **Heavy oil residue** from atmospheric distillation is fed into a tall heated column under vacuum pressure.
2. Components are separated under **reduced pressure** to prevent thermal breakdown.
3. **Vaporized fractions** are sent to a **condenser** and cooled.
4. The now refined **base oil is collected** and used to formulate industrial lubricants.

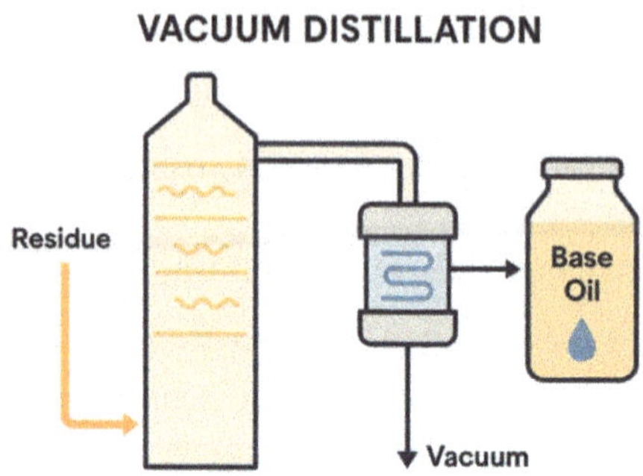

Figure V2 Simple Vacuum Distillation Process

Why Vacuum Distillation Matters for Lubricants

- **Preserves molecular structure** of lubricant-grade hydrocarbons
- Prevents **thermal degradation** that would harm oil performance
- Enables production of **stable, high-viscosity base oils** with good oxidation resistance
- Allows recovery of **valuable components** from otherwise unusable residues

919. Vacuum Grease

Vacuum grease is a specialized lubricant designed to **seal and lubricate components in vacuum systems**, where air and gas pressures are extremely low (down to ultra-high vacuum levels). Table V2 shows different types of vacuum grease available

Key Characteristics:

- **Low vapor pressure** – does not evaporate or outgas under vacuum
- **High thermal stability** – withstands a wide range of temperatures (often –40°C to +200°C or more)
- **Chemically inert** – does not react with gases or system materials
- **Non-migrating** – stays in place and resists creeping or flowing

Common Uses:

- Sealing **ground-glass joints** in lab equipment

- Lubricating **O-rings**, gaskets, and valves in **vacuum chambers**
- Assembling or maintaining **mass spectrometers, electron microscopes,** and **vacuum pumps**

Table V2 Types of Vacuum Grease.

Type	Base Material	Features
Silicone-based	Polydimethylsiloxane	Good thermal stability, non-toxic
Hydrocarbon-based	Petroleum or synthetic oils	Cost-effective, for medium vacuums
Fluorocarbon-based	PFPE (e.g., Krytox®)	Ultra-high vacuum, chemically inert

920. Valve, Directional Control

A **directional control valve** regulates **the path of lubricant flow**, controlling which lubrication points receive oil or grease in a system. It directs flow based on system requirements, load conditions, or preset lubrication cycles.

Key Functions of a Directional Control Valve

- **Controls lubricant flow direction**: Directs oil to specific bearings, gears, or hydraulic actuators.
- **Regulates lubrication timing and distribution**: Used in automated single and dual-line parallel lubrication systems.
- **Prevents lubricant loss and waste**: Ensures lubricant only flows when and where needed.

Types of Directional Control Valves

1. **Manual valves**: Operated by a **lever or knob** to switch lubricant flow.
2. **Solenoid-operated valves**: Electrically controlled for **automated lubrication systems**.
3. **Pilot-operated valves**: Uses system pressure to **activate lubricant flow**.
4. **Flow-divider valves**: Distributes lubricant to **multiple lubrication points evenly**.

Factors Affecting Directional Control Valve Performance

- **Flow rate and system pressure**: Affects **response time and valve function**.

- **Lubricant type (oil/grease)**: Thick lubricants require **higher operating pressures**.
- **Valve actuation method**: Manual, solenoid, or pressure-controlled options affect **system efficiency**.

921. Valve, Flow Control

A flow control valve is a device used to regulate the rate of lubricant flow within a lubrication or hydraulic system. It ensures that lubricant is delivered at a controlled and consistent rate to bearings, gears, and other critical components. Flow control valves play a vital role in preventing overlubrication, underlubrication, and pressure fluctuations, optimizing system performance, and extending equipment life.

Types of Flow Control Valves used in Lubrication Systems

1. **Adjustable flow control valves**:
   - Allows **manual fine-tuning of lubricant flow rate**.
   - Used in **bearing lubrication and hydraulic circuits**.
2. **Fixed-orifice flow control valves**:
   - Contains **a preset opening** that regulates flow based on system pressure.
   - Common in **centralized lubrication systems**.
3. **Pressure-compensated flow control valves**:
   - Maintains a **constant lubricant flow rate**, even if system pressure changes.
   - Used in **high-precision lubrication applications**.
4. **Needle valves**:
   - Uses a **tapered needle-like plunger** to adjust flow.
   - Ideal for **fine control in low-flow lubrication systems**.

922. Valve, Relief

A **relief valve** is a **pressure-regulating device designed to protect lubrication and hydraulic systems from excessive pressure buildup**. It automatically opens when system pressure exceeds a preset limit, allowing excess lubricant or fluid to bypass the system and return to a reservoir or low-pressure area. This prevents damage to pumps, seals, filters, and other components, ensuring safe and efficient operation.

Key Functions of a Relief Valve

- **Prevents overpressure damage**: Opens when pressure exceeds safe operating limits.
- **Ensures continuous lubricant flow**: Regulates system pressure to maintain optimal lubrication.
- **Protects pumps and seals**: Reduces strain on components, preventing premature wear and failure.
- **Enhances system efficiency**: Maintains consistent pressure levels for smooth operation.

How a Relief Valve Works

1. **Normal operation**:
    - The valve remains closed under normal operating pressure, allowing lubricant to flow through the system.
2. **Pressure build-up**:
    - If pressure increases due to a blockage or system demand, the valve senses the rise.
3. **Valve opening**:
    - When the pressure reaches the preset limit, the relief valve opens, redirecting excess fluid.
4. **Pressure stabilization and reset**:
    - Once pressure returns to normal, the valve **closes automatically**, restoring standard operation.

Applications of Relief Valves

- **Lubrication pumps and systems**: Protects against excessive pressure in circulating and centralized lubrication systems.
- **Hydraulic power units**: Prevents overloading in hydraulic circuits and actuators.

- **Filtration systems**: Ensures filters do not rupture due to clogging-induced pressure spikes.
- **Gear and bearing lubrication**: Regulates pressure in high-speed rotating equipment.

923. Valve Lubricant

Valve lubricant is a **specialized lubricant** used to ensure the **smooth operation, sealing, and longevity** of **valves** in various industrial systems—including oil & gas, water, steam, chemical, and food processing applications.

Functions of Valve Lubricant

1. **Lubricates** moving parts (e.g., stem, ball, plug) to reduce friction and wear
2. **Seals** between valve components to prevent leakage
3. **Protects** against corrosion, rust, and contaminants
4. **Aids in valve actuation**—especially important for high-pressure or infrequently operated valves

Key Properties

- **High pressure resistance** (especially in gate and plug valves)
- **Wide temperature tolerance**
- **Chemical compatibility** with process media (e.g., water, gas, hydrocarbons)
- **Non-reactive or inert** for safety in harsh or sensitive environments

Table V3 depicts the recommended grease types for the different types of application valves operate. It is important that the user always match the **lubricant type to the valve type** and **media** (e.g., oxygen, gas, brine, hydrocarbons) the valve operates within.

Table V3 Valve grease types and Their Purpose.

Type	Best For
Petroleum-based greases	General-purpose, low-cost applications
Silicone-based greases	Water, steam, and food-safe environments
Fluorinated greases (PTFE or PFPE)	Harsh chemicals, oxygen, or high-temperature service
Synthetic greases	Wide performance range with excellent stability

924. Vapor Pressure

Vapor pressure is the **pressure exerted by a lubricant's vapor when it is in equilibrium with its liquid phase at a given temperature**. It indicates how easily a lubricant or fluid evaporates under heat. **Lower vapor pressure lubricants resist evaporation**, maintaining stable lubrication, while **higher vapor pressure fluids evaporate more easily**, leading to oil loss, oxidation, and increased wear.

Key Functions of Vapor Pressure in Lubrication

- **Determines evaporation rate**: Affects how quickly a lubricant evaporates under high temperatures.
- **Influences lubricant stability**: Higher vapor pressure can cause **oil thickening, deposit formation, and viscosity changes.**
- **Affects equipment performance**: Excessive evaporation reduces **lubrication effectiveness, leading to increased wear.**
- **Impacts volatility and oil consumption**: Lubricants with high vapor pressure **tend to evaporate faster**, leading to **increased oil consumption** in engines and industrial machinery.

How Vapor Pressure Affects Lubrication Performance

1. **High vapor pressure lubricants**:
    - Evaporate easily at **elevated temperatures**.
    - Lead to **increased oil consumption and potential dry-running conditions.**
    - Common in **low-viscosity fluids or poorly refined mineral oils**.
2. **Low vapor pressure lubricants**:
    - More resistant to **heat and evaporation loss**.
    - Maintain stable **film thickness and viscosity**.
    - Found in **synthetic oils, high-quality mineral oils, and high-performance lubricants**.

Applications of Vapor Pressure Consideration in Lubrication

- **High-temperature lubrication (turbines, compressors, engines)**: Prevents **lubricant loss under extreme heat**.

- **Vacuum systems and aerospace applications**: Ensures lubricants **do not evaporate in low-pressure environments**.
- **Industrial gear and bearing lubrication**: Reduces **oil oxidation and varnish formation**.
- **Refrigeration and compressor lubricants**: Prevents **oil carryover into gas streams**.

Factors Affecting Vapor Pressure in Lubricants

- **Base oil type**: **Synthetic oils (PAO, esters) have lower vapor pressure** than standard mineral oils.
- **Temperature**: **Higher operating temperatures** increase vaporization rates.
- **Lubricant formulation and additives**: Certain **antioxidants and stabilizers** help reduce evaporation losses.

925. Varnish

Varnish is a **thin, insoluble, and sticky residue that forms on internal surfaces of lubrication and hydraulic systems** due to oil degradation, contamination, and oxidation. See Figure V3. It reduces lubrication efficiency, causes overheating, and can lead to equipment failure by restricting oil flow and increasing friction.

Key Functions of Varnish in Lubrication Systems

- **Reduces lubrication efficiency**: Creates **sticky deposits** that prevent proper oil flow and film formation.
- **Increases friction and heat**: Leads to **bearing wear, pump inefficiency, and higher operating temperatures**.
- **Clogs filters and valves**: Causes **restricted oil flow, pressure drops, and system malfunctions**.
- **Promotes equipment failures**: Accelerates **component wear, leading to unplanned downtime**.

Figure V3 Typical varnished components in an engine valve train.

How Varnish Forms in Lubricated Systems

1. **Oxidation reactions**: High temperatures cause oil **molecules to break down**, forming acidic byproducts that react with metal surfaces.
2. **Thermal degradation**: Extreme heat leads to **polymerization of oil additives**, creating varnish-like residues.
3. **Contaminant interaction**: Dirt, moisture, and combustion byproducts mix with degraded oil to form **sticky deposits**.
4. **Electrostatic discharge and spark erosion**: High-voltage conditions in oil cause **micro-pitting and localized varnish formation**.

Common Areas Affected by Varnish

- **Bearings and gears**: Causes **increased friction and overheating**.
- **Hydraulic systems and servo valves**: Blocks **flow pathways, affecting pressure regulation**.
- **Oil coolers and heat exchangers**: Reduces **heat dissipation, leading to higher temperatures**.
- **Filters and reservoirs**: Leads to **clogging and increased pressure differentials**.

Methods for Detecting Varnish in Lubrication Systems

- **Membrane patch colorimetry (MPC Test)**: Measures **varnish potential by analyzing oil color change**.
- **Particle count and ISO cleanliness analysis**: Identifies **insoluble varnish precursors in oil**.
- **FTIR spectroscopy (Fourier transform infrared analysis)**: Detects **oxidation byproducts linked to varnish**.
- **Rotating pressure vessel oxidation test (RPVOT)**: Predicts **oil oxidation resistance and varnish risk**.

Preventing and Removing Varnish in Lubrication Systems

- Use **high-quality lubricants with oxidation resistance**: Synthetic and **premium mineral oils with antioxidant additives** slow varnish formation.
- **Control operating temperatures**: Reduce excessive heat **to minimize oil degradation**.
- **Improve oil filtration and contaminant control**: **Electrostatic oil cleaners, depth filters, and varnish removal units** help eliminate deposits.
- **Regular oil analysis and condition monitoring**: Detect varnish precursors **before they cause damage**.
- **Flushing and chemical cleaning**: Severe varnish buildup may require **chemical solvents or oil flushing systems**.

926. Vegetable-based Oils

Vegetable-based oils are **biodegradable lubricants derived from plant sources such as soybean, rapeseed (canola), sunflower, and palm oil**. They are used as an alternative to petroleum-based lubricants due to their environmental benefits, renewable sourcing, and excellent lubricating properties. These oils are commonly used in food processing, forestry, agriculture, and environmentally sensitive applications.

Key Functions of Vegetable-based Oils in Lubrication

- **Provides high lubricity and film strength**: Ensures **smooth operation and wear protection**.
- **Offers biodegradability and environmental safety**: Breaks down naturally, reducing environmental impact.

- **Resists oxidation with proper additives**: Requires **antioxidants to enhance stability and longevity**.
- **Performs well in extreme pressure conditions**: Provides **good boundary lubrication in high-load applications**.

Types of Vegetable-based Oils in Lubrication

1. **Straight vegetable oils**:
   - Pure plant oils used in **low-load and environmentally sensitive applications**.
   - Examples: **Soybean oil, sunflower oil, rapeseed oil**.
2. **Modified and esterified vegetable oils**:
   - Chemically modified for **better oxidation resistance and thermal stability**.
   - Used in **hydraulic systems, industrial lubrication, and greases**.
3. **Blended bio-based oils**:
   - Mixed with **synthetic esters or mineral oils** to improve **performance and stability**.
   - Common in **biodegradable hydraulic fluids and metalworking lubricants**.

Applications of vegetable-based oils in lubrication

- **Food-grade lubricants (NSF H1 and H3)**: Used in **food processing and pharmaceutical industries**.
- **Hydraulic fluids and industrial lubrication**: Provides **biodegradable alternatives for eco-sensitive areas**.
- **Metalworking fluids**: Offers **natural lubricity for cutting and machining operations**.
- **Chainsaw and forestry lubricants**: Reduces **oil runoff contamination in forest environments**.
- **Agricultural and marine applications**: Used in **tractors, fishing vessels, and equipment in water-sensitive regions**.

Benefits of Vegetable-based Oils

- **Environmentally friendly and biodegradable**: Reduces **pollution risks and regulatory concerns**.

- **Renewable and sustainable resource**: Supports **green initiatives and reduces reliance on fossil fuels.**
- **High lubricity and wear protection**: Provides **excellent film strength and natural friction reduction.**
- **Safe for food-contact applications**: Complies with **FDA and NSF regulations for food processing equipment.**

Key Drawbacks of Vegetable-based Oils in Industry

1. Poor oxidation stability

- **Rapid degradation**: Vegetable oils oxidize faster than synthetic or mineral oils, leading to thickening, sludge formation, and varnish buildup.
- **Shorter service life**: Require frequent monitoring and oil changes, increasing maintenance costs.

2. Limited thermal stability

- **Break down at high temperatures**: Decompose at elevated temperatures (above 80–100 °C), reducing lubrication effectiveness.
- **Risk of polymerization**: Form sticky residues that can lead to filter clogging and component wear.

3. High pour point and poor cold-temperature performance

- **Thicken in cold environments**: Natural vegetable oils have **higher pour points**, making them unsuitable for **cold-weather applications**.
- **Limited flowability**: Can cause start-up issues in hydraulic systems and engines operating in subzero conditions.

4. Hydrolytic instability and water sensitivity

- **Absorb moisture**: Readily react with water, leading to hydrolysis and acid formation.
- **Reduce lubrication quality**: Increased corrosion risk, affecting bearings, gears, and seals.

5. Incompatibility with some seals and materials

- **Cause swelling and degradation**: Certain rubber and elastomeric seals (e.g., nitrile and polyurethane) may soften or deteriorate in contact with vegetable oils.
- **Require compatibility testing**: Some industrial systems may require seal upgrades to prevent leaks and failures.

6. Higher cost compared to conventional oils

- **Expensive production and refining**: When chemical modifications and additives to improve stability are required.
- **Limited availability**: High-quality bio-based lubricants are often more expensive than mineral oils.

7. Risk of microbial contamination and rancidity

- **Bio-degradable but prone to spoilage**: Without stabilizers, vegetable oils can develop odors, degrade, or support microbial growth in humid environments.
- **Require biocides and preservatives**: Need additional additives to prevent microbial contamination.

927. Velocity Factor

In lubrication, the **velocity factor** refers to the influence of **relative surface speed** between moving parts—typically between a **shaft and a bearing or bushing**—on the behavior and performance of the **lubricating film**. It is often considered when selecting the appropriate **lubricant type, viscosity,** and **application method**, especially in dynamic systems like rotating machinery.

Why Velocity Factor Matters:

- As **velocity increases**, a **thinner but more stable oil film** can form (up to a point).
- At **very high speeds**, too viscous a lubricant may cause drag and heat.
- At **low speeds**, a thicker lubricant or special additives may be needed to maintain separation.

Velocity factor is commonly used in **bearing design** and **hydrodynamic lubrication theory**. It's part of the **bearing characteristic number** or **Stribeck curve**, which relates:

- **Load**
- **Viscosity**
- **Speed**
- **Clearance**

Velocity Factor (V) may be expressed as: $V = N \, x \, D \, / \, 60$

Where:

- N = rotational speed (RPM)
- D = shaft diameter (mm or in)
- The result gives **surface speed**, usually in m/s or ft/s

Effect on Lubrication Regimes:

- **Low velocity** → boundary or mixed lubrication
- **Moderate velocity** → full hydrodynamic lubrication (ideal)
- **Very high velocity** → possible lubricant thinning, overheating, or leakage

928. Venturi Cavitation

See "Cavitation."

929. VI

See "Viscosity index."

930. VI Improver

See Viscosity Index (VI) Improver.

931. Virgin Base Oil

Virgin base oil refers to **new, unused base oil that has never been refined, reprocessed, or used in any equipment**. It serves as the foundation for lubricants, hydraulic fluids, and industrial oils, providing a clean and uncontaminated starting point for lubricant formulation. Virgin oils are classified into different base oil groups (I–V) based on their refining process, chemical composition, and performance characteristics.

Key Functions of Virgin Oil in Lubrication

- **Acts as a base stock for lubricants**:
  - Forms the primary component of engine oils, hydraulic fluids, and greases.
  - Used in formulations requiring high purity and stability.

- **Ensures clean and uncontaminated lubrication**:
  - Contains no used oil contaminants, oxidation by-products, or degradation residues.
  - Provides consistent performance and extended lubricant life.
- **Supports additive blending for enhanced performance**:
  - Serves as a foundation for viscosity modifiers, detergents, dispersants, and antiwear agents.
  - Allows for precise formulation to meet industrial and automotive standards.
- **Provides high oxidation stability and thermal resistance**:
  - Resists sludge formation and breakdown under high temperatures.
  - Offers better long-term lubrication compared to rerefined oils.

Types of Virgin Oils Used in Lubricants

- **Group I base oils**:
  - Solvent-refined with higher sulfur content and lower oxidation resistance.
  - Used in general-purpose industrial oils and low-demand applications.
- **Group II base oils**:
- Hydro-processed for better oxidation resistance and thermal stability.
  - Commonly found in modern engine oils, hydraulic fluids, and industrial lubricants.
- **Group III base oils**:
  - Highly refined with superior viscosity control and thermal performance.
  - Used in synthetic-blended and high-performance lubricants.
- **Group IV (PAO) and Group V (Ester) base oils**:
  - Fully synthetic oils with exceptional thermal stability, oxidation resistance, and low-temperature performance.
  - Used in high-end applications like aviation, extreme-pressure lubrication, and long-drain interval oils.

932. Virgin Oil Sample

A **virgin oil sample** is an **unused, uncontaminated lubricant or base oil collected before it is introduced into equipment**. It serves as a reference point for condition monitoring, oil analysis, and contamination detection, allowing maintenance teams to compare in-service oil with its original properties to assess lubricant degradation, contamination, and additive depletion.

Key Functions of a Virgin Oil Sample in Lubrication Analysis

- **Establishes a baseline for oil condition monitoring**:
  - Provides a reference sample to compare with used oil for wear detection, oxidation, and contamination analysis.
  - Helps identify early signs of lubricant degradation or failure.
- **Assists in detecting contamination and degradation**:
  - Compares viscosity, additive concentration, oxidation levels, and particulate contamination.
  - Identifies changes in lubricant chemistry due to moisture, fuel dilution, or acid formation.
- **Ensures correct lubricant specification and purity**:
  - Confirms that the supplied oil meets OEM specifications and industry standards.
  - Detects mixing errors, additive separation, or off-specification products.
- **Supports failure investigations and root cause analysis**:
  - Differentiates between normal lubricant aging and abnormal contamination or breakdown.
  - Helps in diagnosing system failures caused by incorrect lubrication or contamination.

Factors Affecting Virgin Oil Sample Accuracy

- **Storage and handling conditions**: Contaminated storage can **alter oil properties before use**.
- **Sampling methodology**: Must follow **ASTM or ISO standards for reliable comparisons**.

- **Additive settling or separation**: Some additives may **settle over time, affecting sample consistency**.

933. Viscometer

A **viscometer** is a **precision instrument used to measure the viscosity of lubricants, oils, and fluids**. See Figure V4. It determines a fluid's **resistance to flow**, which is critical for assessing lubricant performance, ensuring proper oil selection, and monitoring lubricant degradation over time. Viscometers are essential tools in oil analysis, quality control, and industrial lubrication management

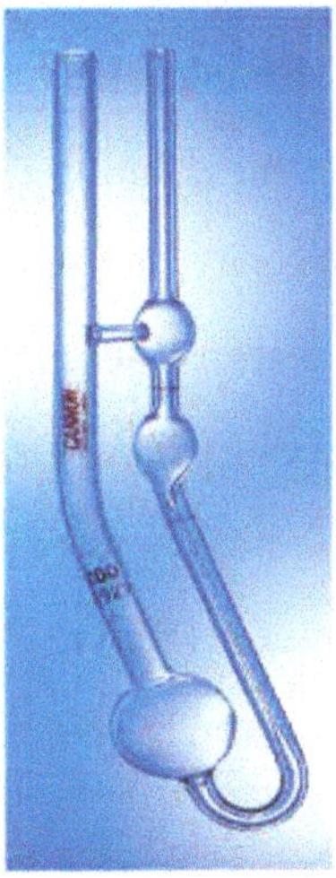

Figure V4 Fischer scientific viscometer.

Key Functions of a Viscometer

- **Measures lubricant viscosity**:
  - Determines **kinematic or dynamic viscosity** to assess fluid thickness and flow properties.
  - Ensures oil meets OEM and industry specifications.
- **Monitors oil condition and degradation**:
  - Detects viscosity changes due to oxidation, contamination, or additive depletion.

- Helps predict lubricant failure before equipment damage occurs.
- **Supports lubricant selection and performance testing**:
  - Confirms that the selected lubricant provides adequate film strength and wear protection.
  - Ensures proper viscosity grades for different operating conditions.

Types of Viscometers Used in Lubrication Analysis

- **Capillary viscometers (Saybolt, Ubbelohde, Ostwald)**:
  - Measures the time it takes for oil to flow through a narrow tube.
  - Used for kinematic viscosity testing in laboratory settings.
- **Rotational viscometers (Brookfield, Couette, Cone-and-plate)**:
  - Uses a rotating spindle immersed in fluid to measure resistance.
  - Ideal for measuring viscosity in greases and non-Newtonian fluids.
- **Falling ball and falling piston viscometers**:
  - Measures the time it takes for a ball or piston to fall through the fluid.
  - Used for on-site and laboratory viscosity testing.
- **Saybolt universal viscometer (SSU)**:
  - Determines the time required for 60 mL of oil to pass through a standard orifice at specific temperatures.
  - Commonly used for measuring fuel and lubricant viscosities.

Factors Affecting Viscometer Accuracy

- **Temperature variations**: Lubricant viscosity **decreases as temperature increases**, requiring precise temperature control.
- **Contamination and oil degradation**: Water, dirt, or oxidation products **can alter viscosity readings**.
- **Shear sensitivity**: Some lubricants, such as **multigrade oils and greases**, may show **shear-thinning or thickening behavior**.

934. Viscometrics

Viscometrics refers to the **study and measurement of viscosity-related properties in lubricants, oils, and industrial fluids**. It involves analyzing **fluid resistance to flow** under varying conditions to ensure optimal lubrication, monitor oil degradation, and improve machine performance. Viscometrics is essential for predictive maintenance, quality control, and lubricant formulation.

Key Functions of Viscometrics in Lubrication

- **Determines lubricant flow behavior**:
  - Measures viscosity changes across different temperatures and pressures.
  - Ensures lubricants provide proper film thickness for wear protection.
- **Monitors oil condition and degradation**:
  - Identifies viscosity breakdown due to oxidation, contamination, or additive depletion.
  - Helps in predictive maintenance by detecting lubrication issues early.
- **Optimizes lubricant selection for specific** Applications:
  - Ensures that lubricant viscosity matches the operating conditions of equipment.
  - Prevents underlubrication (excess wear) or overlubrication (drag and energy loss).

Key Parameters in Viscometrics

- **Kinematic viscosity (measured in centistokes, cSt)**:
  - Determines the time it takes for a fluid to flow under gravity.
  - Commonly measured using capillary viscometers (e.g., Saybolt, Ubbelohde, and Cannon-Fenske).
- **Dynamic viscosity (measured in centipoise, cP)**:
  - Measures the internal resistance to flow under applied force.
  - Tested using rotational and falling ball viscometers.

- **Viscosity index (VI)**:
  - Indicates how much viscosity changes with temperature.
  - Higher VI means better temperature stability and consistent performance.
- **Shear stability**:
  - Assesses how well a lubricant maintains its viscosity under mechanical stress.
  - Important for gear oils, hydraulic fluids, and multigrade engine oils.
- **Viscosity-temperature relationship**:
  - Helps in selecting oils that perform efficiently in extreme cold or high-heat conditions.
  - Affects start-up lubrication and high-load performance.

935. Viscosity

Viscosity is the **measure of a fluid's resistance to flow**. It determines how easily a lubricant can spread, maintain film thickness, and reduce friction between moving parts. Viscosity, defined by the oil's molecule size, is one of the most critical properties in lubrication, influencing equipment performance, wear protection, and energy efficiency. Figure V5 demonstrates how temperature, load, and physical state can affect viscosity behavior.

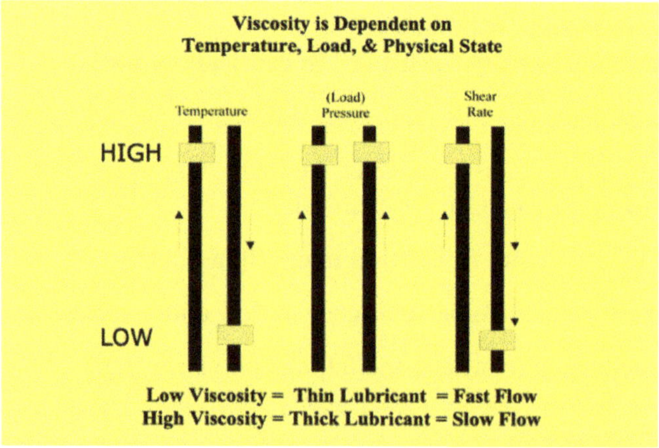

Figure V5 Viscosity behavior chart. Source: Lubrication for Industry, Kenneth E. Bannister, Industrial Press.

Key Functions of Viscosity in Lubrication

- **Controls lubricant flow and film formation**:
  - High viscosity **creates a thicker lubricant film**, protecting surfaces under heavy loads.
  - Low viscosity **flows more easily**, reducing internal drag and energy losses.
- **Reduces friction and wear**:
  - Prevents **metal-to-metal contact by maintaining a protective film**.
  - Enhances **component lifespan in bearings, gears, and engines**.
- **Affects heat dissipation and cooling**:
  - Proper viscosity ensures **efficient heat transfer**, preventing lubricant breakdown.
  - Too thick = **inefficient cooling**, too thin = **increased wear and overheating**.
- **Influences lubricant efficiency and fuel economy**:
  - Lower viscosity oils reduce **energy loss due to fluid drag**, improving efficiency.
  - Higher viscosity oils **withstand heavy loads but may increase resistance** in high-speed applications.

Types of Viscosity in Lubrication

- **Kinematic viscosity (measured in centistokes, cSt)**:
  - Measures the **fluid's resistance to flow under gravity**.
  - Commonly tested using capillary viscometers (e.g., Saybolt, Ubbelohde, and Cannon-Fenske).
- **Dynamic viscosity (Measured in Centipoise, cP)**:
  - Measures the **fluid's internal resistance to movement under force**.
  - Common in rotational and falling ball viscometers.
- **Apparent viscosity**:
  - Applies to **non-Newtonian fluids (greases, polymer-thickened oils)**, which change viscosity under force.

  - Important for high-load applications, food-grade lubricants, and extreme pressure greases.
- **Viscosity index (VI)**:
  - Indicates **how viscosity changes with temperature**.
  - High VI = better stability across temperature variations.

Applications of Viscosity in Lubrication

- **Engine oils**: Ensures **proper start-up lubrication and high-temperature protection**.
- **Hydraulic fluids**: Maintains **consistent operation in pumps, actuators, and valves**.
- **Gear and transmission oils**: Prevents **wear in high-load, high-speed environments**.
- **Industrial lubricants**: Used in **bearings, compressors, and turbine systems for efficient lubrication**.

Factors Affecting Viscosity Performance

- **Temperature**: Heat reduces viscosity, while cold **thickens lubricants**, affecting flow.
- **Contamination (water, dirt, fuel, soot)**: Alters **oil thickness and lubrication efficiency**.
- **Shear stability**: Some lubricants break down under **mechanical stress, reducing viscosity**.
- **Additives and base oil type**: Viscosity modifiers enhance **stability and performance in extreme conditions**.

Conclusion: Viscosity is a fundamental property of lubricants that determines their flow behavior, film strength, and performance under different conditions. By selecting the correct viscosity for specific applications, industries can reduce wear, enhance efficiency, and improve the longevity of both lubricants and machinery.

936. Viscosity – Absolute

See "Viscosity – Dynamic."

937. Viscosity Comparison Chart

A **viscosity comparison chart** provides a visual reference for comparing the viscosity grades of different lubricants (see Figure V6), including ISO VG (International Standards Organization Viscosity Grade), SAE (Society of Automotive Engineers), and AGMA (American Gear Manufacturers Association) classifications. It helps users select the right lubricant by mapping viscosity values across various measurement standards, temperatures, and applications. The chart typically includes kinematic viscosity values in centistokes (cSt) at 40 and 100 °C, and Saybolt Second viscosity (SUS) at 100 and 210 °F highlighting how different oils perform under operating conditions. By comparing engine oils, hydraulic fluids, gear oils, and industrial lubricants, a viscosity chart ensures that maintenance teams and engineers can make informed lubricant choices based on temperature, load, and speed requirements, preventing under or overlubrication, reducing wear, and optimizing efficiency.

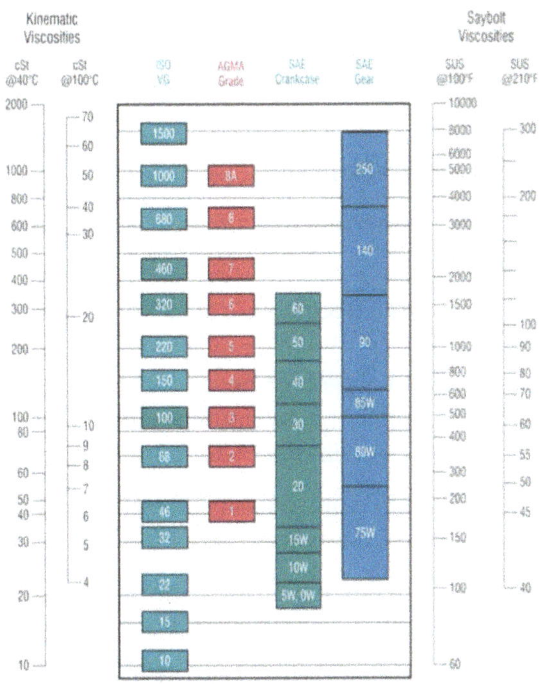

Figure V5 Viscosity comparison chart. Source: Lubrication for Industry, Kenneth E. Bannister, Industrial Press.

938. Viscosity Index

Viscosity index (VI) is a **numerical value that indicates how a lubricant's viscosity changes with temperature**. A higher VI means the oil maintains a more stable viscosity across temperature variations, while a lower VI indicates that viscosity changes significantly with heat or cold. VI is crucial in selecting engine oils, hydraulic fluids, and industrial lubricants, ensuring optimal performance in high-temperature and cold-start conditions.

Key Functions of Viscosity Index in Lubrication

- **Measures viscosity stability across temperature ranges**:
  - High VI lubricants remain **thicker at high temperatures** and **flow better in cold conditions.**
  - Low VI oils experience **greater thinning at high temperatures and thickening at low temperatures.**
- **Enhances lubrication efficiency**:
  - Reduces **friction, wear, and energy loss in varying operating conditions.**
  - Improves **oil film strength and consistency** in **bearings, gears, and engines.**
- **Prevents lubricant breakdown**:
  - High VI oils resist **thermal degradation**, reducing **sludge, varnish, and oxidation.**
  - Maintains **lubricant effectiveness in extreme environments.**

Viscosity index ranges and classification

- **Low VI (<35)**: Common in **conventional mineral oils**, which experience **significant viscosity changes with temperature**.
- **Moderate VI (35–80)**: Found in **basic industrial and automotive oils,** suitable for **stable environments**.
- **High VI (80–110)**: Used in **modern engine oils, hydraulic fluids, and gear oils** requiring **temperature flexibility**.
- **Very high VI (>110)**: Found in **synthetic and high-performance lubricants**, designed for **severe temperature fluctuations**.

The viscosity index, or VI, is a measure of oil's viscosity change due to temperature. Oils with higher VI rating are more desirable as they are more stable under changing temperature conditions reflecting a narrower change in viscosity over a standard temperature range.

As noted above, paraffinic oils have much higher VI ratings making them more stable and desirable where a wide operating temperature range is experienced.

Oils can be grouped and classified by their VI property as depicted in Table V4 – Viscosity Index rating Table.

Table V4 Viscosity Index rating Table.

VI RATING	VI GROUP
< 35	LOW – **LVI**
35–80	MEDIUM – **MVI**
80–110	HIGH – **HVI**
>110	VERY HIGH - **VHVI**

939. Viscosity Index (VI) Improver

Viscosity index (VI) improvers are **additives used in lubricants to enhance their viscosity stability across a wide temperature range**. They help oils maintain optimal flow properties at both high and low temperatures, ensuring effective lubrication under varying operating conditions. VI improvers are essential in engine oils, hydraulic fluids, gear oils, and industrial lubricants where temperature fluctuations affect performance.

Key Functions of VI Improvers

- **Enhance temperature stability**:
  - Prevents excessive thinning at high temperatures and thickening at low temperatures.
  - Ensures consistent lubrication film strength under extreme conditions.
- **Improve cold-start performance**:
  - Reduces oil thickening in cold environments, allowing smoother engine startups.
  - Enhances flowability in hydraulic and transmission systems.

- **Reduce oil consumption and volatility**:
  - Prevents excessive evaporation and lubricant loss at high temperatures.
  - Helps maintain longer oil drain intervals in engines and industrial equipment.
- **Extend equipment life**:
  - Maintains proper lubricant film thickness to minimize wear and friction.
  - Reduces oxidation and deposit formation by keeping oil viscosity stable.

Types of VI Improvers

- **Olefins and copolymers (OCPs)**:
  - Common in engine and hydraulic oils due to their thermal stability and shear resistance.
  - Provides a good balance between performance and cost.
- **Polymethacrylates (PMAs)**:
  - Used in high-performance lubricants and synthetic oils.
  - Offers excellent low-temperature fluidity and oxidation resistance.
- **Styrene-butadiene copolymers (SBCs)**:
  - Used in gear oils and specialty lubricants.
  - Provides superior shear stability and viscosity retention.

Applications of VI Improvers

- **Engine oils**: Ensures **optimal lubrication across hot and cold conditions**.
- **Hydraulic fluids**: Improves **fluidity and system efficiency in extreme temperatures**.
- **Gear and transmission oils**: Reduces **friction and wear in high-load applications**.
- **Industrial lubricants**: Enhances the **performance of compressors, turbines, and heavy machinery**.

Factors Affecting VI Improver Performance

- **Shear stability**: Some VI improvers can **break down under high loads**, reducing effectiveness.
- **Base oil compatibility**: Performance depends on **oil formulation and additive interaction**.
- **Operating conditions**: Extreme temperatures and pressure **affect viscosity retention**.
- **Oil drain intervals**: Over time, VI improvers degrade, requiring **oil analysis and replenishment**.

Benefits of VI Improvers

- **Ensures stable lubrication**: Maintains **viscosity in varying temperatures**.
- **Enhances equipment protection**: Reduces **friction, wear, and overheating**.
- **Improves fuel and energy efficiency**: Reduces **fluid resistance in hydraulic and engine systems**.
- **Extends lubricant service life**: Minimizes **viscosity breakdown and degradation**.

940. Viscosity – Dynamic

Dynamic viscosity, also known as **absolute viscosity**, is a **measure of a fluid's internal resistance to flow under an applied force**. It quantifies how a lubricant responds to shear stress, which is crucial for determining lubricant film thickness, flow characteristics, and energy efficiency in machinery. Dynamic viscosity is measured in poise (P) or centipoise (cP) and is commonly tested using rotational or falling ball viscometers.

Key Functions of Dynamic Viscosity in Lubrication

- **Determines lubricant flow behavior**:
  - Higher dynamic viscosity means the lubricant **flows more slowly and resists motion**, ideal for **high-load applications**.
  - Lower dynamic viscosity allows the lubricant to **flow more easily**, improving **efficiency in high-speed components**.

- **Maintains lubrication film strength**:
  - Prevents **metal-to-metal contact** by ensuring proper oil film thickness.
  - Reduces **friction and wear in bearings, gears, and hydraulic systems**.
- **Influences energy efficiency and heat dissipation**:
  - High dynamic viscosity oils create **more internal resistance**, leading to **higher energy consumption**.
  - Lower viscosity oils improve **fuel efficiency in engines and reduce heat buildup in hydraulic systems**.

Dynamic Viscosity Measurement and Units

- **Measured in Pascal-seconds (Pa·s), Poise (P), or Centipoise (cP)**:
  - 1 P = 0.1 Pa·s | 1 cP = 0.001 Pa·s.
  - Most lubricants have viscosities in the **centipoise (cP) range**.
- **Common testing methods**:
  - **Rotational viscometers**: Measures fluid resistance using **a rotating spindle submerged in oil**.
  - **Falling ball viscometers**: Times how long a ball **falls through the lubricant under gravity**.
  - **Capillary tube viscometers (for Newtonian Fluids)**: Measures how oil flows under **gravity and applied force**.

941. Viscosity – Kinematic

Kinematic viscosity is the **measure of a fluid's resistance to flow under the influence of gravity**. It describes how easily a lubricant can move through a system and is crucial for maintaining proper lubrication film thickness, reducing wear, and optimizing energy efficiency. Kinematic viscosity is measured in centistokes (cSt) and is commonly tested using capillary viscometers, such as the Ubbelohde or Cannon-Fenske viscometer.

Key Functions of Kinematic Viscosity in Lubrication

- **Determines lubricant flow rate and film formation**:
  - Higher kinematic viscosity oils flow slower and provide thicker lubrication films.
  - Lower-viscosity oils flow faster and reduce internal friction for improved efficiency.
- **Ensures proper lubrication in high and low temperatures**:
  - Prevent excessive thinning at high temperatures and thickening at low temperatures.
  - Maintains lubricant integrity in extreme operating conditions.
- **Affects oil circulation and cooling efficiency**:
  - Lighter oils improve cooling and heat dissipation in fast-moving components.
  - Heavier oils offer better protection in high-load and slow-moving machinery.

Kinematic Viscosity Measurement and Units

- **Measured in centistokes (cSt) at standard temperatures**:
  - **40 °C (104 °F)**: Used for industrial and automotive lubricants.
  - **100 °C (212 °F)**: Common for engine and high-temperature applications.
- **Common Testing Methods**:
  - **Capillary viscometers (e.g., Ubbelohde and Cannon-Fenske)**: Measures the time it takes for a fluid to flow through a calibrated tube under gravity.
  - **Saybolt universal seconds (SUS or SSU)**: Converts time-based viscosity measurements into kinematic viscosity values.

942. Viscosity Modifier

See "Viscosity Index (VI) Improver."

943. Viscosity Shear Plane

A **viscosity shear plane** is the **layer within a lubricant where fluid molecules experience varying shear forces due to relative motion between surfaces**. It is a critical concept in lubrication because it determines how lubricant flows, resists shear stress, and maintains film integrity under load and movement. The shear plane plays a key role in hydrodynamic, boundary, and elastohydrodynamic (EHD) lubrication, influencing friction, wear, and lubricant performance.

Key Functions of the Viscosity Shear Plane in Lubrication

- **Controls lubrication film thickness and stability**:
  - The shear plane influences how lubricant layers separate moving surfaces.
  - Proper shear characteristics ensure consistent film strength in gears, bearings, and hydraulic systems.
- **Determines viscosity behavior under load and motion**:
  - In **Newtonian fluids**, viscosity remains constant regardless of shear rate.
  - In **non-Newtonian fluids**, viscosity can **change under shear stress**, affecting flow and lubrication.
- **Affects shear stability and viscosity loss**:
  - High mechanical stress at the shear plane can cause **viscosity breakdown**, reducing protection.
  - Viscosity modifiers help **stabilize lubricant viscosity under shear forces**.

Types of Shear Behavior in Lubrication

- **Hydrodynamic shear plane**:
  - Occurs when full-film lubrication separates two surfaces.
  - Found in bearings, journal surfaces, and fluid couplings.
- **Elastohydrodynamic (EHD) shear plane**:
  - Forms in high-load rolling contacts, such as gears and roller bearings.

  - Lubricant film compresses under pressure but maintains a protective barrier.

- **Boundary lubrication shear plane**:
  - Appears when lubricant film is thin or absent, and friction modifiers or antiwear additives reduce metal contact.
  - Found in slow-moving, high-load applications.

Factors Affecting Shear Plane Performance

- **Lubricant type and viscosity index**: Oils with high VI resist shear thinning under temperature changes.
- **Mechanical stress and load conditions**: Higher loads increase shear stress, affecting viscosity stability.
- **Additive and polymer content**: Viscosity modifiers enhance shear resistance but may degrade over time.

944. Viscous Friction

Viscous friction is the **resistance to motion caused by the internal friction of a fluid as it flows between surfaces**. It occurs when lubricant molecules interact and resist shear forces, affecting energy efficiency, heat generation, and wear protection in mechanical systems. Viscous friction is directly related to lubricant viscosity, with higher viscosity oils creating more resistance but providing better film strength, while lower viscosity oils reduce friction but may offer less protection.

Viscous friction is similar to Fluid friction but differs in that it is a specific type of fluid friction caused by the internal resistance within a fluid due to its viscosity.

Key Functions of Viscous Friction in Lubrication

- **Regulates lubricant film thickness and protection**:
  - Higher viscous friction ensures **stronger film formation**, reducing **metal-to-metal contact**.
  - Lower viscous friction improves **fluid flow but may lead to reduced film strength in high-load applications**.

- **Influences energy efficiency and power loss**:
    - Excessive viscous friction in thick oils **increases drag**, **reducing system efficiency**.
    - Optimal viscosity selection **balances protection with minimal energy loss**.
- **Determines heat generation in lubricated systems**:
    - High-viscosity lubricants cause **more internal friction**, **generating excess heat**.
    - Lower viscosity lubricants flow more easily, **reducing heat buildup** in high-speed applications.

Types of Viscous Friction in Lubrication

- **Hydrodynamic viscous friction**:
    - Occurs when a full lubricant film separates surfaces.
    - Found in bearings, journal surfaces, and high-speed rotating components.
- **Boundary viscous friction**:
    - Happens when lubricant films become thin or break down.
    - Common in slow-moving, high-load applications where additives reduce direct surface contact.
- **Elastohydrodynamic (EHD) viscous friction**:
    - Forms in rolling element bearings and gear contacts under high pressure.
    - Lubricant viscosity temporarily increases under extreme loads to provide protection.

945. Volatility

Volatility is the **tendency of a lubricant or fluid to evaporate under heat or pressure**. It determines how much oil evaporates over time, affecting lubricant stability, oil consumption, and overall performance. High-volatility lubricants tend to thin out, degrade faster, and leave behind deposits, while low-volatility oils maintain film thickness and provide long-lasting protection. Volatility is a key factor in engine oils, hydraulic fluids, and industrial lubricants, especially in high-temperature applications.

Key Functions of Volatility in Lubrication

- **Affects oil evaporation and consumption**:
  - High-volatility oils evaporate faster, leading to oil loss and increased top-ups.
  - Low-volatility lubricants retain viscosity and film strength longer.
- **Influences lubricant stability and performance**:
  - High volatility causes oxidation, sludge formation, and varnish buildup.
  - Low-volatility oils resist thermal breakdown and maintain cleaner engine components.
- **Determines high-temperature lubrication efficiency**:
  - Higher volatility oils thin out quickly at high temperatures, reducing protection.
  - Low-volatility oils remain stable, ensuring consistent lubrication in extreme conditions.

Measurement of Volatility

- **Noack volatility test (ASTM D5800)**:
  - Measures the **percentage of oil lost due to evaporation at 250 °C (482 °F) for 1 hour**.
  - Lower values indicate **better thermal stability and reduced oil loss**.
- **Thermogravimetric analysis (TGA)**:
  - Evaluates weight loss under **controlled heating conditions**.

Factors Affecting Volatility in Lubricants

- **Base oil type**: Synthetic oils (**PAO, esters**) have lower volatility than **conventional mineral oils**.
- **Operating temperature**: Higher temperatures accelerate **evaporation and degradation**.

- **Additives and formulation**: Antioxidants help **stabilize oils and reduce evaporation losses**.

Benefits of Low-volatility Lubricants

- **Reduces oil consumption and maintenance costs**: Minimizes **evaporation losses and top-ups**.
- **Prevents deposit formation and oil thickening**: Avoids **sludge buildup, varnish, and oxidation issues**.
- **Improves high-temperature performance**: Ensures **lubricant longevity in extreme operating conditions**.
- **Enhances equipment protection and efficiency**: Maintains **consistent film thickness and lubrication properties**.

946. Voltammetry

Voltammetry is an **electrochemical predictive analysis technique used to detect and measure antioxidant additives, degradation by-products, and contaminants in lubricants**. It works by applying a varying electrical potential to a lubricant sample and measuring the resulting current response, which provides insights into the lubricant's chemical composition and oxidative stability. This technique is valuable in oil condition monitoring, predictive maintenance, and lubricant formulation analysis.

Key Functions of Voltammetry in Lubrication Analysis

- **Measures antioxidant additive levels**:
  - Detects remaining antioxidant concentration in used oil.
  - Predicts oxidation stability and lubricant life expectancy.
- **Identifies oxidation and degradation by-products**:
  - Tracks the formation of organic acids, varnish precursors, and sludge-forming compounds.
  - Helps prevent premature oil failure and equipment wear.
- **Detects contaminants and wear metals**:
  - Identifies trace metals, water ingress, and foreign substances that can impact lubricant performance.

- ○ Provides early warning of lubricant contamination before system damage occurs.

How Voltammetry Works in Lubrication Testing

1. **Electrochemical cell setup**:
   - The lubricant sample is placed in **a test cell with electrodes**.
2. **Voltage application and current measurement**:
   - A controlled voltage is applied, and the **resulting current response is recorded**.
3. **Data interpretation**:
   - The current peaks correspond to **specific chemical species**, allowing quantification of **antioxidant levels, degradation markers, and contaminants**.

Applications of Voltammetry in Lubrication

- **Turbine and compressor oils**: Tracks **antioxidant depletion in long-life lubricants**.
- **Engine and hydraulic oils**: Monitors **oxidation levels and oil degradation trends**.
- **Transformer and insulating oils**: Assesses **electrochemical stability and contamination risks**.
- **Metalworking fluids**: Detects **chemical changes affecting lubrication performance**.

947. Volumetric Analysis

Volumetric analysis is a **quantitative chemical analysis method used to determine the concentration of specific components in lubricants, fuels, and industrial fluids**. It involves measuring the volume of a reactant solution needed to complete a chemical reaction with a known substance in the lubricant. This method is commonly used in oil condition monitoring, contamination detection, and lubricant formulation quality control.

Key Functions of Volumetric Analysis in Lubrication

- **Determines additive concentration**:
  - ○ Measures detergents, dispersants, antiwear agents, and oxidation inhibitors.

  - Ensures lubricants meet OEM and industry standards.
- **Detects contaminants and degradation by-products**:
  - Identifies acidic or basic compounds, water content, and oxidation residues.
  - Helps assess oil degradation and contamination sources.
- **Measures acid and base numbers (TAN/TBN)**:
  - Total acid number (TAN): Determines acid buildup due to oxidation.
  - Total base number (TBN): Evaluates detergent and alkaline reserve in engine oils.

How Volumetric Analysis Works

1. **Sample preparation**:
   - A precise volume of lubricant is extracted for testing.
2. **Titration process**:
   - A reactant (acid or base) is gradually added until a chemical reaction occurs.
3. **Endpoint detection and calculation**:
   - A color change, electrical signal, or pH shift indicates completion.
   - The amount of reactant used determines the concentration of the measured component.

Types of Volumetric Analysis Used in Lubrication

- **Acid-base titration**:
  - Measures **TAN (acid number) and TBN (base number)** for lubricant condition monitoring.
- **Karl Fischer titration**:
  - Determines **water contamination levels in oils and fuels**.
- **Redox titration**:
  - Identifies **oxidation stability and antioxidant depletion**.

- **Complexometric titration**:
  - Used for **metal analysis in wear debris monitoring**.

Applications of Volumetric Analysis in Lubrication

- **Engine and gear oils**: Assesses **acid buildup, additive depletion, and contamination**.
- **Hydraulic and turbine oils**: Detects **oxidation products and water ingress**.
- **Transformer and insulating oils**: Ensures **proper dielectric properties and contamination control**.
- **Industrial fluids and metalworking oils**: Monitors **chemical stability and performance over time**.

Factors Affecting Volumetric Analysis Accuracy

- **Sample handling and storage**: Contaminated samples **can skew results**.
- **Reagent concentration and purity**: Must use **properly calibrated solutions**.
- **Temperature and environmental conditions**: Affects **reaction rates and endpoint detection**.

See also "Titration."

W

948. W

The letter **"W" in lubrication** commonly refers to **winter-grade viscosity** in engine oils and lubricants, as defined by the **SAE J300 viscosity classification system**. It indicates the oil's cold-temperature performance, ensuring proper lubrication in low-temperature conditions.

Key Meanings of "W" in Lubrication

- **Winter viscosity rating (SAE J300 Standard)** – The **"W"** in oil viscosity grades (e.g., 5W-30 and 10W-40) refers to winter performance, meaning the oil's ability to flow at low temperatures. A lower number before "W" (e.g., 0W and 5W) indicates better cold-start flow properties.
- **Water resistance in lubricants** – **WR (water-resistant) grease**) denote resistance to water contamination or washout.
- **Wheel and rail lubrication (W/R Lubrication)** – In railway applications, **"W"** is associated with wheel flange lubrication, reducing friction and wear between train wheels and rails.

949. Water-based Lubricants

Water-based lubricants are **lubricating fluids that use water as a primary component instead of traditional oil-based formulations**. These lubricants provide effective cooling, low viscosity, and environmental benefits, making them ideal for high-temperature applications, metalworking, and environmentally sensitive industries. They are widely used in hydraulic systems, food-grade lubrication, and metal-cutting operations where heat dissipation and fire resistance are crucial.

Key Functions of Water-based Lubricants

- **Provides superior cooling and heat dissipation**:
  - Water's high thermal conductivity allows for better cooling in high-speed and high-friction applications.

DOI: 10.1201/9788743806608-970

  - Prevents overheating in metalworking and hydraulic systems.
- **Reduces environmental impact and toxicity**:
  - Water-based lubricants are often biodegradable, nontoxic, and eco-friendly.
  - Ideal for food-grade, medical, and environmentally regulated industries.
- **Enhances fire resistance and safety**:
  - Water-based lubricants are nonflammable, reducing the risk of fire hazards in high-heat environments.
  - Used in fire-resistant hydraulic systems and high-temperature processes.

Types of Water-based Lubricants

- **Emulsions (oil-in-water or water-in-oil)**:
  - Contains oil dispersed in water with stabilizing additives.
  - Used in metalworking fluids, hydraulic systems, and industrial cooling.
- **Water-soluble lubricants**:
  - Fully dissolves in water, forming a uniform lubricating film.
  - Found in textile, food-grade, and pharmaceutical lubrication applications.
- **Gelled or thickened water-based lubricants**:
  - Uses polymeric thickeners or additives to create a more viscous lubricant.
  - Applied in biodegradable greases and specialty industrial coatings.

Applications of Water-based Lubricants

- **Metalworking and cutting fluids**: Provides **cooling and lubrication for drilling, milling, and grinding operations**.
- **Hydraulic fluids (fire-resistant)**: Used in **steel mills, underground mining, and aerospace applications**.

- **Food-grade and medical lubrication**: Applied in **processing equipment where incidental food contact occurs**.
- **Paper and textile industry**: Reduces **friction in high-speed rollers and machinery**.

Benefits of Water-based Lubricants

- **Improves cooling and heat management**: Reduces **friction-related temperature increases**.
- **Enhances workplace safety**: Nonflammable and **low-toxicity formulation**.
- **Minimizes environmental impact**: Offers **biodegradability and eco-friendly disposal options**.
- **Reduces fire hazards in high-temperature Operations**: Preferred for **fire-resistant hydraulic fluids**.

950. Water Contamination (In Oil)

Water contamination in oil is the **presence of free, emulsified, or dissolved water in lubricants, hydraulic fluids, and industrial oils**. It is one of the most damaging contaminants, leading to lubricant degradation, corrosion, reduced lubrication effectiveness, and premature equipment failure. Water can enter the system through leaks, condensation, improper storage, or process contamination, making its detection and control critical for machinery reliability and lubricant longevity.

Key Effects of Water Contamination in Oil

- **Reduces lubrication performance and film strength**:
    - Water reduces oil viscosity, leading to thinner lubrication films and increased metal-to-metal contact.
    - Causes premature bearing failure, increased wear, and component damage (see Figure W1).
- **Accelerates oxidation and oil degradation**:
    - Water reacts with oil molecules, forming acids, sludge, and varnish deposits.
    - Leads to higher total acid number (TAN) and reduced lubricant lifespan.

- **Causes corrosion and rust formation**:
  - Water contamination promotes rust and pitting on metal surfaces.
  - Increases the risk of bearing and gearbox failure.
- **Forms emulsions that block oil flow and filters**:
  - Water mixes with oil, creating sticky emulsions that clog filters and narrow passages.
  - Leads to reduced lubrication efficiency and pressure fluctuations in hydraulic systems.
- **Encourages microbial growth and biological contamination**:
  - In industrial lubricants and cutting fluids, water contamination can support bacterial and fungal growth.
  - Causes foul odors, sludge formation, and operational issues.

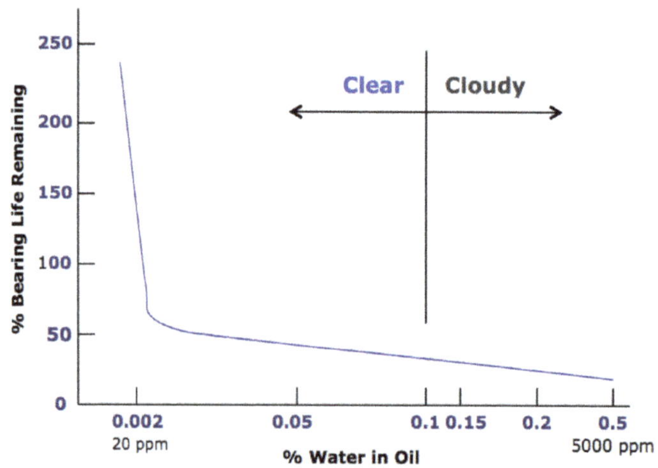

Figure W1 Graph depicting how water contamination impacts bearing life.

Types of Water Contamination in Oil (See Figure W2)

1. **Dissolved water**:
   - Water molecules **are fully mixed within the oil**, appearing **clear**.
   - Difficult to detect without specialized testing.

2. **Emulsified water**:
    - Water droplets are **suspended in the oil**, creating a **cloudy or milky appearance**.
    - Causes lubrication instability and **filter clogging**.
3. **Free water**:
    - Water **separates from the oil**, forming a **visible layer at the bottom of reservoirs or sumps**.
    - The most damaging form, leading to **rapid corrosion and microbial growth**.

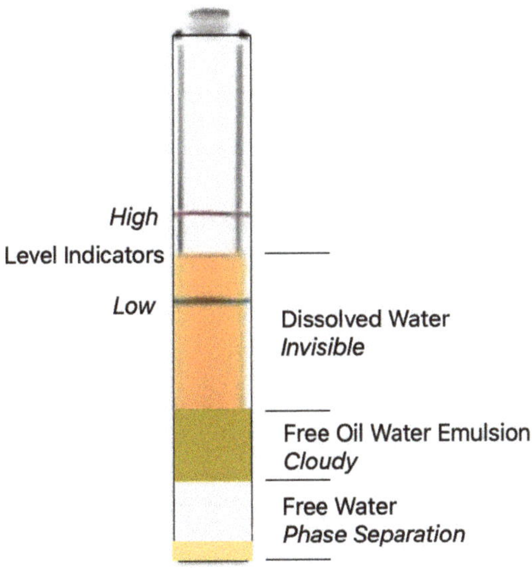

Figure W2 Oil reservoir sight glass showing all three stages of water-contaminated oil. Courtesy: ENGTECH Industries Inc.

Detecting Water Contamination in Oil

- **Crackle test**: A drop of oil is heated on a hot plate; popping or bubbling indicates the presence of water.
- **Karl Fischer titration (ASTM D6304)**: The most accurate method for measuring water content in ppm (parts per million).
- **FTIR spectroscopy (Fourier transform infrared analysis)**: Detects water molecules in oil through infrared absorption peaks.

- **Visual inspection**: Cloudy, milky, or foamy oil suggests emulsified water contamination.

Sources of Water Contamination in Lubricants

- **Seal and gasket leaks**: Water enters through damaged seals in pumps, gearboxes, and hydraulic systems.
- **Condensation in storage tanks and reservoirs**: Temperature changes cause moisture buildup in oil systems.
- **Process and environmental exposure**: Open reservoirs or water-cooled systems introduce water into lubricants.
- **Faulty oil handling and storage**: Poorly sealed containers absorb moisture from the atmosphere.

Methods to Remove Water Contamination from Oil

- **Vacuum dehydration**: Uses low pressure and heat to remove dissolved and free water.
- **Centrifugation and coalescing filters**: Separates water from oil using high-speed spinning or special filter membranes.
- **Desiccant breathers and sealed reservoirs**: Prevents moisture ingress by filtering out airborne water vapor.
- **Water-absorbing filter elements**: Designed to trap and remove free and emulsified water.

951. Water-resistant Lubricant

Water-resistant lubricants refers to a lubricant with the **ability to maintain its protective properties, viscosity, and stability when exposed to water**. Water exposure is a major challenge in industrial environments such as steel mills, marine applications, paper mills, mining, and food processing, where moisture, steam, and direct water contact can lead to lubricant degradation, corrosion, and equipment failure. Lubricants with high water resistance repel moisture, prevent emulsification, and maintain effective lubrication in wet conditions.

Key Functions of Water-resistant Lubricants

- **Prevents emulsification and lubricant breakdown**:
  - Water-resistant lubricants do not mix with water, maintaining their viscosity and lubricating properties.

  - Reduces sludge, foaming, and oxidation issues caused by water ingress.
- **Protects against rust and corrosion**:
  - Forms a hydrophobic (water-repelling) barrier on metal surfaces, preventing rust and pitting.
  - Essential in marine, offshore, and humid industrial environments.
- **Maintains load-carrying and antiwear properties in wet conditions**:
  - Water-resistant greases and oils ensure adequate lubrication even when exposed to moisture.
  - Reduces friction and wear in submerged or high-humidity applications.

Types of Water-resistant Lubricants

- **Hydrophobic (water-repellent) lubricants**:
  - Formulated with additives and base stocks that repel water.
  - Used in marine, offshore, and heavy-duty industrial applications.
- **Greases with high water washout resistance**:
  - Thickened with calcium sulfonate, aluminum complex, or lithium complex for superior water resistance.
  - Used in food-grade lubrication, bearings, and outdoor equipment.
- **Synthetic and ester-based oils with water separation properties**:
  - Designed to shed water quickly (demulsibility) rather than forming emulsions.
  - Used in hydraulic systems, gear oils, and turbine lubricants.

Applications of Water-resistant Lubricants

- **Steel mills and paper mills**: Prevents **lubricant washout in high-humidity and water-prone environments**.
- **Marine and offshore equipment**: Protects **bearings, gears, and seals in saltwater conditions**.
- **Mining and heavy machinery**: Ensures **lubrication in wet, muddy, and high-pressure applications**.

- **Food processing plants**: Used in **H1 food-grade lubricants to prevent contamination in washdown areas**.

Factors Affecting Water Resistance in Lubrication

- **Base oil type and additives**: Synthetic esters and **PAO-based lubricants** offer **higher water resistance** than conventional mineral oils.
- **Thickener type in greases**: **Calcium sulfonate and aluminum complex greases** provide **better water washout resistance** than lithium greases.
- **Operating conditions and water exposure**: Prolonged water exposure requires lubricants with **enhanced corrosion inhibitors and demulsifying properties**.

952. Wax

Wax in lubrication refers to **natural or synthetic waxes used as lubricant additives or as an undesirable by-product of petroleum-based lubricants**. Waxes influence lubricant flow properties, viscosity stability, and cold-temperature performance. While some waxes are intentionally added to enhance lubrication in specialized applications, others, such as paraffinic waxes, can negatively affect lubricant performance, particularly at low temperatures.

Key Functions of Wax in Lubrication

- **Improves lubrication and surface protection**:
  - Some waxes, such as polyethylene wax or PTFE (polytetrafluoroethylene), enhance friction reduction.
  - Used in dry-film lubricants, chain lubricants, and high-temperature applications.
- **Affects low-temperature flow properties**:
  - Waxy deposits in mineral-based lubricants can cause thickening and poor flow at cold temperatures.
  - Lubricants must be treated with pour point depressants (PPDs) to reduce wax crystallization.
- **Enhances wear and corrosion protection**:
  - Waxes can form a protective barrier on surfaces, preventing moisture exposure and oxidation.

  - Used in automotive wax-based lubricants, coatings, and corrosion inhibitors.

Types of Wax in Lubrication

- **Paraffinic wax (petroleum-based wax)**:
  - A by-product in mineral oil refining that can cause oil thickening at low temperatures.
  - Requires pour point depressants to prevent crystallization in cold-weather lubricants.
- **Synthetic waxes (polyethylene and PTFE waxes)**:
  - Used as lubricant additives for friction reduction and wear protection.
  - Common in greases, chain lubricants, and dry-film coatings.
- **Microcrystalline wax**:
  - A more refined petroleum wax with higher viscosity and better film-forming properties.
  - Used in sealants, anticorrosion coatings, and extreme-pressure lubricants.

Applications of Wax in Lubrication

- **Dry film lubricants and chain lubes**: Forms **a protective, low-friction layer for high-dust environments**.
- **Greases and industrial lubricants**: Used in **high-viscosity greases for heavy-load applications**.
- **Automotive and aerospace lubrication**: PTFE-based wax lubricants **reduce wear in high-speed components**.
- **Rust and corrosion prevention coatings**: Wax-based films **protect metal surfaces from oxidation**.

Factors Affecting Wax Behavior in Lubricants

- **Temperature sensitivity**: Waxes solidify at low temperatures, causing **oil thickening and filter blockages**.
- **Solubility and compatibility**: Must be carefully blended to **prevent wax precipitation in lubricants**.

- **Lubricant refining process**: Highly refined oils have **less wax content**, improving **cold-flow properties**.

Conclusion: Wax in lubrication can be beneficial when used as an additive for wear protection, dry lubrication, and corrosion prevention. However, uncontrolled wax content in lubricants, particularly in mineral oils, can negatively impact cold-temperature performance. By selecting the right wax-based lubricant or refining process, industries can optimize lubricant flow properties, surface protection, and long-term equipment reliability.

953. Way (machine)

A **machine way** is a precision-engineered guide or sliding surface that facilitates controlled linear motion between machine components, ensuring stability, accuracy, and minimal friction. Commonly found in lathes, milling machines, precision CNC equipment, and straight-side stamping presses, machine ways provide support and alignment for moving parts, such as carriages, tables, and slides. They are typically hardened and ground surfaces made from materials like steel or cast iron, often coated or lubricated to reduce wear, minimize stick-slip motion, and enhance smooth operation.

Proper lubrication with way oils (ISO 68 or 220) or polymer-based coatings prevents metal-to-metal contact, improves load distribution, and maintains precision in high-load, slow-speed applications.

954. Way Oil

Way Oil, also known as **slideway lubricant**, is a **specialized lubricant designed for machine ways, guides, and sliding surfaces in industrial equipment**. It provides high film strength, antistick-slip properties, and wear protection, ensuring smooth, precise movement in metalworking machines such as lathes, milling machines, and CNC equipment. Way oils are formulated with tackiness additives to prevent oil runoff, maintaining a consistent lubrication film under high loads and slow speeds.

Key Functions of Way Oil in Lubrication

- **Reduces stick-slip motion and improves precision**:
    - Prevents jerky or uneven motion that can affect machining accuracy.
    - Ensures smooth, controlled sliding movement of machine tables and carriages.

- **Provides high film strength and load-carrying capability**:
  - Forms a durable lubricating film to support heavy loads.
  - Prevents metal-to-metal contact and excessive wear on way surfaces.
- **Resists washout from coolants and cutting fluids**:
  - Tackiness additives keep oil in place, even in coolant-heavy environments.
  - Ensures continuous protection despite fluid exposure.
- **Minimizes friction and surface Wear**:
  - Reduces scraping and galling on precision-machined surfaces.
  - Extends the lifespan of machine ways, bearings, and guides.

Common way oil grades and specifications

- **ISO VG 68**: Used in light-to-medium load applications, such as smaller machine tools.
- **ISO VG 220**: Designed for heavier-duty machinery and larger sliding surfaces.

Applications of Way Oil

- **CNC machines and milling machines**: **Ensures smooth table and saddle movement**.
- **Lathes and grinders**: Prevents stick-slip motion in carriage and cross-slide mechanisms.
- **Injection molding machines and presses**: Protects guide rails and linear bearings.
- **Heavy-duty industrial equipment**: Lubricates large, slow-moving ways and slides.

955. Wear

Wear is the gradual removal or deformation of material from a surface due to mechanical interaction with another surface, fluid, or environmental factor. Wear occurs in bearings, gears, hydraulic systems, and rotating machinery, affecting performance, efficiency, and component lifespan. Wear can be caused by friction, impact, corrosion, or contamination, and proper lubrication is essential to minimize wear, reduce energy losses, and extend equipment life.

956. Wear Control

Wear control refers to **strategies and technologies used to minimize material loss and extend the lifespan of mechanical components**. Effective wear control is essential in bearings, gears, hydraulic systems, engines, and rotating machinery, where excessive wear leads to reduced efficiency, increased friction, overheating, and equipment failure. Proper lubrication, material selection, and contamination management are key factors in reducing wear and improving reliability.

Key Functions of Wear Control in Lubrication

- **Prevents metal-to-metal contact**:
    - Lubricants create a protective film that separates moving surfaces.
    - Reduces friction, heat generation, and surface degradation.
- **Minimizes abrasive and adhesive wear**:
    - Keeps contaminants and debris suspended or removed through filtration.
    - Reduces material transfer and surface damage in high-load applications.
- **Enhances load-carrying and antiwear properties**:
    - Additives like antiwear (AW) and extreme pressure (EP) agents chemically react to form protective layers.
    - Ensures long-term component durability under varying speeds and loads.

Methods of Wear Control

- **Proper lubricant selection**:
    - Choosing the correct viscosity and base oil type ensures optimal film thickness and lubrication efficiency.
    - Synthetic oils provide better wear protection in extreme temperatures and high-load applications.

- **Use of antiwear and extreme pressure additives**:
  - AW additives (ZDDP, phosphorus-based compounds) form protective coatings to prevent surface wear.
  - EP additives (sulfur–phosphorus, molybdenum-based compounds) protect against shock loads and high-pressure conditions.
- **Filtration and contaminant control**:
  - High-efficiency filters remove wear particles, dirt, and water contamination.
  - Clean oil ensures longer lubricant life and improved wear resistance.
- **Surface treatments and coatings**:
  - Plating (chrome, nickel, or nitriding) increases surface hardness.
  - Dry film lubricants (PTFE, molybdenum disulfide) reduce friction in extreme environments.
- **Monitoring and predictive maintenance**:
  - Oil analysis and wear particle testing detect early signs of wear.
  - Condition monitoring sensors track lubrication effectiveness and surface degradation.

957. Wear Debris Density

Wear debris density refers to the **concentration and mass of wear particles present in a lubricant or oil sample, typically measured in milligrams per liter (mg/L) or parts per million (ppm)**. It is a critical indicator of machine health, lubricant effectiveness, and component wear severity. High wear debris density suggests excessive surface degradation, contamination, or impending failure, making it a key parameter in condition monitoring and predictive maintenance programs.

Key Functions of Wear Debris Density in Lubrication Analysis

- **Detects early signs of equipment wear and failure**:
  - High-density readings indicate abnormal wear in gears, bearings, and hydraulic systems.
  - Helps prevent catastrophic breakdowns by identifying wear trends early.

- **Evaluates lubricant performance and contaminant levels**:
    - Monitors how well a lubricant protects against friction and load-induced wear.
    - Identifies contaminants such as dirt, water, and oxidation byproducts that accelerate wear.
- **Classifies wear modes and sources of damage**:
    - Distinguishes between normal operating wear and excessive component deterioration.
    - Identifies specific wear mechanisms such as abrasive, adhesive, fatigue, or corrosive wear.

Measurement of Wear Debris Density

- **Analytical ferrography**: Separates and analyzes ferrous (iron-based) and nonferrous particles to determine wear severity.
- **Particle count analysis (ISO 4406/NAS 1638)**: Measures solid particle concentration in a lubricant sample.
- **X-ray fluorescence (XRF) spectroscopy**: Determines **elemental composition** of wear debris to identify specific alloy components.
- **Gravimetric analysis**: Filters out **solid contaminants and measures their total weight per liter of oil**.

Benefits of Monitoring Wear Debris Density

- **Prevents unexpected equipment failures and costly repairs**: Detects abnormal wear before damage escalates.
- **Optimizes lubricant change intervals and maintenance schedules**: Reduces unnecessary downtime and oil wastage.
- **Improves machine reliability and efficiency**: Ensures long-term equipment performance.
- **Enhances predictive maintenance strategies**: Enables data-driven decision-making for proactive maintenance.

958. Wear Inhibitor

See "Antiwear Additive."

959. Wear Mechanisms

There are four common wear mechanisms that cause surface degradation and eventual loss of usefulness in bearing surfaces.

See "Abrasive Wear," "Adhesive Wear," "Fatigue Wear," and "Corrosive Wear."

960. Wear Particle Analysis (WPA)

Wear particle analysis (WPA) is a condition monitoring **technique used to assess machine health by examining wear debris** in lubricants, hydraulic fluids, and greases. It provides insights into component wear rates, failure modes, and contamination levels, allowing for early detection of abnormal wear and predictive maintenance planning. WPA is essential in industrial machinery, automotive engines, turbines, and heavy-duty equipment, where lubrication health directly affects performance and longevity.

For more detailed information, see "Wear Debris Density."

961. Wear Rate

Wear rate is the rate at which material is **lost from a surface due to friction, abrasion, adhesion, or fatigue in a lubricated system**. It is a critical parameter in predictive maintenance, lubrication selection, and equipment longevity analysis. Wear rate is typically measured in volume loss per unit time (e.g., mm^3/hour) or mass loss per cycle and is influenced by lubrication effectiveness, operating conditions, and material properties.

Key Factors Affecting Wear Rate

- **Lubricant film thickness and stability** – A well-maintained oil or grease film reduces direct metal-to-metal contact, lowering wear rate.
- **Lubricant additives and chemistry** – Antiwear (AW), extreme pressure (EP), and boundary lubrication additives help protect surfaces under extreme conditions.
- **Operating load and speed** – High loads increase contact stress, while high speeds can affect lubricant film formation.
- **Contaminants and debris in lubricant** – Dirt, metal particles, and water accelerate wear by acting as abrasive elements.
- **Temperature and oxidation stability** – High temperatures degrade lubricants, reducing their ability to prevent friction and wear.

Methods for Measuring Wear Rate

1. **Wear particle analysis (WPA)**: Examines the size, shape, and composition of wear debris in lubricant samples.
2. **Gravimetric wear measurement**: Weigh components before and after testing to determine mass loss.
3. **Pin-on-disk test (ASTM G99)**: Simulates wear conditions between lubricated sliding surfaces.
4. **Four-ball wear test (ASTM D4172)**: Measures wear scar diameter under controlled lubrication conditions.
5. **Fretting wear analysis**: Evaluates wear caused by micromotion under lubricated contact points

962. Wedge Film Lubrication

Wedge film lubrication is a type of **hydrodynamic lubrication** in which a **tapered or wedge-shaped oil film** forms between two moving surfaces, generating a load-bearing pressure that prevents metal-to-metal contact. This lubrication mechanism is crucial in bearings, gears, and rotating machinery, where it helps reduce friction, minimize wear, and enhance efficiency.

Key Principles of Wedge Film Lubrication

- **Formation of a lubricant wedge**: As one surface moves relative to another, lubricant is drawn into the converging gap, forming a pressure gradient. See Figure W3.
- **Load-carrying capacity**: The pressure generated in the lubricant wedge separates the surfaces, preventing adhesive wear.
- **Self-sustaining lubrication**: The motion of the surfaces continuously replenishes the lubricant film, ensuring consistent protection.
- **Viscosity and speed dependency**: Higher viscosity lubricants and increased relative speed enhance the film thickness and stability.

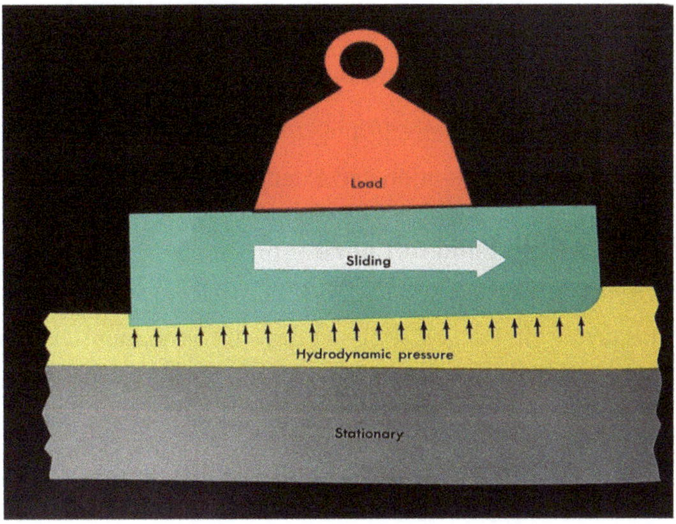

Figure W3 Lubricant wedge formed in hydrodynamic lubrication. Source: Plant Lubrication for Industrial Facilities – Fourth Edition Kenneth E. Bannister.

963. Weld Point

Weld point refers to the load at which a lubricant fails to prevent metal-to-metal contact, causing surfaces to seize or weld together due to extreme pressure and friction. A higher weld point indicates better load-carrying capacity, making it crucial for gears, bearings, and heavy-duty applications where high pressures and shock loads occur.

Key Functions of Weld Point in Lubrication

- **Determines lubricant's extreme pressure performance**:
  - Measures the maximum load a lubricant can withstand before failure.
  - Essential for selecting EP oils and greases for high-load applications.
- **Prevents metal-to-metal contact and seizure**:
  - Ensures lubricant film integrity under heavy loads.

  - Reduces scuffing, galling, and material transfer between surfaces.
- **Enhances gear, bearing, and industrial equipment protection**:
  - Helps in preventing catastrophic failures in high-load components.
  - Ensures smooth operation and extended component life.

Measurement of Weld Point in Lubricants

- **Four-ball EP test (ASTM D2783)**:
  - Measures weld load by applying force on a rotating steel ball against three fixed balls in lubricant.
  - The weld point is the load at which the balls weld together due to lubricant failure.
- **Load wear index (LWI)**:
  - Indicates a lubricant's **ability to prevent wear under increasing loads** before reaching the weld point.
- **Timken OK load test (ASTM D2509)**:
  - Determines **the maximum load a lubricant can handle before scuffing occurs** in a rotating contact environment.

Conclusion: Weld point is a critical measure of a lubricant's ability to withstand extreme pressure, preventing metal surfaces from seizing or welding together. By selecting high-weld-point lubricants with EP additives, industries can protect gears, bearings, and high-load machinery from failure, ensuring long-term reliability and efficiency.

964. Wet Friction

See "Fluid Friction."

965. Wetting Agent

A wetting agent is a **surface-active additive that reduces surface tension, allowing lubricants to spread more effectively over metal surfaces**. Wetting agents are designed to improve lubricant adhesion, penetration, and coverage, ensure better film formation, reduce friction, and enhance cooling in industrial and automotive applications. They are commonly used in metalworking fluids, cutting oils, greases, and industrial lubricants where surface coverage and fluid distribution are critical.

Key Functions of Wetting Agents in Lubrication

- **Enhances lubricant spread and surface coverage**: Improves **fluid distribution across metal surfaces**, reducing dry spots and uneven lubrication.
- **Reduces surface tension for better penetration**: Allows lubricants to **reach tight clearances, bearings, and moving parts**, improving wear protection.
- **Prevents film breakdown and Improves adhesion**: Ensures the lubricant **remains on surfaces under load and high-speed operations**.

Types of Wetting Agents Used in Lubricants

- **Surfactants (surface-active agents)**: Modify fluid properties to **reduce surface tension and improve spreading**.
- **Polar additives**: Contain **polar molecules that bond to metal surfaces**, enhancing **lubrication film retention**.
- **Emulsifiers**: Used in **water-based lubricants and metalworking fluids** to **improve oil–water interaction** for cooling and lubrication.

966. Whale Oil

Whale oil is a **natural lubricant derived from the blubber of whales**, historically used in **industrial machinery**, **automotive applications, and precision instruments** due to its **high lubricity, oxidation resistance, and low freezing point**. The most prized whale for lubricants was the Sperm whale, (see Figure W4) which carried an enormous amount of high-quality oily wax product named Spermaceti in its head cavity causing it to be almost fished out of existence until whaling controls and petroleum based lubricants gave the whale a new lease on life. See Figure W4. Before the development of synthetic and petroleum-based lubricants, whale oil was widely utilized for high-performance lubrication, textile machinery, and early automotive engines. However, due to environmental regulations and the ban on commercial whaling, whale oil has been replaced by synthetic and plant-based alternatives.

Key Functions of Whale Oil in Lubrication

- **Provides superior lubrication and film strength**: Forms a **durable oil film**, reducing friction and wear in **high-load applications**.
- **Resists oxidation and degradation**: Naturally contains **long-chain fatty acids and esters**, preventing **varnish and sludge formation**.
- **Performs well in extreme temperatures**: Offers **low pour points**, ensuring fluidity in **cold environments**, and maintains **stability at high temperatures**.

Historical Uses of Whale Oil in Lubrication

- **Industrial machinery and textile equipment**: Used in **spinning mills, looms, and sewing machines** for smooth operation.
- **Automotive and steam engines**: Applied as **gear and cylinder oils** in early **automobiles, locomotives, and marine engines**. From the 1940s through the early 1970s, automatic transmission fluid (ATF) used whale oil as a rust and corrosion inhibitor.
- **Precision instruments and clocks**: Served as **a stable lubricant for fine mechanical components**.
- **Military and aerospace applications**: Used in **aircraft hydraulic systems and lubricants during World War II**.

Replacement of Whale Oil and Modern Alternatives

- **Synthetic lubricants (PAO, Esters)**: Provide **better oxidation stability and wear protection**.
- **Vegetable-based lubricants (Rapeseed, Castor Oil)**: Offer **biodegradability and similar lubricity properties**.
- **Petroleum-based lubricants**: Became **widely available and more economical** after the mid-20th century.

Factors That Led to Whale Oil Phase-out

- **Environmental and ethical concerns**: Commercial whaling was banned due to endangered species protection laws.
- **Advancements in lubricant technology**: Modern lubricants outperform whale oil in extreme pressure, oxidation resistance, and longevity.
- **Cost and availability**: Synthetic and petroleum-based oils became more accessible and cost-effective.

Figure W4 Sperm whales were especially prized for the high-quality "Spermaceti" oil contained in their heads. Source: National Wildlife Federation.

967. Wheel Bearing Grease

Wheel bearing grease is a high-performance lubricant designed to protect wheel bearings from friction, wear, heat, and contamination. It ensures smooth operation, extends bearing life, and prevents failures in automotive, industrial, and heavy-duty applications.

Key Functions of Wheel Bearing Grease

- **Reduces friction and wear**: Forms a protective lubricating film that prevents metal-to-metal contact.
- **Withstands high temperatures**: Designed to endure heat buildup from braking and wheel rotation.
- **Prevents water and contaminant ingress**: Resists water washout, dirt, and debris, protecting bearings in harsh environments.
- **Maintains stability under load**: Provides consistent lubrication under heavy loads and varying speeds.
- **Prevents rust and corrosion**: Contains anticorrosion and oxidation inhibitors to protect metal surfaces.

Types of Wheel Bearing Grease

- **Lithium complex grease**: Commonly used for automotive and industrial bearings, offering high-temperature stability and water resistance.

- **Calcium sulfonate grease**: Provides exceptional water resistance, corrosion protection, and high-load performance, ideal for marine and heavy equipment applications.
- **Polyurea grease**: Designed for sealed-for-life bearings, offering long-lasting lubrication with high thermal stability.
- **Molybdenum disulfide (MoS_2) Grease**: Used in heavy-duty and off-road applications, reducing wear in high-load and shock conditions.
- **Synthetic grease**: Offers superior performance in extreme temperatures and high-speed applications, ideal for racing, aerospace, and industrial machinery.

968. Wheel Flange Lubrication

Wheel flange lubrication is a **friction-reducing technique used in rail transportation and industrial applications to minimize wear, reduce energy consumption, and extend wheel and rail life**. It involves the application of specialized lubricants or friction modifiers to the wheel flange-rail interface, where high contact stress, lateral forces, and metal-to-metal wear occur. Proper lubrication enhances operational efficiency, prevents excessive rail wear, and reduces noise in curved track sections and high-load applications.

Key Functions of Wheel Flange Lubrication

- **Reduces friction and wear at the wheel-rail interface**: Forms a thin lubricating film, preventing excessive metal loss and surface damage.
- **Extends wheel and rail life**: Minimizes rolling contact fatigue and surface pitting, ensuring longer service life for rail components.
- **Decreases energy consumption and fuel costs**: Reduces rolling resistance and traction force, improving energy efficiency.
- **Lowers noise and vibration in curves**: Prevents wheel squeal and excessive rail contact noise, enhancing passenger comfort and workplace safety.

Types of Wheel Flange Lubrication Methods

- **On-board wheel flange lubrication**: Uses automatic lubricators mounted on the train, applying lubricant directly to the wheel flange during operation. See "Figure W5."

Figure W5 Bijur–Delimon mobile "Railjet" wheel flange lubrication unit. Courtesy: Bijur–Delimon International.

- **Wayside rail lubrication systems**: Fixed trackside applicators dispense lubricant onto rails at strategic locations, typically in curves and high-wear zones.
- **Manual lubrication**: Performed periodically by maintenance crews applying grease or solid lubricants to critical track sections.

Common Lubricants Used in Wheel Flange Lubrication

- **Solid stick lubricants**: Applied to **the wheel flange**, offering **consistent friction reduction** without excessive spread.
- **Graphite and molybdenum-based lubricants**: Provide **high-temperature stability and low wear rates**.
- **Synthetic and biodegradable lubricants**: Used in **environmentally sensitive areas** to minimize environmental impact.
- **Dry film coatings and friction modifiers**: Designed to **reduce wear while maintaining adequate traction** in railway applications.

Applications of Wheel Flange Lubrication

- **Railways and metro systems**: Reduces **track and wheel wear in curves, switches, and crossings**.
- **Heavy haul and freight rail operations**: Protects **high-load train wheels from excessive wear and metal fatigue**.

- **Industrial track systems (mines, ports, and steel plants)**: Ensures **smooth movement of rail-mounted equipment** under extreme conditions.
- **Transit and high-speed rail networks**: Improves **ride comfort and reduces maintenance costs**.

969. WHIMIS

WHMIS (Workplace Hazardous Materials Information System) is **Canada's national hazard communication standard** designed to ensure workers receive the necessary information on hazardous materials used in the workplace. It provides a consistent system for classifying chemicals, labeling hazardous products, and communicating safety precautions. WHMIS aligns with the **globally harmonized system (GHS)** for chemical classification and labeling, ensuring workplace safety and regulatory compliance.

Key Functions of WHMIS in Workplace Safety

- **Standardizes hazard communication**: Ensures consistent classification, labeling, and safety documentation for hazardous products.
- **Protects workers from chemical hazards**: Provides information on safe handling, storage, and emergency response procedures.
- **Enhances workplace safety training**: Employers must train workers on WHMIS symbols, safety data sheets (SDS), and proper chemical handling.

WHMIS Hazard Classes and Categories

WHMIS classifies hazardous materials into **physical, health, and environmental hazards**, with pictograms indicating specific risks:

- **Physical hazards**: Flammable, explosive, oxidizing, corrosive, compressed gases.
- **Health hazards**: Toxic, carcinogenic, respiratory irritants, skin sensitizers.
- **Environmental hazards**: Aquatic toxicity, long-term environmental effects.

Key WHMIS Components

- **Labels and pictograms**: Products must have hazard labels with WHMIS symbols, signal words, and precautionary statements.
- **Safety data sheets (SDS)**: Detailed 16-section documents providing hazard information, handling instructions, and emergency procedures.
- **Worker education and training**: Employers must train employees to identify hazards, read labels, and follow safety protocols.

Applications of WHMIS in Industry

- **Manufacturing and industrial facilities**: Ensures **safe handling of lubricants, chemicals, and process fluids**.
- **Oil and gas sector**: Regulates **chemical storage, transportation, and spill response procedures**.
- **Laboratories and research centers**: Provides **clear labeling and safety measures for hazardous substances**.
- **Construction and automotive services**: Ensures **worker safety when using paints, solvents, and fuel additives**.

970. White Metal Bearing

See "Babbitt."

971. White Room

See "Clean Room."

972. Wick Lubricator

A **wick lubricator** is a **passive lubrication system that uses a capillary-action wick to deliver a controlled amount of lubricant to moving machine parts**. It ensures consistent, automatic lubrication by allowing oil to flow gradually and continuously to bearings, shafts, and other frictional surfaces. Wick lubricators are commonly used in textile machinery, clocks, steam engines (see Figure W6), and light-duty industrial equipment, to reduce wear and extend component life.

Figure W6 Multiple "Wick" lubricators employed on a ship steam engine main bearings. Courtesy: ENG**TECH** Industries Inc.

Key Functions of a Wick Lubricator

- **Provides continuous and controlled lubrication**: The wick draws oil from a reservoir and releases it slowly to the lubrication point, preventing overlubrication.
- **Minimizes friction and wear in moving components**: Maintains a thin oil film on surfaces, reducing metal-to-metal contact and heat generation.
- **Prevents oil waste and contamination**: Uses capillary action to regulate flow, avoiding excess oil loss or dripping.

Components of a Wick Lubricator

- **Oil reservoir**: Stores **lubricating oil**, ensuring a continuous supply.
- **Wick (cotton, felt, or synthetic fibers)**: Draws oil upward through **capillary action**, controlling lubrication rate.
- **Lubrication point**: Directs **oil flow to the bearing, shaft, or moving part needing lubrication**.

- **Cover or housing**: Protects **the wick and oil from contamination or evaporation**.

Factors Affecting Wick Lubricator Performance

- **Wick material and capillary action efficiency**: Determines oil flow rate and consistency.
- **Viscosity of lubricating oil**: Higher viscosity oils flow slower, affecting the lubrication rate.
- **Operating temperature and environment**: Extreme temperatures can dry out wicks or alter oil flow behavior.
- **Contaminant protection**: Proper sealing ensures dust and debris do not clog the wick.

973. Wicking

Wicking refers to the **capillary action by which a lubricant is drawn through a wick material to deliver controlled, continuous lubrication to a surface or component**. It relies on fluid movement through porous or fibrous materials, such as felt, cotton, or synthetic fibers, allowing oil to reach bearings, shafts, and frictional surfaces without the need for pumps or external pressure. Wicking is widely used in wick lubricators, self-lubricating bearings, and automated lubrication systems, ensuring precise and consistent lubrication in light-duty industrial applications.

Key Functions of Wicking in Lubrication

- **Provides continuous and controlled lubricant flow**: Delivers oil gradually and steadily to lubrication points, preventing overlubrication.
- **Enhances wear protection and reduces friction**: Forms a thin, even film of lubricant, minimizing metal-to-metal contact and heat generation.
- **Improves lubrication efficiency in confined spaces**: Ensures oil reaches hard-to-access areas, such as small bearings and precision components.

Common Wick Materials Used in Lubrication

- **Cotton and wool wicks**: Traditional materials with good absorption properties for light-duty applications.
- **Felt wicks**: Used in bearings, gears, and textile machinery due to their durability and stable capillary action.

- **Synthetic fiber wicks**: Offer higher resistance to temperature variations and chemical exposure.
- **Sintered metal wicks**: Found in self-lubricating bearings, providing long-term oil retention and controlled release.

974. Wiper Seal

See "Radial Lip Seal."

975. Wire Rope Lubricant

Wire rope lubricant is a **specialized lubricant designed to protect wire ropes from wear, corrosion, and fatigue while ensuring smooth operation**. It penetrates between the strands and around the exterior surface to reduce friction, prevent rust, and enhance flexibility. Proper lubrication is essential for extending wire rope service life in cranes, elevators, mining equipment, offshore rigs, and industrial hoisting systems.

Key Functions of Wire Rope Lubricant

- **Reduces friction between wire strands**: Minimizes internal wear and strand fatigue, preventing premature failure.
- **Provides corrosion and rust protection**: Forms a moisture-resistant barrier, shielding against saltwater, chemicals, and harsh weather conditions.
- **Maintains wire rope flexibility and strength**: Prevents brittleness and cracking, ensuring safe and efficient operation.
- **Repels contaminants and debris**: Protects against dust, dirt, and abrasive particles that accelerate wear.

Types of Wire Rope Lubricants

- **Penetrating lubricants**: Designed to **seep into the core of the rope**, ensuring **internal strand lubrication**. Ideal for **high-flex applications**.
- **Coating lubricants**: Forms a **protective film on the rope's surface**, shielding against **external contaminants and corrosion**.
- **Biodegradable wire rope lubricants**: Eco-friendly alternatives for **marine, forestry, and environmentally sensitive areas**.

Applications of Wire Rope Lubricants

- **Cranes and hoists**: Reduces **wear in lifting cables subjected to high loads and cyclic bending**.
- **Mining and offshore equipment**: Protects against **saltwater, extreme temperatures, and heavy load stress**.
- **Elevators and cable cars**: Ensures **smooth, reliable movement with reduced noise and vibration**.
- **Bridges and suspension systems**: Prevents **wire fatigue and corrosion in exposed environments**.

Factors Affecting Wire Rope Lubrication Performance

- **Lubricant penetration and adhesion**: Must reach internal strands while resisting washout or runoff.
- **Operating environment**: Requires water-resistant or extreme-pressure formulations for harsh conditions.
- **Lubrication frequency and application method**: **Regular reapplication ensures consistent protection and performance**.

976. Wire Rope Pressure Lubrication System

A **wire rope pressure lubrication system** is a **specialized lubrication method that applies high-pressure lubricant to wire ropes, ensuring deep penetration and full coverage**. Unlike manual or drip lubrication, pressure systems force lubricant into the core and between the strands, reducing internal friction, wear, and corrosion. These systems are essential for heavy-duty applications in cranes, mining, offshore drilling, elevators, and industrial lifting operations, where wire ropes are exposed to high loads, bending fatigue, and harsh environmental conditions.

Key Functions of a Wire Rope Pressure Lubrication System

- **Ensures complete lubricant penetration**: Uses **high-pressure application to reach the inner core of the rope**, improving strand lubrication.
- **Reduces friction and extends wire rope life**: Minimizes **metal-to-metal contact between strands**, preventing premature wear.
- **Provides superior corrosion protection**: Forms a **moisture-resistant barrier** against **water, salt, dust, and chemicals**.

- **Improves lubrication efficiency and reduces waste**: Delivers **precise lubricant amounts**, preventing overapplication and contamination.

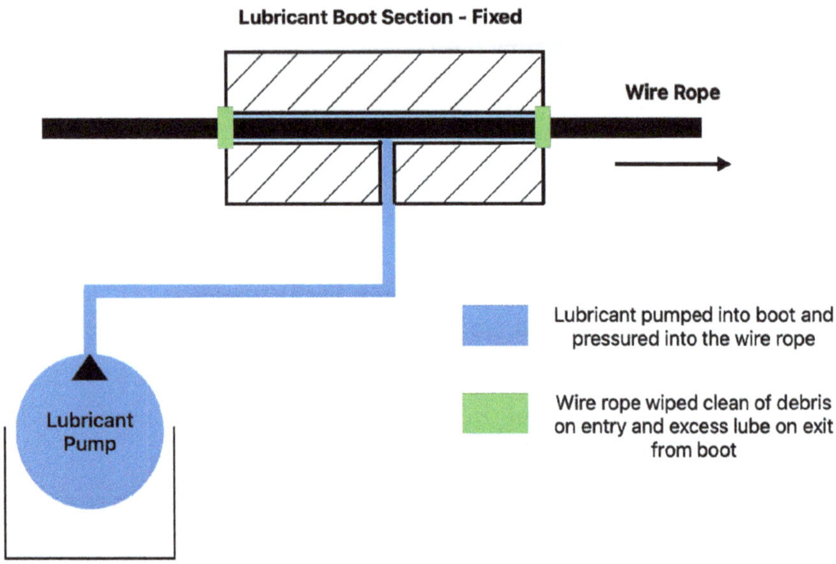

Figure W7 Wire rope lubrication schematic.

Components of a Wire Rope Pressure Lubrication System (See Figure W7)

- **Lubricant reservoir**: Stores the **wire rope lubricant**, ensuring a continuous supply.
- **High-pressure pump**: Delivers **lubricant under pressure**, forcing it into the rope's core.
- **Lubrication applicator head**: Designed to **encase the wire rope and distribute lubricant evenly**.
- **Seals and scraper rings**: Remove **excess lubricant and prevent buildup**, ensuring clean operation.
- **Hoses and control valves**: Regulate **flow rate and pressure for optimal lubricant application**.

977. WO

See "Work Order."

978. Work Hardening

Work hardening (also referred to as **strain hardening**) is the **increase in the hardness and strength of a metal surface due to plastic deformation** caused by repeated mechanical stress, such as friction, impact, or rolling contact. This phenomenon affects lubrication, wear resistance, and material durability, making it a critical factor in bearing surfaces, gears, and tribological applications.

Key Causes of Work Hardening

- **Friction and surface deformation**: Repeated contact between surfaces under high loads alters the metal structure, increasing hardness.
- **Rolling and sliding contact stresses**: Common in bearings, cams, and gear teeth, where metal undergoes continuous deformation.
- **Cold working processes**: Metal forming techniques like forging, rolling, and stamping introduce work hardening to strengthen surfaces.
- **Impact and repeated load cycles**: Shock loads and vibrations contribute to localized strain hardening, improving wear resistance.

Effects of Work Hardening on Lubrication and Wear

- **Increased surface hardness**: Reduces adhesive wear by making the metal surface more resistant to deformation.
- **Reduced ductility and toughness**: Excessive work hardening can lead to brittleness and micro-cracking, increasing the risk of surface fatigue failure.
- **Altered lubrication performance**: Hardened surfaces may require specialized lubricants to ensure adequate film formation and wear protection.
- **Improved wear resistance**: Work-hardened surfaces can enhance abrasion resistance, extending component life.

979. Work Order – Lubrication

A **lubrication work order** is a **formal document used in maintenance management to schedule, track, and execute lubrication tasks on machinery and equipment**. It ensures proper lubricant application, adherence to maintenance schedules, and compliance with operational standards, preventing equipment failures due to improper lubrication through effective work instruction. Lubrication work orders are essential for industrial plants, automotive fleets, manufacturing facilities, and asset management programs, where lubrication is critical for equipment longevity and efficiency.

Key Functions of a Lubrication Work Order

- **Schedules and tracks lubrication tasks**: Provides a clear record of when and where lubrication is performed, ensuring regular maintenance.
- **Ensures proper lubricant selection and application**: Specifies lubricant type, viscosity, and method of application to prevent misapplication.
- **Improves equipment reliability and reduces downtime**: Prevents premature wear, overheating, and mechanical failures due to inadequate lubrication.
- **Enhances compliance with maintenance standards**: Ensures adherence to ISO 55001, ICML 55, and OEM lubrication best practices.

Components of a Lubrication Work Order

- **Work order number and date**: Unique identifier for tracking and documentation.
- **Equipment details**: Specifies machine name, asset ID, and location.
- **Lubricant type and specifications**: Lists oil or grease type, viscosity grade, and additive requirements.
- **Lubrication method**: Defines manual application, automated system refill, or specialized lubrication techniques.
- **Lubrication frequency and schedule**: States daily, weekly, monthly, or condition-based intervals.
- **Technician assigned and completion status**: Records who performed the task and any observations.
- **Special instructions and safety precautions**: Includes lockout/tagout (LOTO) procedures, PPE requirements, and contamination control measures.

980. WR

See "Work Request."

981. Work Request – Lubrication

A **lubrication work request** is a **formal request submitted to maintenance to notify and schedule lubrication tasks for machinery and equipment**. It is used to report lubrication deficiencies, request lubricant replenishment, or schedule preventive maintenance lubrication activities. Lubrication work requests ensure timely lubrication interventions, prevent premature equipment failures, and maintain operational efficiency in industrial plants, automotive fleets, and manufacturing facilities.

Key Functions of a Lubrication Work Request

- **Initiates scheduled or emergency lubrication tasks**: Allows operators, technicians, or engineers to request lubrication servicing when needed.
- **Provides information for the demand lubrication work order**: Provides sufficient details for the planner to develop a quality work order and time requirement for scheduling the work.
- **Ensures proper lubrication and equipment protection**: Prevents friction-related failures, excessive wear, and overheating.
- **Improves maintenance workflow and record-keeping**: Provides a documented process for tracking lubrication needs and request for action to be taken.
- **Enhances communication between operators and maintenance teams**: Ensures critical lubrication tasks are prioritized and executed efficiently.

Components of a Lubrication Work Request

- **Request number and date**: Unique identifier for tracking and documentation.
- **Equipment details**: Specifies machine name, asset ID, and location.
- **Lubrication issue description**: Reports low lubricant levels, abnormal noise, overheating, or improper lubrication.

- **Urgency level and priority**: Indicates whether the request is urgent, routine, or a preventive maintenance requirement.
- **Requester information**: Includes operator or maintenance personnel submitting the request.

982. Worked (Grease)

Worked grease refers to **grease that has been subjected to mechanical stress, shear, or prolonged use in machinery**, **affecting its consistency**, **stability, and lubricating properties**. As grease is worked, it undergoes structural changes, including softening, oil separation, and shear thinning, which impact its performance in bearings, gears, and industrial applications. Proper grease selection and maintenance are essential to ensure long-term lubrication, prevent premature wear, and optimize machinery reliability.

Effects of Mechanical Working on Grease

- **Shear thinning**: Continuous mechanical stress can cause **grease to soften**, leading to migration or leakage from lubrication points.
- **Oil bleed or separation**: Excessive working can result in **base oil separating from the thickener**, reducing lubricating efficiency.
- **Thickening or hardening**: In some cases, prolonged use in high-temperature conditions can cause **thickener depletion**, leading to grease hardening.
- **Contaminant accumulation**: Grease can **trap dirt**, **moisture**, **and debris**, affecting its lubricating properties and increasing wear.

983. Worked Penetration (Grease)

Worked penetration is a **measure of grease consistency and stability**, defined as **the depth (in tenths of a millimeter) that a standard cone penetrates a grease sample after mechanical working**. It is tested using ASTM D217 (Standard Test for Cone Penetration of Lubricating Grease) and helps classify grease into National Lubricating Grease Institute (NLGI) grades, ensuring proper grease selection for industrial and automotive applications.

Key Functions of Worked Penetration in Lubrication

- **Determines grease consistency and grade**: Establishes NLGI classification (000 to 6), where lower penetration values indicate firmer grease.

- **Evaluates shear stability and mechanical work resistance**: Measures how grease softens or remains stable under repeated mechanical stress.
- **Ensures proper application in bearings and machinery**: Helps in choosing the right grease for different speeds, loads, and environmental conditions.

See also "Cone Penetration Test"

984. WPA

See "Wear Particle Analysis."

X

985. X-ray Flourescence (XRF)

X-ray fluorescence (XRF) is a **nondestructive analytical technique used to determine the elemental composition of wear debris, contaminants, and additives in lubricants, greases, and fuels**. It helps in identifying specific metals from machinery components, tracking lubricant additive depletion, and detecting contamination sources, making it an essential tool for predictive maintenance, failure analysis, and quality control in industrial lubrication programs.

Key Functions of X-ray Fluorescence in Lubrication Analysis

- **Identifies wear metals and component degradation**: Detects **iron, copper, aluminum, and other metals** to determine which parts are wearing.
- **Analyzes lubricant additives and formulation changes**: Measures **zinc, phosphorus, calcium, and molybdenum levels** to ensure lubricant performance.
- **Detects contamination from external sources**: Identifies **silicon (dirt), sodium (coolant leaks), and sulfur (fuel contamination)** to prevent failures.
- **Supports predictive maintenance and failure prevention**: Helps track wear trends over time to prevent unexpected breakdowns.

How X-ray Fluorescence (XRF) Works

1. **Excitation**: A high-energy **X-ray beam is directed at the oil or wear debris sample**, causing atoms to become excited.
2. **Fluorescence emission**: The excited atoms emit **secondary X-rays unique to each element's atomic structure**.

DOI: 10.1201/9788743806608-1008

3. **Detection and analysis**: The XRF spectrometer **measures the energy and intensity of emitted X-rays**, identifying **elemental composition and concentration**.

4. **Data interpretation**: The results are **compared to baseline lubricant formulations and wear metal thresholds** to assess equipment health.

Y

986. Y-grade Grease

Y-grade grease is a classification of high-performance industrial **grease designed for extreme pressure (EP), high-load, and high-temperature applications**. It is typically used in heavy-duty machinery, mining equipment, and extreme environments where conventional greases may fail.

Key Characteristics of Y-grade Grease

- **Extreme pressure (EP) protection**: Formulated with EP additives such as sulfur-phosphorus or molybdenum disulfide (MoS_2) to withstand high loads and shock conditions.
- **High-temperature stability**: Can operate in elevated temperatures without breaking down or losing consistency.
- **Water and contaminant resistance**: Designed to resist water washout, dirt, and debris, making it suitable for wet and harsh environments.
- **Superior adhesion and lubrication retention**: Stays in place on bearings, gears, and heavy equipment components, reducing the need for frequent reapplication.
- **Oxidation and corrosion protection**: Contains anticorrosion inhibitors to prevent rust and degradation.

Applications of Y-grade Grease

- **Heavy-duty industrial machinery** – Lubricates bearings, bushings, and joints under extreme operating conditions.
- **Mining and construction equipment** – Used in excavators, loaders, and off-road vehicles for superior wear protection.
- **Steel mills and foundries** – Withstands high temperatures and extreme pressures in metal processing operations.

DOI: 10.1201/9788743806608-1010

- **Marine and offshore equipment** – Protects moving parts exposed to water, salt, and corrosion.
- **Agricultural and forestry machinery** – Ideal for tractors, combines, and logging equipment operating in dusty and wet conditions.

987. Y-channel Oil Flow

Y-channel oil flow refers to a lubrication flow pattern where **oil is distributed through a branching or Y-shaped channel system**. This type of flow configuration is used in hydraulic systems, engine lubrication circuits, and industrial lubrication networks to ensure even distribution of lubricant to multiple components while maintaining optimal pressure and flow rates.

Key Characteristics of Y-channel Oil Flow

- **Divides lubricant flow efficiently**: A single oil supply splits into two or more branches, ensuring multiple components receive lubrication.
- **Maintains controlled pressure and flow**: The design of the Y-channel prevents excessive pressure drops and ensures steady lubricant delivery.
- **Reduces lubrication delays and flow imbalances**: Ensures that critical components receive lubricant simultaneously, reducing wear and overheating risks.
- **Common in complex lubrication systems**: Used in engines, turbines, and industrial gearboxes, where the oil must be directed to multiple bearings or contact points.

Applications of Y-channel Oil Flow

- **Internal combustion engines**: Used in crankshaft lubrication systems, where oil is distributed from a single main gallery to multiple bearing points.
- **Hydraulic and lubrication delivery systems**: Ensures even oil distribution is split to multiple mainline distribution legs.

988. Yaw-induced Lubrication Variations

Yaw-induced lubrication variations refer to changes in lubricant distribution, film thickness, and load-bearing capacity caused by yaw movements in rotating or pivoting systems. These variations are especially significant in applications such as wind turbine yaw bearings, aerospace components, and industrial machinery with rotational motion.

Key Factors Affecting Yaw-induced Lubrication Variations

- **Uneven load distribution**: As yaw movement shifts the direction of applied forces, lubricant film thickness may vary across the contact surface, leading to uneven wear.
- **Shear-induced lubricant flow**: Rotational yaw motion can redistribute lubricant across bearing surfaces, potentially causing localized thinning or excess accumulation in certain areas.
- **Temperature variations and viscosity changes**: Uneven lubrication due to yaw motion may result in localized heating, affecting lubricant viscosity and film stability.
- **Lubrication starvation in extreme angles**: Excessive yaw movements can cause lubricant migration away from critical areas, increasing friction and wear risks.
- **Yaw rate and frequency**: The speed and frequency of yaw movement impact how quickly lubrication adjustments occur, influencing performance and lubricant longevity.

Applications Affected by Yaw-induced Lubrication Variations

- **Wind turbine yaw bearings**: Changes in wind direction cause yaw system movement, affecting lubricant film formation and requiring yaw-specific lubrication strategies.
- **Aerospace control systems**: Aircraft yaw mechanisms experience changing forces that can alter lubrication behavior in rudder actuators and yaw dampers.
- **Industrial rotary equipment**: Large slewing bearings and turntables require yaw-adaptive lubrication to prevent uneven wear.
- **Heavy-duty construction equipment**: Yaw motion in excavators, cranes, and loaders affects lubrication in pivot joints and bearings.

989. Yellow Metal Corrosion Protection

Yellow metal corrosion protection refers to preventing the degradation of copper, brass, bronze, and other copper-containing alloys when exposed to lubricants, hydraulic fluids, and industrial oils. Yellow metals are prone to corrosion and staining due to exposure to sulfur-containing additives, oxidation byproducts, and acidic contaminants.

Key Causes of Yellow Metal Corrosion

- **Sulfur-containing additives**: Extreme pressure (EP) and antiwear (AW) additives with active sulfur can react with yellow metals, leading to sulfidation and tarnishing.
- **Oxidation and acidic contaminants**: Lubricant degradation can produce acids that corrode copper alloys, weakening components.
- **Water contamination**: Moisture accelerates oxidation and galvanic corrosion, increasing yellow metal deterioration.
- **High operating temperatures**: Heat can speed up chemical reactions between the lubricant and yellow metal surfaces, leading to pitting and surface damage.

Methods for Protecting Yellow Metals from Corrosion

- **Use lubricants with copper corrosion inhibitors (ASTM D130 Tested)**: Additives such as triazoles and thiadiazoles form a protective film on copper surfaces.
- **Select lubricants with inactive sulfur compounds**: Some EP and AW additives use nonreactive sulfur, preventing sulfidation of copper alloys.
- **Maintain proper oil filtration and contamination control**: Removing oxidation byproducts, moisture, and acids helps prevent corrosive wear.
- **Choose pH-stable lubricants**: Ensure lubricants have a neutral or slightly basic pH to minimize acidic corrosion risks.
- **Regular oil analysis and corrosion testing**: Conduct ASTM D130 (Copper Strip Corrosion Test) to detect potential lubricant-induced corrosion issues.

990. Yellow Metal Compatibility

See "Yellow Metal Corrosion Protection."

991. Yield

Yield in lubrication terms refers to **the minimum force or stress required to initiate the flow or deformation of a lubricant, particularly greases and**

highly viscous fluids. It is a critical property in grease lubrication, tribology, and fluid dynamics, determining how a lubricant behaves under mechanical stress, temperature changes, and operational loads. Proper understanding of yield characteristics ensures efficient lubricant application, film formation, and wear protection in industrial and automotive applications.

Key Functions of Yield in Lubrication

- **Determines flow behavior of grease and viscous lubricants**: Defines the minimum force needed for grease to transition from a solid-like state to a flowing state.
- **Prevents excessive leakage and migration**: Ensures grease remains in place until mechanical stress is applied, improving retention in bearings and gears.
- **Enhances film formation and load-carrying capacity**: Allows lubricants to form a protective layer under specific loads and pressures, reducing wear.
- **Optimizes lubrication efficiency in extreme conditions**: Prevents premature grease displacement in high-speed, high-load, or high-temperature applications.

Yield Behavior in Different Lubricants

- **Greases**: Have a **defined yield point**, meaning they remain stationary **until enough force is applied to induce flow**.
- **Oils and fluid lubricants**: Typically, **do not have a yield point** but their viscosity determines **resistance to flow**.
- **Thixotropic lubricants**: Exhibit **time-dependent yield behavior**, where repeated mechanical stress **reduces viscosity**, **allowing controlled flow**.

Factors Affecting Yield in Lubrication

- **Thickener type and grease composition**: Determines **structural integrity and resistance to flow**.
- **Operating temperature and pressure**: High temperatures **reduce yield strength**, making grease more fluid, while high pressures **increase resistance to flow**.
- **Shear stress and mechanical agitation**: Continuous movement **lowers grease yield point**, improving distribution in lubrication points.

- **Additives and modifiers**: Viscosity modifiers, tackifiers, and extreme-pressure (EP) additives **influence yield properties for specific applications.**

Applications of Yield Properties in Lubrication

- **Bearings and rolling elements**: **Ensures grease remains in place until mechanical stress initiates flow.**
- **Sealed gearboxes and enclosed lubrication systems**: **Prevents oil or grease migration while maintaining film strength.**
- **High-load industrial machinery**: **Requires lubricants with controlled yield points to withstand extreme pressures.**
- **Automotive and aerospace applications**: **Optimized grease yield properties ensure efficient lubrication in variable-speed and high-impact conditions.**

992. Yield Point

Yield point in lubrication refers to **the minimum shear stress required to initiate the flow of a non-Newtonian lubricant, particularly greases and highly viscous fluids.** It determines when a lubricant transitions from a semi-solid state to a flowing state, making it essential for grease selection, application efficiency, and wear protection in industrial and automotive systems. A proper understanding of the yield point ensures optimal lubricant retention, controlled flow, and effective film formation under load and mechanical stress.

Key Functions of Yield Point in Lubrication

- **Defines the start of lubricant flow**: Ensures grease remains stationary until enough force is applied to initiate movement.
- **Prevents grease migration and leakage**: Maintains lubrication in bearings, gears, and enclosed systems by resisting flow under static conditions.
- **Ensures proper lubrication film formation**: Allows grease to flow only when needed, reducing wear and maintaining protection.
- **Optimizes lubrication for high-load and low-speed applications**: Prevents premature grease displacement in heavy machinery, automotive components, and industrial equipment.

993. Yttrium-based Lubricant Additives

Yttrium-based lubricant additives are specialized lubricant additives that use yttrium (Y), a rare earth metal, to enhance oxidation resistance, improve wear protection, and boost thermal stability in high-performance lubricants. These additives are primarily used in extreme environments, aerospace applications, and advanced industrial lubricants where conventional additives may degrade under extreme heat and pressure.

Key Functions of Yttrium-based Additives

- **Enhances oxidation resistance**: Yttrium compounds help prevent oil breakdown, reducing sludge, varnish, and acid formation.
- **Improves high-temperature stability**: Yttrium-based lubricants are highly resistant to thermal degradation, making them ideal for jet engines, turbines, and space applications.
- **Increases wear resistance**: Acts as an antiwear (AW) and extreme pressure (EP) additive, reducing metal-to-metal contact and wear.
- **Protects against corrosion**: Forms a protective surface layer that resists oxidation and chemical attacks on metal surfaces.
- **Enhances lubricant longevity**: Reduces the need for frequent oil changes by improving lubricant durability and stability.

Applications of Yttrium-based Additives in Lubricants

- **Aerospace and jet engine lubricants**: Used in high-performance turbine oils to withstand extreme temperatures and pressures.
- **High-temperature industrial lubricants**: Applied in metalworking fluids, hydraulic systems, and gear oils that operate in harsh conditions.
- **Nuclear and space applications**: Provides radiation-resistant and high-stability lubrication for critical components.
- **Automotive and racing lubricants**: Found in high-performance motor oils where wear protection and thermal stability are essential.

Z

994. ZDDP

Zinc dialkyldithiophosphate (ZDDP) is a multifunctional antiwear, extreme pressure (EP), and antioxidant additive widely used in engine oils, industrial lubricants, and gear oils. Originally invented by the Castrol Oil company to combat excessive wear in flat tappet engine designs. ZDDP is now a popular antiwear additive used in industrial application oils, usually in concentrations of less than 1%. ZDDP is designed to provide critical protection against metal-to-metal contact, oxidation, and high-pressure wear, making it an essential component in automotive, heavy machinery, and high-performance lubrication systems.

Key Functions of ZDDP in Lubrication

- **Prevents metal-to-metal contact and reduces wear**: Forms a protective phosphate-based film on metal surfaces, minimizing scuffing, galling, and friction.
- **Provides extreme pressure (EP) protection**: Reacts under high-load conditions to prevent gear and bearing surface damage.
- **Acts as an antioxidant to prevent oil degradation**: Inhibits oxidation reactions, reducing sludge, varnish, and deposit formation.
- **Improves lubricant longevity and performance stability**: Ensures consistent viscosity and antiwear protection over extended drain intervals.

How ZDDP Works in Lubrication

1. **Under normal operating conditions**: ZDDP molecules remain dissolved in the oil, enhancing lubricant stability and oxidation resistance.
2. **Under high pressure and heat**: ZDDP chemically reacts with metal surfaces, forming a thin protective film of zinc and phosphate compounds that prevents direct metal contact.

DOI: 10.1201/9788743806608-1019

3. **During boundary lubrication**: When the lubricant film is compromised, ZDDP provides a last-resort protective barrier, reducing severe wear and preventing catastrophic failure.

Factors Affecting ZDDP Performance

- **Operating temperature and load conditions**: Higher temperatures accelerate ZDDP reaction rates and depletion.
- **Base oil compatibility and additive blending**: Works best in well-formulated lubricant packages without additive interference.
- **Phosphorus content restrictions**: Modern regulations limit ZDDP use in some low-emission and catalytic converter-equipped vehicles due to its potential to poison emissions control systems.

995. Zebra Marker

A **zebra strip** is a **visual enhancement tool used on fluid-level sight tubes to improve the visibility of dirty, contaminated, or opaque lubricants**. It features alternating black and white stripes that create a contrast effect, making it easier to detect fluid levels, air bubbles, or contamination buildup in oil reservoirs, hydraulic tanks, and lubrication systems.

Key Functions of Zebra Strips in Fluid Level Monitoring

- **Improves visibility of fluid levels in dirty or opaque oils**: Enhances contrast to make oil levels easier to read, even in dark or contaminated fluids.
- **Aids in quick and accurate lubricant inspections**: Provides a clear reference for operators to assess fluid conditions without opening the system.
- **Helps detect air bubbles and fluid contamination**: Highlights suspended particles, aeration, and water separation in oil.
- **Enhances safety and maintenance efficiency**: Reduces misreading errors that could lead to overfilling, underlubrication, or system failures.

Applications of Zebra Strips in Lubrication and Fluid Management

- **Hydraulic reservoir sight tubes**: Improves fluid level visibility in hydraulic systems operating in harsh conditions.

- **Gearbox and oil tank level indicators**: Assists in monitoring lubricant levels in enclosed gear drives and reservoirs.
- **Machine lubrication systems**: Used in oil-filled sight glasses, sumps, and condition monitoring indicators.
- **Industrial process fluids and coolant tanks: Helps detect fluid clarity changes, contamination, and separation layers.**

Figure Z1 The zebra strip clearly demonstrates the fluid level of the clear water.

Factors Affecting Zebra Strip Effectiveness

- **Striping pattern and contrast intensity**: Higher contrast improves readability in dark or degraded fluids. See Figure Z1.
- **Lighting and viewing angle**: Proper positioning ensures accurate level detection and visibility under various lighting conditions.

- **Fluid type and contaminant load**: Works best with heavily contaminated oil, emulsions, and fluids with poor transparency.

Benefits of Using the Zebra Strip on Fluid Level Tubes

- **Enhances fluid level readability in contaminated lubricants**: Ensures **operators can easily monitor lubricant and hydraulic fluid levels**.
- **Reduces maintenance errors and improves inspection accuracy**: Prevents **misreading levels in dark or aerated oils**.
- **Supports condition monitoring and predictive maintenance**: Allows for **early detection of contamination, leaks, and aeration issues**.
- **Increases safety and operational reliability**: Minimizes **the risk of lubrication-related failures due to incorrect fluid levels**.

996. Zerk Grease Fitting

A **Zerk grease nipple**, also known as a **grease nipple**, is a **small metal valve used to allow grease to be injected into bearings, bushings, and other mechanical components without disassembly**. The Zerk fitting's convex outer shape easily accommodates attachment and detachment of the grease gun tip from many angles. Lubricant pressure forces a sprung-loaded ball to move back against the spring pressure and allows lubricant to flow into the bearing while preventing contaminants from entering the lubrication point. Once the gun is detached, the ball returns to its closed position allowing no lubricant to escape, and little or no dirt inclusion if the grease nipple and gun are wiped cleaned prior to use.

Key Functions of a Grease Nipple

- **Facilitates easy and controlled grease application**: Allows grease to be injected under pressure into bearings and lubrication points.
- **Prevents contaminants from entering bearings and bushings**: Features a spring-loaded check valve that seals the fitting after lubrication.
- **Ensures proper lubricant distribution**: Serves as a direct grease entry point to critical friction points to lubricate, minimize wear, and reduce friction.

Types of Grease Nipples (See Figure Z2)

- **Straight grease nipples**: Standard design for **direct access lubrication points**.

- **Angled grease nipples (45° or 90°)**: Used when **lubrication points are in hard-to-reach locations**.
- **Flush-type grease nipples**: Sit **flush with the surface**, preventing **damage in high-impact areas**.
- **Button head grease nipples**: Designed for **high-volume grease applications** in **heavy-duty machinery**.
- **Metric and imperial (SAE) threaded nipples**: Available in **various thread sizes and fittings to match lubrication system requirements**.

Figure Z2 Zerk fittings come in many configurations. Source: Plant Lubrication For Industrial Facilities – Fourth Edition, Kenneth E. Bannister.

Factors Affecting Grease Nipple Performance

- **Compatibility with grease guns and lubrication tools**: Requires matching couplers and pressure ratings for effective grease delivery.
- **Operating environment and contamination risks**: Sealed grease nipples are preferred in dusty, wet, or corrosive environments.
- **Thread size and installation fit**: Must match the equipment manufacturer's specifications to ensure proper sealing.
- **Maintenance and cleaning practices**: Regular inspection and cleaning prevent blockages and ensure consistent grease flow.

997. Zerk, Oscar U.

Oscar Zerk (1878–1968) was a prolific inventor and innovator in the field of lubrication and mechanical engineering. Born in Austria, Oscar was an engineer and prolific inventor of over 300 patented inventions during his lifetime. Zerk moved to Kenosha WI, United States in 1924 one year after filing for a US patent on the invention he is most famed for, the "**Lubricant Receiving Nipple**" which later became known as the "Zerk" fitting. This simple pioneering invention changed the face of lubrication forever and is even today is likely still the single most sold piece of lubrication equipment. See "Zerk Grease Fitting."

Key Contribution: The Zerk Fitting:

- Invented by Oscar Zerk in 1924, the Zerk fitting is a small, threaded metal fitting with a ball check valve that allows grease to be injected directly into machinery components, such as bearings and joints.
- The fitting prevents grease from leaking out while blocking contaminants like dust and moisture, ensuring optimal lubrication performance.

Legacy: The Zerk fitting remains an essential component in modern lubrication systems, used in countless applications worldwide. Oscar Zerk's ingenuity addressed a critical need for efficient and reliable machinery maintenance during a time of rapid industrialization. His work not only revolutionized lubrication but also laid the groundwork for advancements in machinery design and maintenance practices.

998. Zero-wear Lubrication

Zero-wear lubrication refers to a lubrication condition where friction and wear are minimized to the extent that no measurable material loss occurs over an extended period. This is achieved through optimized lubricant film formation, advanced additive technology, and surface engineering. Zero-wear lubrication is crucial in high-precision machinery, aerospace applications, and ultra-long-life components, where even minimal wear can lead to performance degradation or failure.

Key Principles of Zero-wear Lubrication

- **Full-fluid film formation**: Ensures a continuous lubricant film that completely separates moving surfaces, preventing metal-to-metal contact.

- **Advanced additive chemistry**: Utilizes antiwear (AW), extreme pressure (EP), and friction-reducing additives to enhance surface protection.
- **Optimized lubricant viscosity and stability**: Lubricants are designed to maintain viscosity stability across varying temperatures and loads to prevent film breakdown.
- **Surface engineering and coatings**: Components may be treated with low-friction coatings (e.g., DLC – diamond-like carbon, MoS_2, or ceramic coatings) to further reduce wear.
- **Precision lubrication delivery**: Uses smart, automated lubrication systems that maintain optimal film thickness and prevent lubricant starvation.

999. Zeta Potential

Zeta potential refers to the **electrostatic potential at the boundary layer of dispersed particles in a fluid**, such as contaminants, wear debris, or additives in lubricating oils and greases. It is a key factor in colloidal stability, lubricant filtration efficiency, and contamination control, influencing how particles interact, disperse, or agglomerate in a lubricant.

Key Functions of Zeta Potential in Lubrication

- **Determines particle stability in lubricants**: High zeta potential values (positive or negative) indicate strong electrostatic repulsion between particles, preventing contaminant agglomeration and sedimentation.
- **Enhances dispersion of additives**: Helps stabilize detergent, dispersant, and antiwear additives, ensuring uniform distribution and effectiveness.
- **Affects lubricant filtration and contaminant removal**: Influences particle charge interactions with filter media, impacting filtration efficiency in hydraulic and engine oils.
- **Controls wear debris behavior in oil analysis**: In condition monitoring, zeta potential can indicate whether wear debris and contaminants remain suspended or settle, affecting oil cleanliness levels.
- **Prevents emulsion formation and oil–water separation issues**: Affects how lubricants interact with water contamination, influencing demulsibility and emulsion stability.

Factors Affecting Zeta Potential in Lubrication

- **Lubricant composition and additives**: Certain dispersants and detergents alter surface charge to enhance particle stability.
- **Contaminant and wear particle composition**: Different metals, oxides, and debris may exhibit varied electrostatic behavior, impacting suspension stability.
- **pH and electrolyte concentration**: Variations in acidic or alkaline conditions influence zeta potential and lubricant dispersion stability.
- **Temperature and shear conditions**: High temperatures and mechanical stresses can modify zeta potential by altering additive interactions and particle dispersion.

1000. ZN/P

The **ZN/P ratio** is a **key tribological parameter used to evaluate lubrication performance under different operating conditions**, where:

- **Z = Viscosity** (typically in centistokes, cSt)
- **N = Speed** (typically in revolutions per minute, RPM)
- **P = Load** (typically in pressure or force per unit area, such as psi or N/mm^2)

The **ZN/P ratio** is widely used to determine **the lubrication regime, oil film thickness, and the likelihood of wear or failure** in machine components such as **bearings, gears, and journal surfaces**. See also "Stribeck Curve."

Key Functions of the ZN/P Ratio in Lubrication (See Table Z1)

- **Determines lubrication regime (boundary, mixed, or hydrodynamic lubrication)**:
  - **High ZN/P values** indicate **thicker oil films**, leading to **hydrodynamic lubrication** with minimal wear.
  - **Low ZN/P values** suggest **thin films or boundary lubrication**, increasing wear risk.
- **Optimizes lubricant selection based on load and speed conditions**:
  - Helps in selecting the **correct viscosity grade** for specific **operating speeds and loads**.

- **Prevents excessive wear and surface fatigue in bearings and gears**:
    - Ensures **sufficient lubricant film formation to separate metal surfaces**.
- **Enhances predictive maintenance and condition monitoring strategies**:
    - Used to **calculate lubrication efficiency and adjust maintenance schedules**.

Table Z1 How ZN/P ratio affects lubrication.

ZN/P Ratio	Lubrication Regime	Lubricant Behavior	Wear Risk
High (Thick Film)	Hydrodynamic	Full fluid film separates surfaces	Low
Medium (Partial Film)	Mixed	Partial contact, some film support	Moderate
Low (Thin Film)	Boundary	Direct metal-to-metal contact	High

Factors Affecting ZN/P Performance

- **Lubricant viscosity (Z)**: Higher viscosity increases the **ZN/P ratio**, improving **film thickness but reducing energy efficiency**.
- **Operating speed (N)**: Faster speeds improve **fluid film formation**, reducing **boundary contact risk**.
- **Load (P)**: Higher loads reduce **ZN/P**, increasing **wear potential** if viscosity is insufficient.
- **Temperature and contaminants**: High heat and contamination can **reduce lubricant effectiveness, altering ZN/P values**.

1001. Zone Lubrication

Zone lubrication refers to a **targeted lubrication strategy** where different machine components receive specific lubrication treatments based on operating conditions, load distribution, and motion requirements. It ensures optimal lubricant application by dividing machinery into separate lubrication zones, each with customized lubrication delivery methods, lubricant types, and application frequencies.

Key Principles of Zone Lubrication

- **Lubrication is applied based on operating conditions**: High-load areas may receive heavy-duty greases, while fast-moving components may use low-viscosity oils.
- **Minimizes overlubrication and contamination risks**: Each lubrication zone receives only the necessary amount of lubricant, reducing waste and preventing contamination.
- **Supports automated and centralized lubrication systems**: Many modern lubrication systems can apply lubricants to different zones independently, optimizing efficiency.
- **Improves maintenance and predictive lubrication strategies**: Helps operators monitor wear trends and adjust lubrication based on real-time performance data.

About the Author

Kenneth E. Bannister is a globally recognized expert in industrial machinery lubrication and asset management, known for his innovative approach and authoritative voice in the industry. With decades of experience, Ken has transformed lubrication reliability, condition monitoring, and asset management practices, influencing organizations worldwide through his consulting, writing, and training programs.

- **Prolific author and thought leader:** Kenneth E. Bannister has to date authored and published over 500 books, articles, and white papers, including several highly respected and best-selling books on industrial lubrication and maintenance management. His influential works, such as *"Lubrication for Industry," "Practical Lubrication for Industrial Facilities," and "Lubrication Strategies and Tips – How to Kick-Start Your Lubrication Program,"* are widely used by maintenance professionals globally to implement best practices and enhance equipment reliability.
- **Consultant, educator, and innovator:** As a trusted management consultant and educator, Ken has guided countless companies in developing and refining their lubrication and asset management programs. His real-world, application-focused training sessions empower organizations to extend machinery life, reduce downtime, and achieve cost savings. Known for his innovative solutions and strategic vision, Ken has been instrumental in helping companies optimize their maintenance and lubrication practices.
- **Industry influence and global leadership:** Kenneth E. Bannister's impact on the industry extends beyond his consulting and writing. He has played a pivotal role in writing and developing international maintenance and lubrication standards, including the ICML 55.1® World Lubrication Standards, through his active involvement with industry organizations and councils. His dedication to promoting innovation and best practices in lubrication and reliability has earned his clients' numerous industrial awards and recognitions.

A visionary and ambassador for excellence: Ken's work emphasizes the importance of integrating lubrication management into comprehensive maintenance strategies, using data-driven decision-making and proactive maintenance approaches to achieve operational excellence and reliability. His leadership, authority, and passion for advancing the field continue to inspire the next generation of maintenance professionals and influence industry standards worldwide.

Kenneth E. Bannister is not only a pioneer in lubrication and asset management but also a respected ambassador of innovation and excellence, shaping the future of industrial maintenance and reliability practices.

For Product Safety Concerns and Information please contact our EU
representative GPSR@taylorandfrancis.com
Taylor & Francis Verlag GmbH, Kaufingerstraße 24, 80331 München, Germany

www.ingramcontent.com/pod-product-compliance
Ingram Content Group UK Ltd.
Pitfield, Milton Keynes, MK11 3LW, UK
UKHW022209190926
475702UK00003B/268
* 9 7 8 8 7 7 0 0 4 1 4 0 9 *